Design and Packaging of Electronic Equipment

Design and Packaging of Electronic Equipment

Joel L. Sloan

Guidance and Control Systems Division
Litton Systems, Inc.

VNR VAN NOSTRAND REINHOLD COMPANY
——————— New York ———————

Copyright © 1985 by Van Nostrand Reinhold Company Inc.

Library of Congress Catalog Card Number: 83-23406
ISBN: 0-442-28819-0

Manufactured in the United States of America

Published by Van Nostrand Reinhold Company Inc.
135 West 50th Street
New York, New York 10020

Van Nostrand Reinhold Company Limited
Molly Millars Lane
Wokingham, Berkshire RG11 2PY, England

Van Nostrand Reinhold Company Limited
480 Latrobe Street
Melbourne, Victoria 3000, Australia

Macmillan of Canada
Division of Gage Publishing Limited
164 Commander Boulevard
Agincourt, Ontario M1S 3C7, Canada

15   14   13   12   11   10   9   8   7   6   5   4   3   2   1

Library of Congress Cataloging in Publication Data

Sloan, Joel L.
  Design and packaging of electronic equipment.

  Bibliography: p.
  Includes index.
  1. Electronic packaging.   2. Electronic apparatus and
appliances—Temperature control.   I. Title.
TK7870.S525   1984   621.381   83-23406
ISBN 0-442-28819-0

Preface

The inspiration for this book grew out of a collection of notes and course materials that have been used to broaden the applied skills of practicing packaging engineering personnel and for the design and development of electronic equipment. The material encompassed by this book is geared to the solution of problems that arise in developing modern designs which are required to meet design constraints and environments imposed by the user. Performance and reliability of deployed production equipment substantiates the use of these methods.

The design, development and testing of an electronic assembly requires a broad knowledge of many technical disciplines. The design engineer must be able to integrate the effects of environment, static and dynamic loads and heat transfer with consideration of package size, weight and mounting arrangement, taking into account the materials, components and cabling, into a package that provides the desired performance and reliability. The design engineer must be able to assess component temperatures as a function of the equipment power dissipation; electronic circuit layout and cooling method. He or she may want to know the stresses developed on component leads or attachments due to handling or transport shock and vibration. This text provides easily applied methods that enable the design engineer to perform these tasks and develop a basic understanding of the physical behavior of the equipment.

The methods used are specifically oriented to calculator aided manual computation. Design parameters can be included in the formulations which enable parametric evaluations and optimization of performance relative to configuration characteristics. Either steady state or transient behavior can be determined. The approaches used may be adapted for automated computation for large scale studies that are not practical for manual treatment. Simple representations of the actual design which are amenable to evaluation are developed. Expertise in choosing a representation is developed through experience by repeated application of the techniques described in the text.

Many relationships have been derived that can be used to evaluate the behavior of most design configurations without resorting to matrix manipulations. Matrix methods have also been used to increase the power of the techniques so that configurations of greater complexity can be studied. A thorough development of the matrix methods which are specifically oriented to manual manipulation have been included as an Appendix.

v

Relationships are presented that permit a direct efficient solution for the desired design quantities. Assumptions, limitations and the use of approximations are clearly defined. Few proofs are included, and although some expressions and techniques appear unorthodox, their development is rigorous. Use of any relationship is based on employing consistent units. Although the text material is inclusive, a background in elementary mechanics and strength of materials would be useful to the reader.

The text is designed as a primary reference for packaging electronic equipment. Numerous design examples with in-depth explanations are included. A building block approach is used to introduce each discipline. A unified approach which applies similar mathematical treatment to both heat transfer and structural disciplines is used. The use of English and metric units expands the utility of the text. The material presented is suited to courses in electronic equipment packaging and analysis, the latter including heat transfer, thermoelasticity, structural mechanics and dynamics of structures.

At the onset of a design, the design engineer collects all the facts about the environment, installation, operational requirements, equipment functional partitioning and component complement. Chapter 1 delineates the factors which influence equipment designs that are considered during the design process. These factors could affect component placement and the selection of a structural configuration or cooling approach. The nature of heat transfer in electronic equipment is explored in Chapter 2. Design considerations are discussed. Chapters 3 and 4 expand on the applications of conduction, convection and radiation that occur in electronic equipment. Thermally induced structural loads and stresses are discussed in Chapter 5. Practical methods that can be used to avoid stresses sufficient to cause solder joint failure at component lead attachments are included. Stresses on attached stiffeners and on component attachments due to bonding are also addressed.

Internal forces due to applied external loads in an electronic system are discussed in Chapter 6. The nature of static equipment displacements and methods for their determination are studied in Chapter 7. Chapter 8 expands on those techniques to evaluate the dynamic response of the design to vibration, acoustic noise and shock. The relationship between the fundamental frequencies and the forces and stresses experienced by an assembly are explored. A practical method of evaluating the response to random vibration is discussed. Amplification due to snubbing is also addressed.

I wish to express my sincere appreciation to Dave Steinberg and my wife Bernida for their encouragement and inspiration during the preparation of the manuscript; to Don Beverage for the preparation of the graphic art; to Pearl Fleming for her moral support, her patient and precise typing of the manuscript, and the annotations for the illustrations.

JOEL L. SLOAN

Contents

Design and Packaging of Electronic Equipment

1
Factors Influencing Equipment Design

The sophistication, complexity and packaging density of electronic equipment follow a trend established by circuit and device development. This evolution in technology, materials and techniques has dramatically improved the capability, performance, reliability and availability of systems that perform tasks bounded only by the limits of the human imagination. New uses for compact sophisticated equipment meeting a variety of demands imposed by the user and the environment emerge continually. The diversity of designs and service conditions coupled with the necessity to demonstrate equipment integrity as a condition of acceptance have increased the importance of the design engineer placing additional emphasis on his abilities.

Equipment design is influenced by the effect the service or test environment has on vulnerable elements. The nature of the element environment is also affected by the design itself. Heat build-up in a densely packaged assembly, for example, can result in internal ambient temperatures considerably higher than those of the external surrounding environment. The maximum response stress of a circuit board is a consequence of its structural properties and the behavior of its housing subject to external disturbances. Severe conditions can occur if the behavior of several elements interact, combining effects. The environment also imposes constraints on the design influencing its interface boundary. Shields and insulation may be required, for example, to prevent heating due to radiation and conduction from excessive mounting surface temperatures. An isolation mount may be required, for example, to attenuate high energy amplitudes at critical unit frequencies.

The configuration, cooling and structural philosophy reflected in the design are a consequence of the environment and the physical constraints imposed by the mounting space. Equipment designed for specific requirements may be unsuitable for those applications which impose different constraints. On this basis an environmental-application envelope can be defined for general purpose equipment which may extend its usefulness and service life for a broad class of conditions. Environments commonly encountered in addition to tests and application constraints which influence designs are presented to promote a better understanding for their implementation. Consideration of these factors may improve environmental resistance while minimizing procurement and maintenance costs for military as well as commercial equipment.

1.1 ENVIRONMENTS

The design engineer is responsible for the performance and behavior of equipment in the environments defined for its use. It is necessary to develop an understanding of the effects these conditions have on equipment in order to minimize any failure mechanisms that can occur. The design engineer should always take advantage of observing in progress tests on a variety of designs to gain insight into the nature of failures. Experience would show that failures may be attributed to poor design characteristics, the choice of an incorrect or misuse of a particular part or material, or poor workmanship practices. In any case, the ability of the equipment to resist the effects of the environment would be compromised.

Design environments for military and commercial equipment vary considerably. There is almost as much variation among military services. Equipment used on high speed aircraft, missiles or reentry vehicles experience more severe environments than that on military transports or commercial aircraft. Shipboard military equipment experiences greater shock levels than equipment used on ground support, tracked or wheeled vehicles.

Environments may also be dependent upon geographic location. Equipment designed and used in tropical environments must specifically include consideration for the effects of moisture, fungus and corrosion. Equipment exposed to marine or shore environments will experience salt laden atmospheres. The effects of blowing sand and dust should be included for equipment in desert locations.

Handling and transport environments can exceed those related to service use. Shock and vibration levels are dependent upon equipment size, weight and handling features provided by the design engineer.

Environments may be either natural or induced. Natural environments are conditions generated by forces of nature. Induced environments are generated as a result of vehicle or equipment operation. Table 1.1 lists primary environments and the method of the occurence.

The environment surrounding the equipment or device may change as a result of its mass, interfaces or its operation. The mass ratio of the equipment to its supporting structure may alter the severity of the anticipated vibration or shock environment. Heat dissipated by equipment or devices in confined spaces could increase its ambient temperature, causing higher than predicted device temperatures to occur. The resulting response magnitude for these conditions could cause unexpected malfunction or equipment failure. These and

Table 1.1. Natural and Induced
Environments

Environment	Natural	Induced
Temperature	X	X
Moisture		X
Humidity	X	
Salt-Fog	X	
Sand and Dust	X	
Pressure—Atmospheric	X	
Shock		X
Vibration		X
Acoustic Excitation		X
Acceleration		X
Gravity	X	
Radiation—Solar	X	

similar conditions which develop environments for the specific application can be determined by analysis or carefully designed tests.

The environment may also vary as a function of position and location within an installation. Specified environments generally envelop conditions that could occur at several candidate equipment locations to minimize the possibility of malfunction or failure. Depending on the methods of arriving at the specified level, considerable margins relative to actual vehicle or field environment could be included. This approach certainly increases the availability of the equipment but penalizes the cost and weight of the design while imposing additional design constraints.

A great variety of materials, processes and manufacturing techniques are available to the design engineer to develop products which meet environmental requirements. The selection of a suitable design for a particular application requires:

1. An understanding of essential design and performance characteristics under all service conditions;
2. An ability to evaluate candidate designs in terms of the environment and the response or behavior sought;
3. Selection of economical fabrication processes that do not compromise the design configuration;
4. Performance tests to validate the approach and determine the accuracy of the predicted behavior; and
5. Documenting results and finalizing design documentation for production and quality control.

Table 1.2. Temperature Factors

Mechanism	Effect on Equipment	Accelerating Factors
Increasing Temperature	Loss of strength, reduced stiffness, reduced resonant frequency, softening, distortion aging and creep	Lubricants, rubber parts, plastics, corrosion, fatigue, load intensity, time duration
Reducing Temperature	Increased viscosity, increased stiffness, increased resonant frequency, brittleness, reduced impact resistance	Lubricants, rubber parts, plastics, duration
Thermal Expansion and Contraction	Change in size and shape, buckling, cracking, distortion, loosening	Temperature cycling, temperature range, unequal expansion coefficients, stress concentrations, lack of strain relief

Temperature. Mechanical properties of almost all known materials are affected by changes in temperature. The rate of which chemical reactions take place in joined materials may double for each 18°F (10°C) increase in temperature. Critical combinations of temperature, condensation, humidity and atmospheric contaminants such as salt, fungus, sand, dust and industrial byproducts can cause performance loss in equipment and gradual deterioration of materials and surfaces so that the service life of the unit is prematurely shortened. Table 1.2 lists temperature related factors that can affect performance and service life of electronic equipment. The accelerating factors shown in the third column are materials or conditions that accelerate the effect of the mechanism.

Many materials exhibit a loss in mechanical properties at high temperatures. The degree of deterioration in most cases is time dependent and a function of the load intensity. The useful upper limit of most thermal plastics of 240°F (115.6°C) is related to the heat distortion temperature of the material. Component and mounting surface temperatures above 212°F (100°C) result in a loss of solder joint strength. Operating and nonoperating conditions which cause a variation of 250°F (138.9°C) in component mounting temperatures will eventually lead to joint creep-rupture fatigue failure. Temperature changes in equipment protected with vibration or shock isolators can result in performance loss or failure. Rubber components used in mounts and snubbers exhibit large variations in mechanical properties which affect both resonant frequency and damping. Displacement and acceleration levels at 70°F (21.1°C) increase at elevated ambient and material temperatures, affecting the design envelope and the degree of protection provided to sensitive components.

At temperatures below $-40°F$ ($-40°C$), some nonoperating electronic equipment may experience performance loss or other difficulties at turn-on. These problems may be linked to dimensional changes due to material expansion or contraction. High loads, stresses and poor thermal conductivity can occur at joined interfaces that use materials exhibiting different expansion rates. Mated plastic and metal parts require special attention.

Moisture. Above all, moisture creates an insidious environment which negatively affects performance and deteriorates materials and finishes. Permeation and surface moisture on insulating materials develop loss mechanisms that affect circuit performance. When combined with warm temperatures, moisture accelerates metallic corrosion and fungus growth. Conversely, protection against moisture mitigates the effects of fungus and corrosion. Table 1.3 lists related moisture factors.

Trapped moisture in circuit boards and other nonmetallic materials develops damaging effects and equipment failure when freezing temperatures are encountered. Trapped moisture expands when frozen, causing delamination, cracked solder joints and fractured artwork runs, and permanent distortion in highly stressed assemblies.

Sealed units prevent or retard the entry of moisture. Unless a positive internal pressure is maintained, nonmetallic seals and gaskets offer little resistance to the entry of water vapor. Equipment operating in humid environments experiencing high on-off duty cycles during which internal parts are heated then cooled cause considerable moisture to be pumped into the unit. Nonmetallic seals provide rain protection while permitting access to internal assemblies. The housing, fasteners and gaskets, however, should be properly designed to provide an effective seal against breathing over the temperature and pressure range for which the equipment is designed. Despite all design precautions, gasketed joints may become defective, resulting in trapped moisture.

Table 1.3. Moisture Factors

Mechanism	Effect on Equipment	Accelerating Factors
Material Absorption	Lowers insulation resistance and dielectric strength, increases loss factors	High temperatures, high relative humidity
Surface Absorption	Current leakage, corona, arc tracking, changes insulation resistance and dielectric strength	Temperature cycling, contaminants, corrosion, fungus, salt, dust
Wet/Dry Cycles	Trapped moisture in poorly sealed enclosures	Temperature cycling, on-off equipment, duty cycles

A free breathing open housing design avoids the effects of trapped moisture. Conformal coatings on critical and sensitive circuit elements must be provided in this case to assure adequate protection. The application of protective finishes and coatings should be controlled to minimize pin holes, poor adhesion, scratches, and contaminants such as dust, lubricants and fingerprints.

Moisture induced corrosion can develop high contact resistance in electrical connectors, galvanic effects between joined materials and stress cracking. Materials under stress in the presence of moisture develop anodic corrosion sites, leading to the development of stress cracks and whiskers that hasten equipment deterioration and loss of material strength.

Humidity.   This environment influences the *rate* of moisture absorption and penetration in materials and components. These processes are due to the presence of water vapor, which causes the formation of a water film on surfaces and in the pores of materials. Penetration into nonmetallic materials occurs by various mechanisms; the common factor is that the molecular spacing in these materials exceeds the diameter of the water molecule.

Permeation of moisture or water vapor in nonmetallic materials due to humidity causes the same effects on equipment as discussed under moisture, above. The moisture film can alter electromagnetic characteristics of equipment, initiate corrosive actions, cause dimensional changes and reduce material strength.

Salt-Fog.   This atmosphere is found in marine locations and along bordering land masses. Salt is a rate factor for most deteriorating effects involving corrosion products and galvanic activity at interfaces of dissimilar metallic materials. Related corrosion factors are shown in Table 1.4.

Electrical characteristics and visual appearance of nonmetallic materials are

Table 1.4. Corrosion Factors

Mechanism	Effect on Equipment	Accelerating Factors
Chemical Reaction	Increased contact resistance products of oxidation	Concentration of weak acids and corrosive contaminants
Galvanic	Interface deterioration, loss of mechanical strength, open circuits	High EMF couple, large cathodic area, stress cycling
Pitting	Loss of mechanical strength and component function	Rough finishes, inclusions, discontinuities, surface defects
Migration	Short circuit of insulation, performance loss, corona, whisker development	Unrelieved stresses, accumulating moisture

Table 1.5. Dust Factors

Mechanism	Effects on Equipment	Accelerating Factors
Surface Absorption	Short circuits insulation, performance loss, corrosion	Moisture, circuit sensitivity
Penetration	Increases friction, abrasion and wear, fouls moving parts, deteriorates connector interfaces, lubricants, bearings	Vibration, movement, decreased particle size
Surface Abrasion	Loss of finish, paint and control legends	Increased wind velocity, turbulence

affected by salt. Obvious forms of corrosion are flaking and lifting of paints or finishes and the presence of white, black, red or other colorful residues in the unit.

Sand and Dust. This environment is most prevalent in desert regions. The dust becomes airborne with little wind resistance and usually remains suspended over a long period of time. Under these conditions, dust penetrates into cracks, crevices and enclosure openings settling on internal surfaces, circuit boards and components. Related dust factors are shown in Table 1.5.

The electrically conductive dust may short circuit artwork on boards or become explosive if sufficient concentrations surrounding high voltage circuits are encountered. Removal of panel function control legends by blowing dust may render the equipment inoperable. Screens, filters and other openings provided for the entry of cooling air may become clogged decreasing flow which can lead to excessive component temperatures. When filters are used in the cooling path, the flow rate must be adjusted to account for losses due to clogging. Filter cleaning or replacement under these conditions should be an integral part of a preventative maintenance program for the equipment. Air cooled cold plates that use high fin or pin densities will also require periodic cleaning when exposed to these environments.

Atmospheric Pressure. Pressure and density affect the design of almost all electronic equipment. Pressure should be considered when designing sealed enclosures and components and when defining the equipment cooling philosophy. Variation in pressure due to changes in altitude coupled with temperature changes may induce moisture condensation on equipment. Related pressure factors are listed in Table 1.6.

Changes in differential pressure encountered in aircraft or missile flights may rupture or distort housings or pressurized containers. These effects may cause overheating of equipment or loss of contained fluids or gases. Loss of equipment internal environments may also cause arcing and corona in high

Table 1.6. Pressure Factors

Mechanism	Effects on Equipment	Accelerating Factors
Leakage	Rupture of seals, container rupture or distortion, loss of enclosed environment, stiction higher component temperatures	Variation in temperature or pressure
Voltage Breakdown	Current leakage, corona, arcing, performance loss	High voltage, increasing altitude, decreasing conductor spacing
Outgassing	Deterioration of optical surfaces, variation in chemical and physical material properties	Very low pressures, low density materials
Decompression	Decreased interface conductance, reduced conduction and convection cooling efficiency	Increasing altitude, low contact pressure, surface finishes, natural convection

voltage equipment. Excessive component temperatures may result from the reduction in density and decrease in joint conductances of fastened assemblies.

In space environments, where very low ambient pressures are encountered, some materials out-gas, affecting their chemical and physical properties in addition to depositing opaque films on windows, lenses or other optical surfaces.

Shock.   Failures due to shock occur because the maximum response stress exceeds the ultimate or yield strength of the material. Relatively few stress reversals are involved as compared with vibration induced failures; this distinction as related to the nature of failures experienced by equipment is important.

Equipment handling during test and installation may incur shocks of greater severity than do operational environments. Handling shock is often expressed in terms of an initial velocity imparted to the item whereas operational shocks are described in terms of foundation acceleration or displacement. The effect of these environments on equipment and supported items can be evaluated once the frequency characteristics of the design are determined and the duration and magnitude of the shock motion are known.

Shock spectra have been developed for a wide variety of pulse shapes. For a particular pulse shape, each spectrum defines the frequency content and peak magnitude of the response. The spectrum characterizes the response of an infinite number of single degree of freedom systems. The effects of a shock may easily be assessed in terms of the frequency characteristics of the design.

Generally, all elements of an assembly that have the same resonant frequencies experience comparable peak acceleration responses. The oscillatory motions throughout the assembly quickly dissipate provided damping exists and reinforcing shock pulses do not occur. Oscillation within the system may disturb or excite other responding assemblies that are structurally coupled, the latter exhibiting similar frequency characteristics. The resulting motions are often complex due to these interactions. Related shock factors are detailed in Table 1.7.

Equipment malfunction or damage to supporting attachments caused by shock are related to:

1. Absolute motion of the supported item, which may incite relative motion between assemblies;
2. Relative motion between a supported item and its foundation that could result in support failure or collision between the item and its surroundings.

Increasing the compliance of supports or attachment structure generally decreases the response acceleration but increases absolute or relative displacements. Appropriate materials, components and construction are required to withstand the anticipated response. The severity of shock equipment can withstand without damage is governed by the allowable stresses of the most highly stressed component. Stresses are minimized by achieving a balance between response displacements and acceleration. Weight and cost are minimized by developing a design for which the stress levels of all elements reach their allowables concurrently.

Vibration.   Failures due to vibration occur because the accumulated stress reversals exceed the endurance strength of the materials involved. Steady state excitation at equipment resonance contributes dramatically to the cumulative process. The nature of response amplitudes are dependent upon the characteristics of the design and input excitation. Equipment designs are often cascaded.

Table 1.7. Shock Factors

Mechanism	Effects on Equipment	Accelerating Factors
Fracture	Cracks or catastrophic failure of equipment, chassis, module and component attachments	Brittle materials, materials with low ultimate or yield strength, prestressing, excessive distance between unit C.G. and attachments
Collision	Impact of neighboring parts and assemblies	Compliant structure and support designs, inadequate spacing
Brinelling	Impairment of sliding or rolling contact mechanisms	Preload levels, interface compliance

As an example, consider components attached to circuit boards and modules which are supported by a frame or chassis mounted in a housing. If any of these components have similar resonant frequencies, substantially greater accelerations and displacements will occur. The loads and stresses experienced by vulnerable elements and the influence of supporting structure design on the control of developing amplitudes are of more concern than the nature of excitation at mounting points.

Acceleration applied to the housing mounting points develops response accelerations on each assembly. The probability of damage to any assembly is proportional to the maximum stress developed. An infinitely stiff supporting structure exhibits the same accelerations as the housing, so the maximum stresses which result are proportional to the maximum accelerations of the housing. Compliant chassis and circuit boards experience stresses that are a function of the environment developed by the housing and their design characteristics.

Structures with small damping usually exhibit a dramatic response at resonance. This is characterized by relatively large amplitudes and attendant stresses. The nature of resonance and dynamic interactions that occur are usually the underlying cause of equipment failure. Resonant amplification for sinusoidally excited systems is directly proportional to Q. Equipment structures and circuit boards generally have relatively small damping and consequently large values of Q. Large values of Q are more evident at lower levels of excitation. As the level of excitation is increased, the system behaves somewhat nonlinearly, limiting the response amplitude. The nonlinear characteristics involve small movement at interfaces and greater dissipation so the system exhibits substantially smaller Q at increased levels. The response amplitude for random vibration is directly proportional to $\sqrt{Q}$ within the frequency band considered.

Vibrations may have an infinite variety of spectral distributions. They can be periodic or random and can involve any medium. Vibrations can be either translational or rotational, or a combination of both, the result of a cyclic variation of force, displacement, strain or pressure. Vibration factors are given in Table 1.8.

Vibration leads to a progressive deterioration of the elements whose resonances fall within the excitation envelope. This is due to fatigue which is a result of cumulative exposure to vibratory excitation. Higher amplitudes, which cause greater stresses, are more damaging than lower ones. By the limitation of displacements, component stresses will be reduced correspondingly, increasing the vibration lifetime of the equipment. Although fatigue is indicative of many stress reversals, the time to failure may be short due to the frequencies involved.

Excessive response amplitudes in cascaded or built-up systems can be avoided by uncoupling their natural frequencies. Each resonance should be sep-

Table 1.8. Vibration Factors

Mechanism	Effect on Equipment	Accelerating Factors
Fatigue	Fracture of attachments structural members, artwork, wire interfaces, component leads, loosening of parts, abraded cables and surfaces	Coupled resonant frequencies, excessive displacements, stress concentrations, lack of strain relief, insufficient support or restraint, surface conditions
Collision	Impact of neighboring parts and assemblies	Close spacing of assemblies
Noise	Loosened attachments and parts, fractured elements, intermittent connections	Insufficient tightening, excessive relative displacement, compliance
Fretting Corrosion	Pitting and flaking of surfaces carrying loads	Insufficient preload, insufficient material hardness
Creep-elastic Distortion	Alignment errors in mechanical or optical devices	Structural compliance, damping, redundancy, use of nonmetallic materials

arated from one another by at least one octave or a factor of two above or below the nearest resonance.

Instrument and mechanical systems sometimes include vibratory sources to eliminate error sources and hang-off due to stiction. Vibration may also be useful in some positioning systems.

Equipment containing blowers and motors with unbalanced rotating masses may experience troublesome vibration and noise if the rotation speed is high. Rotating frequencies may be close to the natural frequencies of the structure supporting the device, amplifying the disturbance.

Acoustic Excitation. Failures due to acoustic excitation occur because the accumulated stress reversals exceed the endurance strength of the materials involved. The effect on equipment is the same as for vibration. Acoustic excitations extend to higher frequencies which can not be reproduced by vibration test equipment. Limitations are imposed by the frequency response of the shaker and attenuation of the fixture. Electronic assemblies which have large compliant exposed surfaces with low mass to area ratios and low internal damping or high Q are particularly susceptible to acoustic stimulation.

Acoustic excitation is distributed as random energy from a source. The source may be boundary layer, jet or missile engine noise. Vibrating structural elements tend to modify the transmission of random excitation. The input is transformed at element resonances to simultaneous sinusoidal waveforms of fluctuating amplitudes. For purposes of analysis, the fluctuating amplitudes are assumed to be a stationary process. Stated succinctly, the rms pressure or acceleration is independent of time.

Equipment designed for vibration usually withstands acoustic environments. Noise levels internal to a housing are reduced from 6 to 30 db by absorption and transmission losses associated with the materials and cascaded construction. Greater attenuation is obtained with coatings of rubber, plastic or paper materials on exposed surfaces.

Acceleration. Acceleration loads act through the center of gravity of an assembly developing bending, flexural and shear stresses in the supporting structure and at attachment interfaces. Stresses are minimized by reducing the distances between attachments and center of mass of each assembly. Accelerations do not develop sufficient stress reversals, so fatigue is not a consideration. Failure occurs because the load exceeds the ultimate or yield strength of the materials used. Maximum design loads, displacements and stress occur when acceleration exists concurrently with vibratory excitation.

Usually acceleration is referenced to the dimensionless quantity g where one g represents 32.17 ft/s^2 (9.81 m/s^2). If the load in an element at static conditions is known, the maximum load due to acceleration is the product of g and the static load.

The effects of acceleration on equipment are similar to those described for Table 1.7.

1.2 EVALUATION AND WORKMANSHIP TESTS

Evaluation and workmanship tests for electronic equipment are pseudoenvironments developed to simulate severe service or field conditions. These tests are normally accelerated by increasing the level of severity to obtain an indication of capability and performance in a reasonable period of time. Some tests are conducted on equipment prototypes to demonstrate design feasibility before investment in hard tooling and costly manufacturing processes is made. Tests which demonstrate the equipment's ability to meet imposed requirements define the environments that influence design.

Workmanship tests are performed on production equipment to assure consistency and uncover faulty materials or assembly practices without aging the equipment. Tests are performed for short durations at reduced levels or near those actually anticipated in service. These tests are invaluable in uncovering faulty electrical connections, loose fasteners, loose particles in devices and inadequately installed items. Failures of this type usually occur at the onset of equipment use. Persistently recurring difficulties may require corrective design changes. Inability to tighten inaccessible fasteners or cold solder joints due to excessive local conduction may appear as a problem in workmanship but actually is design dependent. Workmanship tests are not meant to determine improper use of materials or to uncover faulty structural configurations, or to evaluate component arrangement.

Environmental design requirements are established empirically or analytically. When measurements are performed, they are usually limited to a few points that characterize vehicle or system behavior. Specific levels for field or service conditions at candidate equipment locations are rarely determined. Analytical criteria is based on models which represent the behavior of the environment and the system under study. Design levels for equipment are limited by the refinement of the model and the degree to which the actual system is represented. Specific data representing the environment at equipment locations can be used to optimize designs for a particular application.

Usually, design requirements are established in a general sense since equipment locations are not known in advance. Measured or analytically determined characteristics are enveloped to generalize the behavior. Where there had been notches or valleys due to a quiescent system state, there appears an environment equal to or greater than peak levels occuring elsewhere or at another time or frequency in the system. Often the levels represented by the enveloped environment are then increased to account for uncertainties relative to system behavior, model simplifications, production variables or service environments.

Considerable variations may exist between military and commercial requirements. Most military electronic equipment is designed to operate when exposed to frigid arctic as well as hot desert climates, in vented or unvented compartments which may be subjected to noise, acceleration, shock, and vibration. Commercial and industrial equipment, on the other hand, is designed for sheltered environments which limit the variations and severity of climatic and induced dynamic conditions. The controlling design factors for commercial equipment are cost and reliability. Except for transport and handling loads, dynamic service conditions are benign. Thermal environments and attendant operating temperatures for components and assemblies are usually studied and evaluated in order to enhance system reliability of commercial equipment. Similar considerations apply to industrial and military equipment which operates over greater environmental temperature variations and may include service accelerations or other dynamic loads.

Sheltered environments permit the use of unsealed units and the application of natural cooling techniques to a greater extent than can be accommodated in exposed conditions. Some industrial applications require operation in dirty environments, which necessitate closed or sealed designs. Under these conditions, the evaluations of design and performance of industrial and military equipment are approximately comparable.

Equipment evaluation factors that influence design are listed in Table 1.9. The degree of severity of tests designed to examine equipment suitability in

Table 1.9. Environmental Evaluation Factors

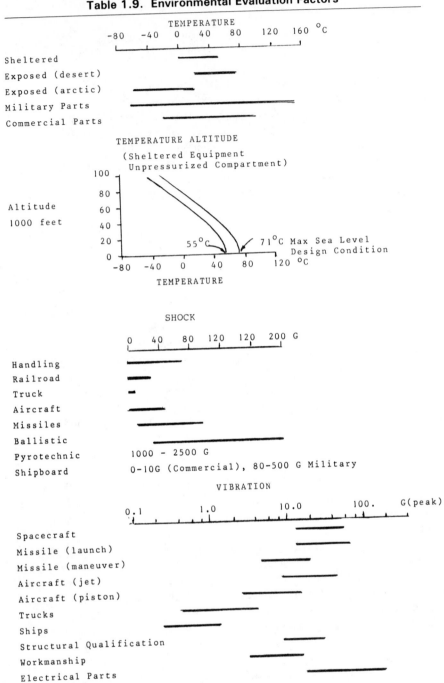

TEMPERATURE

| | -80 | -40 | 0 | 40 | 80 | 120 | 160 °C |

Sheltered
Exposed (desert)
Exposed (arctic)
Military Parts
Commercial Parts

TEMPERATURE ALTITUDE
(Sheltered Equipment
Unpressurized Compartment)

Altitude
1000 feet

100
80
60
40
20
0

55°C   71°C Max Sea Level
                Design Condition

-80 -40   0   40   80   120 °C

TEMPERATURE

SHOCK

0   40   80   120   120   200 G

Handling
Railroad
Truck
Aircraft
Missiles
Ballistic
Pyrotechnic   1000 - 2500 G
Shipboard   0-10G (Commercial), 80-500 G Military

VIBRATION

0.1   1.0   10.0   100.   G(peak)

Spacecraft
Missile (launch)
Missile (maneuver)
Aircraft (jet)
Aircraft (piston)
Trucks
Ships
Structural Qualification
Workmanship
Electrical Parts

14

Table 1.9. Environmental Evaluation Factors (*Continued*)

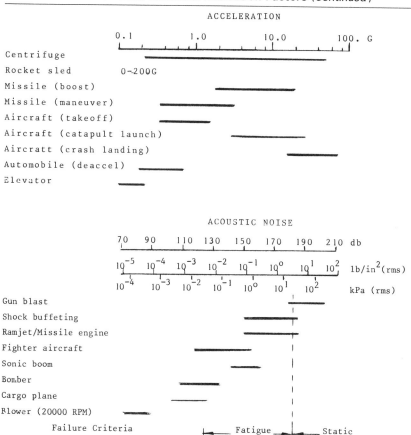

ACCELERATION

	0.1	1.0	10.0	100. G
Centrifuge				
Rocket sled	0-200G			
Missile (boost)				
Missile (maneuver)				
Aircraft (takeoff)				
Aircraft (catapult launch)				
Aircraft (crash landing)				
Automobile (deaccel)				
Elevator				

ACOUSTIC NOISE

70 90 110 130 150 170 190 210 db

10^{-5} 10^{-4} 10^{-3} 10^{-2} 10^{-1} 10^{0} 10^{1} 10^{2} lb/in^2(rms)

10^{-4} 10^{-3} 10^{-2} 10^{-1} 10^{0} 10^{1} 10^{2} kPa (rms)

Gun blast
Shock buffeting
Ramjet/Missile engine
Fighter aircraft
Sonic boom
Bomber
Cargo plane
Blower (20000 RPM)

Failure Criteria   Fatigue   Static

service depends on the contraction used to determine test duration, amplitudes and cyclical conditions. The contraction is based on the physical nature of the actual environments being simulated. Usually, it is not practical to obtain a suitable sample or a complete definition of the service environment. Variation in test duration or amplitudes from representative service data may be determined in terms of construction practices and materials generally used in electronic equipment.

The resonant amplification Q of a system is related to the stress σ for the materials used by

$$Q = C_o \sigma^{2-n} \qquad (1.1)$$

where the experimentally determined value C_o is a function of material and construction. The value of n, which defines the slope of specific damping energy versus the ratio of reversed stress to fatigue strength ranges between 2 and 3 for a variety of commonly used materials for device and structural applications. The mean slope has the value of 2.4.

Since the response acceleration can be expressed as the product of the system amplification and input acceleration

$$\ddot{x} = Q\ddot{x}_f \tag{1.2}$$

the nature of the input can be developed in terms of the resultant stress produced by substituting (1.1) into (1.2), giving

$$\ddot{x}_f = \frac{A\sigma^{n-1}}{MC_o} \tag{1.3}$$

where $F = M\ddot{x}$ using $F = \sigma A$ with A and M representing the area and mass respectively of a simplified model. An input acceleration which develops a member stress at the endurance level becomes

$$\ddot{x}_{fe} = \frac{A\sigma_e^{n-1}}{MC_o} \tag{1.4}$$

A stress acceleration relationship based on an average slope of 2.4 is determined as the ratio of (1.3) to (1.4):

$$\left(\frac{\ddot{x}}{\ddot{x}_e}\right)_f = \left(\frac{\sigma}{\sigma_e}\right)^{1.4} \tag{1.5}$$

The fatigue behavior of materials, which is a measure of ability to withstand repeated cyclical stress, may be expressed in terms of the cycles to failure. The slope of the resulting stress-cycle curve ranges between 7 and 13 for a variety of materials commonly encountered in equipment design. The mean slope has a value of 9.0. A general relationship describing the fatigue-cycle behavior becomes

$$\frac{N}{N_e} = \left(\frac{\sigma_e}{\sigma}\right)^{9.0} \tag{1.6}$$

Combining (1.5) and (1.6), an expression which relates input acceleration amplitude to fatigue cycles is obtained as

$$\frac{N}{N_e} = \left(\frac{\ddot{x}_e}{\ddot{x}}\right)_f^{6.4} \tag{1.7a}$$

Since the number of cycles, N, is directly related to the duration of vibration, t, (1.7a) may be used to evaluate test parameter interrelationships of duration and acceleration. In this case, (1.7a) becomes

$$\frac{t}{t_e} = \left(\frac{\ddot{x}_e}{\ddot{x}}\right)_f^{6.4} \tag{1.7b}$$

As an example, the contraction of a one hour swept vibration test at the levels of Figure 1.1(a) to a ¼ hour swept vibration test requires the increased test levels of Figure 1.1(b) to maintain comparable damage criteria.

Equation (1.7) may also be used to assess the relative severity of environments; this assessment enables the design engineer to determine test conditions comparable to those used to develop the design. Consider the random vibration design spectrum shown as the solid line in Figure 1.2. A new application requires that the equipment meet a higher level indicated by the broken line, above. The two levels can be compared by normalizing the duration of the design spectrum to that of the new requirement. A variation of (1.7b) which enables the design engineer to construct a graphic representation of the contracted spectrum becomes

$$\frac{t}{t_e} = \left(\frac{W_e}{W}\right)_f^{3.2} \tag{1.7c}$$

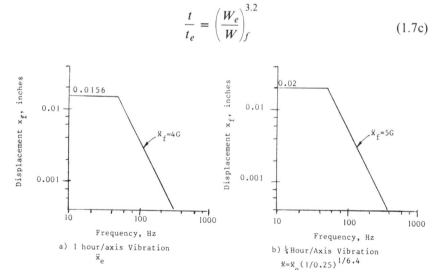

Figure 1.1. Contraction of Vibration Test Time.

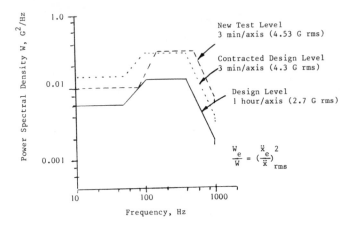

Figure 1.2. Comparison of Random Vibration Requirements.

for constant values of mean squared acceleration density W at a defined frequency. The result of applying (1.7c) to the solid line spectrum is shown in Figure 1.2 as a dotted line. The dotted line spectrum represents a damage level at 3 minutes/axis comparable to the original design spectrum at 1 hour/axis. The contraction indicates that the new environment is not as severe for design frequencies <100 Hertz and only slightly more severe at frequencies >400 Hertz. Equation (1.7c) may be used to determine the overall G rms level for the contraction envelope if desired. The relationship between the spectral density values and the rms accelerations for any two identically contoured spectrums may be determined from (1.7b) and (1.7c) as

$$\frac{W_e}{W} = \left(\frac{\ddot{x}_e}{\ddot{x}}\right)^2_{rms} \tag{1.7d}$$

Equation (1.7) is based on the assumption that the accumulated damage as a result of exposure to a prolonged service environment can be represented by the damage accumulated by a more intense environment over a shorter period of time. On this basis relationships for contracting noise environments can also be developed. With the assumption that the accumulated damage in each narrow band changes proportionately, the change in the sound pressure level (SPL) as a function of test duration for the average material stress-cycle slope becomes

$$\log\left(\frac{t}{t_e}\right) = \frac{9}{20}(SPL_e - SPL) = \frac{9}{20}\Delta SPL \text{ db} \tag{1.8a}$$

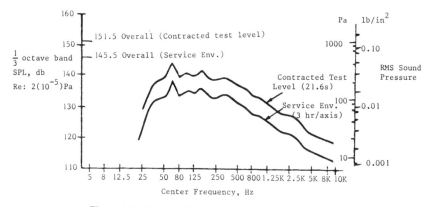

Figure 1.3. Missile and Aircraft Acoustic Noise Spectrum.

The ratio of test time for an increase of 6db in the overall sound pressure level for the service environment of Figure 1.3 using (1.8a) is 501:1. Thus equipment exposed to a noise source for 3 hours can be tested at a level 6db higher for only 21.6 seconds. The contraction of the corresponding pressure spectral density may be determined using the relationship

$$\frac{t}{t_e} = \left(\frac{W_{pe}}{W_p} \right)^{4.5} \tag{1.8b}$$

which with the aid of (1.8a) can be expressed as

$$\frac{W_{pe}}{W_p} = 10^{(\Delta SPL/10)} \tag{1.8c}$$

The effect on the rms pressure due to a change in the acoustic test level may be determined from

$$\left(\frac{P_e}{P} \right)_{rms} = 10^{(\Delta SPL/20)} \tag{1.8d}$$

Equations (1.8c) and (1.8d) were used to develop the contracted acoustic test level spectrum and rms pressure from the service environment of Figure 1.4.

Design test levels are often developed using (1.7) and (1.8) after measured amplitudes and durations for the service environments have been increased to account for uncertainties. The resulting requirements develop greater displacements and loads on equipment and at attachment interfaces. Comparative

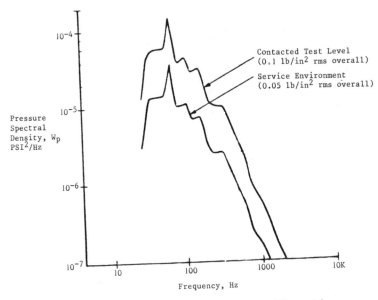

Figure 1.4. Acoustic PSD for Environments of Figure 1.3.

damage at increased intensities for shorter exposure periods is dependent only on stress-cycle relationships for the materials and construction typically found in electronic hardware. Equations (1.7) and (1.8) do not apply if the greater intensities introduce nonlinearities in the behavior of the equipment.

The responses of an assembly or component may either increase more or less than a similar increase in excitation. The mechanisms associated with this behavior are complex. This is due to the non-linear influence of damping, the redistribution of loads and stresses, and elastic modification. Some component boards, structural elements and panels become more compliant as deflection increases and others stiffen. The resonant amplification factor Q tends to follow changes in frequency but not in constant proportion. Increasing the acceleration excitation at a resonant frequency causes a decrease in Q. Generally, Q tends to decrease more in highly damped systems for a corresponding increase in excitation.

Higher acceleration levels due to contraction can cause a part which experiences stress levels below the endurance level in the service environment to fail prematurely. Greater intensities due to contraction may also cause reversion of stress margins at critical locations from positive to negative. A design exhibiting positive margins wherein the yield stress is never reached in one environment can incur negative margins, deformation and premature failure in a more intense environment.

Other effects which alter behavior can also occur. Larger displacements due to increased acceleration levels may exceed established spacing on adjacent assemblies. Impact and abrupt velocity changes that result could develop high local acceleration levels that may exceed the ultimate strengths of surrounding components and attachments, causing the development of fractured components, leads and solder joints that would not occur at the original service test level.

Use of a chamber for conducting free air temperature tests may also influence the behavior of equipment. Chamber cavity circulation developed by an integral fan can create conditions of greater severity at low temperatures and lower severity at high temperatures. The radiant thermal energy exchange between the equipment and chamber walls is an additional factor.

The simulation of free air conditions for heat dissipating equipment implies that forced chamber circulation is not permitted. Additionally, a minimum spacing between the equipment and chamber walls must be maintained. The spacing which develops approximately free air conditions is a function of equipment volume and maximum surface heat dissipation, as shown in Figure 1.5. A greater wall spacing permits higher unit surface heat densities.

Hardware evaluation tests assure a satisfactory level of reliability when the equipment is introduced into service. Reliability is determined through a prediction of the fatigue life and load carrying capacity of the structural elements and the ability of the equipment to function properly. Both analysis and testing provide the needed information. Analysis is generally used where relatively simple mathematical models provide a reasonable characterization of hardware behavior. Tests are used where idealized models fail to account for certain significant parameters which involve configuration, interface or environmental complexities. It is generally desirable to perform both analysis and tests to com-

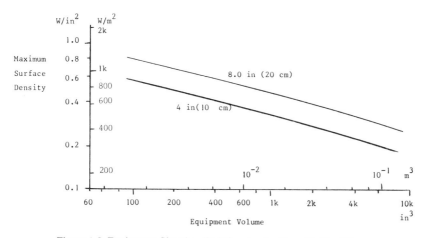

Figure 1.5. Equipment-Chamber Wall Spacing for Free Air Conditions.

plement each other, adding a higher degree of understanding relative to equipment behavior and the techniques involved in idealizing physical systems. Analysis provides basic insight into the physical character of hardware that can not always be obtained from test. Data obtained from idealized models may also be used to optimize sensor locations for a test configuration.

1.3 APPLICATION CONSTRAINTS

Equipment designs are influenced by the environments to which they are exposed, the tests which evaluate capability and the constraints imposed by the installation or operation. Installation constraints are those factors which contact or penetrate an imaginary boundary immediately around the equipment. These include limitations imposed by the electrical, mechanical or man-machine interfaces and limitations regarding configuration, size, weight, center of gravity, or orientation. Operational constraints include performance, response or reaction rates, repeatability and reliability. Related applications factors are shown on Tables 1.10 and 1.11 for installation and operational constraints respectively.

Installation, location and the local environment into which equipment is placed can adversely affect equipment life and performance. Provisions must be provided in the mounting space for equipment cooling. Naturally cooled equipment uses either convection or conduction to remove generated heat. Natural convection depends on a free exchange of the surrounding air at design temperature conditions. Conduction depends on a suitable attachment interface and the ability of the mated surface to remove generated heat without

Table 1.10. Application Factors
(Installation Constraints)

Factor	Effect on Equipment
Electrical Interfaces	Size and location of cables/connectors, test/troubleshooting provisions, grounding, radiated/conducted interference
Mechanical Interfaces	Forced or conductive cooling provisions, mounting/ attachment provisions, clearances for dynamic displacements, noise sensitivity, alignment and keying provisions
Man-Machine Interfaces	Handling provisions, control selection and placement, access for test, repair or replacement, tool requirements, lighting, visibility, function selection
Spatial	Configuration, size, weight, mounting approach, connector location, cooling technique, weight/balance criteria
Orientation	Cooling efficiency, sensitivity to direction dependent loads, gravity

Table 1.11. Application Factors
(Operational Constraints)

Factor	Effect on Equipment
Performance	Sensitivities to resonant amplifications, noise, and displacements, thermal control criteria, dynamic response criteria
Reaction Rates	Cold start stability criteria, heater requirements, insulation/thermal capacity constraints
Repeatibility	Mechanical interface stability, constraints on structural redundancy
Reliability	Load factor criteria, component temperature constraints, performance margins, fatigue, life criteria, power dissipation

experiencing an inordinate temperature rise. Force-cooled equipment requires clearances at inlet and exhaust openings to prevent choking and loss of design flow rates. Design provisions should also be included in event of short term loss of cooling or excessive temperature variations.

Operation at temperatures below the dew point at conditions of high humidity can cause moisture condensation within equipment and on components, circuit boards and connectors. Airborne equipment is extremely susceptible due to a wide variation in altitude, temperature and humidity that often occurs during flight. Condensation within equipment or on sensitive electronic circuits can be avoided by isolating the cooling air by confining it in a cold plate or plenum.

Heat dissipated by electronic equipment is inversely proportional to their thermal efficiencies in obtaining a useful output. Equipment with large surface areas having low power dissipation rarely exhibit temperature related problems. Compact high power density designs require special attention to assure effective heat removal and attainment of reliable operating temperatures. For any design, power-wasteful components should be minimized and packaging techniques employed to distribute heat sources and develop controlled heat removal paths to an available sink.

Maintenance provisions should permit test, removal and replacement of assemblies for an operational system without affecting the cooling effectiveness of other assemblies. This includes filters and screens that may become blocked.

The behavior of the equipment attachment interface may include components of both translational and rotary motion. The compliance of the mounting structure for electronic equipment may influence alignment and dynamic stability of sensitive systems. Lack of planarity and tolerance accumulation can introduce prestressing that can affect behavior. Similar conditions can develop because of structural redundancy within the equipment.

2
Cooling Techniques

2.1 FUNDAMENTAL ASPECTS OF HEAT TRANSFER

Most electronic equipment generates heat as a function of obtaining a useful output. In this case, heat develops within electronic parts having low thermal efficiencies. In general, the operating temperature depends on the heat applied to the system. One objective of the package design is to provide effective heat removal paths which reduce part temperatures and thereby increase the product's life. All modes of heat transfer are involved in determining the paths and the temperature of any part. These modes are conduction, convection, radiation and heat storage.

Heat transfer by *conduction* is through the substance of the element, analogous to electrical conduction. Heat applied to a mounting surface from an electronic part conducts through the material to all adjoining interfaces, seeking a cooler reference. Heat transferred by *convection* relies on the relative motion of a viscous media. If the motion of air adjacent to a heated surface is due to a reduction in density as a result of heating, then heat is transferred from the surface to the air by convection. In *radiant* heat transfer, there is a transfer of energy between distant bodies by electromagnetic radiation, analogous to light radiation. A surface exposed to the sun is heated by thermal radiation. Components mounted to a heated surface which is adjacent to a cooler structure loses heat by radiation. *Heat storage* relates to a change in the content of thermal energy in the medium along any heat transmission path. The quantity of heat stored is a function of time, the rate of temperature change and the thermal capacity of the materials involved.

The design engineer has the prerogative of selecting any of the heat transfer modes in developing a configuration. The modes selected should be compatible with the constraints imposed by the equipment interfaces and service environment to assure that performance objectives are not compromised. Equipment designed for forced convection requires provisions for coolant inlets and exhausts which should not be restricted by adjacent equipment surfaces or compartment walls. Even though one mode of heat transfer may be used as the primary means of heat removal from a package, a combination of the remaining heat transfer processes are still involved.

24

The electronic package is a specially constructed enclosure which provides the means of removing generated heat enabling devices to operate reliably in accordance with their design rating characteristics. The electronics package may also protect electronic components from the effects of severe external environments which may be due to large variations in temperature or exposure to solar radiation.

The effectiveness of natural convection, forced convection and radiation are compared for a cylindrical component in Figure 2.1. These results provide an indication on the relative merit of these heat transfer modes for cooling any assembly.

The choice of cooling mode influences the package design, reliability, weight and cost. The objective of limiting part temperatures becomes increasingly difficult to accomplish if the dissipated power is not managed effectively during circuit and system design. Limitations imposed on equipment size and orientation affects the ability to remove generated heat and the means available to control part temperatures.

Cooling by natural convection or radiation is inherently reliable and low cost since no special equipment such as fans or pumps are required. Mechanical parts required to move fluids or gases can malfunction or wear out. Designs which use natural convection and radiation are also less susceptible to failure or loss of performance due to airborne contaminants.

Part temperatures are a consequence of the power they dissipate, their physical dimensions and the temperature of their immediate surroundings. The steady state temperature rise of any heat dissipating element depends on the balance between heat generated and lost to its surroundings. The heat loss is due to the contributive effects of heat transferred by radiation, convection and

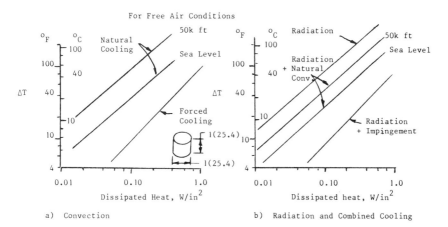

Figure 2.1. Temperature Rise of a Typical Electronic Part.

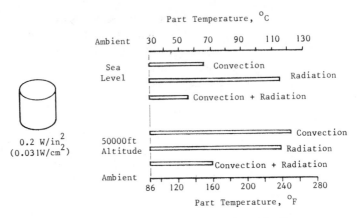

Figure 2.2. Part Temperature Due to Natural Cooling.

conduction. Figure 2.1(a) describes the temperature rise of a typical component considering heat transfer by convection only. A component dissipating 0.2W/in^2 (0.031W/cm^2) would operate at $63°\text{F}$ ($35°\text{C}$) above ambient temperature at sea level. The same component would experience a $164°\text{F}$ ($91.1°\text{C}$) increase in temperature at an altitude of 50,000 ft. Use of Figure 2.1 based on an ambient temperature of $86°\text{F}$ ($30°\text{C}$) results in the component temperatures shown in Figure 2.2.

As seen from Figure 2.1, forced convection cooling is a more effective method of heat removal than free convection. The higher velocities encountered in forced convection decreases the resistance to heat transfer across the developed boundary layers. At relatively low flow rates, thermal resistances about half of those obtained with free convection and radiation can be obtained. In general, thermal resistances of the order of $18°\text{F/W} \cdot \text{in}^2$ ($1.55°\text{C/W} \cdot \text{cm}^2$) can be achieved with practical mass flow rates.

The characteristics of equipment design as influenced by the heat transfer modes are summarized in Table 2.1. Conduction, convection and radiation, the three natural modes of heat transfer, may occur individually or in any combination.

Conduction. The basic law of heat conduction at steady state in its most simple form is

$$Q = \frac{kA \, \Delta T}{L} \tag{2.1}$$

given Q = rate of heat transfer;
       k = thermal conductivity of the material;

Table 2.1. Related Design Characteristics

Heat Transfer Modes	Design Characteristic
Natural Convection	• Applicable over moderate ambient temperatures • Nonlinear function of size, shape, orientation and temperature • Cools moderate to low component densities • Vented designs require louvers, screens or perforated surfaces • Improves with increasing surface area normally extended by fins, pins or decreased component density • Effectiveness decreases ∼1.5% per 1000 ft increase in altitude
Forced Convection	• Requires supplied coolant or self-contained fan or pumps which could contribute to heat load • Applicable over moderate to high ambient temperatures • Cools moderate to high component densities • Requires the establishment of flow paths and baffles • Requires consideration of pressure loss and choking in selecting coolant prime mover • Requires consideration of acoustic noise
Conduction	• Cools moderate to high component densities • Requires a dedicated high pressure interface contact area • Requires use of high conductivity materials, short lengths and large cross sections
Radiation	• Requires consideration of surface finishes and coatings • Effectiveness independent of altitude increasing dramatically with greater temperature differences
Thermal Storage	• Requires materials with high thermal capacity • Requires large mass—which increases equipment weight

A = cross sectional area perpendicular to the direction of heat flow;
L = length of the conducting path;
ΔT = temperature difference between the two ends of the path length.

If heat is dissipated into a material by a component, a steady flow occurs to cooler surroundings. The simplest flow is one dimensional, where the direction of flow is parallel throughout the cross section. A simple example is heat flow in a strip of constant cross section from one point along its length to another. For fixed dissipation, Q, the temperature of a part may be reduced with the aid of (2.1) by decreasing the distance to an available sink or increasing the cross sectional area or conductivity of the path material. Thermal conductivity, k, is a thermodynamic property of the material. The value of k increases with temperature for gases and nonmetallic materials. The conductivity of liquids generally decreases with increasing temperature. Metallic materials exhibit varied behavior with typical values given in Appendix C.

Convection. The basic equation for convection is

$$Q = h_c A \, \Delta T \tag{2.2}$$

where Q = rate of heat transfer;
h_c = coefficient of heat transfer;
A = surface area normal to heat flow;
ΔT = temperature difference between the surface and coolant bulk.

The temperature gradient ΔT is confined to a very thin fluid layer immediately adjacent to the surface. At a distance greater than the thickness of this boundary layer, an isothermal condition exists due to the mixing motion of the coolant. The boundary layer thickness is a consequence of coolant velocity. Two conditions relative to the coolant velocity are laminar and turbulent flow. Coolant flow across rough surfaces can develop turbulence, which increases the heat transfer rate relative to a smooth surface exhibiting a laminar condition.

Turbulence defines a condition in which a smooth flow is broken up into eddies and cross currents. Such conditions exist for flow across circuit board components or assemblies which present an uneven rough surface. For duct flow, there is a velocity that is critical. Below the critical velocity, the flow is uniform and smooth. Above the critical velocity, the flow is turbulent.

Most indirectly cooled equipment using ducted flows exhibits *laminar* flow conditions. Friction losses for laminar flows do not increase as rapidly as those for turbulent conditions. This minimizes the requirements for the coolant prime mover.

For a given flow path, the friction losses due to turbulence increase in proportion to the square of the flow rate, while the fan power required increases with the cube. Designs for forced convection require careful consideration of availability and power dissipation of the fan and the effects of the noise generated. Additionally, the flow path and its attendant losses together with the equipment inlet and exhaust configurations must be accounted for during the design phase. Fan positioning within the inlet or exhaust plenum is another factor that can degrade performance of the convective flow system.

Natural Convection. Heat transfer by natural convection is influenced by many factors which include coolant properties, flow conditions, surface characteristics and temperature. The resulting nonlinear relationship is

$$Q = 0.00414 \, \frac{ACP^{0.5}}{L_c^{0.25}} \, \Delta T^{1.25} \text{ watts} \tag{2.3}$$

Table 2.2. Shape Factors and Characteristic Length for Various Configurations

Surface or Body	Orientation	L_c	C
Rectangular Plane ($L \times W$)	Horizontal (heated side up)	$2LW/(L + W)$	0.54
Rectangular Plane ($L \times W$)	Horizontal (heated side down)	$2LW/(L + W)$	0.27
Rectangular Plane ($L \times W$)	Vertical	Vertical dim. < 24 in	0.59
Misc. Plane	Vertical	Area/horizontal width	0.59
Circular Plane	Vertical	$0.785 \times$ diameter	0.59
Cylinder (long)	Horizontal	Diameter	0.53
Cylinder	Vertical	Vertical dim. < 24 in	0.55
Sphere	Any	Diameter/2	0.63
Parallel Planes[a]	Horizontal (heated Bottom plane)	Spacing	0.21
Parallel Planes	Vertical (length/space or $R > 3$)	Spacing[b]	$0.195/R^{1/9}$
Parallel Planes	Vertical (length/space or $R < 3$)	Spacing[b]	$0.28/R^{1/4}$
Small Parts	Any (height H, width W)	$HW/(H + W)$	1.45

[a]Conduction across interspace is used for horizontal planes when the temperature of the upper surface is greater than the lower.
[b]Spacing between surfaces should be sufficient to prevent boundary layer choking. Only conduction across the space occurs for spacings which inhibit convective flow.

given Q = rate of heat transfer;
A = surface area normal to flow, in^2;
L_c = characteristic length (Table 2.2);
C = shape factor (Table 2.2);
P = altitude to sea level pressure ratio;
ΔT = ambient to surface temperature difference, °C.

Heat transfer in natural convection is due to the circulation of a coolant that experiences a reduction in density of the boundary layer in contact with a heated body. Usually the coolant is air but other gases may be used. Natural convection accounts for an appreciable portion of the heat transferred from electronic devices at sea level. At higher altitudes, the effectiveness of this mode decreases, as shown in Figure 2.3. A decrease of about 1.5 percent per 1000 ft (0.3 km) occurs for altitudes up to 2000 ft (0.61 km).

EXAMPLE 2.1. Determine the temperature of a 200 milliwatt 0.75 in (19 mm) long by 0.2 in (5 mm) diameter component mounted horizontally within the plain surfaced

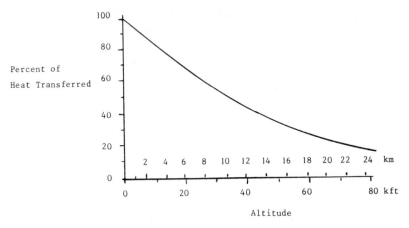

Figure 2.3. Effectiveness of Natural Convection Versus Altitude.

sealed housing (Figure 2.4) which dissipates 50 watts by natural convection to an ambient of 122°F (50°C) at sea level.

The heat dissipated by the housing is assumed to be distributed on all surfaces except the bottom. The quantity of heat dissipated by each surface is given by (2.3).

Top surface: Area $A = 7 \times 10 = 70$ in^2; characteristic length from Table 2.2, L_c $= 2(7)(10)/(7 + 10) = 8.24$ in; and shape factor $C = 0.54$.

$$q_{top} = 0.00414 \frac{AC}{L_c^{0.25}} \Delta T^{1.25} = \frac{0.00414(70)(0.54)}{8.24^{0.25}} \Delta T^{1.25}$$

$$q_{top} = 0.0924 \, \Delta T^{1.25}$$

Vertical surfaces: Area $A = (7 \times 7 + 7 \times 10) \times 2 = 238$ in^2; $L_c = 7$ and $C = 0.59$.

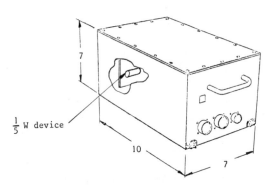

Figure 2.4. Natural Convection Cooled Component.

$$q_{\text{vert}} = 0.00414 \frac{AC}{L_c^{0.25}} \Delta T^{1.25} = \frac{0.00414(238)(0.59)}{7^{0.25}} \Delta T$$

$$q_{\text{vert}} = 0.357 \, \Delta T^{1.25}$$

The total heat dissipated by the top and vertical surfaces equals the heat dissipated by convection from the housing. This may be expressed as

$$q_{\text{housing}} = q_{\text{top}} + q_{\text{vert}} = 50 \text{ watts}$$

or

$$0.0924 \, \Delta T^{1.25} + 0.357 \, \Delta T^{1.25} = 50$$

$$\Delta T = \left(\frac{50}{0.0924 + 0.357} \right)^{1/1.25} = 43.4\,°\text{C}$$

The average surface temperature of the housing is

$$T_{\text{surface}} = T_{\text{ambient}} + \Delta T = 50 + 43.4$$

$$T_{\text{surface}} = 93.4\,°\text{C}$$

The internal convective ambient is equal to the surface temperature plus the air to surface temperature difference which is about the same as the external wall ΔT; therefore the 200 milliwatt device dissipates to an ambient of 136.8°C. Since the device is a small part, its characteristic length from Table 2.2 becomes

$$L_c = HW/(H + W) = 0.75(0.2)/(0.75 + 0.2) = 0.16 \text{ in}$$

and the shape factor $C = 1.45$. Rearranging (2.3), the solution for the device temperature rise becomes

$$\Delta T = \left(\frac{241.54 L_c^{0.25} Q}{AC} \right)^{0.8} \tag{2.4}$$

$$\Delta T = \left[\frac{241.54 \, (0.16)^{0.25} \, (0.2)}{(0.2\pi)(0.1 + 0.75)(1.45)} \right]^{0.8} = 18.9\,°\text{C}$$

The device temperature is

$$T_{\text{device}} = T_{\text{ambient}} + \Delta T = 136.8 + 18.9$$

$$T_{\text{device}} = 155.2\,°\text{C}$$

Q.E.D.

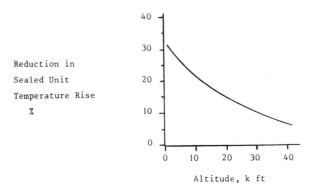

Figure 2.5. Effect of Altitude on a Vented Unit.

When environmental conditions permit, louvers or other openings may be used to reduce the temperature rise of components. The buoyant force of heated air leaving the housing through openings at the top of the housing draws cooler ambient air into the unit through openings near the bottom. The temperature rise of the heated column of air must be accounted for in determining device temperatures. Larger parts should be located to avoid baffling the coolant from smaller hotter components. At sea level, vented housings are about 33 percent cooler than a sealed unit. The effectiveness of venting decreases at altitude, as shown in Figure 2.5.

Sloped surfaces are not as effective as vertical surfaces in venting. The loss in efficiency is due to a redistribution of the coolant boundary layer. Equation (2.3) is modified to account for moderate deviations from conditions which determine the shape factor and characteristic length (from Table 2.2):

$$Q = 0.00414AC \left(\frac{\cos \alpha}{L_c} \right)^{0.25} P^{0.5} \Delta T^{1.25} \tag{2.5}$$

The effectiveness of plain surfaces may be increased by the addition of fins, ribs, pins or corrugations which increases the convective surface area. For a given dissipation, (2.4) illustrates the reduction in temperature rise expected for an increase in surface area provided the new area is at the same temperature as the original surface. A temperature gradient along the extended surface is the result of conductive or spreading impedances, which decreases the effectiveness of the added surface area. Referring to this loss in terms of a fin efficiency η, (2.3) becomes

$$Q = 0.00414 \frac{\eta ACP^{0.5}}{L_c^{0.25}} \Delta T^{1.25} \tag{2.6}$$

A design example which treats the effect of an extended surface is discussed in Section 4.2.

Radiation. The total energy radiated electromagnetically from a perfect radiator with a reflectivity of zero at any particular temperature is equal to σT^4, which is known as the Stefan-Boltzmann equation. Materials or finishes which act as a perfect radiator or black body are nonexistent. *Emissivity*, which is the ratio of the emissive energy of an actual surface to that of a black body, is used to define the energy emitted by any surface as $\epsilon\sigma T^4$. The materials and finishes commonly employed in electronics equipment and parts have the ability to absorb all incident radiation, a property which is dependent on temperature as well. Many materials exhibit the property when their emissivity is nearly the same as their absorptivity. The basic equation for radiant exchange between surfaces exhibiting these characteristics, namely *gray bodies*, is given by

$$Q = 3.485(10^{-12})\epsilon F(T_1^4 - T_2^4) \ \text{W/in}^2 \qquad (2.7a)$$

which in metric units is

$$Q = 5.704(10^{-8})\epsilon F(T_1^4 - T_2^4) \ \text{W/m}^2 \qquad (2.7b)$$

given Q = rate of heat transfer per unit area;
   ϵ = emissivity characteristic of the surfaces;
   F = shape factor dependent upon surface orientation and geometry;
   T = absolute surface temperatures, $°R$ (2.7a), $°K$ (2.7b).

Parameters associated with (2.7) may be determined using Figure 2.6 or Figure 4.9 as illustrated by Example 4.3. When cooling consists of a combination of several heat transfer modes, the use of Figure 2.7, which employs the heat transfer coefficient h_r in the form of (2.2), may be a time saving aid.

EXAMPLE 2.2.   A circuit dissipating 10W is distributed over a 50 in^2 (0.032 m^2) area with an average emissivity of 0.4. This surface, which radiates to a 90°C surface, has a geometric shape factor of 1. A conductive path having a conductance of 0.5W/°C connects the heated surface to a 40 in^2 (0.026 m^2) surface which radiates to a 50°C chassis surface. This surface exhibits an emissivity of 0.5 with a geometric shape factor of 1. A schematic representation of the heat transfer configuration is shown in Figure 2.8(a). Determine the temperature of the heated component surface.

An initial estimate for the temperature differences for the two surfaces is used to evaluate the corresponding radiative conductances, $h_rA\epsilon F$. Based on a 25°C temperature difference, the conductance to the 90°C surface in terms of Figure 2.7 becomes

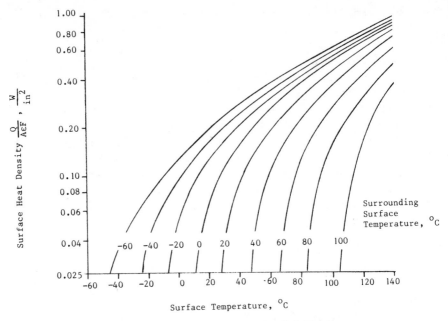

Figure 2.6. Characteristics of Radiating Surfaces.

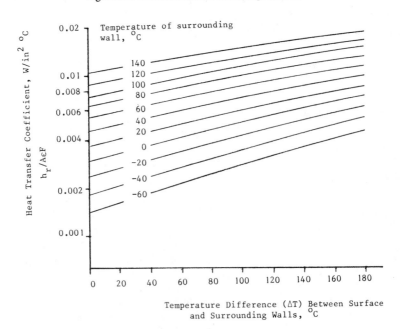

Figure 2.7. Radiation Coefficient of Heat Transfer.

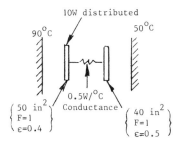
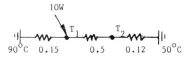

a) Schematic of Heat Transfer Geometry   b)   Heat Transfer Network

Figure 2.8. Idealized Network for Radiation Heat Transfer.

$$(h_r A \epsilon F)_1 = 0.0075(50)(0.4)(1) = 0.15 \text{W}/^\circ\text{C}$$

The heat transfer coefficient in metric units is determined from (2.7) and Figure 2.7 as

$$h_r = 0.0075 \frac{5.704(10^{-8})}{3.485(10^{-12})(1.8^\circ R/^\circ K)^4} = 11.69 \text{W}/\text{m}^2 \cdot^\circ\text{C}$$

The conductance in metric units becomes

$$(h_r A \epsilon F)_1 = 11.69(0.032)(0.4)(1) = 0.15 \text{W}/^\circ\text{C}$$

When 50°C is used for the initial estimate for the temperature difference to the 50°C surface, the conductance determined in a similar manner gives the thermal network shown in Figure 2.8(b).

Solution for temperature T_1 follows the procedure of Appendix A.1 as illustrated in Example A.6.1. The matrix representation becomes

$$\begin{bmatrix} 0.62 & -0.5 \\ -0.5 & 0.65 \end{bmatrix} \begin{Bmatrix} T_2 \\ T_1 \end{Bmatrix} = \begin{Bmatrix} 0.12(50) \\ 10 + 0.15(90) \end{Bmatrix}$$

Following the reduction procedure of Appendix A.6, the reduced matrix determining T_1 and T_2 becomes

$$\begin{bmatrix} 1 & -0.806 \\ 0 & 1 \end{bmatrix} \begin{Bmatrix} T_2 \\ T_1 \end{Bmatrix} = \begin{Bmatrix} 9.677 \\ 114.73 \end{Bmatrix}$$

The first equation represented by the reduced matrix determines T_2 to be

$$T_2 = 9.677 + 0.806 T_1 = 9.677 + 0.806(114.73^\circ\text{C})$$

$$T_2 = 102.15^\circ\text{C}$$

The values of T_1 and T_2 are used to determine improved values for the assumed temperature differences. The resulting updated matrix gives closer approximations to T_1 and T_2. Since the results obtained confirm the initial estimate for the temperature differences, the temperature of the heated component mounting surface is 114.73°C.   Q.E.D.

Thermal Storage.   Any part of an electronics assembly which is along a heat transmission path will store heat. The basic relationship is

$$Q = \rho C_p V \frac{\Delta T}{\Delta t} \qquad (2.8)$$

where Q = rate of heat transfer;
     ρ = material density;
    ΔT = change in temperature of the material;
    Δt = change in time;
    C_p = specific heat of the material.

Materials with a high thermal capacity (C_p) can absorb more heat than materials with a low capacity for a given temperature rise. Capacities for various materials are listed in Appendix C. Materials which experience a change in state from a solid to a liquid or from a liquid to a gas can absorb a large quantity of heat during the transition. Fusibles, composed of materials such as paraffins or organic salts, undergo reversible changes in state from a solid to a liquid when heated and a liquid to a solid when cooled. These materials are commonly used for energy storage in the absence of adequate thermal sinks.

The thermal capacity of the assembly affects its response to transients, which may be due to changes in environment or operational conditions. A condition of thermal equilibrium exists when temperatures do not change with time. A transient upsets this equilibrium, requiring the system to establish an energy balance on the new conditions. The temperature distribution varies with time until this balance is obtained. The rate of change of the assembly temperatures are governed by the distribution of thermal capacity throughout the configuration.

2.2 STEADY STATE TEMPERATURE DISTRIBUTION

Steady state conditions exist when temperatures throughout the assembly do not change with time. When this occurs, the thermal energy in the system and at any location is in balance. Equilibrium at any location is simply stated as *heat in equals heat out.*

Sources and Sinks. The temperature distribution of an assembly is a function of its geometry, the placement of heat dissipating components and the nature of heat transfer internal and external to the configuration. Thermal energy seeks paths of least resistance on route to a cooler reference, or sink. Heat will flow from a component assembly of high temperature to one of lower temperature even though both dissipate heat. The ultimate sink for heat dissipated by an assembly is the surrounding environment, which may include its mounting interface structure.

Any component, assembly or unit that dissipates heat is a thermal source. The motor of a fan which provides forced convection cooling is an example of a source of thermal energy. Diodes, transistors, resistors, capacitors, transformers and chokes are also some of the many heat sources found in electronic assemblies. These sources may be packaged individually as discrete components or as part of an integrated assembly in a dual-in-line package (DIP), leadless carrier or as part of a hybrid assembly. The manner in which sources are packaged may inhibit the transfer of heat to its surroundings.

Concentrated sources refers to individual or small clusters of heat dissipating components. The junctions of an LSI chip are an example of concentrated sources used in determining the temperature distribution within the interconnecting wafer. A *uniformly distributed heat source* may consist of many nearly equivalent thermal sources distributed over a defined area. *Distributed sources* are analogous to a blanket heater.

Graphical Methods. The temperature distribution of many component mounting surfaces encountered in electronics packaging may be intuitively estimated after a review of the distribution of sources and available thermal sinks. If desired, a sketch to the approximate scale of the surface may be constructed using lines of constant temperature and/or lines depicting the heat flow. Lines of constant temperature tend to hug concentrated sources and acutely shaped boundaries of known temperature. Isothermal lines spread out in the field of distributed sources and at obtusely shaped boundaries of known temperature. Isothermal lines are also perpendicular to boundaries which are adiabatic where no heat is transferred.

A rough sketch outlining the primary heat flow paths can be used to develop an equivalent network for determining particular temperatures on the surface. An estimate of the temperature distribution may be obtained directly from the sketch if the system of isotherms and heat flow lines are drawn so that they intersect everywhere at right angles to form little quadrilaterals. The quadrilaterals should approximate squares if possible. The temperature or heat flow may be obtained using a variation of (2.1) which becomes

$$Q = Nkb \, \Delta T^* \qquad (2.9)$$

given N = number of heat flow paths
$\quad\quad\quad b$ = material thickness
$\quad\quad\quad \Delta T^*$ = temperature difference between isotherms

The temperature distribution and heat flow can be estimated for surfaces of any shape using this technique. When several sources are present, superposition must be used. An isothermal plot for each source acting independently with nulled boundary temperatures is required in addition to a separate plot for the temperature distribution due only to the boundary temperatures. The temperature of any point on the surface is obtained by adding the values of corresponding points on each plot. Accuracies to within 10 percent of exact values may be obtained. Unfortunately, the graphical determination of the temperature distribution is slow and tedious.

One Dimensional Idealization. Approximate temperature distributions of complex surface configurations or three dimensional systems can be obtained rapidly using idealizations based on a network of discrete conductances, finite elements or finite differences. Exact distributions based on the solution of governing differential equations are practical for only the simplest configurations. Differential equations are often used to develop equivalent one dimensional conductances for two or three dimensional systems. In this way, the temperature distribution of complex heat transfer systems can be resolved rapidly and accurately.

Consider, for example, the temperature distribution of a sector of an annular surface as shown in Figure 2.9(a), with line heat source q along the inner radius and temperature T_o on the outer radius. Concentric lines of constant temperature hug the line source, spreading out so they approach the outer boundary as shown in Figure 2.9(b), due to the increasing cross sectional area as the outer boundary is approached.

The differential equation for the temperature distribution is obtained by performing a heat balance on the differential element, Figure 2.9(a), which becomes

$$\frac{1}{\theta r}\frac{d}{dr}\left(\theta rkb\,\frac{dT}{dr}\right) = 0 \tag{2.10}$$

subject to the boundary conditions

$$\theta rkb\,\frac{dt}{dr} = -q \quad\quad \text{at } r = r_i$$

$$T = T_o \qu\quad\quad \text{at } r = r_o$$

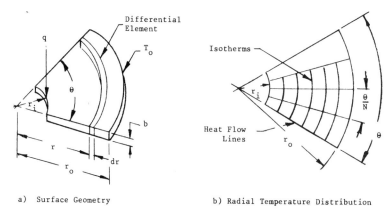

a) Surface Geometry   b) Radial Temperature Distribution

Figure 2.9 Annular Sector.

The solution obtained by integration and substitution of the appropriate boundary conditions is

$$T = \frac{q}{\theta k b} \ln \left(\frac{r_o}{r} \right) + T_o \qquad (2.11)$$

A pseudo one dimensional conductance based on the relationship $Q = G \, \Delta T$ for the annular configuration using (2.11) becomes

$$G = \frac{\theta k b}{\ln \left(\dfrac{r_o}{r_i} \right)} \qquad (2.12)$$

Many useful pseudo one dimensional conductances have been derived (Table 2.3) to facilitate network characterization of hardware configurations for the determination of the temperature distribution.

EXAMPLE 2.3.   A 5W power transistor is mounted on a 0.06 in. (1.52 mm) thick bracket which is supported on two sides by a chassis at a temperature of 120°F (49°C), Figure 2.10(a). Determine the transistor-bracket interface temperature using (a) the graphical technique and (b) an idealized network approach.

(a) *Graphical Solution.* On Figure 2.10(b), isotherms are sketched concentrically around the circular source spreading out as they approach the constant temperature boundaries as in Figure 2.9(b). Heat flow lines are drawn and the isotherms are corrected so that little quadrilaterals are formed. Advantage is taken of symmetry. Since

Table 2.3. Equivalent Conductance

Geometry	Conductance, G

$$\dfrac{\theta k b}{\ln \left(\dfrac{r_o}{r_i}\right)} \qquad (2.12)$$

$$\dfrac{k b \ln \left(\dfrac{r_o}{r_i}\right)}{\theta} \qquad (2.13)$$

$$\dfrac{k b (Z - W)}{L \ln \left(\dfrac{Z}{W}\right)} \qquad (2.14)$$

$$\dfrac{2 k b}{\theta_2 - \theta_1} \qquad (2.15)$$

$$\dfrac{k b B}{H} \qquad (2.16)$$

$$\dfrac{2 \pi k L}{\ln \left(U + \sqrt{U^2 - 1}\right)} \qquad (2.17)$$

$$U = \dfrac{d^2 - (r_1^2 + r_2^2)}{2 r_1 r_2}$$

Parallel Surface

$$\dfrac{2 \pi k L}{\ln \left(\dfrac{H}{r} + \sqrt{\left(\dfrac{H}{r}\right)^2 - 1}\right)} \qquad (2.18)$$

$$\ln \dfrac{2 \pi k L}{(2 H / r)} \qquad H \gg 2 r$$

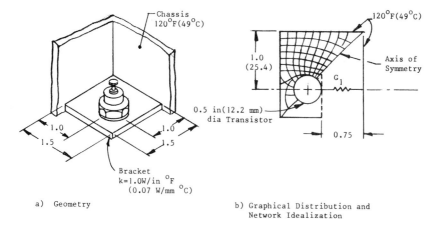

a) Geometry

b) Graphical Distribution and
Network Idealization

Figure 2.10 Transistor Bracket Temperature Distribution.

there are 16 heat flow paths connecting the source to the chassis, the isothermal temperature difference using (2.9) becomes

$$\Delta T^* = \frac{q}{Nkb} = \frac{5}{16(1.0)(0.06)} = 5.21°F$$

In metric units $\Delta T^* = \dfrac{5}{16(.07)(1.52)} = 2.94°C$

The transistor mounting temperature using 8 isotherms between the source and chassis becomes

$$T_{transistor} = 7(5.21) + 120 = 156.5°F$$

$$= 7(2.94) + 49 = 69.6°C$$

(b) *Idealized Network Solution.* One conductance G_1, Figure 2.10(b), represents one-half of the symmetrical surface. The conductance for a trapezoidal surface (2.14) with the short base tangent to the component gives

$$G_1 = \frac{1.0(0.06)(0.75)}{0.75 \ln\left(\dfrac{1.5}{0.75}\right)} = 0.0865 W/°F$$

In metric units

$$G_1 = \frac{0.07(1.52)(19.05)}{19.05 \ln\left(\dfrac{38.1}{19.05}\right)} = 0.153 W/°C$$

Since half the generated heat is transferred through G_1, the transistor mounting temperature becomes

$$T_{transistor} = 120 + \Delta T = 120 + \frac{q/2}{G_1}$$

$$= 148.9°F$$

This is a good approximation since isotherms tangent to the transistor are nearly parallel to the short base of the trapezoidal area used. The exact solution gives the transistor mounting temperature as $153°F$ $(67.2°C)$.   Q.E.D.

2.3 TRANSIENT TEMPERATURE BEHAVIOR

Transient behavior occurs when the temperature anywhere within the equipment changes with time. The transient may be due to any or a combination of events or conditions related to a change in environment or generated heat. A transient occurs when power is first applied to the equipment or when an assembly is exposed to a change in ambient temperature. When power is applied to equipment which has not been operating for some time, the heat generated causes an immediate increase in temperature at each source. Initially, very little heat is transferred to the surroundings. Heat transfer to neighboring areas and to the surrounding ambient develops slowly as temperature differences increase. The source temperatures are limited by the sink or ambient temperatures surrounding the equipment. Temperatures considerably above the ambient occur within equipment for low conductance (high impedance) heat transfer paths. Conversely, temperatures slightly above ambient result for high conductance (low impedance) paths. A knowledge of the transient temperature distribution provides the basis of determining component temperatures and heat rates for varying dissipations or environmental conditions that may influence performance or limit equipment life.

Time Constant Methods.   Equipment and various assemblies do not respond instantly to a change in generated heat or environment because of inertia. A measure of the thermal inertia of any part of an assembly is given by its time constant. The time constant is the elapsed time required for the temperature to change $(1/e)$th of its original value. The relationship between the system transfer function and the time constant is discussed in Appendices B.1 and B.5. Many assemblies and components may be represented by the model illustrated in Figure B.1-8. The steady state temperature T_{ss} for the assembly of Figure B.1-8 generating heat q is obtained by setting $WC_p = 0$ giving

$$T_{ss} = T_s + q/U \tag{2.19}$$

Equation (2.19) shows that the steady state temperature is a function of the sink temperature T_s and the temperature difference due to the heat transfer through conductance U. The time constant τ is determined from (B.1-32b) in terms of the time ϕ required to obtain temperature T with known initial temperature T_i giving

$$\tau = \frac{\phi}{\ln \left(\dfrac{T_{ss} - T_i}{T_{ss} - T} \right)} \tag{2.20}$$

The temperature at any increment of the time constant is found by writing (2.20) in the form

$$T = (1 - e^{-\phi/\tau})(T_{ss} - T_i) + T_i \tag{2.21}$$

Equation (2.21) represents the response of the model in Figure B.1-8 to step changes in sink temperature and generated heat that may occur at any time. When these conditions vary with time, a fractional increment of the time constant is used in (2.21). In this case, T_i represents the assembly temperature at the onset of each increment and T_{ss} represents the average steady state temperature during the increment. For those cases where the sink temperature varies linearly, the transient response may be obtained using

$$T = (1 - e^{-\phi/\tau})(T_{si} - T_i + q/U - m\tau) + T_i + m\phi \tag{2.22}$$

where T_{si} represents the initial sink temperature and m represents the slope. For an assembly which experiences a combination of linear or step changes in constraints, the response may be obtained by using (2.21) or (2.22) at the appropriate time.

EXAMPLE 2.4.   An assembly which dissipates 10W is operated in an 80°F (26.7°C) ambient after being stored at 40°F (4.4°C). Determine (a) the steady state temperature of the assembly if the conductance to ambient is 0.2 W/°C, (b) the time constant if 131°F (55°C) is obtained after 10 minutes of operation, and (c) the temperature history of the assembly.

(a) *The Steady State Temperature.* Since the sink temperature is 26.7°C during operation, the steady state temperature using (2.19) becomes

$$T_{ss} = 26.7 + 10/0.2 = 76.7°C \ (170°F)$$

(b) *The Time Constant.* Since the assembly reaches 55°C from an initial temperature of 4.4°C in 10 minutes, the time constant using (2.20) becomes

$$\tau = \frac{10}{\ln\left(\dfrac{76.7 - 4.4}{76.7 - 55}\right)} = 8.31 \text{ minutes}$$

(c) *The Transient Response.* The transient behavior may be plotted quickly using points determined on the basis of the number of time constants. Figure B.1-9 illustrates the procedure for heating or cooling. The assembly temperature at a half-time constant ($\phi = \tau/2$) or 4.16 minutes using (2.21) is

$$T_{\tau/2} = (1 - 1/\sqrt{e})(76.7 - 4.4) + 4.4 = 32.8°\text{C}$$

At one time constant (8.31 minutes) and two time constants (16.62 minutes), the assembly temperatures are

$$T_{1\tau} = (1 - 1/e)(76.7 - 4.4) + 4.4 = 50.1°\text{C}$$

$$T_{2\tau} = (1 - 1/e^2)(76.7 - 4.4) + 4.4 = 66.9°\text{C}$$

The temperature at successive time constants may also be expressed in terms of the preceding time constant temperature, i.e.,

$$T_{2\tau} = (1 - 1/e)(T_{ss} - T_{1\tau}) + T_{1\tau}$$

The transient response developed by sketching a line through the above points is shown on Figure 2.11(a).   Q.E.D.

EXAMPLE 2.5.   Determine the temperature history of the assembly in Example 2.4 for an ambient changing linearly in accordance with $T_s = 1.7\phi + 4.4°\text{C}$ accounting for a drop of 5W in the power dissipated when the assembly temperature exceeds 60°C.

The transient response is determined from (2.22) with $m = 1.7°\text{C/min}$ and $T_{si} = 4.4°\text{C}$. At $\tau/2$, the assembly temperature is

$$T_{\tau/2} = (1 - 1/\sqrt{e})[4.4 - 4.4 + 10/0.2 - 1.7(8.31)] + 4.4 + 1.7(8.31/2)$$

$$T_{\tau/2} = (1 - 1/\sqrt{e})35.87 + 11.46 = 25.6°\text{C}$$

at one and two time constants the assembly temperatures are

$$T_{1\tau} = (1 - 1/e)35.87 + 4.4 + 1.7(8.3) = 41.2°\text{C}$$

$$T_{2\tau} = (1 - 1/e^2)35.87 + 4.4 + 1.7(2)(8.3) = 63.7°\text{C}$$

The sketched response in Figure 2.11(b) is used to determine the elapsed time of 15.1 minutes required for the assembly to reach 60°C. New initial conditions at 15.1 minutes become

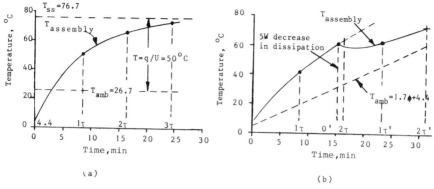

Figure 2.11. Transient Behavior of an Assembly. (a) Step input; (b) Linear ambient with a step change in power dissipation.

$$T_{s1} = 1.7(15.1) + 4.4 = 30.1°C$$

$$T_i = 60°C$$

The response progresses from this point with time ϕ reinitialized. The temperature at $\tau/2$ (an accumulated time of 19.26 minutes) is

$$T_{\tau/2} = (1 - 1/\sqrt{e})[30.1 - 60 + 5/0.2 - 1.7(8.31)] + 60 + 1.7(8.31/2)$$
$$T_{\tau/2} = (1 - 1/\sqrt{e})(-19.03) + 67.06 = 59.6°C$$

At one time constant (23.41 accumulated minutes) and at two time constants (31.72 accumulated minutes), the assembly temperatures are

$$T_{1\tau} = (1 - 1/e)(-19.03) + 60 + 1.7(8.31) = 62.1°C$$
$$T_{2\tau} = (1 - 1/e^2)(-19.03) + 60 + 1.7(2)(8.31) = 71.8°C$$

The composite response is shown in Figure 2.11(b).   Q.E.D.

The time constant for an application with a different conductance U_a can be determined from the original configuration with time constant τ and conductance U using the relationship

$$\tau_a = \tau(U/U_a) \qquad (2.23)$$

Empirical or measured transient response temperatures may be used to determine the time constant of the test configuration. Any portion of the response should be divided equally into three increments. The selected increment should be as large as possible. The temperature at successive increments are designated T_1, T_2 and T_3. When $\Delta\phi$ represents the selected increment in time, the time constant becomes

$$\tau = \frac{\Delta\phi}{\ln\left(\dfrac{T_2 - T_1}{T_3 - T_2}\right)} \tag{2.24}$$

The steady state temperature for the measured response can be predicted using

$$T_{ss} = \frac{T_2^2 - T_1 T_3}{2T_2 - T_1 - T_3} \tag{2.25}$$

EXAMPLE 2.6.   Determine the time constant and steady state temperature for the response shown in Figure 2.11(a).

Using the increment $\Delta\phi = 8.3$ minutes and the temperatures corresponding to the divisions at 1τ, 2τ and 3τ in (2.24) gives

$$\tau = \frac{8.3}{\ln\left(\dfrac{66.9 - 50.1}{73.1 - 66.9}\right)} = 8.32 \text{ minutes}$$

The steady state temperature using these temperatures in (2.25) becomes

$$T_{ss} = \frac{66.9^2 - 50.1(73.1)}{2(66.9) - 50.1 - 73.1} = 76.7\,^\circ\text{C}$$

The results compare closely to the results obtained in developing the response in Example 2.4 with the same parameters.
Q.E.D.

Numerical Techniques.   The thermal behavior of the assembly of Example 2.4 was based on the assumption that the generated heat and the thermal capacity of the structure are uniformly distributed. This assumption permitted the use of the simple representation of Figure B.1-8 to investigate the assembly temperature response. The representation is based on lumping the assembly so that the model of Figure B.1-8 is descriptive of the actual characteristics of its interfaces and thermal geometry. This modeling approach can be extended to include locations, components and assemblies constituent to an item of equipment or adjoining interfaces, neighboring equipment and the surrounding environment. The temperature history of any defined lumped representation consisting of more than two temperatures requires the use of numerical techniques.

In many cases, equipment may be modeled requiring temperature behavior at many locations. The analytical solution of the resulting large set of differential equations by the methods of Appendix B is not practical. In these cases, the use of numerical techniques may be used to obtain solutions. The techniques used are readily adapted to computer use if the developed equations are too numerous to be solved by manual methods.

Euler Method. The Euler method reduces a set of differential equations to a set of algebraic equations. The required temperatures are found directly using multiplication and addition. Since the thermal response for equipment and assemblies are developed around known initial values, the temperatures after some increment in time $\Delta\phi$ are required. Time is then incremented using

$$\phi^{\nu+1} = \phi^{\nu} + \Delta\phi \qquad (2.26)$$

The simplest way to estimate the temperature distribution at $\phi^{\nu+1}$ is to determine the value of the derivative $dt/d\phi$ at ϕ^{ν} and move ahead in time using

$$T^{\nu+1} = T^{\nu} + \left(\frac{dT}{d\phi}\right)^{\nu} \Delta\phi \qquad (2.27)$$

The temperature history is developed by evaluating $T^{\nu+1}$ at the end of increment $\Delta\phi$ using T^{ν} as the initial value from the previous increment. The defined initial conditions are used to establish T^{ν} for the first step. Equation (2.27) can be generalized for a system of equations based on matrix representations of the quantities involved.

The derivative $dT/d\phi$ represents the connectivity, heating and heat transfer conditions imposed on each lumped representation. Both linear and nonlinear conditions may be used. Nonlinear conditions result whenever natural convection or radiation are used. Parameters of $dT/d\phi$ which vary as a function of time are easily accommodated during the incremental solution process. The increment $\Delta\phi$ may be adjusted to correspond to the event when the event does not occur at a specific step. The adjusted step should be $<\Delta\phi$ to preserve accuracy. The original increment may be resumed after incorporating the event if desired.

The temperature history given by (2.27) progresses via the derivative evaluated at the old time. This condition contributes some error at each step. Greater accuracy results when small increments are used. In practice, larger increments are desirable in order to minimize the computational effort required. An increment $\Delta\phi \simeq \tau_{min}/5$ will minimize the loss of accuracy as the solution progresses.

EXAMPLE 2.7.   Determine the temperature history of the assembly of Example 2.5 using Euler's method.

The differential equation based on a heat balance for the assembly as illustrated in

Appendix B.1 accounting for a linearly varying ambient and dissipation Q is

$$WC_p \frac{dT}{d\phi} + UT = (1.7\phi + 4.4)U + Q \qquad (2.28a)$$

Expressing (2.28a) in terms of the derivative gives

$$\frac{dT}{d\phi} = (1.7\phi + 4.4 - T)\frac{U}{WC_p} + \frac{Q}{WC_p} \qquad (2.28b)$$

where the quantities Q, U and WC_p may be determined from Example 2.4, i.e.,

$$WC_p = U\tau = 0.2(8.31)$$

$$WC_p = 1.662 \ \text{W} \cdot \text{min}/^\circ\text{C}$$

Substituting quantities and writing (2.28b) in the form of (2.27) gives

$$T^{\nu+1} = T^\nu + (6.546 + 0.2046\phi - 0.1203T^\nu)\,\Delta\phi \qquad (2.29)$$

A suitable increment $\Delta\phi$ is found using

$$\Delta\phi \simeq \tau_{\min}/5 = 8.31/5$$

$$\Delta\phi = 1.66 \ \text{min}$$

An increment $\Delta\phi = 1.5$ min will be used. The assembly temperature after each 1.5 min interval using (2.29) becomes

$$T^{\nu+1} = 0.819T^\nu + 0.307\phi^\nu + 9.819 \qquad (2.30)$$

The temperature $T^1 = 13.425\,^\circ\text{C}$ is determined after 1.5 min using an initial temperature $T = 4.4\,^\circ\text{C}$ and $\phi = 0$ minutes. The temperature $T^2 = 21.282$ after 3.0 min is found by evaluating (2.30) using $T^1 = 13.425\,^\circ\text{C}$ and $\phi^1 = 1.5$ minutes. Subsequent temperatures are determined using quantities from the previous step in the manner described. The results obtained from Example 2.5 are illustrated in Figure 2.11(b).

When the assembly reaches $60\,^\circ\text{C}$, the dissipated heat decreases to 5W. This event occurs after approximately 15 minutes. Temperatures subsequent to this point are determined using (2.28b) and (2.27) with $Q = 5\text{W}$. The Euler relationship may be expressed as

$$T^{u+1} = 0.819T^u + 0.307\phi^u + 5.307 \qquad (2.31)$$

where $u = 0$, $\phi = 15$ min and $T = 60\,^\circ\text{C}$. The temperatures at successive 1.5 minute increments subsequent to this point are determined from (2.31) as

ϕ min	15	16.5	18.0	19.5	21	22.5	24	25.5	27
T °C	60.00	59.08	58.79	59.00	59.64	60.63	61.89	63.39	65.08

These results are shown in Figure 2.11(b).   Q.E.D.

Semiexplicit Method. The temperature history developed by the Euler method is based on a direct application of the quantities at each increment. The semiexplicit method is based on using the actual closed form solution for each temperature. The terms which couple related temperatures to the differential equation of interest are grouped as a pseudo heat load. The pseudo heat load is adjusted iteratively using average temperatures determined from the solution of coupled differential equations within each increment $\Delta\phi$. When changes in the temperature diminish to a negligible amount within a given increment, the correct temperature distribution has been achieved. The solution then progresses to the next increment using quantities determined for the previous increment as initial conditions.

Consider the temperature distribution given by the set of coupled differential equations

$$
\begin{bmatrix}
(WC_p)_1 & & & & \\
& (WC_p)_2 & & & \\
& & \cdot & & \\
& & & \cdot & \\
& & & & \cdot \\
& & & & & (WC_p)_n
\end{bmatrix}
\begin{Bmatrix}
\dot{T}_1 \\
\dot{T}_2 \\
\cdot \\
\cdot \\
\cdot \\
\dot{T}_n
\end{Bmatrix}
$$

$$
+
\begin{bmatrix}
G_{11} & G_{12} & \cdots & G_{1n} \\
G_{21} & G_{22} & \cdots & G_{2n} \\
\cdot & \cdot & & \\
\cdot & \cdot & & \\
\cdot & \cdot & & \\
G_{n1} & G_{n2} & & G_{nn}
\end{bmatrix}
\begin{Bmatrix}
T_1 \\
T_2 \\
\cdot \\
\cdot \\
\cdot \\
T_n
\end{Bmatrix}
=
\begin{Bmatrix}
Q_1 \\
Q_2 \\
\cdot \\
\cdot \\
\cdot \\
Q_n
\end{Bmatrix}
\qquad (2.32)
$$

Any temperature T_j of (2.32) can be found from the solution of

$$
(WC_p)_j \frac{dT_j}{d\phi} + G_{jj}T_j = Q_j - \sum_{\substack{i=1 \\ i \neq j}}^{n} G_{ji}T_i
\qquad (2.33)
$$

The semiexplicit solution of (2.33) is

$$T_j = \frac{\overline{Q}_j}{G_{jj}}(1 - e^{-\Delta\phi/\tau_j}) + T_{j0}e^{-\Delta\phi/\tau_j} \qquad (2.34)$$

where

$$\overline{Q}_j = Q_j - \Sigma\, G_{ji}\overline{T}_i \qquad \text{(pseudo heat load)}$$

$$\tau_j = (WC_p)_j/G_{jj}$$

The barred quantities in (2.34) indicate that average values for the increment $\Delta\phi$ are used. T_{j0} refers to the temperature T_j at the beginning of the increment. Time ϕ moves ahead according to (2.26). The exponential quantities can be determined once for a particular increment. Equation (2.34) is numerically stable and provides greater accuracy than could be obtained with the Euler method for the same increment.

Although the semiexplicit method produces highly accurate solutions with linear averaging of the temperatures, some improvement may be realized using a higher order method. Higher order averaging methods, however, do increase the computational effort. Various averaging methods which apply to uniform increments are given in Table 2.4.

Use of (2.35b) and (2.35c) requires known temperatures at two and three time increments respectively. Equation (2.35a) must always be used for the first increment. Parabolic averaging may be used during the second increment.

EXAMPLE 2.8.   Determine the temperature response of a centrally located 0.2W component and the corresponding point on the circuit board using (a) the semiexplicit method and (b) Euler's method. The schematic of the board assembly, Figure 2.12(a), may be represented by the model of Figure 2.12(b).

(a) *Semiexplicit Solution.* The two differential equations for the model of Figure 2.12(b) developed with the aid of Appendix A.1 in the form of (2.32) are

Table 2.4. Averaging Methods

Case	Characteristic	Equation	
I	Straight Line	$T = (T_n + T_{n+1})/2$	(2.35a)
II	Parabolic	$T = (-T_{n-1} + 8T_n + 5T_{n+1})/12$	(2.35b)
III	Cubic	$T = (T_{n-2} - 5T_{n-1} + 19T_n + 9T_{n+1})/24$	(2.35c)

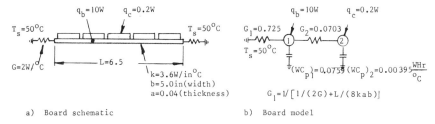

a) Board schematic   b) Board model

Figure 2.12. Transient Circuit Board Model.

$$
\begin{bmatrix} (WC_p)_1 & 0 \\ 0 & (WC_p)_2 \end{bmatrix} \begin{Bmatrix} \dot{T}_1 \\ \dot{T}_2 \end{Bmatrix}
$$

$$
+ \begin{bmatrix} G_1 + G_2 & G_2 \\ -G_2 & G_2 \end{bmatrix} \begin{Bmatrix} T_1 \\ T_2 \end{Bmatrix} = \begin{Bmatrix} q_b + G_1 T_s \\ q_c \end{Bmatrix} \qquad (2.36)
$$

The exact solution of (2.36) may be obtained using the transfer function technique outlined in Appendix B, particularly B.2 and B.5. These techniques are not practical for large models with many differential equations. The differential equations (2.33) developed from (2.36) are

$$(WC_p)_1 \dot{T}_1 + (G_1 + G_2) T_1 = G_2 T_2 + q_b + G_1 T_s$$

$$(WC_p)_2 \dot{T}_2 + G_2 T_2 = G_2 T_1 + q_c \qquad (2.37)$$

The equations which determine the temperature distribution for each increment are developed from (2.34) using (2.37) as

$$T_1 = \frac{(G_2 \overline{T}_2 + q_b + G_1 T_s)}{G_1 + G_2} (1 - e^{-\Delta\phi/\tau_1}) + T_{10} e^{-\Delta\phi/\tau_1}$$

$$T_2 = \frac{(G_2 \overline{T}_1 + q_c)}{G_2} (1 - e^{-\Delta\phi/\tau_2}) + T_{20} e^{-\Delta\phi/\tau_2} \qquad (2.38)$$

where $\qquad \tau_1 = (WC_p)_1/(G_1 + G_2)$

$$\tau_2 = (WC_p)_2/G_2$$

Substituting quantities from Figure 2.12 into (2.38) and simplifying gives

$$T_1 = (58.154 + 0.884\overline{T}_2)0.21 + 0.7899 T_{10} \qquad (2.39a)$$

$$T_2 = (2.845 + \overline{T}_1)0.3296 + 0.67036 T_{20} \qquad (2.39b)$$

where $\Delta\phi = 2\tau_2/5 = 0.0225$ hr. The initial temperatures $T_{10} = T_{20} = 50°C$ for the first increment. The solution is started using $\overline{T}_2 = T_{20} = 50°C$ giving

$$T_1 = [58.154 + 0.0884(50)]0.21 + 0.7899(50) = 52.635°C$$

The average value of $\overline{T}_1$ is

$$\overline{T}_1 = (52.635 + 50)/2 = 51.318°C$$

This value is used in (2.39b) to determine T_2 as

$$T_2 = (2.845 + 51.318)0.3296 + 0.67036(50) = 51.37°C$$

The average value of $\overline{T}_2$ is

$$\overline{T}_2 = (51.37 + 50)/2 = 50.685°C$$

A second evaluation of (2.39a) gives

$$T_1 = [58.154 + 0.0884(50.685)]0.21 + 0.7899(50) = 52.648°C$$

$$\overline{T}_2 = (52.648 + 50)/2 = 51.324°C$$

A second evaluation of (2.39b) gives

$$T_2 = (2.845 + 51.324)0.3296 + 0.67036(50) = 51.372°C$$

Since a negligible change to T_2 has occurred, the temperatures after 0.0225 hr are $T_1 = 52.648°C$ and $T_2 = 51.372°C$. These quantities are used as initial values for the next $\Delta\phi$ increment. The equation for determining T_1 at $\phi = 0.045$ hour is therefore

$$T_1 = (58.154 + 0.0884\overline{T}_2)0.21 + 0.7899(52.648)$$

where $\overline{T}_2 = 52.648$ for the first evaluation.

The temperature history up to 0.09 hour is shown compared to the exact solution in Table 2.5. The exact solution is developed using the method of Appendix B.

(b) *Euler's Method.* Substituting quantities into (2.37) and rearranging gives the expressions for evaluating the derivatives $\dot{T}_1$ and $\dot{T}_2$ as

$$\dot{T}_1 = 609.35 + 0.9262T_2 - 10.478T_1 \qquad (2.40)$$

$$\dot{T}_2 = 50.569 - 17.775T_2 + 17.775T_1$$

The temperature distribution at each increment is developed using (2.27) and (2.40) as

Table 2.5. Transient Circuit Board Response

Temperature	Method	Time, ϕ hours			
		0.0225	0.045	0.0675	0.09
	Euler	52.791	54.992	56.740	58.137
T_1	Semi explicit	52.652	54.769	56.476	57.857
	Exact	52.654	54.780	56.493	57.878
	Euler	51.320	53.107	54.994	56.794
T_2	Semi explicit	51.373	53.078	54.852	56.550
	Exact	51.421	53.151	54.937	56.639

$$\begin{Bmatrix} T_1 \\ T_2 \end{Bmatrix}^{r+1} = \begin{bmatrix} 0.8821 & 0.0104 \\ 0.20 & 0.8 \end{bmatrix} \begin{Bmatrix} T_1 \\ T_2 \end{Bmatrix}^{r} + \begin{Bmatrix} 6.855 \\ 0.5689 \end{Bmatrix} \qquad (2.41)$$

where $\Delta\phi = \tau_2/5 = 0.01125$ hour.

Temperatures for the first increment are determined for initial conditions $T_1^0 = T_2^0 = 50°C$ using (2.41) as

$$\begin{Bmatrix} T_1 \\ T_2 \end{Bmatrix}^{1} = \begin{bmatrix} 0.8821 & 0.0104 \\ 0.20 & 0.8 \end{bmatrix} \begin{Bmatrix} 50 \\ 50 \end{Bmatrix} + \begin{Bmatrix} 6.855 \\ 0.5689 \end{Bmatrix} = \begin{Bmatrix} 51.480 \\ 50.569 \end{Bmatrix} \qquad (2.42)$$

The temperatures for the first increment from (2.42) are used as initial conditions in (2.41) to determine the temperatures for the second increment ($\phi = 0.0225$ hour). The temperature at each additional increment are found in a similar manner using the results of the previous increment as initial conditions in (2.41). The temperature history is shown in Table 2.5.   Q.E.D.

2.4 HEAT REMOVAL SYSTEMS

All heat generated within electronic equipment by components or assemblies seeks the surrounding environment. Heat transfer from every source occurs by all three methods—conduction, convection and radiation—along tedious routes to cooler surroundings. Heat removal systems are implementations that shorten heat transfer paths to the cooler reference. Various media which bring the heat dissipating sources closer to a cooler reference include low impedance heat sinks, cold plates, heat pipes and impingement cooling techniques.

Effective cooling of electronic components, the major sources of heat in electronic equipment, requires a knowledge of the functional circuit partitioning and dissipations of the components used. Lower operating temperatures are achieved if, during circuit design, components with lower dissipations are selected. Equipment operating temperatures can be minimized by planning the

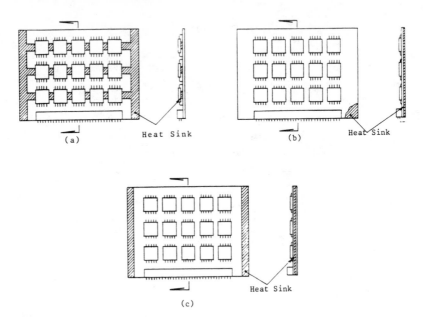

Figure 2.13. Circuit Card Heat Sink Configurations. (a) Conductive frame; (b) Conductive back plate; (c) Conductive core.

equipment layout with effective cooling systems in mind. The approaches used must be compatible with any geometric or interface constraints, other environmental requirements, the maintenance philosophy and cost.

Heat Sinks. Low impedance heat sinks, Figure 2.13, are often used on circuit boards to reduce component operating temperatures by minimizing the component-to-sink temperature difference. Conductively cooled circuit boards use either strips or plates fabricated from copper or aluminum to transfer the generated heat to the card interface. These heat sinks, Figure 2.13(a), are usually bonded to the board surface either beneath rows of similarly shaped components or on the unpopulated opposite surface, Figure 2.13(b).

Printed wiring boards with high generated thermal power densities may be fabricated using an aluminum plate integrally bonded between thin layers of copper clad circuit board material, Figure 2.13(c). Components can be attached to this assembly using the techniques developed for any two-sided circuit boards. Higher packaging densities are achieved by bonding assembled component boards to each side of an aluminum plate. In this case, the heat sink must transfer the heat generated by both assemblies.

Built-up assemblies consisting of copper or aluminum and epoxy glass fiberboard may distort or warp when the equipment is operated. Changes in temperature may cause severe stresses in the attaching media because copper and aluminum experience about twice the expansion of the epoxy glass material. These effects can be reduced by the use of symmetrical sections with equal thicknesses of epoxy glass on each side of the metal center. Thermally induced distortion of nonsymmetric assemblies can be minimized by fabricating them at a temperature approximately midway between their expected high (operating) and low (storage or transport) service temperatures. Chapter 5 describes some methods for evaluating the magnitude and location of peak stresses in assemblies of this type.

Aluminum mounting surfaces are often used to transfer heat generated by high power devices. Designs of this type are found in power supplies. Many high power components must be electrically insulated from conductive cooling surfaces. These components may experience an inordinate temperature rise if the insulating material cracks, distorts or cold flows after many on-off operational cycles or in the presence of vibration. Sometimes poor performance or failure in the presence of high component temperatures may be due to a loosening of the attachment interface as a result of temperature-induced deterioration of the insulating materials.

Conductive surfaces which are attached to electronic assemblies reduce local hot spots by distributing the component generated heat to surrounding areas. The heat sink also provides an effective path for transferring generated heat to an adjoining assembly or to the environment. Conductive sinks provide a means of increasing the convective surface area of an assembly through the use of fins or pins. The reduction of component temperatures and the temperature distribution of heat-sink-cooled assemblies may be assessed using the methods of Chapter 3.

Cold Plates.   High power electronic components or high heat density assemblies that result from miniaturization often require a more effective heat removal system than is offered by conductive heat sinks. The cold plate combines the effects of a conductive heat sink with convection heat transfer to reduce the impedances between the generating heat sources and the thermal sink. Because of the availability of air from either the surrounding environment or conditioned sources, it is commonly used as the sink. The cold plate isolates the cooling air from the circuits and components being cooled, avoiding difficulties attributed to entrained moisture and airborne contaminants.

The configuration and design of the cold plate may take on different forms and sizes depending on equipment packaging and thermal requirements. Specific flow paths depend on the nature and distribution of the assemblies requiring cooling, intermediate conductive paths between the thermal sources and the cold plate interface, and equipment structural, and maintenance considerations.

Cold plates are usually fabricated from aluminum as welded or brazed assemblies. These assemblies can be used as chassis walls, Figure 2.14(a), or

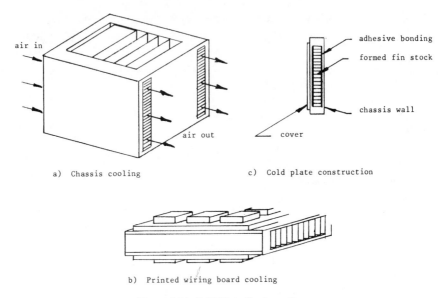

a) Chassis cooling   c) Cold plate construction

b) Printed wiring board cooling

Figure 2.14. Cold Plate Configurations.

for the indirect cooling of high powered electronic assemblies such as that illustrated in Figure 2.14(b). Cold plates can also be integrated into the chassis design via the use of formed finned stock assembled into a prepared cavity, as shown in Figure 2.14(c). Adhesives retain the finned stock, preventing flow short circuiting. The wall-to-air temperature difference along the flow path can be minimized by the selection of a suitable fin density. Higher fin densities also increase the resistance to flow which may impact on the selection of a fan or otherwise constrain the application of the equipment.

A cold plate technique for cooling circuit boards is shown in Figure 2.15. Cooling air flows between unpopulated board surfaces. The heat generated by the components must be conducted to the board and across its thickness to the

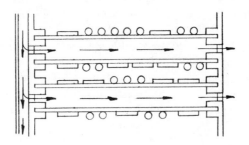

Figure 2.15. Indirect Cooling of Printed Wiring Boards.

cooling air. Care must be taken to assure proper air flow distribution and the prevention of leakage.

Temperature related performance can be enhanced by minimizing operational thermal cycling, which may be the result of variations in equipment generated heat or changes in sink or ambient temperatures. Cold plates equipped with heat pipes will also reduce temperature gradients within the cold plate, developing a uniform temperature sink for attached assemblies.

Hardware design using cold plate techniques achieves an orderly arrangement of conductive heat transfer paths and air flow paths. Components or assemblies with high heat generation should be located either close to the cold plate wall or the coolant inlet. Conductive attachments should be designed to minimize temperature differences. The net result is a configuration whose thermal performance and flow characteristics can be predicted accurately.

In most cases, this data can be developed before hardware fabrication is initiated. Design approaches that will provide a basis of evaluating cold plate configurations in terms of their temperature and flow loss characteristics are described in Chapter 4.

Direct Impingement. In this method, components are cooled directly by blowing air. Flow channels are developed in the equipment design exposing heat generating components to the moving cooling air mass, Figure 2.16. This

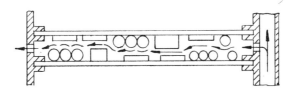

a)   Impingement cooled circuit cards

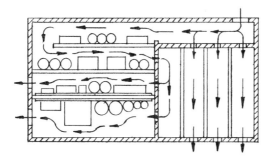

b)   Series - parallel cooling flow paths

Figure 2.16. Forced Convection Cooling.

heat removal method develops low component-to-air temperature differences since the components are directly exposed to the thermal sink.

Component spacing and distribution become a prime design factor in order to assure proper cooling of all components in the flow channel. Coolant routing and the flow channel configuration must assure that sufficient cooling is directed and distributed over high heat generating devices. A flow system of this nature usually develops large flow losses due to viscous drag and turbulence. This condition necessitates balancing of the pressure losses in different flow branches to assure adequate cooling throughout the equipment.

Hot spots on component boards or other assemblies due to concentrations of heat generating sources result in a uneven distribution of heat which increases the difficulty of obtaining effective cooling. Variations in component size and shape add an additional degree of complexity, Figure 2.16. In general, smaller parts experience lower case-to-coolant temperature differences due to end effects which increase its conductance. When determining conductances as a result of flow directly over components, one should include the actual surface area of the parts. Coolant temperature must also be accounted for. Variations in coolant temperature along surfaces of heat producing parts are a consequence of the duct configuration and component spacing at the location of interest. For this reason, hot spots within a given flow configuration determine the flow requirements for that branch. This approach generally results in conservative operating temperatures for low-heat generating components on the same assembly.

Airborne applications require consideration of varying air densities and flow losses due to changes in altitude. Air that cools airborne equipment can become moisture laden and may transport vehicle-associated contaminants which can cause intermittent performance or short-circuit printed circuit boards and connector interfaces. Exposed surfaces and components should be protectively coated to minimize the effects of moisture and other airborne contaminants.

The process of developing impingement cooled equipment requires

- Determination and description of heat generating components.
- Definition of the allowable maximum temperature for different classes of circuits or components used.
- Selection of candidate flow paths accounting for equipment configuration, interfaces and assembly arrangement in terms of generated heat distribution, coolant temperature rise and allowable component temperatures.
- Determination of flow rates and pressure losses along each flow path.
- Determination of component operating temperatures and equipment cooling requirements, e.g. inlet air temperature, flow rate and flow loss.

Heat Pipes. High density electronic packaging is limited by several factors, among which is the ability to remove generated heat efficiently to avoid exceeding allowable component temperatures. Heat pipes offer a simple solution for assemblies with high heat concentration. These devices can eliminate hot spots at high power density locations and reject the excess heat to cooler locations or to the equipment sink without requiring external power for operation. The heat pipe can be integrated into the design of circuit cards and chassis structure or attached as an external component.

Heat pipes are semiactive devices that employ an evaporation-condensation cycle. Heat is transferred within the device by vapor phase transport from a heat source (evaporator) to a heat sink (condensor). The condensate within the closed system returns to the evaporator by means of capillary wick pumping. The heat pipe which provides a large effective conductance at near isothermal conditions is capable of operation in nearly any orientation. For a given temperature rise, the heat pipe conductance is several orders of magnitude greater than a solid copper rod of the same configuration. These devices are sealed, noiseless, lightweight nonmechanical components that may be used to control equipment or assembly temperatures which exhibit varying heat generation rates or equipment exposed to varying thermal environments.

Heat pipe performance is a function of the size of the evaporator and condensor areas, wick construction, fluid media and pipe orientation. A schematic illustrating operational characteristics of a typical heat pipe is shown in Figure 2.17.

The operating temperature and energy transfer performance of a heat pipe is a function of its working fluid. Performance is also a function of the area of the evaporator and condensor sections within the pipe. Fluids that have a high latent heat of vaporization λ, high surface tension σ and a low viscosity ν are considered viable candidates. The relative performance of a fluid in terms of

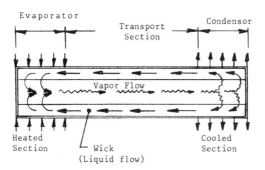

Figure 2.17. Heat Pipe Schematic.

Table 2.6. Heat Pipe Fluid Characteristics

Property	Ammonia	Water	Freon 21	Sulfur Dioxide	Mercury
Operating Range (°F)	−50 to 125	40 to 450	−20 to 200	15 to 110	400 to 820
Surface Tension σ, lb/ft	0.00124	0.005	0.00062	0.001	0.0322
Heat of Vaporization λ, BTU/lb	508.	980.	62	149	128
Kinematic Viscosity ν, ft^2/hr	0.014	0.038	0.0076	0.011	0.00414
Performance Factor p_f, $\dfrac{\text{BTU hr}}{\text{ft}^3}$	45.	129.	5.	13.	996.

its ability to optimize flow can be assessed using the relationship

$$p_f = \frac{\sigma\lambda}{\nu} \qquad (2.43)$$

Characteristics of a few common fluids are shown in Table 2.6.

The energy transferred at the evaporator in terms of the wick flow rate $\dot{m}_{\text{wick}}$ is

$$Q_{\text{evap}} = \lambda\dot{m}_{\text{wick}} \qquad (2.44)$$

The flow rate depends upon the cross-sectional wick area and porosity in addition to the density and capillary diffusion rate of the fluid. Porosity also influences pipe performance at different orientations. A variety of wick structures may be used, including screen or woven wire meshes, sintered powders or extruded grooves along the inside length of the pipe wall. Designs that increase the flow rate experience an attendant increased capability for thermal energy transfer.

The thermal performances of a grooved half-inch diameter heat pipe based on two different fluids are compared in Figure 2.18.

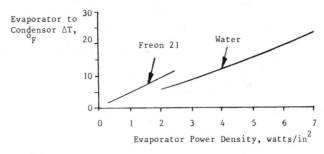

Figure 2.18. Effect of Fluid Media on Heat Pipe Performance.

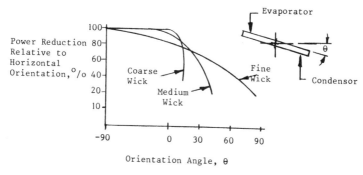

Figure 2.19. Heat Pipe Performance at Different Orientations.

An orientation that causes the heat pipe to work against gravity will degrade its performance. Condensed liquid flows back to the evaporator by capillary action. A design which places the condensor below the evaporator reduces the flow rate through the wick. This decreases the transfer of energy by the pipe. A fine wick, which in general does not permit good flow rates, will provide degraded performance at severe pumping orientations. A coarse wick, which provides good flow rates for horizontal applications, cannot move fluid at other orientations. This is shown graphically in Figure 2.19.

The manufacture and assembly of heat pipes require special care to avoid irreversible damage to vapor transport systems. Incompatibilities between fluid, wick or pipe materials lead to the eventual evolution of noncondensible gases which form a thermal barrier at the condensers. This condition reduces the efficiency or may inhibit operation of the pipe. Heat pipes must be assembled under vacuum conditions to avoid trapped gases. Proper cleansing is essential. Contaminants that diffuse into materials before assembly leach out during operation, inhibiting capillary flow. Special sealing methods are also required to avoid internal contamination of the closed cycle system.

All heat pipes should be tested before assembly into equipment. The tests may be accelerated by operating the pipes vertically with the evaporator down while maintaining higher than required operating temperatures. If contaminants are present, degraded performance will be detected after several weeks of operation.

2.5 EQUIPMENT MODELING

A model is an imaginary physical system that possesses the attributes or salient features of an actual system. The model is more amenable to analysis because it is an idealized simplification of complex hardware configurations. The process of constructing the model involves many approximations which should be clearly stated in a sketched representation of the actual part or assembly under study. In addition, the conditions at the boundaries of the model should properly represent the interaction between the real system and its environment.

In developing a thermal model, it is often assumed that the temperature of the attachment interface is constant. It is expected that the interface temperature is not constant but the designer perceives that a small variation will not appreciably affect the temperature distribution. Approximations of this nature simplify the representation, permitting a reasonably accurate and timely determination of equipment thermal behavior which in turn enables decisions regarding the actual design to be made.

Usually a number of successively more refined analyses follow wherein key approximations and assumptions are checked for their validity. The ability to construct a model appropriately representing the real hardware based on simplifying viable approximations requires a degree of intuitive judgement. Experience with actual hardware in test environments is a necessary prerequisite in developing astute approximations that lead to a rapid, reasonably accurate prediction of equipment behavior. Approximations which are applicable for most designs involve

- Neglecting small effects;
- Assuming the effect of the equipment on its surrounding is negligible;
- Assuming that lumped characteristics represent distributed characteristics;
- Assuming that physical and material properties do not vary with time;
- Assuming linear actions between variables.

Network Representation. The approximation of an actual item of equipment by a network representation results in a major mathematical simplification. A network model contains branch conductances (or impedances) intersecting at nodes. Each node represents a specific volume of material which corresponds to a portion of the actual equipment, its surrounding environment or an adjoining interface. The model of Figure 2.12(b) uses two nodes and two conductances and capacitances to determine the maximum temperature of a circuit board and a correspondingly located component. Many more nodes and branch conductances and capacitances could have been used to represent the system more accurately. Theoretically, the greater the number of nodes, the better the correspondence between the physical model and the actual equipment. The increase in complexity and effort, however, does not necessarily justify the degree of improvement in predicted performance achieved. The designer must exercise judgement in selecting an appropriate network that will provide useful and timely performance data which supports packaging decisions regarding the actual design.

Virtually any hardware configuration may be represented by a network. The

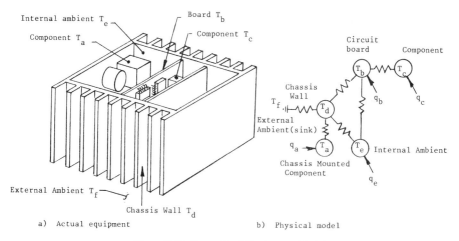

Figure 2.20. Generalized Equipment Model.

configuration may be planar or occupy three dimensional space. Conductances for commonly encountered geometries may be cataloged for future use. Conductance relationships for heat transfer sections often found in hardware configurations are listed in Table 2.3. Each conductance branch of a network represents a heat transfer path within the equipment or assembly being modeled. Conductances across circuit card retainers and bolted interfaces which rely on the contact pressure developed by the device are discussed in Chapter 3.

The temperature distribution of components, assemblies, attachment surfaces and internal ambients for the equipment configuration of Figure 2.20(a) may be determined from a network. Each assembly can be modeled as a network which is linked together with appropriate conductances representing heat transfer paths that actually exist. The resulting network will require automated methods to obtain corresponding node temperatures. The effort and time required to develop a detailed model as described are extensive. This activity is usually initiated late in the hardware design phase in order to validate thermal performance in terms of the equipment design requirements. Hand calculations based on simplified networks such as that shown in Figure 2.20(b) may be used to arrive at viable design concepts. This approach can provide an evaluation of assembly or module configurations and their locations within the equipment with respect to component operating temperatures. Simple networks facilitate parametric evaluations that may be used to optimize a design parameter such as thickness, weight or material.

Equipment or assemblies cooled by forced convection may also be modeled using a network. In this case, branch conductances are defined which represent surface-to-coolant characteristics in addition to those which describe coolant transport properties and heat flow paths within the assembly. The conductive

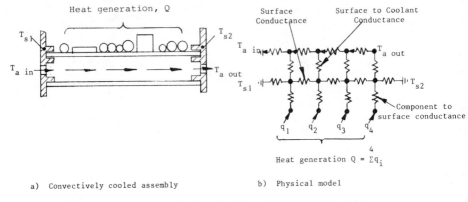

a) Convectively cooled assembly   b) Physical model

Figure 2.21. Idealization of a Convectively Cooled Assembly.

and convective cooling of the component board in Figure 2.21(a) is described by the network of Figure 2.21(b).

Node Considerations.   Generally, nodes are used at locations where temperatures are required. Nodes may also be necessary at locations of heat generation, material interfaces, geometric transitions or at boundaries and thermal sinks. A greater number of nodes are used at locations of high heat concentration to more accurately predict maximum temperatures. Either a linear or non-linear distribution may be used.

Capacitances are used only for transient analysis. The capacitance is proportional to the nodal volume, material density and specific heat. Capacitances which are small compared to others in the network may be combined with larger values or neglected. The nodes which do not have capacity are eliminated using the method of Appendix A.7. The temperature distribution at the selected time interval is determined using the reduced matrix. The temperatures of the remaining nodes may be recovered at any time interval using the products of matrix reduction. This technique is illustrated in Chapter 3.

The effort required in determining the temperature distribution from the network can be minimized with proper node numbering. The numbering sequence used affects only the work necessary to determine the nodal temperatures, not their values. The optimal sequence is accomplished by systematically numbering the nodes to minimize the numerical difference between assigned node numbers.

Planar component mounting surfaces with concentrated or distributed sources may be modeled using finite elements, which develops an analogous network as shown in Figure 2.22. Node-to-source (component) correspondence is not necessary since the quantity of heat generated by components may be allocated in proportion to their distances from surrounding nodes.

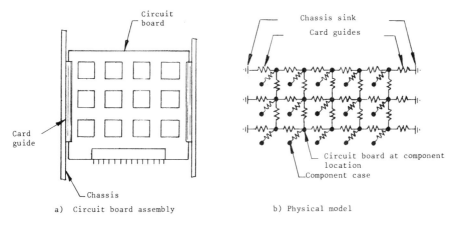

a)   Circuit board assembly   b) Physical model

Figure 2.22. Planar Surface Model.

Symmetry.   Many equipment configurations exhibit symmetry, which may be used advantageously when evaluating the behavior of a design. Where symmetry exists, the size of the network and the attendant effort required to obtain a solution may be reduced. This advantage also provides a method of checking a solution which is obtained using the full configuration.

In order for thermal symmetry to exist, there must be symmetry with respect to geometry, boundary conditions and heat generation. Geometric symmetry requires that the heat transfer configuration be identical about an axis of symmetry. The same requirement is imposed on thermal boundaries and source distribution. Configurations that do *not* satisfy symmetry requirements are shown in Figure 2.23.

An axis of symmetry may be regarded as an insulated boundary. In this case, heat flow is prohibited from crossing the boundary. Instead, equal quantities of heat flow on either side of the axis of symmetry. A planar surface exhibiting double symmetry is illustrated in Figure 2.24(a). A conductance network that is based on splitting the geometry along axes of symmetry representing the shaded portion is shown in Figure 2.24(b).

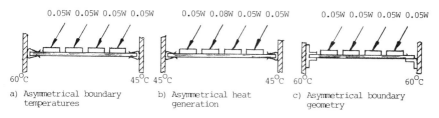

a) Asymmetrical boundary   b) Asymmetrical heat   c) Asymmetrical boundary
   temperatures   generation   geometry

Figure 2.23. Asymmetrical Circuit Board Configurations.

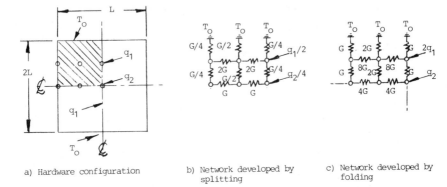

a) Hardware configuration   b) Network developed by   c) Network developed by
                                splitting   folding

Figure 2.24. Equivalent Representations of a Symmetric Quarter Section.

A comparable impedance network is obtained by inverting conductance quantities. Along the axes of symmetry, impedance quantities are doubled whereas conductance quantities are halved. Heat sources coinciding with these axes are halved since the generated heat will flow away from the boundary in equal amounts. Because the heat generated at the intersection of symmetric axes flows outward uniformly, only one quarter of the generated heat is used for the split portion. The resulting network maintains a geometric correspondence with the comparable section of the actual configuration. Only conductance (or impedance) and heat generation quantities along axes of symmetry are modified in this approach.

A similar network, Figure 2.24(c), may also be developed by folding the configuration along axes of symmetry. In this case, conductance (or impedance) and heat generation quantities along symmetric axes are not modified, but branch conductances are doubled or impedances halved throughout the remaining network. Since conditions of symmetry exist on two axes, the network configuration for Figure 2.24(a) is folded first along one axis and then along the other. The result, shown in Figure 2.24(c), has conductances which are doubled along symmetric axes. The remaining quantities are four times the unfolded network conductances. Use of symmetry reduces the network to 6 nodes based on a 15 node representation of the actual configuration.

Properly developed symmetric networks give the same results as those obtained from the original network. Care must be exercised, however, when determining either the effective conductance or heat flow from a network developed using the splitting approach. This condition is demonstrated in determining the heat flow between T_{s1} and T_{s2} from the network of Figure 2.25(a). The quantity of heat flowing in this network using an effective conductance G becomes

$$Q = G(T_{s1} - T_{s2}) \qquad (2.45)$$

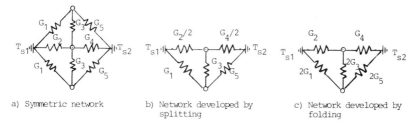

a) Symmetric network

b) Network developed by splitting

c) Network developed by folding

Figure 2.25. Network Equivalence.

The equivalent network developed by splitting along the single axis of symmetry, Figure 2.25(b), physically relates to one-half of the actual geometry. The quantity of heat flowing in the network of Figure 2.25(b) may be determined as

$$Q_a = G_a(T_{s1} - T_{s2}) = \frac{G}{2}(T_{s1} - T_{s2}) \qquad (2.46)$$

The effective conductance of the equivalent network developed by folding is identical to that of the original network configuration. The nodal temperature distribution for any of the network configurations of Figure 2.25 is the same.

Equivalent Networks. The construction of a network representing actual equipment may entail the use of nodes for which thermal response characteristics are not required. It is convenient to use nodes at points of geometric transitions or at interfaces between different materials when developing network representations. If the temperature at these locations is not necessary, the corresponding nodes may be systematically removed using methods of Appendix A-7. The fewer the number of nodes and loops a network contains, the less the effort required to determine its thermal characteristics.

Unnecessary nodes may be removed by employing branch equivalences shown in Figure 2.26. In general, n impedances $R_1, R_2 \ldots R_n$ which are connected in series may be added to give

$$R = R_1 + R_2 + \cdots + R_n \qquad (2.47)$$

A series of conductances may be added in an analogous way using $G = 1/R$ to give

$$\frac{1}{G} = \frac{1}{G_1} + \frac{1}{G_2} + \cdots + \frac{1}{G_n} \qquad (2.48)$$

If n impedances are connected in parallel, their total impedance becomes

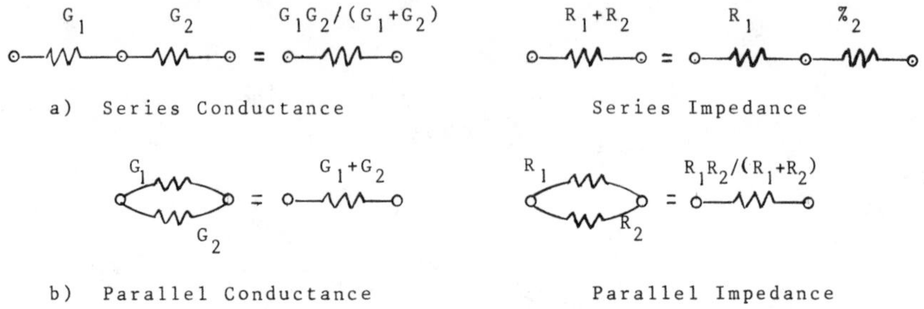

a)   Series Conductance

Series Impedance

b)   Parallel Conductance

Parallel Impedance

Figure 2.26. Branch Equivalence.

$$\frac{1}{R} = \frac{1}{R_1} + \frac{1}{R_2} + \cdots + \frac{1}{R_n} \tag{2.49}$$

The total conductance for parallel connected conductances may be determined analogously as

$$G = G_1 + G_2 + \cdots + G_n \tag{2.50}$$

Unnecessary nodes and loops can be removed by using equivalent relationships between three interconnected branch impedances arranged in either a delta or star configuration. Expressions that enable conversion between these two configurations are shown in Figure 2.27.

Methods developed in Appendix A.7 are also applicable to the removal of network nodes and loops.

EXAMPLE 2.9.   Determine the node temperatures for the impedance network shown in Figure 2.28(a) using (a) an equivalent network based on symmetry and (b) an equivalent network based on removing the unwanted temperature represented by node 3.

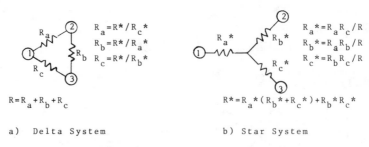

a)   Delta System

b)   Star System

Figure 2.27. Relationships between Delta-Star Impedance Connected Systems.

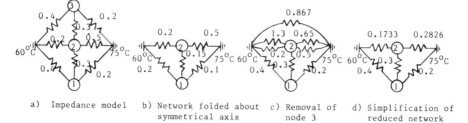

a) Impedance model   b) Network folded about   c) Removal of   d) Simplification of
                   symmetrical axis   node 3   reduced network

Figure 2.28. Equivalent Methods of Reducing Network Size.

(a) *Solution Based on Symmetry.* The equivalent network shown in Figure 2.28(b) is developed by folding the given model along the symmetrical axis. Nodal temperatures 1 and 2 may be determined by applying a loop-flow approach directly to the impedance network or by inverting the branch quantities and using the nodal conductance method.

Since the number of loop equations which equals the number of branches minus the number of nodes (3.59) is greater than the number of nodal equations, the conductance method is used. The impedance quantities inverted, the solution form using methods of Appendix A is

$$\begin{bmatrix} 13.667 & -6.667 \\ -6.667 & 21.667 \end{bmatrix} \begin{Bmatrix} T_2 \\ T_1 \end{Bmatrix} = \begin{Bmatrix} 450 \\ 1050 \end{Bmatrix} \tag{2.51}$$

Upon inverting the conductance matrix of (2.51), the solution becomes

$$\begin{Bmatrix} T_2 \\ T_1 \end{Bmatrix} = \begin{bmatrix} 0.0861 & 0.0265 \\ 0.0265 & 0.0543 \end{bmatrix} \begin{Bmatrix} 450 \\ 1050 \end{Bmatrix} = \begin{Bmatrix} 66.57 \\ 68.94 \end{Bmatrix} \tag{2.52}$$

(b) *Solution Based on Removing Unwanted Node 3 Temperature.* The equivalent network shown in Figure 2.28(c) is obtained by employing a star-to-delta conversion using the relationships delineated in Figure 2.27. The resulting parallel impedance branches may be combined into a single element as shown in Figure 2.26(b) or by using (2.49). An additional simplification is provided by eliminating the 0.867 impedance branch which directly couples both sinks. Heat flow in this branch does not effect the nodal temperatures. The simplified network is shown in Figure 2.28(d). The solution using methods of Appendix A on the conductance representation becomes

$$\begin{bmatrix} 12.642 & -3.333 \\ -3.333 & 10.833 \end{bmatrix} \begin{Bmatrix} T_2 \\ T_1 \end{Bmatrix} = \begin{Bmatrix} 611.613 \\ 525. \end{Bmatrix}$$

$$\begin{Bmatrix} T_2 \\ T_1 \end{Bmatrix} = \begin{bmatrix} 0.0861 & 0.0265 \\ 0.0265 & 0.1004 \end{bmatrix} \begin{Bmatrix} 611.613 \\ 525. \end{Bmatrix} = \begin{Bmatrix} 66.572 \\ 68.948 \end{Bmatrix} \quad (2.53)$$

The results obtained demonstrate the equivalence of the different network simplifications used.　　　　　　　　　　　　　　　　　　　　　　　　Q.E.D.

2.6 TEMPERATURE CONTROL METHODS

Temperature control methods are used to control the energy exchange between an electronic component, an assembly or an item of equipment and its environment in a way that will keep operating temperatures within specified allowable limits. These limits are established to assure reliable performance and to extend the useful service life of the equipment.

Thermal control techniques can be grouped into three categories: (1) passive, (2) semipassive and (3) active.

Passive temperature control is achieved through geometrical and thermophysical considerations alone. This technique does not require additional input power or the use of rotating mechanical devices or forced coolant motion. Conduction, natural convection and radiation heat transfer constitute passive means of controlling temperatures. Examples include the use of coatings, insulations, heat sinks and phase change materials.

Semipassive techniques use rotating devices, forced coolant motion or require additional input power to achieve temperature control. Liquid cooling and forced convection heat transfer are semipassive techniques. Thermal electric devices, heat pipes and heaters are examples of semipassive temperature control components.

Active techniques involves refrigeration using either a Carnot or absorption cycle heat pump to control temperature. Either of these active approaches uses a radiator which rejects heat through condensation of a fluid. The radiator may operate at a temperature which is higher than the controlled temperature of the component or assembly. Active techniques require additional power and involve the use of rotating components and forced coolant motion to achieve temperature control.

Passive Design Aspects.　Passive control techniques are usually preferred for design applications because moving components, flowing coolants or increased electrical power are not required for implementation. Because of this, passive control is usually more reliable; there are no elements in the heat transfer system that could fail.

Passive cooling using conductive heat transfer provides paths which are effective for disposing of heat by radiation. These paths assure lower component temperatures due to conduction to surrounding sinks and a related inherent thermal capacity which constrains the temperature rise. Conductive heat sinks are used for temperature control of electronic packages which experience cyclical variations in power dissipation or heat absorption from the environment.

Passive control may also be obtained from thermal insulators. These materials are placed between the component and a hot source or cold sink to resist the transfer of heat to or from the component. Coatings placed on surfaces of components or assemblies control the temperature by inhibiting thermal radiation. Coatings must be selected with the proper ratio of absorptivity to emissivity to achieve the heat balance necessary to control surface temperatures.

Phase change materials that store heat by a change in enthalpy can be an effective method of thermal control of temperature sensitive or heat generating assemblies where the total time-integrated operational heat load is small. The enthalpy change of melting and solidification of materials that have a melt point close to the design environment of electronic equipment exhibits a range of about 30 watt·hr/lb (66 watt·hr/kg). These materials, which provide approximately an order of magnitude higher capacity for equal weight compared with other packaging materials, absorb heat on melting or supply heat when solidifying.

Conceptual design aspects for passive phase change temperature control are shown in Figure 2.29. The heat generated in the electronic package in Figure 2.29(a) is absorbed by the melting of the solid fusible material in addition to some sensible heat absorption by the melted liquid material. The thermal interface with the fusible material remains essentially at a constant temperature which is determined by the material used. After all of the fusible material has changed phase, the temperature of the assembly rises at a rate governed by its thermal mass or inertia. The configuration shown in Figure 2.29(b) uses the

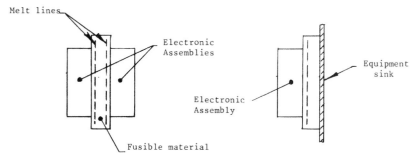

a) Cold plate configuration   b) Temperature control configuration

Figure 2.29. Phase Change Hardware Configurations.

Table 2.7. Properties of Paraffin Fusible Materials

$$C_f \begin{cases} 0.15 \text{ watt} \cdot \text{hr/lb } ^\circ\text{F} \\ 0.59 \text{ watt} \cdot \text{hr/kg} ^\circ\text{C} \end{cases} \quad \rho_f \begin{cases} 0.028 \text{ lb/in}^3 \\ 780. \text{ kg/m}^3 \end{cases}$$

Material	Formula	Melt Point °F	Melt Point °C	Heat of Fusion Watt·hr/lb	Heat of Fusion Watt·hr/kg
Eicosane	$C_{20}H_{42}$	98	36.7	31.1	68.4
Heneicosane	$C_{21}H_{44}$	105	40.6	26.9	59.4
Docosane	$C_{22}H_{46}$	112	44.4	31.4	69.1
Tricosane	$C_{23}H_{48}$	117	47.2	29.3	64.5
Tetracosane	$C_{24}H_{50}$	124	51.1	31.9	70.3
Pentacosane	$C_{25}H_{52}$	129	53.9	29.9	65.8
n-Hexacosane	$C_{26}H_{54}$	133	56.1	32.2	71.0
Heptacosane	$C_{27}H_{56}$	138	58.9	29.6	65.2
Octacosane	$C_{28}H_{58}$	142	61.1	31.9	70.3
Nonacosane	$C_{29}H_{60}$	147	63.9	30.2	66.5
Triacontane	$C_{30}H_{62}$	150	65.6	31.6	69.7
Dotriacontane	$C_{32}H_{66}$	157	69.4	32.5	71.6

fusible material between the heat generating components and the equipment thermal sink. The heat transferred to or from the sink is limited by this technique provided the fusible element remains in transition. This configuration maintains temperature control even if the sink temperature exceeds the allowable component temperature for brief periods.

The effects of the molten layer of material that develops adjacent to heated surfaces may be evaluated using transient analysis. The use of waffling, fins or conductive grids that distribute the molten layer throughout the fusible is preferred for containers with a large volume of material. Expansion cavities must be included in the fusible container to accommodate the larger volume of the molten material. About 15 percent of the solid fusible volume is required for expansion.

Thermal properties of the paraffin group C_nH_{2n+2} are listed in Table 2.7. Paraffins are noncorrosive, nontoxic and low cost. They are easily packaged in containers of simple or complex geometry that can be adapted to most electronic equipment requiring temperature control.

The weight of fusible material required may be determined using

$$W_f = \frac{Q\tau - (T_m - T_i)\overline{C}_e}{q_f + (T_m - T_i)C_f} \tag{2.54}$$

given Q = assembly power dissipation, W;
  τ = period requiring temperature control, hr;
  T_i = initial assembly temperature;

T_m = fusible melt point;
q_f = heat of fusion;
C_f = specific heat of fusible;
$\overline{C}_e$ = thermal capacity of the assembly excluding fusible material (includes fusible container).

EXAMPLE 2.10.   All of the heat generated from a 10 watt assembly initially soaked at 80°F (26.7°C) must be absorbed by its mass for an operational period of one hour without exceeding 150°F (65.6°C) at the cold plate. If the specific heat and weight of the assembly are 0.07 watt·hr/lb°F (0.28 watt·hr/kg°C) and 1.0 lb (0.45 kg) respectively, determine (a) the weight and volume of a packaged fusible satisfying the requirements and (b) the weight and volume of an aluminum plate that limits the cold plate temperature to 150°F (65.6°C) after one hour.

(a) *Packaged Fusible Configuration.* Assuming that a 0.25 lb (0.11 kg) aluminum container would be sufficient, the thermal capacity of the assembly not including the fusible material becomes

$$\overline{C}_e = (WC)_e = 0.07(1.0 + 0.25) = 0.0875 \text{ watt·hr/°F}$$

$$= 0.28(0.45 + 0.11) = 0.157 \text{ watt·hr/°C}$$

The allowable maximum temperature can be met using Triacontane. The weight of fusible required is determined from (2.54) using values from Table 2.7 as

$$W_f = \frac{10(1.0) - (150 - 80)(0.0875)}{31.6 + (150 - 80)(0.15)} = 0.092 \text{ lb}$$

In metric units this becomes

$$W_f = \frac{10(1.0) - (65.6 - 26.7)(0.157)}{69.7 + (65.6 - 26.7)(0.59)} = 0.042 \text{ kg}$$

The total weight of the packaged fusible which includes the fusible material and container is

$$W = 0.25 + 0.092 = 0.342 \text{ lb}$$

$$= 0.11 + 0.042 = 0.152 \text{ kg}$$

The volume of the package fusible accounting for 15% material expansion is

$$V = 0.25/0.1 + 0.092(1.15)/0.028 = 6.3 \text{ in}^3$$

which in metric units is

$$V = 0.11/2768. + 0.042(1.15)/780. = 10^{-4} \text{ m}^3$$

The weight assumed for the fusible container appears adequate for the required material volume. If the weight of the container should be insufficient, the computations would be repeated using a new estimate.

(b) *Aluminum Plate Configuration.* The maximum allowable increase of the cold plate temperature is obtained from given data as

$$\Delta T_{cp} = 150 - 80 = 70°F \ (38.9°C)$$

The allowable assembly heat rate to achieve the maximum cold plate temperature after one hour is

$$\dot{T} = \Delta T_{cp}/\tau = 70./1.0$$

$$\dot{T} = 70°F/hr \ (38.9°C/hr)$$

The thermal capacity required to obtain the allowable heat rate becomes

$$\overline{C}_e = Q/\dot{T} = 10/70$$

$$\overline{C}_e = 0.143 \ watt \cdot hr/°F \ (0.257 \ watt \cdot hr/°C)$$

The difference between the required thermal capacity and that of the assembly determines the requirements for the aluminum plate as

$$\overline{C}_p = 0.143 - 0.07 = 0.073 \ watt \cdot hr/°F \ (0.131 \ watt \cdot hr/°C)$$

The weight and volume of the plate required are determined using material properties from Appendix C as

$$W_p = \overline{C}_p/C = 0.073/0.07 = 1.04 \ lb \ (0.473 \ kg)$$

$$V_p = W_p/\rho_p = 1.04/0.1 = 10.4 \ in^3 \ (1.7 \times 10^{-4} \ m^3)$$

Phase change materials offer an attractive alternative when weight and volume are important considerations. This is illustrated by comparing the two configurations. Transient response characteristics for each configuration are also compared in Figure 2.30. The weight and volume advantages provided by a fusible are offset by a higher average sink temperature over the operational period.   Q.E.D.

Semipassive Design Aspects.   Semipassive temperature control techniques require the use of certain electro-mechanical components that are designed to provide a hot or cold thermal reference or develop coolant motion. With the exception of heat pipes, most of these components consume power when performing their intended functions. The heat generated by semipassive control techniques must be removed by the equipment cooling system, depending on the design mechanization used.

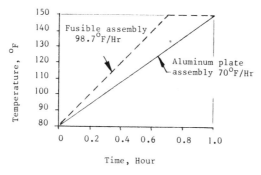

Figure 2.30. Response Characteristics for Two Passive Temperature Control Techniques.

A heat pipe is a component that controls temperature using saturated vapor convection and condensation to transfer heat over large distances with small temperature differentials. Heat pipes have been developed for a variety of electronic packaging applications. Circuit card heat pipes, for example, provide uniform and efficient cooling, permitting greater flexibility of component arrangements with higher power densities than could be achieved with passive control techniques. Components along a heat-pipe-cooled surface operate at essentially the same interface temperature regardless of position because of the high thermal conductance provided by the pipe. Simple flat heat pipes have been used at the attachment interface of transformers to control winding temperatures. Heat pipes have also been integrated into chassis to transfer heat to a cold plate or sink, as shown in Figure 2.31. Heat pipes may be straight or contain multiple bends; they may also be rigid or flexible.

The cooling developed by thermoelectric devices is due to an energy exchange of electrons flowing across a junction of two materials that have different available electron energy levels. The junction known as a Peltier couple consists of two semiconductor elements that are doped to form a p and an n type material. The Peltier couple is shown in Figure 2.32(a).

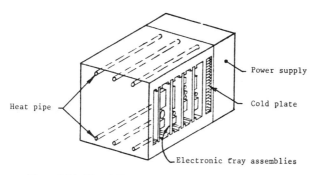

Figure 2.31. Chassis Temperature Control Using Heat Pipes.

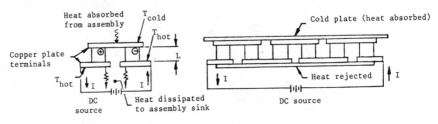

a) Peltier couple   b) Module consisting of multiple couples

Figure 2.32. Thermoelectric System.

A DC current applied to the junction passes from the n to the p semiconductor material, reducing the temperature of the common copper plate terminal. This occurs as a result of electrons passing from the lower energy level p material to the higher energy n material. Heat absorbed at the cold junction is pumped to the hot junction at a rate proportional to the current applied to the couple. Heat can be supplied at the common terminal by merely reversing the direction of current flow.

A thermoelectric module consists of numerous individual couples interconnected electrically in series and thermally in parallel. These devices can provide a suitable operating temperature for components or assemblies even though the sink or ambient may be at a higher temperature. Thermoelectrics can provide precise temperature control by selecting an appropriate current. They can also maintain temperature control inside sealed compartments.

At steady state conditions, a heat balance on the cold junction of Figure 2.32(a) gives

$$Q_{\text{cold}} = Q_p - Q_j/2 - Q_{\text{cond}} \qquad (2.55)$$

where Q_p is the rate of heat absorption due to the Peltier effect, Q_j is the rate of Joule heating in the couple of which only half flows to the cold junction, and Q_{cond} is the rate of heat transferred by conduction from the hot to the cold junction. With appropriate substitutions (2.55) becomes

$$Q_{\text{cold}} = \alpha T_c I - \tfrac{1}{2}I^2 R_e - \frac{\Delta T}{R_t} \qquad (2.56)$$

given ΔT = temperature difference $(T_h - T_c)$;
  T_c = absolute temperature of cold junction, $^\circ$K;
  T_h = absolute temperature of hot junction, $^\circ$K;
  I = current flow, amp;

Table 2.8. Material Properties of Thermoelectric Couples

Material State	Resistivity ρ, ohms	Power α, volt/°C	Conductivity K, watt/ cm°K	Figure of Merit Z, °K^{-1}
Bi$_2$Te$_3$ (pure)	0.00267	425(10^{-6})	0.00785	0.00231
Bi$_2$Te$_3$ — Bi$_2$Se$_3$ — Sb$_2$Te$_3$	0.00201	388(10^{-6})	0.00627	0.00312
Bi$_2$Te$_3$ — AgI	0.00277	442(10^{-6})	0.01038	0.00177

R_e = electrical resistance, $\rho L/A$ ohms;
R_t = thermal resistance, L/KA, °C/watt;
α = thermoelectric power, volts/°C;
A = cross sectional area of thermoelectric material, cm^2;
K = thermal conductivity.

Bismuth tellaride (Bi$_2$Te$_3$) is used as a thermoelectric material in either a pure or modified form. Assuming that the geometry of both p and n legs of the couple are the same, the electrical and thermal resistances may be combined. The resultant material properties are listed in Table 2.8.

The thermoelectric couple develops the greatest ΔT when there is no external heat load or $Q_{cold} = 0$. The temperature difference at this condition may be determined from

$$\Delta T_{max} = \frac{1}{2Z}(1 - \sqrt{1 + 2ZT_h})^2 \text{ °C} \tag{2.57}$$

The current that optimizes the cold junction heat removal Q_{cold} in (2.56) is determined from the condition $dQ_{cold}/dI = 0$. Maximum current and corresponding maximum pumping power that can be developed by the couple are determined at $\Delta T = 0$ as

$$I_{max} = \alpha T_h/R_e \text{ amps} \tag{2.58}$$

$$Q_{cold\ (max)} = \frac{R_e}{2} I_{max}^2 \text{ watts} \tag{2.59}$$

Internal junction resistances increase the maximum current in (2.58) an additional 10 percent. The application of the thermoelectric couple for the control of assembly or component temperatures establishes its performance characteristics. These characteristics are defined in terms of (2.57)–(2.59) using

$$\xi_{\Delta T} = \xi_I(2 - \xi_I) - \xi_Q \tag{2.60}$$

where ξ_I is the fractional value of characteristic I relative to the maximum obtainable by the couple, i.e. $\xi_{\Delta T} = \Delta T/\Delta T_{max}$. The coefficient of performance which is defined as the rate of cooling divided by the input power becomes

$$\phi = \frac{Q_{cold}}{IV} = \frac{2\xi_I - \xi_I^2 - \xi_{\Delta T}}{2\xi_I^2 + \xi_I\xi_{\Delta T}ZT_c} \tag{2.61}$$

for an applied voltage V to the couple. The coefficient of performance which is maximum for a particular application becomes

$$\phi_{max} = \frac{T_c}{\Delta T} \left[\frac{\sqrt{1 + Z(T_h + T_c)/2} - T_h/T_c}{\sqrt{1 + Z(T_h + T_c)/2} + 1} \right] \tag{2.62}$$

EXAMPLE 2.11. The interface temperature of an assembly dissipating 10W must be limited to 40°C in a 50°C environment. It is assumed that all of the generated heat will be removed by thermoelectrics and that heat absorbed from the environment is negligible. The interface temperature difference between the assembly and the thermoelectric can be held to 2°C. The temperature difference between the thermoelectric and ambient can be held to 8°C. The bismuth telluride element used has a length of 0.3 cm and a cross-area of 0.01 cm^2. Determine the size and performance characteristics of the thermoelectric temperature control device.

The maximum performance characteristics of the thermoelectric couple are determined from (2.57) through (2.59) using

$$T_h = 331°K (58°C)$$

$$R_e = \rho L/A = 0.00267(0.3)/0.01 = 0.08 \text{ ohm}$$

as

$$\Delta T_{max} = \frac{1}{2(0.00231)} [1 - \sqrt{1 + 2(0.00231)(331)}]^2 = 75.4°C$$

$$I_{max} = 1.1(425)(10^{-6})331/0.08 = 1.93 \text{ amp}$$

$$Q_{cold (max)} = 1.93^2(0.08)/2 = 0.149 \text{ watt}$$

accounting for internal junction resistances. A multiple of Peltier couples are required since the assembly dissipates more heat than a single couple can pump. Using twice the assembly dissipation as a guide, the number of couples is approximately determined as

$$N = 2Q/Q_{cold (max)} = 2(10)/0.149 \simeq 134$$

The configuration that results may be packaged as one or more modules, each conceptually similar in cross-section to Figure 2.32(b). A 38°C cold plate is required to assure a 40°C interface temperature. The thermoelectric must maintain a 20°C temperature difference to satisfy the design requirement. The fractional temperature characteristic becomes

$$\xi_{\Delta T} = \Delta T/\Delta T_{\max} = 20/75.4$$

$$\xi_{\Delta T} = 0.265$$

(2.63)

The fractional heat pumping characteristic using 134 couples is

$$\xi_Q = Q/Q_{\text{cold (max)}} = 10/(134)(0.149)$$

$$\xi_Q = 0.5$$

(2.64)

The input current to the thermoelectric module is determined from (2.60) using (2.63) and (2.64) as

$$I = \xi_I I_{\max} = 0.515(1.93) = 1 \text{ amp}$$

The coefficient of performance of the selected module configuration for this design application is determined using (2.61) as

$$\phi = \frac{2(0.515) - 0.515^2 - 0.265}{2(0.515^2) + 0.515(0.265)(0.00231)(38 + 273)}$$

$$\phi = 0.795$$

The input power

$$P_{\text{in}} = Q_{\text{cold}}/\phi = 10/0.795 = 12.6 \text{ watts}$$

in addition to the 10 watts generated by the assembly must be removed from the hot side of the thermoelectric by the environment. The heat transfer method must be capable of transferring 22.6 watts to a 50°C environment at a temperature difference of 8°C. The voltage required to operate the module is

$$V = P/I = 12.6/1.0 = 12.6 \text{ volts}$$

A lower coefficient of performance and a higher input current would result if the number of couples used were reduced. The module size is approximately three times the element area. Since each couple is composed of two elements, the module area becomes

$$A_{\text{module}} = 2(0.01)(134) \times 3 = 8.04 \text{ cm}^2$$

which can be accommodated in a package measuring 2.84 cm on a side.   Q.E.D.

Thermoelectric module construction requires insulation between copper terminals on the hot and cold sides of each couple and between adjoining cold plates or structure. Surfaces should be flat to insure maximum contact reducing thermal gradients within the assembly. Care should be exercised when fastening cold plate to hot plate structures to assure that undesirable heat flow paths that may short circuit the thermoelectric system are minimized. A thermal shunt will reduce the heat pumping capacity of the assembly. Modules should not be used as structural members. Thermoelectric modules are available from suppliers as packaged assemblies including cold plates and heat sinks.

The temperature difference obtainable from a couple is limited by the figure of merit Z of available materials as shown in (2.57). Larger ΔTs may be obtained by a cascade arrangement where the cold surface of one couple serves as a heat sink for another. A 4 to 1 stacking factor has been found to yield good results. In this arrangement, four couples are used in the higher temperature level for each couple in the adjacent lower temperature level. More than three levels are impractical since the heat pumping capacity of a couple decreases at lower base temperatures.

Thermoelectric modules require a DC source with a ripple of less than 10 percent. A higher ripple current may be used but the performance is degraded. Ripple results in a reduction of ΔT_{max} for a given input current.

A difficult temperature control problem materializes when the component or assembly temperature varies above and below the control temperature. Either heating or cooling can be accomplished by reversing the direction of current flow through the thermoelectric module. Since current switching is not desirable, heaters may be added which provide makeup heat when needed. If necessary, thermoelectric cooling can be shut down when the heaters are activated.

The effectiveness of semipassive temperature control is enhanced when it is considered during concept development and in the initial hardware design phase. In general, optimal temperature control, heat removal and structural design can result when properly considered during the process of selecting, locating and orienting components in the product configuration.

3
Mechanics of Conduction

Heat conduction in electronic equipment, like other modes of heat transfer, takes place only when the temperatures at various locations of an assembly differ. The rate of conduction heat flow is proportional to the area perpendicular to the direction of flow and to the temperature difference between any two points. The rate of flow is also inversely proportional to the distance between these points as expressed by (2.1).

Conduction processes may take place between different parts of a homogeneous body, at interfaces between mated parts and between closely spaced assemblies. Heat flow from the surface of one body to another or to the surrounding media can be considered as conduction through a thin fluid layer within which no convection exists. In gases, conduction occurs by molecular and atomic diffusion, and in liquids and some dielectrics, by means of elastic waves.

A line joining points in a body whose temperatures are identical defines an *isothermal surface*. Isothermal surfaces do not intersect since a common point can not be at two different temperatures. Isotherms were used to develop a graphical visualization of heat flow in Figures 2.9 and 2.10. They either end at boundaries of the body or close on themselves within it.

3.1 CONCEPT OF CONDUCTION COOLING

The effort required for evaluation of the temperature distribution and rate of conduction heat transfer of an assembly or part may be reduced by considering the geometrical characteristics of heat flow. Heat flow in a circuit board whose thickness is small compared to its other dimensions is primarily two dimensional. Further, where the component-generated heat distribution can be considered uniform, heat flow can be studied using a one dimensional approach. Many components and hardware assemblies may be modeled as one instead of two or three dimensional systems.

Differential equations that describe steady state heat conduction in one, two and three dimensional systems are listed in Table 3.1. Equation (2.10) is an instance of applying Case III of Table 3.1 with $Q = 0$ to an annular sector.

Table 3.1 Differential Equation for Conduction Heat Transfer

Case	Differential Equation	General Solution	Remarks
One Dimensional			
I	$\dfrac{d^2 T}{dx^2} + \dfrac{Q}{k} = 0$	$T = C_1 x + C_2 x - \dfrac{Q}{2k} x^2$	Heat source, $Q\,\dfrac{W}{in^3}\left(\dfrac{W}{m^3}\right)$
II	$\dfrac{d^2 T}{dx^2} - M^2 T = 0$	$T = C_1 \sinh Mx + C_2 \cosh Mx$	Surface-to-ambient exchange
III	$\dfrac{1}{r}\dfrac{d}{dr}\left(r\dfrac{dT}{dr}\right) + \dfrac{Q}{k} = 0$	$T = C_1 \ln r + C_2 - \dfrac{Q}{k} r^2$	Cylindrical system with heat source Q
Two Dimensional			
IV	$\dfrac{\partial^2 T}{\partial x^2} + \dfrac{\partial^2 T}{\partial y^2} = 0$	$T = (C_1 \sin \xi x + C_1 \cos \xi x) \times (C_3 e^{\xi y} + C_4 e^{-\xi y})$	Planar system $T = Ce^{\pm \xi x} e^{\pm j\xi y}$
V	$\dfrac{\partial^2 T}{\partial r^2} + \dfrac{1}{r}\dfrac{\partial T}{\partial r} + \dfrac{1}{r^2}\dfrac{\partial^2 T}{\partial \phi^2} = 0$	$T = (C_1 r^{\xi} + C_2 r^{-\xi}) \times (C_3 \sin \xi\phi + C_4 \cos \xi\phi)$	Cylindrical system
Three Dimensional			
VI	$\dfrac{\partial^2 T}{\partial x^2} + \dfrac{\partial^2 T}{\partial y^2} + \dfrac{\partial^2 T}{\partial z^2} = 0$	$T = C \exp(\pm \xi_1 x) \exp(\pm \xi_2 y) \exp(\pm j \sqrt{\xi_1^2 + \xi_2^2}\, z)$	Cartesian coordinates
VII	$\dfrac{\partial^2 T}{\partial r^2} + \dfrac{1}{r}\dfrac{\partial T}{\partial r} + \dfrac{1}{r^2 t}\dfrac{\partial^2 T}{\partial \phi^2} + \dfrac{\partial^2 T}{\partial z^2} = 0$	$T = A(r)(C_1 \sin \xi_1 \theta + C_2 \cos \xi_1 \theta) \times (C_3 e^{\sqrt{\xi_2} z} + C_4 e^{-\sqrt{\xi_2} z})$	Cylindrical systems $A(r)$ expressed as Bessel functions

Analytical solutions for two and three dimensional systems can be obtained only for relatively simple geometries. Numerical techniques are used to obtain approximate solutions for complex configurations. Conductances that may be used to model many two dimensional geometries are listed in Table 2.3.

Conduction heat transfer for complex assemblies may be conveniently studied using thermal conductance or resistance parameters. Thermal conductance using (2.1) is defined as

$$G = kA/L \qquad (3.1)$$

so that the rate of heat transfer between a hot location at temperature T_h and a cold location at temperature T_c becomes

$$Q = G(T_h - T_c) \qquad (3.2)$$

If the thermal resistance R is defined as the reciprocal of thermal conductance, then the rate of heat flow can be written as

$$Q = (T_h - T_c)/R \qquad (3.3)$$

Many important hardware configurations that are planar or axisymmetric may be considered as one dimensional systems. The one dimensional approach is used when the rate of heat flow is in one direction, which leads to a direct application of the general solution of Table 3.1.

The Influence of Attachment and Joining Methods. Rectangular component mounting surfaces that are attached to supporting structures along opposite edges are commonly used in electronic equipment. Such assemblies are referred to as *circuit boards, trays* or *modules*. These packaging configurations facilitate functional partitioning of the circuits, testing, repair and replacement. The heat on these surfaces may be discrete or distributed, with the heat removed by conduction at the edge supports.

Edge supports are available in a variety of configurations. To obtain effective conductive heat transfer, the edge support should provide a large contacting surface area and exert a high pressure on the edges of the circuit board. Support configurations that develop low pressure or serve only as clearance guides result in high temperature differences across the interface. In these cases, conduction heat transfer occurs across an effective air gap. Thermal resistances in terms of the actual support length used are given for several common configurations in Table 3.2.

Wedge-lock devices, Table 3.2(D), can develop a pressure in excess of 200 lb/in^2 (1.4 MPa) at the interface, which develops a 60 percent clamped struc-

Table 3.2 Thermal Resistances of Common Edge Restraints
°C · in / W (°C · mm / W)

Condition	(A)	(B)	(C)	(D)
0–50 kft (0–15.2 km)	12.0 (305.)	8.0 (203.)	6.0 (153.)	1.8 (46.)
~100 kft (30.5 km)	15.5 (394.)	10.5 (267.)	8.0 (203.)	1.9 (48.)

tural interface. For applications that do not require the plug in convenience of edge restraints of the type shown in Table 3.2, screws may be used to secure the assembly.

Screws and other threaded fasteners can develop sufficient pressure to produce low resistance thermal paths between mated parts. The resistance obtained depends on the interface materials, the nature of the contacting surfaces, the pressure and the interface fluid media. Flat, parallel, smooth, hard, stiff mating surfaces develop lower resistances. Interface materials that are thermally conductive can reduce the developed resistance relative to that obtained by air still more. Foil fillers may be used to compensate for rough uneven mating surfaces. Rough surfaces have finishes greater than 80 rms whereas smooth finishes are generally less than 20 rms.

Increasing pressure developed by larger screws with greater torque will decrease the interface thermal resistance. Smoother surfaces give lower resistances at lower pressures, and are not appreciably affected by increasing pressure. The resistance developed by rough surfaces, however, is affected dramatically by increasing pressure. Thermal resistances for mated surfaces as a function of contact pressure and roughness at sea level conditions are shown in Table 3.3.

Table 3.3 Interface Resistance at 27°C
in² · C / W (mm² · C / W)

Interface Materials	Roughness rms	Contact Pressure lb/in² (MPa)			
		250 (1.7)	500 (3.4)	750 (5.2)	1000 (6.9)
Copper/Copper	175	0.49 (317)	0.25 (158)	0.12 (78)	0.06 (42)
	100	0.10 (63)	0.08 (53)	0.06 (42)	0.05 (32)
	55	0.03 (18)	0.01 (9)	0.01 (9)	0.01 (9)
Aluminum/Aluminum	175	0.69 (447)	0.54 (352)	0.40 (261)	0.33 (211)
	100	0.60 (387)	0.45 (292)	0.35 (229)	0.29 (187)
	55	0.11 (70)	0.10 (63)	0.09 (60)	0.09 (56)
Aluminum/Copper	175	0.42 (273)	0.36 (233)	0.28 (181)	0.20 (130)
	55	0.08 (55)	0.07 (46)	0.06 (38)	0.05 (30)

Table 3.4 Interface Resistance of Riveted Joints in$^2 \cdot$°C/W (mm$^2 \cdot$°C/W)

Interface Materials	Temperature °F (°C)		
	100 (38)	200 (93)	300 (149)
Aluminum/Aluminum	1.09 (705)	0.90 (581)	0.82 (528)

Thermal resistances of riveted joints are dependent on the same material, surface and interface properties that affect bolted or screw fastened parts. Variations in resistances are due to preparation and assembly process parameters that are difficult to control even with automated techniques. Typical values are listed in Table 3.4 as a function of temperature for joined aluminum surfaces.

Rectangular One Dimensional Systems. The application of one dimensional systems to component mounting surfaces with heat removal on opposite edges is illustrated by the model in Figure 3.1. Concentrated sources q_1, q_2, q_{a1} and q_{a2} expressed in watts represent the heat generated along a line across the width of the surface at location ai. Distributed sources $\dot{q}_{a3}$ and $\dot{q}_{a4}$ are expressed in watt/in (w/mm) of length. Distributed sources may be added or subtracted to obtain nonuniform heat distributions representing component generated heat conditions along the length of the surface. Figure 4.29 illustrates the development of nonuniform heating by superposition of uniform heat distributions.

Steady state heat conduction for a one dimensional system with a single source is described by Case I of Table 3.1. The differential equation representing the configuration of Figure 3.1 is determined by superposition to be

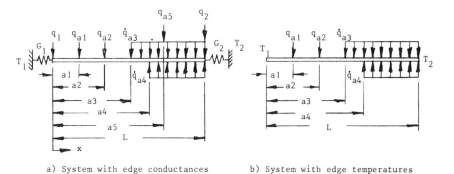

a) System with edge conductances   b) System with edge temperatures

Figure 3.1. One Dimensional Conduction Cooled Component Mounting System (thickness a, width b and conductivity k).

$$kab \frac{d^2 T}{dx^2} = -\dot{Q}(x) \tag{3.4}$$

The general solution of (3.4) subject to boundary conditions (4.65) for the configuration of Figure 3.1(a) becomes

$$T(x) = T(0) + \{G_1[T(0) - T_1] - q_1\} \frac{x}{kab} - \sum_i^n \langle T_i(x) \rangle_{ai} \tag{3.5}$$

The angular brackets $\langle \rangle$ are used to describe a function singularity with the meaning

$$\langle f(x) \rangle_{ai} \quad \begin{matrix} = 0 & x < ai \\ = f(x) & x \geq ai \end{matrix} \tag{3.6}$$

Temperature functions $T_i(x)$ which depend on whether concentrated or distributed heating occurs at location ai are determined from

Concentrated heating, q_{ai}: $T_i(x) = q_{ai}(x - ai)/kab$ (3.7a)

Distributed heating, $\dot{q}_{ai}$: $T_i(x) = \dot{q}_{ai}(x - ai)^2/2kab$ (3.7b)

Continuously varying heating, $q_{bi}(x)$: $T_i(x) = \frac{1}{kab} \int_0^x q_{bi}(x) \, dx$ (3.7c)

The surface temperature at $x = 0$ is obtained from

$$T(0) = \frac{\left(1 + \dfrac{G_2 L}{kab}\right)(G_1 T_1 + q_1) + G_2 T_2 + q_2 + G_2 \sum_1^n T_i(L) + Q}{G_1 + G_2 + \dfrac{G_1 G_2 L}{kab}} \tag{3.8}$$

where Q represents the total heat generated by the assembly excluding edge sources q_1 and q_2 which may be expressed as

$$Q = \sum_{i=1}^n [q_{ai} + \dot{q}_{ai}(L - ai) + q_{bi}(L)] \tag{3.9}$$

The maximum surface temperature for particular configurations often found in electronic equipment is given in Table 3.5.

Table 3.5 Maximum Surface Temperatures of Heated Edge Cooled Assemblies

Case	Configuration	Equation	
I		$T_{max} = T_s + \dfrac{Q}{2G} + \dfrac{QL}{8kab}$	(3.10)
II		$T_{max} = T_s + \dfrac{Q}{G} + \dfrac{QL}{2kab}$	(3.11)
III		$T_{max} = T_s + \dfrac{Q}{2G} + \dfrac{QL}{4kab}$	(3.12)
IV		$T_{max} = T_s + \dfrac{Q}{2G} + \dfrac{QL}{16kab}$	(3.13)
V		$T_{max} = T_s = \dfrac{Q}{G} + \dfrac{QL}{3kab}$	(3.14)

Component mounting surfaces with known edge temperatures, shown in Figure 3.1(b), represent a special case of (3.5) with $G_1 = G_2 \rightarrow \infty$. The temperature is given by

$$T(x) = \left[T_2 - T_1 + \sum_{i}^{n} T_i(L) \right] \left(\frac{x}{L}\right) + T_1 - \sum_{i}^{n} \langle T_i(x) \rangle_{ai} \quad (3.15)$$

Maximum surface temperature for the configurations in Table 3.5 with known edge temperature T_s may be found by setting $Q/G = 0$ in (3.10) through (3.14).

EXAMPLE 3.1. Components generating 15 watts are mounted to a 0.09 in (2.3 mm) thick, 5 in (127 mm) wide aluminum plate which is fastened to an aluminum chassis at one end as shown in Figure 3.2(a). Three screws develop an average pressure of 250 psi (1.7 MPa) over an area of 4.5 in^2 (2903 mm^2) at the chassis interface. Determine the temperature distribution for (a) uniform heating, (b) 10W uniformly distributed and 5W concentrated in components mounted 4 in (101.6 mm) from the chassis interface and (c) 10W linearly distributed with $\dot{q}_{max}$ at $x = 0$ adjacent to the chassis interface and 5W concentrated in components as in part (b), above.

(a) 15W Uniform Heating. The plate-to-chassis interface resistance is determined from Table 3.3 as 0.60 in^2·C/W (387 mm^2·C/W) based on the defined attachment

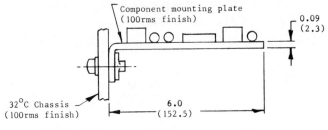

a) Edge cooled component assembly

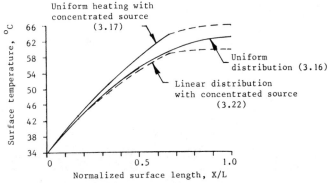

b) Temperature distribution with different heat distributions

Figure 3.2. Conduction Cooled Component Assembly.

pressure, interface materials and surface finish. For the given interface area, the conductance is

$$G_1 = \frac{4.5 \text{ in}^2}{0.6 \text{ in}^2 \cdot \text{C/W}} = \frac{2903 \text{ mm}^2}{387 \text{ mm}^2 \cdot \text{C/W}} = 7.5 \text{ W/}^\circ\text{C}$$

Since the plate surface is uniformly heated over its entire length, $a1 = 0$ giving a temperature function from (3.7b) as

$$T_0(x) = \frac{(2.5 \text{W/in}) x^2}{2(3.5 \text{W/in} \cdot \text{C})(0.09 \text{ in})(5 \text{ in})} = 0.79 x^2 \, ^\circ\text{C}$$

or in metric units

$$T_0(x) = \frac{(0.098 \text{W/mm}) x^2}{2(0.138 \text{W/mm} \cdot \text{C})(2.3 \text{ mm})(127 \text{ mm})} = 0.00122 x^2 \, ^\circ\text{C}$$

The mounting surface temperature at the chassis interface may be determined from (3.8). An alternative approach to (3.8) can be used for this configuration, however, since all the generated heat flows through conductance G_1, seeking the cooler $32\,°C$ chassis. In this case, the temperature at $x = 0$ becomes

$$T(0) = T_1 + Q/G_1 = 32 + 15/7.5$$

$$T(0) = 34\,°C$$

The temperature distribution from (3.5) is

$$T(x) = 34 + 7.5(34 - 32)\,\frac{x}{3.5(0.09)5} - 0.79x^2 \,°C$$

$$T(x) = 34 + 9.52x - 0.79x^2 \,°C \tag{3.16a}$$

In metric units

$$T(x) = 34 + 7.5(34 - 32)\,\frac{x}{0.138(2.3)127} - 0.00122x^2 \,°C$$

$$T(x) = 34 + 0.372x - 0.00122x^2 \,°C \tag{3.16b}$$

The temperature distribution is shown graphically in Figure 3.2(b).

(b) 10W Uniform with 5W Concentrated Heating. Conductance G_1, determined in part (a), remains the same. The temperature function for 10W uniform heating with $a1 = 0$ is given by (3.7b) as

$$T_0(x) = \frac{1.67x^2}{2(3.5)(0.09)(5)} = 0.53x^2 \,°C$$

which in metric units is

$$T_0(x) = \frac{0.0656x^2}{2(0.138)(2.3)(127)} = 8.14(10^{-4})x^2 \,°C$$

The temperature function for the 5W concentrated heating at $a2 = 4$ in (101.6 mm) using (3.7a) is

$$T_4(x) = \frac{5(x - 4.0)}{(3.5)(0.09)(5)} = 3.174(x - 4.0)\,°C$$

In metric units

$$T_4(x) = \frac{5(x - 101.6)}{(0.138)(2.3)(127)} = 0.124(x - 101.6)\,^{\circ}\text{C}$$

The temperature distribution from (3.5) becomes

$$T(x) = 34 + 9.52x - 0.53x^2 - \langle 3.174(x - 4.0) \rangle_{4.0} \,^{\circ}\text{C} \qquad (3.17a)$$

which in metric units is

$$T(x) = 34 + 0.372x - 8.14(10^{-4})x^2 - \langle 0.124(x - 101.6) \rangle_{101.6} \,^{\circ}\text{C} \qquad (3.17b)$$

The temperature distribution for this heating is depicted in Figure 3.2(b).

(c) 10W Linearly Distributed with 5W Concentrated Heating. Since the total heat flowing to the chassis is unchanged from the configuration studied in part (a), the conductance remains the same. Also, the temperature function for the concentrated heating determined in part (b) applies since its location is the same. The linear heat distribution may be stated as

$$\dot{q}(x) = \dot{q}_{max} \left[1 - \left(\frac{x}{L} \right) \right] \qquad (3.18)$$

The total heat at location x due to this distribution is

$$q(x) = \int_0^x \dot{q}(x)\,dx = \dot{q}_{max}x \left[1 - \frac{1}{2}\left(\frac{x}{L} \right) \right] \qquad (3.19)$$

Since heating over the surface length has been defined as

$$\dot{q}_{max} = \frac{2Q}{L} \qquad (3.20)$$

where $Q = q(L)$.

The temperature function is determined by substituting (3.19) and (3.20) into (3.7c), giving

$$T_0(x) = \frac{2Q}{kab} \int_0^x \left(\frac{x}{L} \right) \left[1 - \frac{1}{2}\left(\frac{x}{L} \right) \right] dx = \frac{QL}{kab} \left(\frac{x}{L} \right)^2 \left[1 - \frac{1}{3}\left(\frac{x}{L} \right) \right] \qquad (3.21)$$

Substituting defined quantities (3.21) becomes

$$T_0(x) = \frac{10(6)}{(3.5)(0.09)(5)} \left(\frac{x}{6} \right)^2 \left[1 - \frac{1}{3}\left(\frac{x}{6} \right) \right] = 38.09 \left(\frac{x}{6} \right)^2 \left[1 - \frac{1}{3}\left(\frac{x}{6} \right) \right]$$

which in metric units is

$$T_0(x) = \frac{10(152.5)}{(0.138)(2.3)(127)} \left(\frac{x}{152.5}\right)^2 \left[1 - \frac{1}{3}\left(\frac{x}{152.5}\right)\right]$$

$$= 37.83 \left(\frac{x}{152.5}\right)^2 \left[1 - \frac{1}{3}\left(\frac{x}{152.5}\right)\right]$$

Using (3.5), the temperature distribution is

$$T(x) = 34 + 9.52x - 38.09 \left(\frac{x}{6}\right)^2 \left[1 - \frac{1}{3}\left(\frac{x}{6}\right)\right]$$

$$-\langle 3.174(x - 4.0)\rangle_{4.0} \ ^\circ C \quad (3.22a)$$

which in metric units becomes

$$T(x) = 34. + 0.372x - 37.83 \left(\frac{x}{152.5}\right)^2 \left[1 - \frac{1}{3}\left(\frac{x}{152.5}\right)\right]$$

$$- \langle 0.124(x - 101.6)\rangle_{101.6} \ ^\circ C \quad (3.22b)$$

The temperature distribution given by (3.22) is compared to those obtained from parts (a) and (b) above in Figure 3.2(b).   Q.E.D.

In addition to conduction, heated surfaces may also be cooled by natural or forced convection. One dimensional systems that are cooled by conduction and forced convection are studied in Section 4.3. The temperature distribution for one dimensional systems cooled by conduction combined with natural convection may be determined from the general solution for Case II of Table 3.1. The steady state temperature distribution for configuration of Figure 3.1(a) subject to convection to an ambient at temperature T_∞ and boundary conditions (4.65) is

$$T(x) = T(0) \cosh Mx + [\xi_1 + T(0)\xi_2] \sinh Mx$$

$$- T_\infty(\cosh Mx - 1) - \sum^{n} \langle T_i(x)\rangle_{ai} \quad (3.23)$$

given $N_s = \dfrac{A_h}{bL} = \dfrac{\text{Total heat transfer area}}{\text{Surface area}}$;

h = convection heat transfer coefficient;

T_∞ = ambient;

$$M^2 = hN_s/ka;$$

$$\xi_1 = -\frac{g_1 + G_1 T_1}{Mkab};$$

$$\xi_2 = \frac{G_1}{Mkab}.$$

Temperature functions $T_i(x)$ which depend on whether concentrated or distributed heating occurs at location ai are determined from

Concentrated heating, q_{ai}: $T_i(x) = \dfrac{Mq_{ai}}{N_s hb} \sinh M(x - ai)$ (3.24a)

Distributed heating, $\dot{q}_{ai}$: $T_i(x) = \dfrac{\dot{q}_{ai}}{N_s hb} [\cosh M(x - ai) - 1]$ (3.24b)

The surface temperature at $x = 0$ is obtained from

$$T(0) = \frac{(q_1 + G_1 T_1 + G_2 T_\infty)\cosh ML + (kabMT_\infty - G_2\xi_1)\sinh ML + q_2 + G_2(T_2 - T_\infty) + G_2 \sum^n T_i(L) + kab \sum^n \dot{T}_i(L)}{(G_1 + G_2)\cosh ML + \left(N_s hb + \dfrac{G_1 G_2}{kab}\right)\dfrac{\sinh ML}{M}}$$

(3.25)

One dimensional surfaces with known edge temperatures shown in Figure 3.1(b) represent a special case of (3.23) with $G_1 = G_2 \to \infty$. The temperature distribution for known edge temperatures becomes

$$T(x) = \frac{(T_1 - T_\infty)\sinh M(L - x) + \left[T_2 - T_\infty + \sum^n T_i(L) \right] \sinh Mx}{\sinh ML} + T_\infty - \sum^n \langle T_i(x) \rangle_{ai}$$

(3.26)

EXAMPLE 3.2.   The 15W uniformly heated component mounting surface described in Example 3.1 is also cooled by natural convection to an ambient of $46°C$ in the intended service environment. The total convective heat transfer surface area is about four times the available component mounting area. Determine the surface temperature distribution assuming $h = 0.00367 \ W/in^2°C$ ($5.67 \times 10^{-6} W/mm^2°C$).

Equation (3.25) representing the configuration shown in Figure 3.2(a) simplifies to

$$T(0) = \frac{G_1 T_1 \cosh ML + kabMT_\infty \sinh ML + kab \sum\limits^{n} \dot{T}_i(L)}{kabM \sinh ML + G_1 \cosh ML} \tag{3.27}$$

Using (3.24b), the derivative for uniform heating over the entire length with $a1 = 0$ is

$$\dot{T}_0(L) = \frac{\dot{q}M}{N_s hb} \sinh ML \tag{3.28}$$

For given quantities

$$M \quad = \left[\frac{0.00367(4)}{3.5(0.09)}\right]^{1/2} = 0.216 \text{ in}^{-1};$$

$$kab \quad = 3.5(0.09)5 = 1.575 \text{W} \cdot \text{in}/°\text{C};$$

$$\cosh ML = 1.964;$$

$$\sinh ML = 1.691;$$

$$\dot{T}_0(L) \quad = \frac{(2.5)(0.216)(1.691)}{4(0.00367)(5)} = 12.44 °\text{C/in};$$

$$T(0) \quad = \frac{7.5(32)1.964 + 1.575(0.216)(46)1.691 + 1.575(12.44)}{1.575(0.216)1.691 + 7.5(1.964)} = 33.8 °\text{C}.$$

Since the heat conducted to the chassis must cross the interface which is represented by conductance G_1, the heat removed by convection may be determined from

$$q_{conv} = Q - G_1[T(0) - T_1] = 15 - 7.5[33.8 - 32.]$$

$$q_{conv} = 1.5 \text{W}$$

The convective heat transfer can be increased with the addition of fins if lower surface temperatures are required. Lower component attachment temperatures would also result if the ambient temperature surrounding the assembly could be reduced. This would necessitate some modification to the electronic box cooling scheme.

Substituting quantities and parameters in (3.23) and simplifying give the temperature distribution as

$$T(x) = 39.68 \sinh Mx - 46.26 \cosh Mx + 80.06 \tag{3.29}$$

This result, which includes the effect of natural convection, is compared to the conductively cooled assembly (3.16) in Figure 3.3.   Q.E.D.

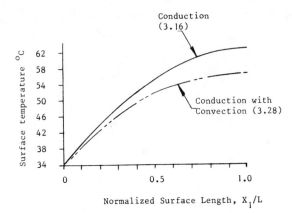

Figure 3.3. Comparative Temperatures of a Uniformly Heated Assembly.

Combined Conduction and Convection. Analytic solutions for one dimensional planar configurations listed in Table 3.1 were derived for constant width, thickness and conductivity. The complexity of analytical solutions increases dramatically if any of these quantities vary over the surface length. Changes in geometry or thermal properties can be accommodated more readily with a network representation.

To develop a network for conductive and convective cooling, the length of the configuration should be partitioned into constant property segments that can be represented by fixed conductance quantities. The network shown in Figure 3.4(a) represents a particular segment of length ℓ with constant geometrical and thermal properties. The convective ambient $T_{\infty\ell}$ in addition to conductances G_ℓ and G_ℓ^* provides the flexibility which enables the development of element strings that describe changes in system geometry and thermal properties. Figure 3.4(b) is an example of a three element string that represents varying properties over the length of a heated assembly.

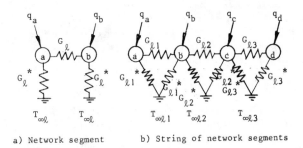

Figure 3.4. Network Representation for Conductive and Convective Cooling.

The conductances of Figure 3.4(a) may be determined for any equipment configuration from

$$G_\ell = \frac{MkA_x}{\sinh M\ell} \tag{3.30}$$

$$G_\ell^* = G_\ell(\cosh M\ell - 1)$$

given $M^2 = \dfrac{hA_h}{\ell kA_x}$;

ℓ = segment length;

A_h = convective heat transfer area;

A_x = conductive heat transfer area (cross-sectional area normal to the direction of heat flow).

EXAMPLE 3.3.   Determine the temperature distribution of the component assembly described in Example 3.2 using a network approach. Compare the results to those obtained using the exact method.

The geometric and thermal properties over the length of the component assembly are constant. The length is partitioned into four segments however, to permit a reasonable distribution of the uniform generated heat. Figure 3.5(a) schematically represents the segmented configuration. Network quantities based on the given configuration for a segment length of 1.5 in (38.1 mm) using (3.30) are

$$kA_x = 1.575 \text{W} \cdot \text{in}/^\circ\text{C};$$

$$A_h = 4(5)(1.5) = 30 \text{ in}^2;$$

$$M = \left[\frac{0.00367(30)}{1.5(1.575)}\right]^{1/2} = 0.216 \text{ in}^{-1};$$

$$G_\ell = \frac{1.575(0.216)}{\sinh (0.324)} = 1.032 \text{W}/^\circ\text{C};$$

$$G^*{}_\ell = 1.032[\cosh (0.324) - 1] = 0.0546 \text{W}/^\circ\text{C}.$$

In metric units these quantities become

$$kA_x = (0.138)(2.3)127 = 40.3 \text{W} \cdot \text{mm}/^\circ\text{C};$$

$$A_h = 4(127)(38.1) = 19354.8 \text{ mm}^2;$$

$$M = \left[\frac{5.67(10^{-6})(19354.8)}{38.1(40.3)}\right]^{1/2} = 0.00845 \text{ mm}^{-1};$$

$$G_\ell = \frac{40.3(0.00845)}{\sinh (0.322)} = 1.039 \text{W}/^\circ\text{C};$$

$$G^*{}_\ell = 1.039 [\cosh (0.322) - 1] = 0.0543 \text{W}/^\circ\text{C}.$$

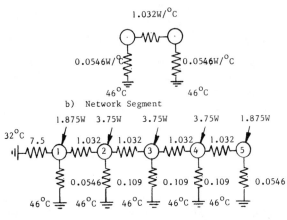

a) Schematic of Component Mounting System

b) Network Segment

c) Heat transfer network

Figure 3.5. Network Representation of a Component Mounting System.

The corresponding segment network shown in Figure 3.5(b) is used to construct the system network of the component mounting configuration shown in Figure 3.5(c). Using the methods of Appendix A, the solution form may be written as

$$
\begin{bmatrix}
8.567 & -1.032 & 0 & 0 & 0 \\
-1.032 & 2.173 & -1.032 & 0 & 0 \\
0 & -1.032 & 2.173 & -1.032 & 0 \\
0 & 0 & -1.032 & 2.173 & -1.032 \\
0 & 0 & 0 & -1.032 & 1.087
\end{bmatrix}
\begin{Bmatrix}
T_1 \\ T_2 \\ T_3 \\ T_4 \\ T_5
\end{Bmatrix}
=
\begin{Bmatrix}
244.39 \\ 8.76 \\ 8.76 \\ 8.76 \\ 4.39
\end{Bmatrix}
\quad (3.31)
$$

The solution of (3.31) compared to the exact distribution (3.29) becomes

	Surface Temperature, °C				
Length from Chassis in (mm)	0.0	1.5 (38.1)	3.0 (76.2)	4.5 (114.3)	6.0 (152.4)
Equation (3.31)	33.90	44.60	51.52	55.39	56.62
Equation (3.29) (Exact)	33.80	44.43	51.29	55.11	56.28

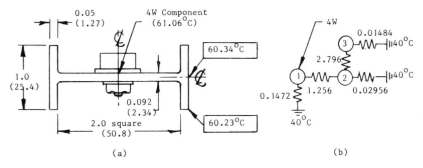

(a)   (b)

Figure 3.6. Component Heat Sink Design. (a) Component heat sink; (b) Network based on component heat sink.

Even though five nodes were used to represent the distributed heating, good correlation was obtained.   Q.E.D.

EXAMPLE 3.4. The aluminum convective heat sink shown in Figure 3.6(a) is used to cool a 4W component in an ambient of $40\,^{\circ}$C. Determine the component mounting temperature for $h = 0.0037\text{W/in}^2\,^{\circ}$C ($5.73 \times 10^{-6}\text{W/mm}^2\,^{\circ}$C).

The network approach based on (3.30) assumes the generated heat is distributed along a line at the center of the heat sink. With symmetry, only one-half of the configuration will be considered.

Network quantities for the 0.09 in (2.34 mm) thick surface with a segment length of 1.0 in (25.4 mm) using (3.30) are

$$kA_x = 3.5(0.09)(2.0) = 0.63\text{W}\cdot\text{in}/\,^{\circ}\text{C};$$

$$A_h = 2(1.0)(2.0) = 4\text{ in}^2;$$

$$M = \left[\frac{0.0037(4)}{1.0(0.63)}\right]^{1/2} = 0.153\text{ in}^{-1};$$

$$G_\ell = \frac{0.63(0.153)}{\sinh(0.153)} = 0.628\text{W}/\,^{\circ}\text{C};$$

$$G^*_\ell = 0.628\,[\cosh(0.153) - 1] = 0.00736\text{W}/\,^{\circ}\text{C}.$$

Network quantities for the 0.05 in (1.27 mm) thick surface with a segment length of 0.5 in (12.7 mm) using (3.30) are

$$kA_x = 3.5(0.05)(2.0) = 0.35\text{W}\cdot\text{in}/\,^{\circ}\text{C};$$

$$A_h = 2(0.5)(2.0) = 2\text{ in}^2;$$

$$M = \left[\frac{0.0037(2)}{0.5(0.35)}\right]^{1/2} = 0.206\text{ in}^{-1};$$

$$G_\ell = \frac{0.35(0.206)}{\sinh (0.103)} = 0.699 \text{W}/^\circ\text{C};$$

$$G^*_\ell = 0.699 \left[\cosh (0.103) - 1\right] = 0.00371 \text{W}/^\circ\text{C}.$$

The configuration shown in Figure 3.6(a) has two axes of symmetry which can be used to reduce the size of the network. The reduced network representing the entire heat sink is shown in Figure 3.6(b). Using the methods of Appendix A, the node temperatures may be determined from the solution of

$$\begin{bmatrix} 1.403 & -1.256 & 0 \\ -1.256 & 4.082 & -2.796 \\ 0 & -2.796 & 2.811 \end{bmatrix} \begin{Bmatrix} T_1 \\ T_2 \\ T_3 \end{Bmatrix} = \begin{Bmatrix} 9.888 \\ 1.184 \\ 0.594 \end{Bmatrix} \qquad (3.32)$$

The temperatures determined from (3.32) are shown superposed on the heat sink configuration in Figure 3.6(a).   Q.E.D.

Circular One Dimensional Systems.   For some applications it is more advantageous to use circular component mounting geometries or configurations that are segments of a circle. The temperature distribution for these one dimensional conduction-cooled configurations may be determined from the solution for Case III of Table 3.1. The steady state temperature distribution for the generalized radially cooled configuration of Figure 3.7 is

$$T(r) = \xi_1 G_2 \ln \left(\frac{r_0}{r}\right) + \xi_0 G_1 \ln \left(\frac{r}{R}\right)$$
$$- \theta k a (\xi_0 + \xi_1) - \sum^{n} \langle T_i(r) \rangle_{ri} \qquad (3.33)$$

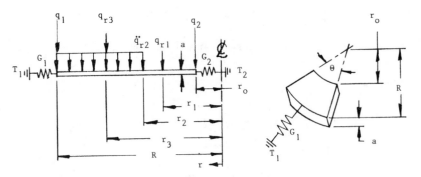

Figure 3.7. Radially Cooled Circular Component Mounting System.

subject to the boundary conditions

$$\theta R ka\dot{T}(R) = q_1 - G_1[T(R) - T_1] \tag{3.34}$$

$$\theta r_0 ka\dot{T}(r_0) = -q_2 + G_2[T(r_0) - T_2]$$

given

$$\xi_0 = \frac{q_2 + G_2 T_2}{G_1 G_2 \ln \left(\dfrac{r_0}{R}\right) - \theta ka(G_1 + G_2)}$$

$$\xi_1 = \frac{Q + q_1 + G_1 \left[\sum T_i(R) + T_1\right]}{G_1 G_2 \ln \left(\dfrac{r_0}{R}\right) - \theta ka(G_1 + G_2)} \tag{3.35}$$

where Q is the total generated heat on the surface excluding edge heat loads q_1 and q_2. Temperature functions $T_i(r)$ which depend on whether concentrated or distributed heating occurs at location ri are determined from

Concentrated heating, q_{ri}: $\quad T_i(r) = \dfrac{q_{ri}}{\theta ka} \ln \left(\dfrac{r}{ri}\right) \quad 0 < ri < R \quad (3.36a)$

Uniform heating, $\ddot{q}_{ri} = \dfrac{2q_{ri}}{\theta(R^2 - ri^2)}$:

$$T_i(r) = \frac{\ddot{q}_{ri}}{ka} \left[\frac{r^2 - ri^2}{4} - \frac{ri^2}{2} \ln \left(\frac{r}{ri}\right)\right] \quad 0 < ri < R$$

$$\tag{3.36b}$$

$$T_0(r) = \frac{\ddot{q}_0 r^2}{4ka} \qquad ri = 0$$

$$\tag{3.36c}$$

The edge temperature at $r = r_0$ and $r = R$ may be determined from

$$T(r_0) = \frac{\theta ka}{G_2} (G_1\xi_0 - G_2\xi_1) + \frac{q_2}{G_2} + T_2 \tag{3.37a}$$

$$T(R) = \frac{\theta ka}{G_1} (G_2\xi_1 - G_1\xi_0) + \frac{Q + q_1}{G_1} + T_1 \tag{3.37b}$$

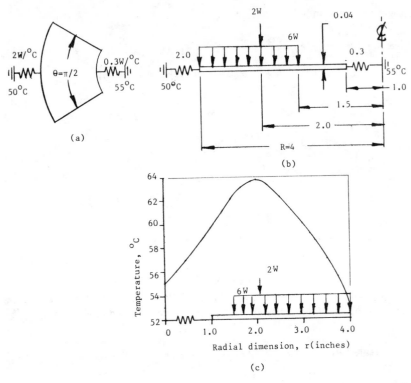

Figure 3.8. Radially Cooled Circuit Module. (a) Geometry; (b) Radial thermal model; (c) Temperature distribution.

For a particular uniformly heated configuration with $r_0 = G_2 = 0$ and $G_1 = \infty$, i.e. T_1 is the edge temperature, simplifies (3.33) to

$$T(r) = T_1 + \frac{Q}{2\theta ka}\left[1 - \left(\frac{r}{R}\right)^2\right]$$ (3.38)

EXAMPLE 3.5.   The circuit module shown in Figure 3.8 is used in a cylindrical equipment configuration. An aluminum heat sink is required to provide both structural support and conductive cooling. Determine the radial temperature distribution for a heat sink thickness of 0.04 in (1 mm) with the boundary constraints defined in Figure 3.8(a) using the heat distribution shown in Figure 3.8(b).

Based on the conductivity of aluminum as given in Appendix C, the surface conductance is

$$ka = (3.5\,\text{W/in} \cdot °\text{C})(0.04 \text{ in}) = 0.14\,\text{W/}°\text{C}$$

The distributed surface heating $\ddot{q}_{1.5}$ and the related temperature function $T_{1.5}(r)$ for a $\theta = \pi/2$ circular segment are determined from (3.36b) as

$$\ddot{q}_{1.5} = \frac{2(6)}{\frac{\pi}{2}(4^2 - 1.5^2)} = 0.56 \text{ W/in}^2$$

$$T_{1.5}(r) = \frac{0.56}{0.14)}\left[\frac{r^2 - 2.25}{4} - \frac{2.25}{2}\ln\left(\frac{r}{1.5}\right)\right] = \left[r^2 - 2.25 - 4.5\ln\left(\frac{r}{1.5}\right)\right]$$

$$(3.39)$$

The temperature function for concentrated heating at $ri = 2$ in from (3.36a) is

$$T_2(r) = \frac{2}{\frac{\pi}{2}(0.14)}\ln\left(\frac{r}{2}\right) = 9.09\ln\left(\frac{r}{2}\right) \qquad (3.40)$$

Computational constants ξ_o and ξ_1 determined from (3.35) for $q_1 = q_2 = 0$ and $Q = 8$ watts are

$$\xi_o = \frac{0.3(55)}{2(0.3)\ln(\frac{1}{4}) - \frac{\pi}{2}(0.14)(2.3)} = -12.34°\text{C}^2/\text{W}$$

$$\xi_1 = \frac{8 + 2(15.64 + 50)}{2(0.3)\ln(\frac{1}{4}) - \frac{\pi}{2}(0.14)(2.3)} = -104.18°\text{C}^2/\text{W}$$

where from (3.39) and (3.40) for $r = R = 4$ inches

$$\sum T_i(R) = T_{1.5}(R) + T_2(R) = 15.64°\text{C}$$

Using given and computed quantities, the edge temperature at the inside radius at $r_o = 1$ inch is determined from (3.37a) as

$$T(r_o) = \frac{\frac{\pi}{2}(0.14)}{(0.3)}[-12.34(2) + 104.18(0.3)] + 55 = 59.82°\text{C}$$

Substituting quantities in (3.33) and simplifying give the temperature distribution as

$$T(r) = -31.254\ln\left(\frac{1}{r}\right) - 24.68\ln\left(\frac{r}{4}\right) + 25.62 - \langle T_{1.5}(r)\rangle_{1.5} - \langle T_2(r)\rangle_2$$

where the temperature functions are given by (3.39) and (3.40). The temperature distribution is shown graphically in Figure 3.8(c). Q.E.D.

Two Dimensional Systems. The general solution for two dimensional systems given in Table 3.1 apply when the rate of heat flow on planar configuration occurs in more than one direction. Analytical solutions for relatively simple geometries may be obtained from the general solution. In most cases, however, the geometry or boundary conditions of planar component mounting configurations are too involved to apply the general solution directly. Instead, variational or numerical techniques that give approximate temperature distributions must be used.

Uniformly heated planar component mounting configurations are commonly used as a basis of estimating the temperature distribution of hardware assemblies early in the development cycle. Many production configurations do possess this property which occurs when each component on the assembly dissipates approximately the same amount of heat. The differential equation for a uniformly heated planar configuration of thickness a at steady state may be written as

$$\frac{\partial^2 T}{\partial x^2} + \frac{\partial^2 T}{\partial y^2} + \frac{Q}{ka} = 0 \tag{3.41}$$

The variational form of (3.41) applied to a rectangular system with dimensions b (width) and L (length) is of the form

$$I = \frac{1}{2} \int_A \left[\left(\frac{\partial T}{\partial x}\right)^2 + \left(\frac{\partial T}{\partial y}\right)^2 - \frac{2QT}{kabL} \right] dA \tag{3.42}$$

Consider the uniformly heated configuration of Figure 3.9 with all edges at temperature T_O. Algebraic or trigonometric functions that generally describe

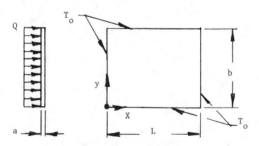

Figure 3.9. Rectangular Uniform Heated Surface.

the temperature distribution while satisfying the conditions at the edges provide the basis of the solution. An algebraic function satisfying this requirement for the configuration in Figure 3.9 is

$$T(x,y) = C \left[\frac{x}{L}\left(1 - \frac{x}{L}\right) \right] \left[\frac{y}{b}\left(1 - \frac{y}{b}\right) \right] + T_o \qquad (3.43)$$

where the constant C will be determined by minimizing the functional I in (3.42). The functions used in (3.43) describe a parabolic temperature distribution indicative of uniform heating. Substituting (3.43) into (3.42) and integrating gives

$$I = \frac{1}{2} \int_o^b \int_o^L \left(\frac{C}{L}\right)^2 \left(1 - 2\frac{x}{L}\right)^2 \left[\frac{y}{b}\left(1 - \frac{y}{b}\right)\right]^2$$

$$+ \left(\frac{C}{b}\right)^2 \left(1 - 2\frac{y}{b}\right)^2 \left[\frac{x}{L}\left(1 - \frac{x}{L}\right)\right]^2$$

$$- \frac{2QC}{kabL} \left[\frac{x}{L}\left(1 - \frac{x}{L}\right)\right] \left[\frac{y}{b}\left(1 - \frac{y}{b}\right)\right] dx\, dy$$

$$I = \frac{1}{2} \left[\frac{C}{90}\left(\frac{b}{L} + \frac{L}{b}\right) - \frac{QC}{18ka} \right] \qquad (3.44)$$

Equation (3.44) is minimized by taking the derivative with respect to C and setting the result equal to zero, which may be stated as

$$\frac{dI}{dC} = 0 \qquad (3.45)$$

Applying (3.45) to (3.44) and solving for the unknown constant C gives

$$C = \frac{5Qbl}{2ka(b^2 + L^2)} \qquad (3.46)$$

Equation (3.43) with C determined from (3.46) represents a simple expression for determining the temperature at any location. Temperature distribution of uniformly heated rectangular component mounting surfaces with insulated or linearly varying edge temperatures may be determined using Table 3.6. Equation (3.15) for uniform heating reduces to (3.50) for the particular configuration represented by Case IV of Table 3.6.

Table 3.6 Temperature Distribution of Uniformly Heated Rectangular Surfaces

Configuration	Temperature Distribution
Case I	$$T(x,y) = \frac{5QbL}{20ka(b^2+L^2)}\left[\frac{x}{L}\left(1-\frac{x}{L}\right)\right]\left[\frac{y}{b}\left(1-\frac{y}{b}\right)\right] + \frac{xy}{bL}(T_0+T_2$$ $$- T_1 - T_3) + \frac{x}{L}(T_3-T_0)+\frac{y}{b}(T_1-T_0)+T_0 \qquad (3.47)$$
Case II	$$T(x,y) = \frac{5QbL}{2ka(4b^2+L^2)}\left[\frac{x}{L}\left(1-\frac{x}{L}\right)\right]\left[\frac{y}{b}\left(1-\frac{y}{b}\right)\right]+\frac{x}{L}(T_1-T_0)$$ $$+ T_0 \qquad (3.48)$$
Case III	$$Y(x,y) = \frac{5QbL}{8ka(b^2+L^2)}\left[\frac{x}{L}\left(2-\frac{x}{L}\right)\right]\left[\frac{y}{b}\left(2-\frac{y}{b}\right)\right]+T_0 \qquad (3.49)$$
Case IV	$$T(x,y) = \frac{QL}{2kab}\left[\frac{x}{L}\left(1-\frac{x}{L}\right)\right]+\frac{x}{L}(T_1-T_0)+T_0 \qquad (3.50)$$
Case V	$$T(x,y) = \frac{QL}{2kab}\left[\frac{x}{L}\left(2-\frac{x}{L}\right)\right]+T_0 \qquad (3.51)$$

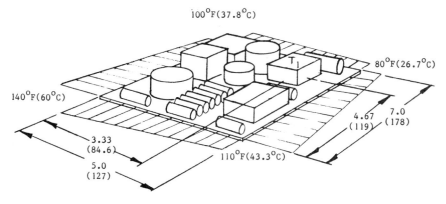

Figure 3.10. Component Assembly with Edge Temperature Distribution.

EXAMPLE 3.6. A 12W uniformly heated assembly experiences linearly varying edge temperatures as shown in Figure 3.10 due to conductive gradients developed by the attachment structure. Determine the attachment temperature of transformer $T1$ for an 0.08 in (2 mm) thick aluminum mounting surface.

The temperature distribution of Figure 3.10 may be represented by Case I of Table 3.6. Quantities used in (3.47) for $T1$ located at $x = 3.33$ in (84.6 mm) and $y = 4.67$ in (119 mm) are

$$\frac{5QbL}{2ka(b^2 + L^2)} = \frac{5(12 \text{ W})(7 \text{ in})(5 \text{ in})}{2(1.97 \text{ W/in}°\text{F})(0.08 \text{ in})[(7 \text{ in})^2 + (5 \text{ in})^2]} = 90.°\text{F (English)}$$

$$= \frac{5(12 \text{ W})(178 \text{ mm})(127 \text{ mm})}{2(0.139 \text{ W/mm}°\text{C})(2 \text{ mm})[(178 \text{ mm})^2 + (127 \text{ mm})^2]}$$

$$= 51.°\text{C (metric)}$$

$$\frac{xy}{bL}(T_o + T_2 - T_1 - T_3) = \frac{(3.33 \text{ in})(4.67 \text{ in})}{(5 \text{ in})(7 \text{ in})}(140 + 80 - 100 - 110)$$

$$= 4.4°\text{F (English)}$$

$$= \frac{(84.6 \text{ mm})(119 \text{ mm})}{(127 \text{ mm})(178 \text{ mm})}(60 + 26.7 - 37.8 - 43.3)$$

$$= 2.5°\text{C (metric)}$$

$$\frac{x}{L}(T_3 - T_o) = \frac{(3.33 \text{ in})}{(5 \text{ in})}(110 - 140) = -20.°\text{F (English)}$$

$$= \frac{(84.6 \text{ mm})}{(127 \text{ mm})}(43.3 - 60) = -11.1°\text{C (metric)}$$

Substituting these quantities into (3.47) gives

$$T_{T1} = 90 \left[\frac{3.33}{5} \left(1 - \frac{3.33}{5} \right) \right] \left[\frac{4.67}{7} \left(1 - \frac{4.67}{7} \right) \right] + 4.4$$

$$- 20. + \frac{4.67}{7} (100 - 140) + 140 = 102.1°F$$

The temperature of $T1$ in metric units is

$$T_{T1} = 51 \left[\frac{84.6}{127} \left(1 - \frac{84.6}{127} \right) \right] \left[\frac{119}{178} \left(1 - \frac{119}{178} \right) \right]$$

$$+ 2.5 - 11.1 + \frac{119}{178} (37.8 - 60) + 60 = 39.1°C$$

Q.E.D.

Numerical Methods for One or Two Dimensional Systems. Component or assembly mounting surfaces with nonuniform heat distributions and irregular geometries can be conveniently studied using numerical techniques. These approximate methods, which are derived from the applicable differential equations, can accommodate many more constraints than can be included in an analytical solution. Procedures using approximate methods are generally much more straightforward than analytical methods and may be successfully applied to a broader set of configurations.

Any numerical approach requires the replacement of the physical surface by a pattern of discrete points or nodes. Instead of finding a solution that varies continuously over the surface, numerical techniques give approximate temperatures at nodal locations. Temperatures within the nodal pattern may be found by interpolation when additional information is necessary.

Numerical treatment using either finite differences or finite elements are generally favored due to their suitability to automated computation. The finite element as a discrete entity representing a substitute portion of the surface with defined boundaries satisfies the governing differential equation. This contrasts with the finite difference method that uses a numerical approximation to the governing differential equation.

Finite Difference Method. The pattern of discrete points that may form rectangular, triangular or other geometric shapes define a *network* or a *finite difference mesh*. Consider the rectangular mesh for the surface shown in Figure 3.11. A heat balance on the network at node 0 of Figure 3.11(b) may be written as

$$q_0 + q_1 + q_2 + q_3 + q_4 = 0 \qquad (3.52)$$

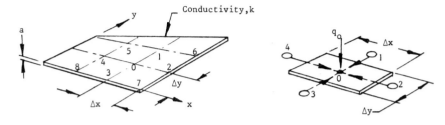

a) Rectangular node pattern on a   b)   Heat balance at node 0
   heat transfer surface

Figure 3.11. Finite Difference Development.

Conduction heat transfer from node 1 to node 0 of Figure 3.11(b) is

$$q_1 = \frac{ka\,\Delta x}{\Delta y}(T_1 - T_0) \qquad (3.53)$$

Writing expressions similar to (3.53) for each network branch intersecting at node 0 and substituting into (3.52) using the normalization $\Delta x = \alpha\Delta y$ gives

$$4T_0 - \frac{2}{\alpha^2 + 1}(T_4 + T_2) - \frac{2\alpha^2}{\alpha^2 + 1}(T_1 + T_3) = \frac{2q_0\alpha}{(\alpha^2 + 1)ka} \qquad (3.54)$$

The coefficients of (3.54) may be structured to correspond with the nodes surrounding the centrally located node as shown in Figure 3.12(a). Equation (3.54) or the geometric coefficient form of Figure 3.12(a) is completely general

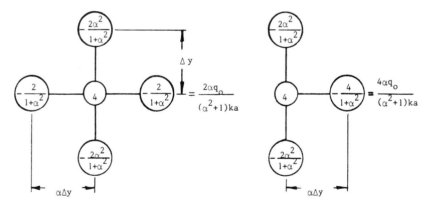

a)   Interior node   b) Edge node along the y axis

Figure 3.12. Finite Difference Coefficient Forms.

and may be applied at any node on the surface mesh. Along edges or corners, the correct nodal coefficient form may be developed from a heat balance as in (3.52) or by applying conditions of symmetry to the form of Figure 3.12(a). Since heat does not flow across edges or lines of symmetry, the appropriate nodal form in Figure 3.12(b) is developed by folding the network of Figure 3.12(a) along the corresponding axis.

An algebraic equation for the temperature of the central node results for each node in the network. This equation is developed by superposing the central coefficient of the cluster of Figure 3.12 at a particular node while assemblying the product of the corresponding nodal temperatures and coefficients for all the nodes in the cluster. The algebraic equation for the temperature of node 2 in Figure 3.11 using Figure 3.12(b) is

$$4T_2 - \frac{2\alpha^2}{1 + \alpha^2}(T_6 + T_7) - \frac{4}{1 + \alpha^2}T_0 = \frac{4\alpha q_0}{(\alpha^2 + 1)ka} \tag{3.55}$$

The simultaneous equations representing a finite difference mesh may be solved for the unknown temperatures using methods of Appendix A or any other known method.

Finite Element Method. The finite element representation of an actual surface resembles the application of the variational approach (3.42), in which the temperature distribution is approximated by the sum of functions, each multiplied by an unknown constant. The unknown quantities are determined by minimizing the variation of energy relative to equilibrium of the selected functions. In the finite element method, the distribution of temperature within an individual element is characterized instead of the entire surface. The temperature distribution of the actual surface is determined from a substitute structure consisting of discrete elements connected together at their nodal points.

Although elements of various geometrical shapes have been studied, a straightforward approach which gives good results may be obtained with the elements shown in Figure 3.13. The elements may be used individually or collectively when developing the substitute surface. The triangular element may be used to fit curved boundaries or edge to edge, forming a rectangular element. The rectangular element of Figure 3.13(b), however, is inherently more accurate due to the use of a higher order polynomial in its development. The edge conductances of the rectangular element, Figure 3.13(b), may become negative for long and narrow configurations. Algebraic development of the conductance matrix (Appendix A) or simultaneous equations based on a nodal heat balance (3.52) follow the defined procedures independent of the sign. Conductance values including the appropriate sign are incorporated after the desired solution form is developed.

Geometry

Element

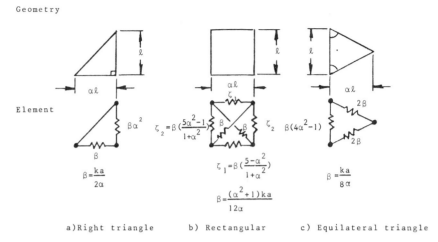

a)Right triangle   b) Rectangular   c) Equilateral triangle

Figure 3.13. Finite Element Conductance Configurations.

EXAMPLE 3.7. Determine the attachment temperature of transformer $T1$ of Example 3.6 using the finite element of Figure 3.13(b).

Figure 3.10 is redrawn in plan and subdivided into nine rectangular elements as shown in Figure 3.14(a). The nodal intersections are numbered so that node 4 corresponds to the location of $T1$. The nodal pattern was selected as a matter of convenience since any pattern could have been used. Interpolation may be required if nodal locations differ from points where specific temperatures are desired. Nodal edge temperatures are determined using a linear relationship from Figure 3.10. The element

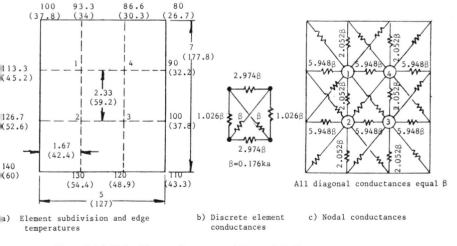

a) Element subdivision and edge   b) Discrete element   c) Nodal conductances
   temperatures   conductances

Figure 3.14. Finite Element Treatment of Figure 3.10 Component Assembly.

aspect ratio is determined as

$$\alpha = \frac{\Delta x}{\Delta y} = \frac{1.667}{2.333} = 0.714 \qquad (3.56)$$

Since 12W is uniformly distributed over the entire surface, the proportion of heat injected into each of the four nodes is

$$q = \frac{1.67(2.33)12}{7(5)} = 1.33\text{W (English)}$$

$$q = \frac{42.4(59.2)12}{177.8(127)} = 1.33\text{W (metric)} \qquad (3.57)$$

The generalized finite element shown in Figure 3.14(b) is developed using Figure 3.13(b) with (3.56). The edge conductances along intersecting elements are added, giving the conductance configuration shown in Figure 3.14(c). Each conductance of Figure 3.14(c) is multiplied by β.

The equivalent heat load injected into a node is composed of the proportion of the distributed heating from (3.57) in addition to the effects of the surrounding edge temperatures. The equivalent heat load for node 1 using Figures 3.14(a) and (c) becomes

$$q_1 = q + [(86.6 + 100 + 126.7)(1) + 93.3(2.052) + 113.3(5.948)]\beta$$

$$q_1 = 1.33 + 1178.7\beta \text{ W (English)}$$

$$q_1 = 1.33 + 459.3\beta \text{ W (metric)}$$

The equivalent heat load for nodes 2, 3 and 4 are determined along similar lines. Using these results and the nodal conductances of Figure 3.14(c), the conductance matrix solution following the procedure given in Appendix A becomes

$$\beta \begin{bmatrix} 20 & -2.052 & -1 & -5.948 \\ -2.052 & 20 & -5.948 & -1 \\ -1 & -5.948 & 20 & -2.052 \\ -5.948 & -1 & -2.052 & 20 \end{bmatrix} \begin{Bmatrix} T_1 \\ T_2 \\ T_3 \\ T_4 \end{Bmatrix}$$

$$\text{English} \qquad\qquad \text{metric}$$

$$= \begin{Bmatrix} 1.33 + 1178.7\beta \\ 1.33 + 1393.7\beta \\ 1.33 + 1171.0\beta \\ 1.33 + 986.3\beta \end{Bmatrix} = \begin{Bmatrix} 1.33 + 459.3\beta \\ 1.33 + 578.6\beta \\ 1.33 + 455.1\beta \\ 1.33 + 352.2\beta \end{Bmatrix} \qquad (3.58)$$

where $\beta = 0.176(1.97)(0.08) = 0.0277\text{W}/°\text{F}$ (English)

$\beta = 0.176(0.140)(2.03) = 0.05\text{W}/°\text{C}$ (metric)

Using reduction methods of Appendix A, the attachment temperature for $T1$ from (3.58) is

$$T_{T1} = 0.121/\beta + 97.75 = 102.1°\text{F (English)}$$

$$T_{T1} = 0.121/\beta + 36.52 = 38.9°\text{C (metric)}$$

These results are virtually the same as those obtained in Example 3.6.   Q.E.D.

The nodal pattern selected for manual computation by the equipment designer should minimize the number of nodes while maintaining a general correspondence to the physical system. It is not necessary to have the nodal pattern replicate each geometric feature or component. Internodal characteristics can usually be accounted for by adjusting conductances or nodal heat loads. Symmetrical conditions should be exploited as an effective means of reducing the number of nodes. The use of nonuniform rectangular elements, Figure 3.15(a), for some assembly configurations can also reduce the number of nodes (elements) needed to characterize the surface. Figure 3.15 illustrates the collective use of both elements shown on Figure 3.13. The application of surface and edge conductances emitting from the model is also illustrated in Figure 3.15(b). Conductances that augment the finite element configuration are used to characterize interfaces, adjacent assemblies and component heat flow paths to the surface. Capacitances may be added to the appropriate nodes if transient response characteristics are required.

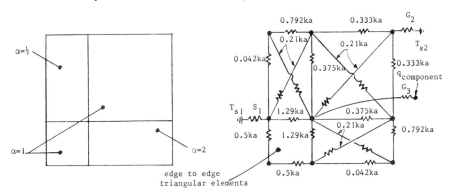

a) Non-uniform element   b) Assembly conductance configuration
    distribution

Figure 3.15. Application of Finite Elements and Interface Conductors.

Usually any lumped idealization including a finite difference or finite element representation gives conservative results. Temperature and transient response characteristics of the substitute system converge to actual values with increasing nodal (element) density. Systems with high nodal densities require automated computation. Idealizations with fewer than 12 nodes have been successfully evaluated by the manual techniques of Appendix A.

3.2 EFFECTIVENESS OF EQUIPMENT MOUNTING AND POSITIONING

The reliability of electronic equipment improves as the operating temperature of all components is reduced. When high power generating devices are used, particular emphasis is given to their location relative to an available sink and to attachment methods which minimize thermal gradients. Often circuit connectivity and component spatial requirements limit possible locations requiring consideration of other cooling schemes. Surrounding low power and passive components are directly affected by high power devices. Since heat flows from a higher temperature location to a cooler one, passive and low power components are subjected to elevated attachment temperatures that may produce higher than expected operating temperatures and consequently, lower equipment reliability.

Electronic circuits are usually partitioned and packaged according to the function they perform. At the functional level, circuit cards provide a convenient and producible component assembly. The size, shape, orientation and attachment of each of these card assemblies is dependent upon the spatial constraints imposed by the next assembly, their power dissipation and the repair and replacement objectives for the equipment. The operating temperature that each card assembly experiences is dependent upon its location relative to other heat generating assemblies and the equipment heat sink to which all generated energy is transferred.

The sink temperature of each assembly in a conduction cooled system is a function of its distance from the equipment attachment interface, the ultimate sink for all generated energy. If all components have approximately the same manufacturers' rating, the higher-dissipating assemblies should be located as close to the ultimate sink as possible. The basis for this goal is demonstrated using two possible distributions of heat generating assemblies shown in Figure 3.16. In each case the higher heat generating assembly (10W) is located closest to the sink. If the 10W assembly at node 1 and the 5W assembly at node 2 were interchanged so that the higher heat generating assembly is positioned further from the equipment sink, the mean temperature of the assemblies in the series distribution and parallel distribution would increase 18 and 23 percent respectively. The effect of position on the temperature of the interchanged assemblies is even more pronounced.

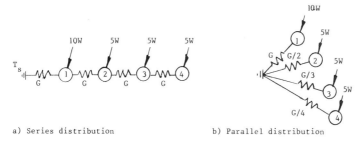

a) Series distribution   b) Parallel distribution

Figure 3.16. Possible Assembly Distributions for Electronic Equipment.

Manufacturers' ratings for power components may be higher than those for signal devices. Under these conditions, power supplies and other power component assemblies may be located further from the equipment sink since they can operate reliably at higher temperatures. In general, however, lower mean component temperatures result when higher heat generating devices are located as close to the ultimate sink as possible at each level of assembly. Within a circuit board assembly or module, for example, the higher heat generating components should be positioned in the vicinity of the attachment interface.

Heat removal by conduction across bolted equipment interfaces requires special attention. Mating surfaces should be large, flat, parallel, smooth, and stiff to obtain an adequate conductance to minimize surface-to-surface temperature difference. Conduction-cooled equipment can operate at any orientation.

Orientation does affect equipment designed for natural convection cooling. Convective flow paths and the nature of surface heat removal can be affected adversely by reorientation. Component and mounting surface temperatures depend on the free exchange of the convective cooling medium. Equipment reorientation alters the internal and external convective flow geometry and the orientation of heated components and component mounting surfaces. The heat transfer from a square surface is 7.3 percent more effective for a vertical orientation than for a horizontal orientation with the heated side facing up. If reorientation causes a vertically cooled surface to lie horizontally with its heated side down, a temperature increase of about 87 percent can be expected on that assembly. Higher temperatures on any assembly can affect others within the enclosure.

The installation of procured equipment or equipment designed for specific applications requires an understanding of the internal and external heat flow mechanization and structural design sufficient to assure reliable performance in the new environment. Equipment might outwardly appear to be conduction-cooled but may actually depend upon natural convection internally. Reorientation of equipment can also affect the displacements experienced by the enclosure and internal assemblies as well as the forces at their attachments.

3.3 IMPEDANCE TECHNIQUES

Impedance techniques offer an attractive alternative to the conductance approach for many network idealizations of actual hardware heat transfer processes. The amount of effort required to determine the temperature distribution in a modeled system where the number of nodes is greater than the loop meshes can be reduced considerably using impedance techniques. The number of loop meshes is related to the number of nodes by

$$N_{\text{loops}} = N_{\text{branches}} - N_{\text{nodes}} \qquad (3.59)$$

The number of branches N_{branches} in (3.59) are the minimum remaining after combining simple series or parallel paths using (2.47) and (2.50).

The system of equations developed from a heat balance on each node is the basis of the conductance approach defined in Appendix A. Once the network has been defined, however, the conductance matrix and the matrix form (A.2) can be written by inspection. The diagonal coefficients of the conductance matrix are the sums of those conductances intersecting at the corresponding nodes. The off-diagonal coefficients are the negatives of those conductances connecting any two corresponding nodes of the network. The popularity of (A.2) is due to the ease of coding this format for automated computation. The nodal temperatures must be found by inversion or reduction, which limits the network size suitable for manual computation.

Matrix Characteristics. The temperatures can be determined directly using impedances since the equations are written from network relationships in the form

$$\{T\} = [G]^{-1}\{Q\} = [Z]\{Q\} \qquad (3.60)$$

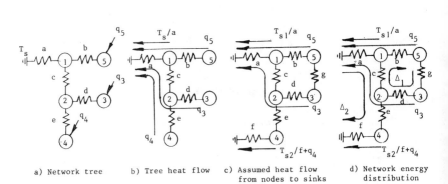

a) Network tree   b) Tree heat flow   c) Assumed heat flow from nodes to sinks   d) Network energy distribution

Figure 3.17. Impedance Network Configurations.

The impedance matrix [$\mathbf{Z}$] for the model of Figure 3.17(a) can be written by inspection since $N_{\text{branches}} = N_{\text{nodes}}$. Heat transfer networks with this characteristic form tree-like structures emitting from a single sink, T_s. The diagonal coefficients of the impedance matrix with $N_{\text{loops}} = 0$ in (3.59) are equal to the sum of the impedances from the sink T_s to the corresponding node. The off-diagonal coefficients are equal to the sum of those impedances common to the corresponding nodes. The impedance matrix solution for the network of Figure 3.17(a) for $T_s = 0$ is

$$[\mathbf{Z}]$$

$$
\begin{Bmatrix} T_1 \\ T_2 \\ T_3 \\ T_4 \\ T_5 \end{Bmatrix} =
\begin{bmatrix}
a & & & & \\
a & a+c & & \text{Symmetrical} & \\
a & a+c & a+c+d & & \\
a & a+c & a+c & a+c+e & \\
a & a & a & a & a+b
\end{bmatrix}
\begin{Bmatrix} 0 \\ 0 \\ q_3 \\ q_4 \\ q_5 \end{Bmatrix}
$$

$$(3.61a)$$

Since only three nodes are subjected to heat generation, (3.61a) may be stated compactly as

$$
\begin{Bmatrix} T_1 \\ T_2 \\ T_3 \\ T_4 \\ T_5 \end{Bmatrix} =
\begin{bmatrix}
a & a & a \\
a+c & a+c & a \\
a+c+d & a+c & a \\
a+c & a+c+e & a \\
a & a & a+b
\end{bmatrix}
\begin{Bmatrix} q_3 \\ q_4 \\ q_5 \end{Bmatrix} \qquad (3.61b)
$$

The temperature of node 3 can be determined directly from (3.61b) as

$$T_3 = (a + c + d)q_3 + (a + c)q_4 + aq_5$$

Many impedance networks of actual hardware heat transfer paths will contain some loops as determined by (3.59). For these configurations, the impedance matrix for a tree-like structure is developed which contains all the modeled nodes, and the appropriate impedance branches. The choice of the tree branches is arbitrary, but it is desirable that the smallest impedances be included. The ignored branch impedances (larger values) are added using methods of Appendix A.6(3).

Heat Flow Method. An attractive alternative to the development of the impedance matrix is using heat flow developed from the network energy distribution. The heat flow method provides a degree of simplification of the impedance technique for the manual solution of modeled actual equipment.

Heat flow in a network is due to heat generation at nodes. The temperature of each sink adjoining a network may be equated to an equivalent nodal heat load at adjacent nodes. Nodal heat is assumed to flow from the source node to a sink along any arbitrary simple path. Simple paths are those that route directly to a sink, usually traversing a minimum number of branches. The heat flow for the network of Figure 3.17(a) is shown in Figure 3.17(b). By means of this method, actual heat flow distributions for tree-like network structures occur. In more complex networks, corrective loop heat flows are required in addition to projected nodal heat flows to obtain the true energy distribution. An assumed nodal heat flow for a network containing a loop as determined by (3.59) is shown in Figure 3.17(c).

Nodal temperatures in a tree network may be determined progressively using the heat flow through various branches. The sink is used as the reference for initiating this procedure. The heat flow in branch a of Figure 3.17(a) is

$$q_a = T_s/a + q_3 + q_4 + q_5 \qquad (3.62)$$

The temperature of node 1 using (3.62) with $T = RQ$ is

$$T_1 = aq_a = T_s + a(q_3 + q_4 + q_5) \qquad (3.63)$$

Since the temperature of node 1 has been determined, the temperatures of coupled nodes 2 and 5 in Figure 3.17(b) become

$$T_2 = T_1 + cq_c = T_1 + c(q_3 + q_4)$$
$$T_5 = T_1 + bq_b = T_1 + bq_5$$

Using (3.63), the temperatures of these nodes may be expressed in terms of the sink temperature T_s as

$$T_2 = T_s + (a + c)(q_3 + q_4) + aq_5$$
$$T_5 = T_s + (a + b)q_5 + a(q_3 + q_4)$$

The temperature of the remaining nodes of Figure 3.17(b) may be determined similarly.

An additional corrective heat flow is required for each loop of a complex heat transfer network before the nodal temperatures can be determined progressively. The quantity of loops (corrective heat flows) in the model may be determined from (3.59). Since the network of Figure 3.17(c) has 5 nodes and 7 branches, only two corrective heat flows are required. Corrective flows are associated with the loops formed in the network. Flows are assigned arbitrarily with heat flowing in a selected direction. Corrective heat flows for the network of Figure 3.17(c) are illustrated in Figure 3.17(d).

Once the corrective heat flows are known, the nodal temperatures may be determined. As before, a sink is used as the reference for initiating the procedure. The heat flows in branches a and f of Figure 3.17(d) are

$$q_a = T_{s1}/a + q_3 + q_5 - \Delta_2 \qquad (3.64a)$$

$$q_f = T_{s2}/f + q_4 + \Delta_2 \qquad (3.64b)$$

The temperatures of nodes 1 and 4 using (3.64) are

$$T_1 = aq_a = T_{s1} + a(q_3 + q_5 - \Delta_2) \qquad (3.65a)$$

$$T_4 = fq_f = T_{s2} + f(q_4 + \Delta_2) \qquad (3.65b)$$

The temperature of any node that is coupled to either node 1 or node 4 can now be determined. Two different expressions that define the temperature of node 2 can be written. One of these may be used to check the result obtained with the other if desired. The temperatures of nodes 2 and 5 using (3.65) are

$$T_2 = T_4 + eq_e = T_{s2} + fq_4 + (e + f)\Delta_2$$

$$T_5 = T_1 + bq_b = T_{s1} + a(q_3 - \Delta_2) + (a + b)q_5 - b\Delta_1$$

Other network nodal temperatures may be determined similarly. Unknown corrective heat flow quantities Δ_i may be evaluated from the energy distribution in the associated network branches. The equation which determines Δ_1 of Figure 3.17(d) becomes

$$b(\Delta_1 - q_5) + c(\Delta_1 + q_3 - \Delta_2) + d(\Delta_1 + q_3) + g\Delta_1 = 0 \quad (3.66)$$

Equation (3.66) is equivalent to setting the sum of the temperature differences around loop Δ_1 equal to zero. An expression similar to (3.66) may be written for Δ_2. Δ_1 and Δ_2 are coupled because each flows through a common branch in Figure 3.17(d). The unknown corrective flows may be determined from the matrix equation

$$
\begin{bmatrix} b + c + d + g & -c \\ -c & a + c + e + f \end{bmatrix} \begin{Bmatrix} \Delta_1 \\ \Delta_2 \end{Bmatrix}
$$

$$
= \begin{Bmatrix} bq_5 - (c + d)q_3 \\ T_{s1} - T_{s2} + (a + c)q_3 + aq_5 - fq_4 \end{Bmatrix} \quad (3.67)
$$

The relationship for Δ_1 in (3.67) is identical to (3.66). Equation (3.67) may be written directly from the network by inspection without the necessity of developing loop balance equations analogous to (3.66). The diagonal coefficients are equal to the sum of those impedances through which branches Δ_i flow. The off-diagonal coefficients are the negative of the sums of those impedances for branches common to Δ_i and Δ_j. The temperature coefficients corresponding to Δ_i on the right side of (3.67) are the sum of the products $R_i q_j$ for the path of Δ_i. Branch flows opposite to the sense of Δ_i are considered positive.

The heat flow method reduces the effort of obtaining the nodal temperatures of many network configurations of actual equipment. Additional reductions in effort are obtained through the use of symmetry and delta-star conversions shown in Figure 2.27.

EXAMPLE 3.8.   A 0.09 in (2.3 mm) thick aluminum bracket shown in Figure 3.18(a) provides a mounting surface for four power transistors. The interface temperature of an adjoining structure is maintained at 131°F (55°C). Determine the attachment temperature of each transistor using (a) the heat flow method, and (b) the heat flow method in conjunction with the delta-star conversion.

(a) *Heat Flow Method.* Right triangle elements of Figure 3.13 were used to construct the impedance model shown in Figure 3.18(b). Unwanted nodes were eliminated by reducing parallel paths into equivalent impedances connecting the four nodes repre-

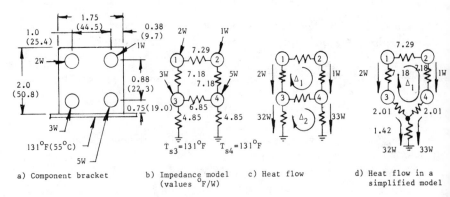

a) Component bracket   b) Impedance model   c) Heat flow   d) Heat flow in a
                           (values °F/W)   simplified model

Figure 3.18. Component Bracket.

senting component locations. The assumed heat flow configuration is shown in Figure 3.18(c). The flow from node 3 to the sink for example becomes

$$q_{3/Ts3} = T_{s3}/4.85 + q_1 + q_3 = 32W \quad \text{(English)}$$

$$q_{3/Ts3} = T_{s3}/2.69 + q_1 + q_3 = 32W \quad \text{(metric)}$$

Using Figure 3.18(c), the matrix equation for determining unknown corrective heat flows becomes

$$\begin{bmatrix} 7.18(2) + 7.29 + 6.85 & -6.85 \\ -6.85 & 4.85(2) + 6.85 \end{bmatrix} \begin{bmatrix} \Delta_1 \\ \Delta_2 \end{bmatrix} = \begin{Bmatrix} 7.18(2) - 7.18(1) \\ 4.85(32) - 4.85(33) \end{Bmatrix}$$

The solution of using methods of Appendix A becomes

$$\begin{Bmatrix} \Delta_1 \\ \Delta_2 \end{Bmatrix} = \begin{bmatrix} 0.03896 & 0.01613 \\ 0.01613 & 0.06709 \end{bmatrix} \begin{Bmatrix} 7.18 \\ -4.85 \end{Bmatrix} = \begin{Bmatrix} 0.202 \\ -0.209 \end{Bmatrix} \quad (3.68)$$

The temperatures of nodes 3 and 4 using Figure 3.18(c) and (3.68) are

$$T_3 = (32 - \Delta_2)4.85 = 156.2°F$$
$$T_4 = (33 + \Delta_2)4.85 = 159.0°F$$

(3.69)

The temperatures of nodes 1 and 2 using Figure 3.18(c), (3.68) and (3.69) are

$$T_1 = T_3 + (2 - \Delta_1)7.18 = 169.1°F \quad (3.70a)$$

$$T_2 = T_4 + (1 + \Delta_1)7.18 = 167.6°F \quad (3.70b)$$

Convenient checking is also provided by the heat flow method. The temperature of node 2 can be determined from T_1 using Figures 3.18(c), (3.68) and (3.70a) as

$$T_2 = T_1 - 7.29\Delta_1 = 167.6°F$$

(b) *Heat Flow Method with a Delta-Star Conversion.* Figure 3.18(d) illustrates the result of applying the delta-star conversion to the model of Figure 3.18(b). The conversion has replaced a loop with a dummy node. The single corrective flow, Δ_1, is determined as

$$\Delta_1 = \frac{(32 - 33)(2.01) + (2 - 1)7.18}{7.18(2) + 2.01(2) + 7.29} = 0.201 \quad (3.71)$$

Applying the heat flow indicated in Figure 3.18(d) and (3.71), the temperatures of nodes 3 and 4 are calculated as

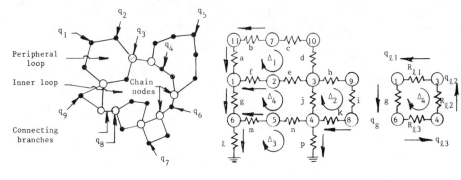

a) Reducible chained loop system   b) A particular chained   c) Simplified network
                                      loop configuration   for determining Δ_4

Figure 3.19. Chained Loop Configurations.

$$T_3 = (32 + 33)1.42 + (32 - \Delta_1)2.01 = 156.2°F$$

$$T_4 = (32 + 33)1.42 + (33 + \Delta_1)2.01 = 159.0°F$$

The delta-star conversion produces the results obtained in part (a) with a reduction in effort. When the delta-star conversion affects sinks, the equivalent nodal heat load is developed prior to its application.   Q.E.D.

Simply Connected Loops.   Corrective heat flows may be determined directly without resorting to a matrix equation if network loops are simply connected as in Figure 3.19(a). A simply connected configuration consists of peripheral loops containing any number of branches chained together with connecting branches to form an inner loop. Any part of the configuration may contain a heat sink. Heat may be injected into any of the nodes. A network configuration consisting of simply connected loops is shown in Figure 3.19(b).

Equivalent impedances of all peripheral loops $R_{\ell i}$ are determined in order to construct the simplified network of Figure 3.19(c). The equivalent impedance of loop 1, which is defined by chain nodes 1 and 3, is

$$R_{\ell 1} = \frac{(a + b + c + d)(f + e)}{a + b + c + d + e + f} = \frac{(a + b + c + d)\,(f + e)}{Z_{\ell 1}} \quad (3.72a)$$

The equivalent impedances of loops 2 and 3 are determined similarly as

$$R_{\ell 2} = \frac{j(h + i + k)}{h + i + j + k} = \frac{j(h + i + k)}{Z_{\ell 2}} \quad (3.72b)$$

$$R_{\ell 3} = \frac{(m + n)(\ell + p)}{(m + n + \ell + p)} = \frac{(m + n)(\ell + p)}{Z_{\ell 3}} \quad (3.72c)$$

The heat flowing in each branch of the simplified network figure 3.19(c) is determined from the flow in the corresponding loop of Figure 3.19(b).The heat flowing in $R_{\ell 1}$ using Figure 3.19(b) becomes

$$q_{\ell 1} = \frac{(aq_a + bq_b)(e + f) + fq_f(a + b + c + d)}{Z_{\ell 1}R_{\ell 1}} \tag{3.73a}$$

Loop flows opposite to the sense of the assumed inner loop flow Δ_4 are considered positive. Heat flows in branches $R_{\ell 2}$ and $R_{\ell 3}$ of Figure 3.19(c) are determined similarly as

$$q_{\ell 2} = -\frac{kq_k(j)}{Z_{\ell 2}R_{\ell 2}} \tag{3.73b}$$

$$q_{\ell 3} = \frac{-mq_m(\ell + p) + (\ell q_\ell - pq_p)(m + n)}{Z_{\ell 3}R_{\ell 3}} \tag{3.73c}$$

Relationships (3.72) are developed from the sum of the products of the flow branch impedance of one loop impedance leg and the impedance of the other leg divided by the loop and equivalent impedances.

The corrective heat flow may be determined from Figure 3.19(c), using (3.72) and (3.73), as

$$\Delta_4 = \frac{R_{\ell 1}q_{\ell 1} + R_{\ell 2}q_{\ell 2} + R_{\ell 3}q_{\ell 3} + gq_g}{R_{\ell 1} + R_{\ell 2} + R_{\ell 3} + g} \tag{3.74}$$

Corrective flows for any peripheral loop may be determined using the known Δ_4 from (3.74). With known corrective heat flows, the temperature distribution may be determined progressively as described earlier.

EXAMPLE 3.9.   The network of Figure 3.18(c) representing the component bracket described in Example 3.8 consists of two simply connected loops. Determine the temperature distribution using the method described for networks of this configuration.

Either loop of Figure 3.18(c) may be considered as an inner loop. Defining loop 1 as the inner loop, the equivalent impedance of loop 2 becomes

$$R_{\ell 2} = \frac{6.85(4.85 + 4.85)}{Z\ell 2} = 4.015°\text{F/W}$$

where $Z_{\ell 2} = 6.85 + 2(4.85) = 16.55°\text{F/W}$. The network for determining Δ_1 is shown in Figure 3.20(a). Heat flow in branch $R_{\ell 2}$ due to the flows in loop 2 is

$$q_{\ell 2} = \frac{32(4.85)6.85 - 33(4.85)6.85}{Z_{\ell 2}R_{\ell 2}} = -\frac{2.01}{R_{\ell 2}} \text{W} \tag{3.75}$$

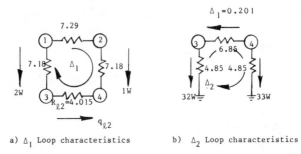

a) Δ_1 Loop characteristics   b) Δ_2 Loop characteristics

Figure 3.20. Loop Flow Determination of Figure 3.18.

The corrective heat flow Δ_1 from Figure 3.20(a) based on (3.75) becomes

$$\Delta_1 = \frac{2(7.18) - 1(7.18) - 2.01}{7.29 + 2(7.18) + 4.015} = 0.201\text{W} \tag{3.76a}$$

This result and the flow geometry shown in Figure 3.20(b), developed from Figure 3.18(c), determines Δ_2 as

$$\Delta_2 = \frac{6.85(0.201) + (32 - 33)4.85}{2(4.85) + 6.85} = -0.209\text{W} \tag{3.76b}$$

The results obtained in (3.76) are virtually the same as (3.68) of Example 3.8. Progressive determination of nodal temperatures is illustrated in that example.   Q.E.D.

EXAMPLE 3.10.   Determine the nodal temperatures for the network of Figure 3.21(a).

The number of loops from (3.59) are depicted by their corrective heat flows on Figure 3.21(a). This network configuration is not simply connected. The four equations for determining Δ_i may be arranged in a matrix and treated by the methods of Appendix A.

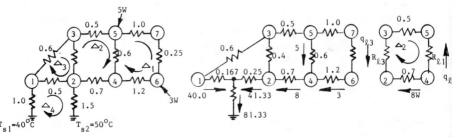

a) Heat transfer network   b) Δ_4 Loop elimination   c) Δ_2 Loop Characteristics

Figure 3.21. Heat Transfer Network of Example 3.10.

A simply connected loop configuration can be developed, however, by applying a delta-star conversion to loop 4 as shown in Figure 3.21(b). The network for determining Δ_2 for the inner loop is illustrated in Figure 3.21(c). Equivalent impedances for loops 1 and 3 are

$$R_{\ell 1} = \frac{0.6(1.0 + 0.25 + 1.2)}{Z_{\ell 1}} = 0.482\,°C/W \qquad (3.77a)$$

$$R_{\ell 3} = \frac{0.4(0.6 + 0.25 + 0.167)}{Z_{\ell 3}} = 0.287\,°C/W \qquad (3.77b)$$

where $Z_{\ell 1} = 1.0 + 0.25 + 1.2 + 0.6 = 3.05\,°C/W$

$Z_{\ell 3} = 0.6 + 0.25 + 0.167 + 0.4 = 1.417\,°C/W$

Heat flows through branches $R_{\ell 1}$ and $R_{\ell 3}$ are determined from Figure 3.21(a) and (b) as

$$q_{\ell 1} = \frac{-3(1.2)(0.6) - 5(0.6)(1.0 + 0.25 + 1.2)}{Z_{\ell 1} R_{\ell 1}} = -\frac{3.12}{R_{\ell 1}}\,W \qquad (3.78a)$$

$$q_{\ell 3} = \frac{40.0(0.167)(0.4) - 41.33(.25)(0.4)}{Z_{\ell 3} R_{\ell 3}} = -\frac{1.03}{R_{\ell 3}}\,W \qquad (3.78b)$$

Using Figure 3.21(c) with (3.77) and (3.78) determines Δ_2 to be

$$\Delta_2 = \frac{-3.12 - 1.03 - 0.7(8)}{0.482 + 0.7 + 0.287 + 0.5} = -4.952W \qquad (3.79a)$$

The corrective flows for loops 1 and 3 based on this result and Figure 3.21(b) become

$$\Delta_1 = \frac{-3.(1.2) + (5 - 4.952)0.6}{Z_{\ell 1}} = -1.171W \qquad (3.79b)$$

$$\Delta_3 = \frac{-4.952(0.4) - 41.33(0.25) + 40.(0.167)}{Z_{\ell 3}} = -3.975W \qquad (3.79c)$$

Nodal temperatures are determined progressively using the sink as a reference. In accordance with Figure 3.21(b) and (3.79), the nodal temperatures become

$$T_1 = 81.33(0.5) + (40. + 3.975)0.167 = 48\,°C$$

$$T_2 = 81.33(0.5) + (41.33 - 3.975)0.25 = 50\,°C$$

$$T_3 = T_1 - 0.6\Delta_3 = 48 + 0.6(3.975) = 50.38\,°C$$

$$T_4 = T_2 + (8 + \Delta_2)0.7 = 52.13\,°C$$

$$T_5 = T_3 - 0.5\Delta_2 = 52.85\,°C$$

$$T_6 = T_4 + (3 + \Delta_1)1.2 = 54.32\,°C$$

$$T_7 = T_6 - 0.25\Delta_1 = 54.03\,°C$$

The temperature distribution may be checked by calculating node temperatures using a different node as a reference. The temperature of node 7 is also determined by node 5 as

$$T_7 = T_5 - \Delta_1(1.0) = 54.02\,°C \qquad \text{Q.E.D.}$$

Forced Convection Methods. Forced convection heat transfer may be required when conduction at an assembly or equipment interface is insufficient to maintain component allowable temperatures. Convectively cooled assemblies have been designed to cool high heat densities while minimizing device-to-coolant temperature differences. Many planar forced convection cooled assemblies or surfaces may be evaluated using the methods of Section 4.3. Equipment configurations that use numerous conductive paths in addition to forced convection cooling may be studied using an impedance network containing designated convective flow branches.

Figure 3.22(a) illustrates a string of directed flow branches interconnecting four coolant nodes. Heat along adjacent surfaces is transferred by convection to the coolant nodes. Each node in the coolant string acts as a heat controlled sink. Only downstream heat flow along directed coolant branches can increase the temperature of a node. Heat injected into any node by convection can only flow downstream to the next node (sink). The function of the directed branches is illustrated in Figure 3.22(b) for q_1, q_3 and q_4 of Figure 3.22(a). Node temperatures of Figure 3.22(b) based on the function of directed flow branches are determined progressively as

$$T_1 = T_{a1} + q_1/WC_p$$

$$T_2 = T_1 \qquad (3.80)$$

$$T_3 = T_2 + q_3/WC_p = T_{a1} + (q_1 + q_3)/WC_p$$

$$T_4 = T_3 + q_4/WC_p = T_{a1} + (q_1 + q_3 + q_4/WC_p$$

a) Coolant flow string   b) Projected heat flow

Figure 3.22. Flow Characteristics of Convective Heat Transfer Branches.

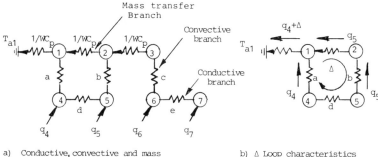

a) Conductive, convective and mass
   transfer branches

b) Δ Loop characteristics

Figure 3.23. Network of a Convectively Cooled Assembly.

The temperature of nodes 1 and 2 is determined by q_1 since heat entering upstream node 3 cannot flow downstream beyond node 2. Equipment temperature distributions using networks containing directed flow branches follow the techniques described for tree and loop configurations.

A hardware network idealization containing a loop and a tree is shown in Figure 3.23(a). The number of loops may be determined from (3.59). Branches a, b and c are surface-to-coolant convective impedances $(1/h_c)$. Nodal conduction is represented by branches d and e. Temperatures of nodes 3, 6 and 7 may be determined directly once the temperature of node 2 is known. As an example, the temperature of node 6 is

$$T_6 = T_2 + (q_6 + q_7)(1/WC_p + c)$$

The corrective heat flow Δ of Figure 3.23(b) must be evaluated before the temperatures of nodes 1, 2, 4 and 5 can be determined. The projected flows shown in Figure 3.23(b) adhere to the characteristics of directed branches previously described. The corrective flow for the loop of Figure 3.23(b) becomes

$$\Delta = \frac{q_5(1/WC_p + b) - q_4 a}{1/WC_p + a + b + d} \tag{3.81}$$

Based on Figure 3.23(b) and (3.81), the temperatures of nodes 1, 2 and 4 are given by

$$T_1 = T_{a1} + (q_4 + \Delta)/WC_p \tag{3.82a}$$

$$T_2 = T_1 + (q_5 - \Delta)/WC_p = T_{a1} + (q_4 + q_5)/WC_p \tag{3.82b}$$

$$T_4 = T_1 + (q_4 + \Delta)a = T_{a1} + (q_4 + \Delta)(1/WC_p + a) \tag{3.82c}$$

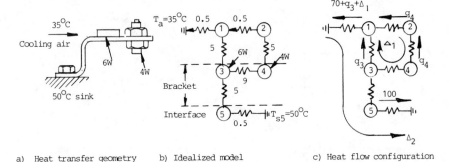

a) Heat transfer geometry   b) Idealized model   c) Heat flow configuration

Figure 3.24. Conductive and Convection Cooled Component Bracket.

The temperature of node 2 (3.82b) could have been determined without evaluating Δ since the coolant temperature rise at this point is a consequence of all downstream heat flows, i.e. $q_4 + q_5$.

EXAMPLE 3.11.   Forced convection in addition to conduction is used to cool the component bracket shown in Figure 3.24(a). An impedance model representing internal and interface conductive processes including surface convective cooling is shown in Figure 3.24(b). Determine the component attachment temperatures at nodes 3 and 4.

The matrix equation for determining Δ_i may be written directly from Figure 3.24(c). Each row corresponds to a loop accounting for coupled flows in peripheral branches. The matrix equation becomes

$$\begin{bmatrix} (5 + 5 + 0.5 + 9) & -5 \\ (-5 + 0.5) & (0.5 + 5 + 5 + 0.5) \end{bmatrix} \begin{Bmatrix} \Delta_1 \\ \Delta_2 \end{Bmatrix}$$

$$= \begin{Bmatrix} 4(5 + 0.5) - 6(5) \\ 76(0.5) + 6(5) - 100(0.5) \end{Bmatrix}$$

The solution for Δ_1 and Δ_2 using methods of Appendix A is

$$\begin{Bmatrix} \Delta_1 \\ \Delta_2 \end{Bmatrix} = \frac{1}{187} \begin{bmatrix} 11 & 5 \\ 5.5 & 19.5 \end{bmatrix} \begin{Bmatrix} -8 \\ 18 \end{Bmatrix} = \begin{Bmatrix} 0.0107 \\ 1.642 \end{Bmatrix}$$

The temperatures determined progressively are

$$T_1 = (76 + 0.0107 - 1.642)0.5 = 37.18°C$$

$$T_2 = T_1 + (4 - 0.0107)0.5 = 39.18°C$$

$$T_3 = T_1 + (6 + 0.0107 - 1.642)5 = 59.02°C$$

$$T_4 = T_3 + 0.0107(9) = 59.12°C$$

The amount of heat transferred by convection may be determined by evaluating

$$q_{convection} = \frac{T_2 - T_a}{WC_p} = 2.09W$$

The balance of the generated heat (7.91W) is transferred to the 50°C sink by conduction.   Q.E.D.

In the absence of forced convection, corrective heat flows Δ_i are coupled by common impedance branches. Nodal coupling in addition to branch coupling can occur in systems containing directed flow branches that represent forced convection. An example of nodal coupling is illustrated by the directed branch geometry of Figure 3.25. In this configuration, the heat flow in loop 2 does not affect the temperatures of nodes 1, 3 and 5 since Δ_2 of Figure 3.25(b) does not flow in branches c and d. The temperatures of nodes 2, 4 and 6, however, are dependent on the temperatures of the loop 2 interface at nodes 1 and 5. The corrective flow for loop 1 can be determined independently as

$$\Delta_1 = \frac{(q_3 + q_{Ta1})a + q_3c - q_{Ta5}b}{a + b + c + d} = \frac{T_{a1} - T_{a5} + q_3(a + c)}{a + b + c + d}$$

With known Δ_1, the temperature of any loop 1 node may be determined. Using known quantities for T_1 and T_5, the corrective flow for loop 2 becomes

$$\Delta_2 = \frac{T_1 - T_5 + q_4(a + e)}{a + b + e + f}$$

$$= \frac{T_{a1} - T_{a5} - \Delta_1(a + b) + q_3a + q_4(a + e)}{a + b + e + f}$$

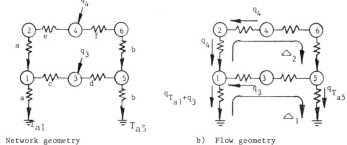

a) Network geometry   b) Flow geometry

Figure 3.25. Nodal Coupled Directed Flow Branches.

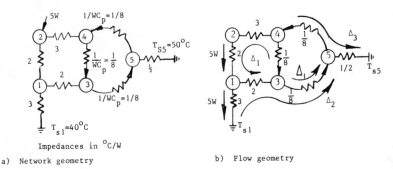

Impedances in °C/W

a) Network geometry   b) Flow geometry

Figure 3.26. Idealization of an Internally and Externally Convectively Cooled Assembly.

where
$$T_1 = T_{a1} + (q_3 - \Delta_1)a$$
$$T_5 = T_{a5} + \Delta_1 b$$

The recovery of loop 2 nodal temperatures follows directly.

EXAMPLE 3.12.   An electronic assembly is cooled by conduction in addition to internally recirculating forced convection. The idealized heat transfer model for this configuration is shown in Figure 3.26(a). Determine the loop heat flow quantities Δ_i of Figure 3.26(b), the temperature of the assembly at node 2 and the decrease in temperature provided by forced convection.

The equations that represent the sum of the temperature differences for each Δ_i flow of Figure 3.26(b) become

$$7.125\Delta_1 + 5(2) + 2\Delta_2 = 0 \tag{3.83a}$$

$$5.125\Delta_2 + 2.125\Delta_1 - T_{s1} + T_5 - 5(3) = 0 \tag{3.83b}$$

$$0.625\Delta_3 - T_4 + T_{s5} = 0 \tag{3.83c}$$

Unknown temperatures T_4 and T_5 in (3.83) are required to determine heat flows Δ_i. These temperatures are defined in terms of Δ_i and T_{si} by selecting flow paths originating at a sink. The resulting expression that determines T_i must be different from those which define Δ_i. In accordance with Figure 3.26(b), T_4 and T_5 are determined from

$$T_5 = T_{s5} + 0.5\Delta_3 \tag{3.84a}$$

$$T_4 = T_3 - 0.125\Delta_1 = T_{s5} + 0.5\Delta_3 + 0.125\Delta_2 \tag{3.84b}$$

Substituting (3.84b) into (3.83c) gives $\Delta_3 = \Delta_2$ so that (3.83) using (3.84) becomes

$$\begin{bmatrix} 7.125 & 2 \\ 2.125 & 5.625 \end{bmatrix} \begin{Bmatrix} \Delta_1 \\ \Delta_2 \end{Bmatrix} = \begin{Bmatrix} -10 \\ 5 \end{Bmatrix} \tag{3.85}$$

given $T_{s1} = 40°C$ and $T_{s5} = 50°C$. The solution of (3.85) using methods of Appendix A gives $\Delta_1 = -1.849W$ and $\Delta_2 = \Delta_3 = 1.587W$. The temperature of the assembly represented by node 2 is determined using a progressive process as $T_2 = 56.54°C$. Compared to an assembly that is only conduction cooled, the internally recirculating forced convection results in a 8.46°C lower assembly temperature.   Q.E.D.

Transient Solutions.   Impedance flow methods may also be used to study the transient response of any idealized network. The techniques used to develop the approximate temperature distribution at each increment of time are similar to those used to study conduction and forced convection models, described previously. Transient networks, however, contain additional impedance branches that represent nodal capacity. The simplicity of the heat flow approach is maintained even though nodes without capacity (massless nodes) are included.

Conductance methods for similar models containing massless nodes are comparatively more complex. This approach requires the removal of massless nodes by a reduction technique. The reduced combined conductance-capacitance matrix for the equivalent network must be inverted to obtain the temperature distribution at selected increments in time. The temperature of the massless nodes may be determined at any selected increment of time by solving the steady state equivalent model developed by the reduction process.

Transient impedance flow methods use the same network developed for the steady state temperature distribution with the addition of node to sink branches representing the impedance $\Delta\phi/(WC_p)_i$ at nodes with mass. The sink temperature for each of these new branches corresponds to the initial temperature of the mass represented by the node. At the onset of each new time increment, these sink temperatures correspond to the node temperature for the previous increment. At each increment of time $\Delta\phi$, the temperature distribution may be determined by the progressive process described previously.

This transient flow method is similar to the Euler technique (2.27) for a conductance network. The approximate temperature distribution developed successively from the preceding distribution is always stable and convergent for any time increment. No restrictions on the magnitude of $\Delta\phi$ are imposed; however, large increments suppress the effects of small thermal capacities or rapidly changing injected heat loads or boundary constraints. Smaller increments generally improve the accuracy of the response.

EXAMPLE 3.13.   In order to determine the temperature history of an assembly represented by the network of Figure 3.27(a), a new branch representing the capacitive

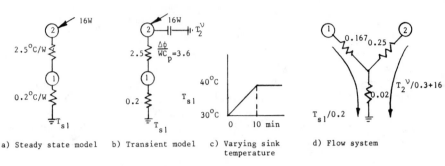

a) Steady state model   b) Transient model   c) Varying sink   d) Flow system
                                                     temperature

Figure 3.27. Model Development for Transient Response.

impedance of node 2 is added as shown in Figure 3.27(b). Determine the temperature history for an initial assembly temperature (node 2) of 30°C when (a) the sink temperature is held at 30°C and (b) the sink temperature varies as shown in Figure 3.27(c).

(a) *Fixed 30°C Sink Temperature.* The flow projections are developed using the methods described previously. The network of Figure 3.27(b) may be simplified by using the delta-star transformation of Figure 2.27. The result is shown in Figure 3.27(d) for a time increment $\Delta\phi$ = ½ hour. The temperature of node 1 is determined from Figure 3.27(d) as

$$T_1^{\nu+1} = \left(\frac{T_{s1}}{0.2} + \frac{T_2^{\nu}}{0.3} + 16\right)0.02 + \frac{0.167}{0.2} T_{s1}$$

$$T_1^{\nu+1} = 0.32 + 0.935 T_{s1} + 0.0667 T_2^{\nu}$$

(3.86a)

The temperature of node 2 is developed similarly as

$$T_2^{\nu+1} = 4.32 + 0.10 T_{s1} + 0.9 T_2^{\nu}$$

(3.86b)

For constant T_{s1} = 30°C, (3.86) becomes

$$T_1^{\nu+1} = 28.37 + 0.0667 T_2^{\nu}$$

$$T_2^{\nu+1} = 7.32 + 0.9 T_2^{\nu} \quad (30°C)$$

(3.87)

The temperature of nodes 1 and 2 after ½ hour is determined by substituting the initial temperature of node 2 at ϕ = 0, (30°C) into (3.87), giving T_1^1 = 30.37 and T_2^1 = 34.32°C. The temperature distribution at ϕ = ⅙ hour is determined by substituting T_2^1 into (3.87), giving T_1^2 = 30.66°C and T_2^2 = 38.21°C. This process may be continued until steady state quantities are reached. Figure 3.28 illustrates the temperature history of the assembly at node 2.

Since node 1 is massless, its temperature may be determined knowing the temper-

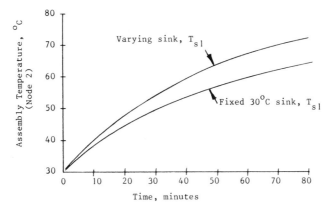

Figure 3.28. Temperature History of Assembly (Node 2).

ature of node 2 at any time ϕ. This relationship is developed by substituting (3.86b) into (3.86a) as

$$T_1 = 0.927 T_{s1} + 0.0741 T_2 \qquad (3.88)$$

Equation (3.88) may be used to determine T_1 if at $\phi = 0$, $T_{s1} \neq T_2$.

(b) *Varying Sink Temperature.* Linear or nonlinear constraints may be imposed at nodes or network sinks. These may be a function of time, temperature or heat flow. These effects may be accounted for by developing an expression that varies dependently upon selected variables or by altering the network at the appropriate increment. The sink temperature of Figure 3.27(c) may be described by

$$T_{s1} = 30 + 60\phi \, ^\circ C \quad 0 \leq \phi \leq 0.167 \qquad (3.89a)$$

$$T_{s1} = 40 \, ^\circ C \quad 0.167 \leq \phi \qquad (3.89b)$$

Relationships valid for $0 \leq \phi \leq 0.167$ are determined by substituting (3.89a) into (3.86) giving

$$T_1^{\nu+1} = 28.37 + 56.22\phi + 0.0667 T_2^{\nu}$$
$$T_2^{\nu+1} = 7.32 + 6.0\phi + 0.9 T_2^{\nu} \qquad (3.90)$$

The temperature distribution using (3.90) at $\phi = \frac{1}{12}$ and $\frac{1}{6}$ hour becomes

$\phi = 0$	$\frac{1}{12}$	$\frac{1}{6}$	hr
$T_1 = 30.$	35.06	40.06	°C
$T_2 = 30.$	34.82	39.66	°C

Relationships valid for $\phi > 0.167$ hr are determined by substituting (3.89b) into (3.86) giving

$$T_1^{n+1} = 37.72 + 0.0667 T_2^n$$

$$T_2^{n+1} = 8.32 + 0.9 T_2^n$$

(3.91)

Using (3.91), the temperature distribution at $\phi = \frac{1}{4}$ hr becomes $T_1^3 = 40.37\,°C$ and $T_2^3 = 44.014\,°C$. This process may be continued until steady state quantities are reached. Figure 3.28 illustrates the assembly (node 2) response to the varying sink temperature of Figure 3.27(c).   Q.E.D.

4

Mechanics of Convection and Radiation

For most applications, both radiation and natural convection occur naturally and often simultaneously. The heat transfer efficiency of these methods depends on the configuration, orientation and temperature of the surfaces involved.

4.1 RADIATION

Radiation acts in a manner similar to light. Traveling at the speed of light, radiation striking a body may be absorbed, reflected or transmitted. The fractional amounts of energy associated with each mechanism depend on the material, body geometry, surface characteristics and temperature. The sum of the fractional quantities always equals 1. An ideal *black body* absorbs all energy falling on it. The absorptivity, α, therefore equals 1. Likewise, a perfect *white body* reflects all energy falling on it so its reflectivity, ρ, equals 1. When all the energy striking a body is transmitted, the transmissivity, τ, equals 1.

A black body emits the maximum possible radiation for a given temperature. *Emissivity*, ϵ, is defined as the ratio of the amount of radiant energy emitted by a body to that of a black body. The emissivities of typical materials used in hardware depend on temperature, surface finish and condition. Emissivities for plastics and other nonmetallic materials are generally higher than those of the metals. Polished metal surfaces which are free of oxidation have very low emissivities. Surface oxidation, paint or roughness dramatically increases the emissivity of metals. Even a thin coat of lacquer 0.0005 in (0.0127 mm) thick can increase the emissivity of polished copper from 0.03 to about 0.80. The emissivity and absorptivity of various materials are shown in Appendix C.

Materials used in electronic hardware may be classified as *gray*, diffusely reflecting opaque surfaces. These surfaces reflect the same fraction of energy over all wavelengths in all directions. A body which radiates and simultaneously absorbs an identical fraction of radiant energy in the absence of other forms of energy exchange is in a state of thermal equilibrium. Under these conditions $\epsilon = \alpha$ and the temperature of the body remains constant. For gray surfaces $\rho + \epsilon = 1$. The net rate of radiation leaving an opaque surface of a system with area A_i and emissivity ϵ_i at uniform temperature T_i is

$$q_i = \frac{\epsilon_i}{1 - \epsilon_i} A_i(\sigma T_i^4 - J_i) = \frac{\sigma T_i^4 - J_i}{R_i} \tag{4.1}$$

where σ is the Stephan-Boltzmann constant and J_i is the associated radiosity which is defined as the sum of the radiation emitted and reflected per unit surface area and R_i is the radiative impedance. The net rate of radiation exchange between two surfaces emitting radiation at rates J_i and J_j becomes

$$q_{ij} = (J_i - J_j)A_i F_{ij} = \frac{J_i - J_j}{R_{ij}} \tag{4.2}$$

where F_{ij} is the geometric shape factor that defines the fraction of energy incident on surface j from surface i. The total radiant exchange between surfaces A_i and A_j depends on the fraction of energy emitted by surface i and the fraction absorbed by surface j and reflected back to i and absorbed, and so on. The total radiant energy may be expressed as

$$Q = \frac{\sigma(T_1^4 - T_2^4)}{R_{net}} \tag{4.3}$$

where the impedance R_{net} is a function of the view the surfaces have of each other and their emitting and absorbing characteristics. If the surface is irregular, the projected area should be used. This is the area of a planar image of the surface normal to the axis of illumination. Finned surfaces, which increase the effective heat transfer area for convection, contribute little to the radiative area of an enclosure. The projected area of the inclined surface $abcd$ of Figure 4.1 is $abc'd'$. Recessed and finned surfaces are treated analogously.

Equations (4.1) and (4.2) provide the basis for determining the total radiant exchange between bodies by means of an equivalent network. The schematic representation of Figure 4.2 is used to find R_{net} in (4.3) for the radiant energy exchange between parallel surfaces A_1 and A_2 of Figure 4.1. The overall impedance between the two surfaces with temperatures T_1 and T_2 is

$$R_{net} = R_1 + R_2 + R_{12} \tag{4.4a}$$

With (4.1) and (4.2) defining element impedances, (4.4a) becomes

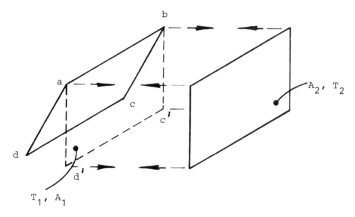

Figure 4.1. Projected Area of an Inclined Surface.

$$R_{net} = \frac{1 - \epsilon_1}{\epsilon_1 A_1} + \frac{1 - \epsilon_2}{\epsilon_2 A_2} + \frac{1}{A_1 F_{12}} \qquad (4.4b)$$

where, for example, $R_1 = (1 - \epsilon_1)/\epsilon_1 A_1$ and $R_{12} = 1/A_1 F_{12}$. Substituting (4.4b) into (4.3) and simplifying gives an expression for the total radiant exchange between two surfaces as

$$Q = \frac{A_1 \sigma \left(T_1^4 - T_2^4 \right)}{\left(\dfrac{1}{\epsilon_1} - 1 \right) + \left(\dfrac{1}{\epsilon_2} - 1 \right) \left(\dfrac{A_1}{A_2} \right) + \dfrac{1}{F_{12}}} \qquad (4.5)$$

For parallel infinite surfaces A_1 and F_{12} equal 1 which reduces (4.5) to the relationship

$$Q = A\epsilon\sigma(T_1^4 - T_2^4) = \frac{A_1 \sigma(T_1^4 - T_2^4)}{\dfrac{1}{\epsilon_1} + \dfrac{1}{\epsilon_2} - 1} \qquad (4.6)$$

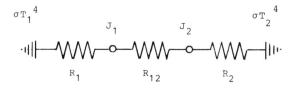

Figure 4.2. Model for Parallel Plate Radiant Exchange.

Table 4.1. Emissivity for Various Configurations

Configuration ($F_{ij} = 1$)	ϵ
I. Infinite parallel surfaces or one surface completely enclosed by a slightly larger surface	$\dfrac{1}{\dfrac{1}{\epsilon_1} + \dfrac{1}{\epsilon_2} - 1}$
II. One surface completely surrounded by another; $A_1 < A_2$	$\dfrac{1}{\dfrac{1}{\epsilon_1} + \dfrac{A_1}{A_2}\dfrac{1}{\epsilon_2} - 1}$
III. One surface completely enclosed by a much larger surface	ϵ_1

This method may be extended to any number of bodies provided the geometric shape factors F_{ij} are known. Table 4.1 lists emissivity ϵ of (4.6) for cases where $F_{ij} = 1$.

Shape Factor Determination. The sum of the fractions of the energy leaving body j equals the total energy, thus

$$\sum_{i=1}^{n} F_{ij} = 1 \qquad (4.7)$$

In addition, the fraction of incident energy on one surface conveyed from another may be expressed in terms of the reciprocity that exists as

$$A_i F_{ij} = A_j F_{ji} \qquad (4.8)$$

Equation (4.8) is convenient when one shape factor is easier to determine than the other. In order to determine the shape factor of a new configuration in terms of a known factor, the simplification

$$G_{ij} = A_i F_{ij} \qquad (4.9)$$

is used. With this definition, (4.7) and (4.8) become

$$\sum_{i=1}^{n} G_{ij} = A_j \qquad (4.10)$$

and

$$G_{ij} = G_{ji} \qquad (4.11)$$

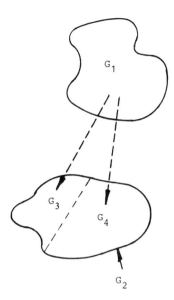

Figure 4.3. Partitioning of Radiating Surfaces.

The relationship of portions of a surface to the whole for radiation from another body, Figure 4.3, can be stated as

$$G_{12} = G_{1(34)} = G_{13} + G_{14} \tag{4.12}$$

Suppose that G_{12} and G_{14} are known and G_{31} remains to be found. Then from (4.11) and (4.12)

$$G_{31} = G_{12} - G_{14} \tag{4.13}$$

With the substitution of equivalent representations of (4.9) into (4.13), the desired shape factor becomes

$$F_{31} = \frac{A_1}{A_3}(F_{12} - F_{14}) \tag{4.14}$$

Equation (4.12) may be expanded to include divisions of each body:

$$G_{(12)(34)} = G_{1(34)} + G_{2(34)} = G_{13} + G_{14} + G_{23} + G_{24} \tag{4.15}$$

One of the most difficult problems in treating radiative transfer between surfaces is determining the geometric shape factor. The evaluation requires inte-

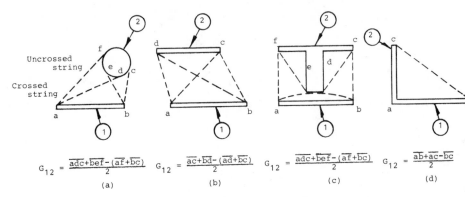

$$G_{12} = \frac{\overline{adc} + \overline{bef} - (\overline{af} + \overline{bc})}{2}$$

(a)

$$G_{12} = \frac{\overline{ac} + \overline{bd} - (\overline{ad} + \overline{bc})}{2}$$

(b)

$$G_{12} = \frac{\overline{adc} + \overline{bef} - (\overline{af} + \overline{bc})}{2}$$

(c)

$$G_{12} = \frac{\overline{ab} + \overline{ac} - \overline{bc}}{2}$$

(d)

Figure 4.4. String Technique Shape Factors.

grations of the radiative interchange over the finite areas involved in the exchange process. Many complex relationships for different configurations have been derived and are available in the literature.

A method due to Hottel evaluates G_{ij} for the interchange between two surfaces per unit length, using a string technique. The determination of G_{ij} becomes

$$G_{ij} = \frac{\Sigma \text{ crossed string length} - \Sigma \text{ uncrossed string length}}{2} \qquad (4.16)$$

The length of the crossed strings includes any portion wrapped around a configuration reaching the termination adjoining the uncrossed strings. The uncrossed strings bound the limits of the radiated surfaces of both bodies. Shape factors for several configurations are defined in Figure 4.4. Accurate shape factor values are obtained for surface lengths normal to the plan that are much greater than the developed length of any string in Figure 4.4.

EXAMPLE 4.1. Determine the geometric shape factors F_{12} and F_{21} for the configuration of Figure 4.5.

The shape factor is given by Figure 4.4(d) as

$$G_{12} = G_{21} = \frac{\overline{ab} + \overline{ac} - \overline{bc}}{2} = \frac{3 + 4 - \sqrt{3^2 + 4^2}}{2}$$

$$G_{12} = 1.0 \text{ in}^2/\text{in of depth} = 8.0 \text{ in}^2$$

The required factors are determined using (4.8) and (4.9) as follow:

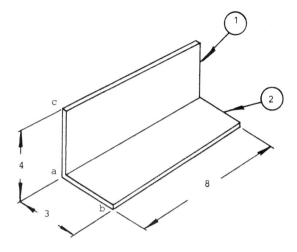

Figure 4.5. Example Radiant Configuration.

$$F_{12} = \frac{G_{12}}{A_1} = \frac{8}{4(8)} = \frac{1}{4}$$

$$F_{21} = \frac{A_1}{A_2} F_{12} = \frac{4(8)}{3(8)} (0.25) = \frac{1}{3} \qquad \text{Q.E.D.}$$

The integal solution for the configuration of Figure 4.4(d) is shown on Figure 4.6(a). The shape factor for parallel directly opposed identical rectangles is shown on Figure 4.6(b).

Shape factors for surfaces frequently used in electronic equipment may be derived and cataloged for reuse.

EXAMPLE 4.2. Finned surfaces are often used to increase the rate of heat transfer from component parts and electronic equipment. Determine the shape factor for a finned surface based on the configuration of Figure 4.7(a) for $H/L > 5$ in a black body ambient.

The shape factor may be evaluated from (4.2) once the overall impedance between the surface and ambient has been determined. Assuming the fins and base are at the same temperature, the radiant energy transfer to the black body ambient is illustrated by the network of Figure 4.7(b). In this case, energy is not reflected from the ambient to the surface giving $\epsilon_{\text{ambient}} = 1$.

The impedances between the fins and the ambient R_6 and R_8 are determined from reciprocity (4.8) as

$$R_6 = R_8 = \frac{1}{A_3 F_{32}} = \frac{1}{A_3 F_{31}} = \frac{1}{A_1 F_{13}} \qquad (4.17)$$

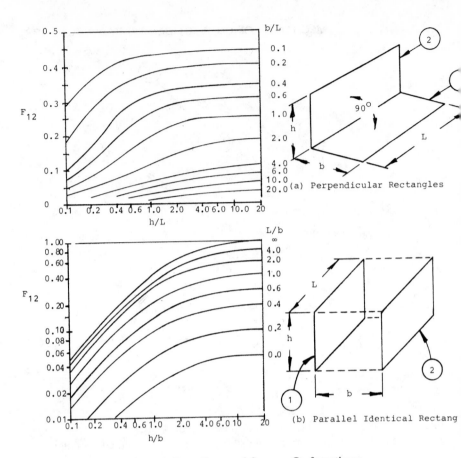

Figure 4.6. Shape Factors of Common Configurations.

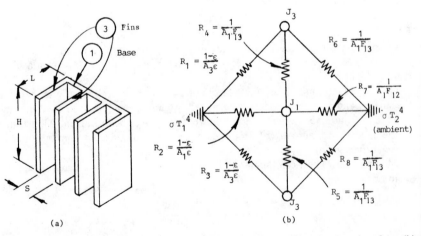

Figure 4.7. Radiative Model of a Finned Surface. (a) Characteristics of a finned surface; (b) Finned surfaces network.

The overall impedance is easily determined by using the symmetry of Figure 4.7(b) as

$$R_{net} = \frac{R_6 R_7}{R_a} + \frac{\left(R_1 + \dfrac{R_4 R_6}{R_a}\right)\left(R_2 + \dfrac{R_4 R_7}{R_a}\right)}{R_1 + 2R_2 + \dfrac{R_4}{R_a}(R_6 + 2R_7)} \tag{4.18}$$

where $R_a = R_4 + R_6 + 2R_7$. For configuration ratios $H/L > 5$, the crossed string method may be used to define F_{12} and F_{13}. For long parallel plates of width S spaced a distance L apart, the shape factor becomes

$$F_{12} = \sqrt{1 + \left(\frac{L}{S}\right)^2} - \left(\frac{L}{S}\right) \tag{4.19}$$

The shape factor for two long orthogonally intercepting plates of widths L and S is

$$F_{13} = \frac{1}{2}\left[\left(\frac{L}{S}\right) + 1 - \sqrt{1 + \left(\frac{L}{S}\right)^2}\right] \tag{4.20}$$

This result may be verified by substituting $S = 4$ and $L = 3$ and solving for F_{12} of Example 4.1. In (4.19) and (4.20), S is the width of surface 1 of Figure 4.7(a). Substituting $A_1 = HS$, $A_3 = HL$, (4.19) and (4.20), into (4.18) and simplifying gives

$$R_{net} = \frac{1}{HS}\left[\frac{1}{F_a} + \frac{\dfrac{1}{\epsilon}\left(\dfrac{1-\epsilon}{\epsilon} + \dfrac{1}{F_a}\right)}{\dfrac{1}{\epsilon} + 2\left(\dfrac{L}{S}\right)\left(\dfrac{1-\epsilon}{\epsilon} + \dfrac{1}{F_a}\right)}\right] \tag{4.21}$$

given $F_a = \sqrt{1 + (L/S)^2} - (L/S) + 1$. Using (4.2), the shape factor of a finned surface becomes

$$F_{12} = \frac{1}{A_s R_{net}} = \frac{1}{(2L + S)HR_{net}} \tag{4.22a}$$

$$F_{12} = \frac{1}{\left[2\left(\dfrac{L}{S}\right) + 1\right]\left[\dfrac{1}{F_a} + \dfrac{\dfrac{1}{\epsilon}\left(\dfrac{1-\epsilon}{\epsilon} + \dfrac{1}{F_a}\right)}{\dfrac{1}{\epsilon} + 2\left(\dfrac{L}{S}\right)\left(\dfrac{1-\epsilon}{\epsilon} + \dfrac{1}{F_a}\right)}\right]} \tag{4.22b}$$

Equation (4.22b) is graphically portrayed on Figure 4.8.   Q.E.D.

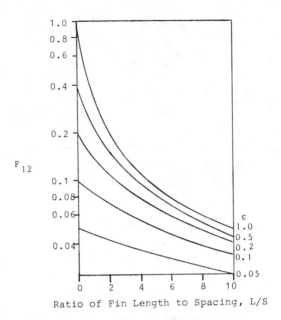

Figure 4.8. Shape Factor for Finned Surfaces.

Influence of Design Characteristics. Radiation transfer depends on the exposed surfaces and surface characteristics between the source and the sink. Intervening opaque surfaces at temperatures similar to the source which mask any part of the sink as seen by the source may adversely affect the rate of transfer and the temperatures of the dissipating parts. Surfaces of an enclosure, cabinet or housing which are at temperatures considerably below those of component assemblies, modules and circuit boards provide an acceptable heat transfer sink. Effective radiative transfer requires that each dissipating assembly have a full view of the walls. Practical and spatial limitations usually preclude this approach, however, with the result that only a small portion of the heated surface area is in view of an external wall surface. Card trays and racks develop high density packages limiting radiative transfer to the exposed circuit cards at each end of the tray. Parts located along the perimeter of each card will lose heat by radiation to the adjacent structure which is in view.

The design engineer may exercise control over the portion of heat transferred by radiation by selective positioning of critical parts and choice of finishes. A painted wall with a large value of emissivity will absorb a higher percentage of the incident radiation, reflecting a smaller quantity back to the part. Low emissivity finishes on the parts will reduce heating by radiation from neighboring parts by reflecting incident radiation, but they will also emit less of its generated heat. A heated surface with an emissivity of 1 is an ideal emitter radiating energy at the maximum possible rate per unit area for any given temperature. A sink with emissivity of 1 will absorb all radiant energy incident upon it. The amount of heat radiated per unit area of exposed surface may be expressed by the Stefan-Boltzmann equation.

$$Q = 3.485(10^{-12})\epsilon F_{12}(T_1^4 - T_2^4) \text{ watts/in}^2 \quad (4.23)$$

given T in $°R$. Figure 4.9 is the graphical representation of (4.23) which may be used to assess the required emissivities of the radiating surfaces to obtain an acceptable surface temperature. The emissivity of (4.23) represents a combined value as given by Table 4.1 for several configurations.

EXAMPLE 4.3. A leadless carrier dissipating 0.20W is mounted on a circuit board that radiates to a finned housing wall shown in Figure 4.10. The finned wall dissipates 10W to an ambient of $120°F$ ($48.9°C$) by radiation.

a. Determine the wall temperature for an external finish with an emissivity of 0.8.

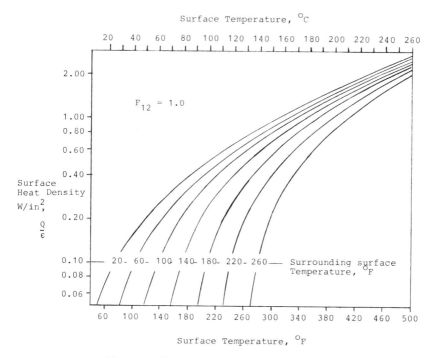

Figure 4.9. Characteristics of Radiating Surfaces.

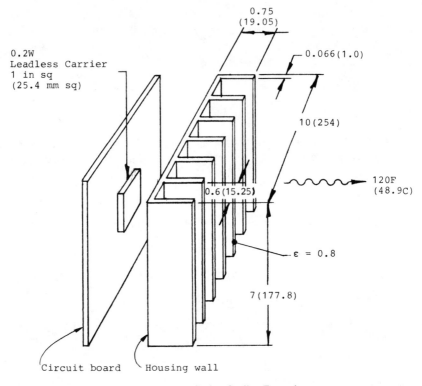

Figure 4.10. Radiation Cooling Example.

The wall temperature may be determined from (4.3) once the overall impedance R_{net} from the finned wall to the ambient has been determined. The overall impedance is determined from (4.22a) as

$$R_{net} = \frac{1}{N(2L + S)HF_{12}} \qquad (4.24)$$

where N = number of finned channels of width S. With a ratio of fin length to spacing given as 1.25, the shape factor from Figure 4.8 for an emissivity of 0.8 is 0.27. With $N = 15$, (4.24) gives

$$R_{net} = \frac{1}{15(2 \times 0.75 + 0.6)7(0.27)} = 0.0167 \text{ in}^{-2}$$

or in metric units

$$R_{net} = \frac{1}{15(2 \times 19.05 + 15.25)(177.8)(0.27)} = 2.6(10^{-5}) \text{ mm}^{-2}$$

Written in a convenient form for determining the finned surface temperature, (4.3) becomes

$$T_1^4 = \frac{QR_{net}}{\sigma} + T_2^4 \qquad (4.25)$$

where $\sigma = 3.485(10^{-12})$ W/in$^2 \cdot R^4$

$$= 5.669(10^{-14})$$ W/mm$^2 \cdot K^4$

substituting parameters as defined in the problem statement

$$T_1^4 = \frac{10(0.0167)}{3.485(10^{-12})} + (120 + 460)^4 = 1.6108(10^{11})$$

$$T_1 = 633.5°R \quad \text{or} \quad t_1 = 173.5°F$$

In metric units this becomes

$$T_1^4 = \frac{10(2.6 \times 10^{-5})}{5.669(10^{-14})} + (48.9 + 273)^4 = 1.532(10^{10})$$

$$T_1 = 351.8°K \quad \text{or} \quad t_1 = 78.8°C$$

b. Determine the component/wall emissivity required to assure a leadless carrier case temperature less than 239°F (115°C). The bare component case has an emissivity of 0.08.

The surrounding wall temperature for the leadless carrier as determined in (a) is used in conjunction with the given maximum case temperature of 239°F (115°C) in Figure 4.9 to obtain an allowable surface heat density Q/ϵ of 0.27. The combined emissivity becomes

$$\epsilon = Q/0.27 = 0.2/0.27 = 0.74$$

From Figure 4.10, the surface areas of the leadless carrier and the back face of the finned wall are

$$A_1 = 1.0 \text{ in}^2 (645.16 \text{ mm}^2)$$

$$A_2 = 10(7) = 70 \text{ in}^2 (45161 \text{ mm}^2)$$

so that

$$A_1/A_2 = 1/70$$

Configuration II of Table 4.1 may be used to determine the required emissivities. Since A_1/A_2 is a small value, Case III applies. The emissivity of the leadless carrier should be

$$\epsilon_1 = \epsilon = 0.74$$

This emissivity may be obtained by applying a thin layer of lacquer or conformal coating to the case. Q.E.D.

4.2 NATURAL CONVECTION

The nature of equipment use and environmental considerations usually influence whether the housing can freely ventilate or must be enclosed. Since forces developed by buoyancy are responsible for maintaining the air flow needed for heat transfer, the open or enclosed housing design has a dramatic effect on internal part temperatures. The use of lower surface power densities by distributing heat sources or employing extended surfaces can reduce part temperatures, insuring a viable design in either case.

The surfaces of components can contribute substantially to the heat transfer area of an assembly. This contribution is maximized when parts are well spaced and the component layout precludes flow baffling by large devices. The orientation and spacing of component boards require consideration to avoid choking and to develop freely circulating air flow paths. Open and closed convective flow patterns are illustrated in Figure 4.11. Distributed sources in closed systems develop heated flows along surfaces which develop cooled down drafts in the interspace, Figure 4.11(A.b). The heated flow due to concentrated heating flows to the interspace with the flow cooled along walls as in Figure 4.11(A.a). Heated horizontal surfaces form ascending plumes of heated air with accompanying recirculating cooler plumes on large freely ventilating surfaces.

Uniform or Approximately Uniform Temperature Distribution. The heat transfer by natural convection from heated surfaces can be evaluated using (2.6). Natural convection heat transfer from surfaces of an enclosure or module occurs in parallel with rates dependent upon characteristics of the surface. The temperature difference between surfaces and the surrounding ambient can be developed from (2.3) and (2.6) as

$$\Delta T_{\text{parallel}} = \left(\frac{241.54Q}{P^{0.5} \sum\limits^{n} \dfrac{1}{Z_i}} \right)^{0.8} \tag{4.26}$$

where

$$Z_i = (L_c^{0.25}/\eta AC)_i$$

A⟩ Closed Systems

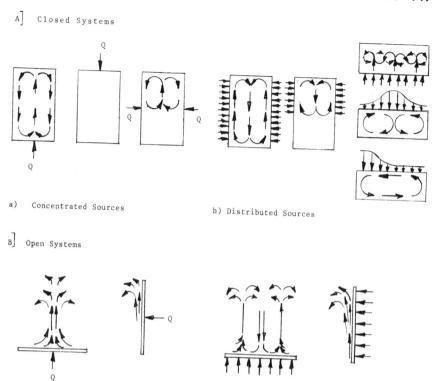

a) Concentrated Sources   b) Distributed Sources

B⟩ Open Systems

a) Concentrated Sources   b) Distributed Sources

Figure 4.11. Natural Convection Flow Patterns.

In a similar fashion, the temperature difference between a heated surface and the surrounding ambient with intervening surfaces serially interposed becomes

$$\Delta T_{\text{series}} = \left(\frac{241.54Q}{P^{0.5}}\right)^{0.8} \sum^{n} z_i^{0.8} \qquad (4.27)$$

Parallel and series configurations are schematically represented in Figure 4.12. Where both series and parallel convective cooling occurs, (4.26) and (4.27) may be combined to give

$$\Delta T = \left[\frac{241.54Q}{P^{0.5} \sum\limits_{i}^{m}\left(\dfrac{1}{\sum\limits_{j}^{n} z_j^{0.8}}\right)_i^{1.25}}\right]^{0.8} \qquad (4.28)$$

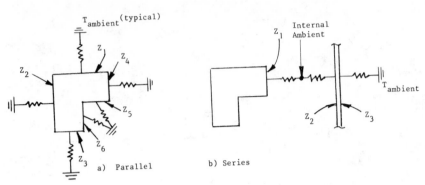

Figure 4.12. Convective Cooling Idealizations.

EXAMPLE 4.4. A 5W electronic assembly dissipates heat by natural convection from three surfaces (Figure 4.13) to an ambient of $40°C$. Determine the average temperature of the heated surfaces.

The surface characteristics using (4.26) and Table 2.2 are summarized in Table 4.2.

The surface-to-ambient temperature difference using (4.26) and Z from Table 4.2 becomes

$$\Delta T = \left[\frac{241.54(5)}{\left(\dfrac{1}{0.127} + \dfrac{1}{0.114} + \dfrac{1}{0.084} \right)} \right]^{0.8} = 20°C$$

The average surface temperature is

$$T_{\text{surface}} = T_{\text{ambient}} + \Delta T$$

$$T_{\text{surface}} = 40 + 20 = 60°C \qquad\qquad \text{Q.E.D.}$$

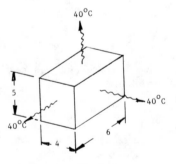

Figure 4.13. Example of a Parallel Cooled Assembly.

Table 4-2. Convective Surface Characteristics

Surface	Orientation	Area, in^2	Characteristic Length L_c, in	η	Z
4 × 5	Vertical—side	20	5.0	1	0.127
4 × 6	Horizontal—top	24	4.8	1	0.114
5 × 6	Vertical—side	30	5.0	1	0.084

Series-parallel convective cooling paths are used to represent an assembly mounted within an enclosure. A typical path is shown in Figure 4.12(b). An intermediate node is used to define the internal ambient temperature developed by the adjoining surfaces.

EXAMPLE 4.5.   The module of Example 4.4 is mounted in an enclosure. Enclosure walls face the three heated surfaces of the assembly as shown in Figure 4.14. Determine the average temperature of the heated assembly surfaces if the external enclosure ambient is 40°C.

The surface to external ambient temperature difference is determined using (4.28) and Z from Table 4.2 and Figure 4.14. The denominator of (4.28) for each wall of Figure 4.14 becomes

$$\left(\frac{1}{\sum\limits_{3} Z_j^{0.8}}\right)_1^{1.25} = \left(\frac{1}{0.084^{0.8} + 0.020^{0.8} + 0.063^{0.8}}\right)^{1.25} = 4.68$$

$$\left(\frac{1}{\sum\limits_{3} Z_j^{0.8}}\right)_2^{1.25} = \left(\frac{1}{0.127^{0.8} + 2(0.088^{0.8})}\right)^{1.25} = 2.52$$

$$\left(\frac{1}{\sum\limits_{3} Z_j^{0.8}}\right)_3^{1.25} = \left(\frac{1}{0.114^{0.8} + 0.025^{0.8} + 0.0164^{0.8}}\right)^{1.25} = 2.613$$

The surface-to-ambient temperature difference using (4.28) is

$$\Delta T = \left[\frac{241.54(5)}{4.68 + 2.52 + 2.613}\right]^{0.8} = 47°C$$

The average surface temperature of the assembly becomes

$$T_{\text{surface}} = T_{\text{ambient}} + \Delta T$$

$$T_{\text{surface}} = 40 + 47 = 87°C \qquad \text{Q.E.D.}$$

Nonuniform Temperature Distribution.   Equations (4.26), (4.27) and (4.28) apply to heated surfaces which are uniform in temperature. The way in which

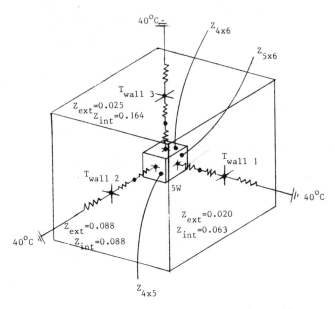

Figure 4.14. Example of Series—Parallel Cooled Assembly.

a surface is heated or cooled may develop temperature gradients which decrease the heat transfer effectiveness of the surface. Finned surfaces, covers or access panels are typical examples. Assuming the base of a finned surface is at a uniform temperature, T_b, the temperature of the fin decreases from T_b to a value approaching ambient at the fin tip, T_L. The minimum temperature at the fin tip is determined as

$$T_L = \frac{T_b}{\text{Cosh}\left(L \sqrt{\frac{2h}{ka}} \right)} \tag{4.29}$$

where k = fin conductivity;
 a = fin thickness;
 h = heat transfer coefficient;
 L = fin length.

It can be seen from (4.29) that a short thick fin of high conductivity would be nearly isothermal. Finned surfaces which have a significant effect on reducing part temperatures are usually shaped differently. A convenient representation for the impedances used in describing the fin effectiveness uses

$$R_s = \frac{1}{hLH} \qquad (4.30a)$$

$$R_{sq} = \frac{1}{ka} \qquad (4.30b)$$

Equations (4.30) represent convective and conductive fin impedances for the geometries described in Figure 4.15. The thermal profile, minimum temperature and fin efficiency is given for three common configurations encountered in hardware design. Either T_{min} or the profile temperature function permits an evaluation of the minimum temperature of the surface. The fin efficiency has been plotted in Figure 4.16 as a function of fin impedances (4.30) for rapid assessment. Convective heat transfer from the surfaces of Figures 4.15(I) and (II) are from a single side analogous of a cover or panel with an external flow. Heat transfer from both sides can be accommodated by doubling h in (4.30a).

Case III of Figure 4.15 requires a determination of the minimum fin spacing which prevents choking inhibiting the buoyant flow of air.

$$S = 0.86 \left(\frac{H}{\Delta T}\right)^{1/4} (1.0087)^{0.45\Delta T} \text{ in} \qquad (4.31)$$

The fin spacing should never be less than that obtained from (4.31). Channels formed by the fins are assumed smooth. Card guides or discrete components mounted within the flow channel develop an obstruction which increases the wall-to-ambient temperature difference. The increased temperature varies in proportion to the cube root of the resulting ratio of channel head loss.

A heat transfer coefficient for natural convection can be developed by comparing (2.2) with (2.3), which gives

$$h_L = 0.00414 CP^{0.5} \left(\frac{\Delta T}{L_c}\right)^{0.25} \qquad (4.32)$$

Use of (4.32) requires an estimate for ΔT which is reevaluated and updated iteratively until the value converges. At this point, the parameters of the design have been determined.

EXAMPLE 4.6.   The aluminum enclosure of Figure 4.17 dissipates 50W to an ambient of 50°C at sea level. The enclosure is to be cooled by natural convection from all surfaces except the bottom. In order to limit part temperatures, the surfaces that interface with electronic assemblies must be limited to 70°C. If these surfaces are

CASE   CONFIGURATION   GOVERNING RELATIONSHIPS

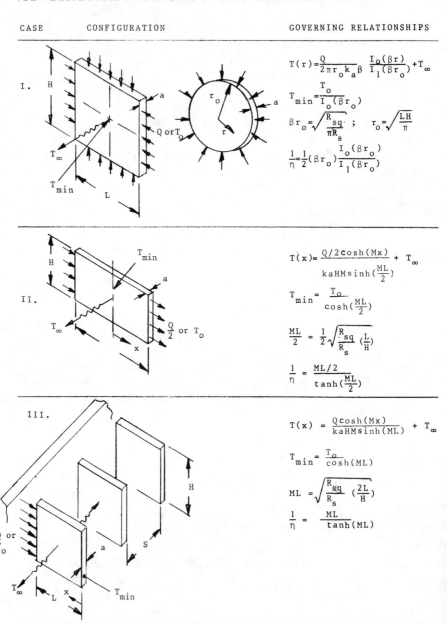

$$T(r) = \frac{Q}{2\pi r_o k_a \beta} \frac{I_o(\beta r)}{I_1(\beta r_o)} + T_\infty$$

$$T_{min} = \frac{T_o}{I_o(\beta r_o)}$$

$$\beta r_o = \sqrt{\frac{R_{sq}}{\pi R_s}} \quad ; \quad r_o = \sqrt{\frac{LH}{\pi}}$$

$$\frac{1}{\eta} = \frac{1}{2}(\beta r_o) \frac{I_o(\beta r_o)}{I_1(\beta r_o)}$$

$$T(x) = \frac{Q/2 \cosh(Mx)}{k_a H M \sinh(\frac{ML}{2})} + T_\infty$$

$$T_{min} = \frac{T_o}{\cosh(\frac{ML}{2})}$$

$$\frac{ML}{2} = \frac{1}{2}\sqrt{\frac{R_{sq}}{R_s}}\left(\frac{L}{H}\right)$$

$$\frac{1}{\eta} = \frac{ML/2}{\tanh(\frac{ML}{2})}$$

$$T(x) = \frac{Q\cosh(Mx)}{k_a H M \sinh(ML)} + T_\infty$$

$$T_{min} = \frac{T_o}{\cosh(ML)}$$

$$ML = \sqrt{\frac{R_{sq}}{R_s}}\left(\frac{2L}{H}\right)$$

$$\frac{1}{\eta} = \frac{ML}{\tanh(ML)}$$

Figure 4.15. Characteristics of Typical Fin Geometrics.

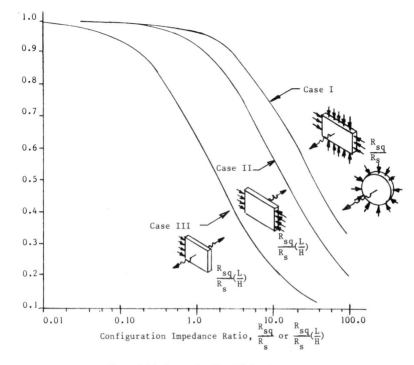

Figure 4.16. Convective Fin Efficiencies.

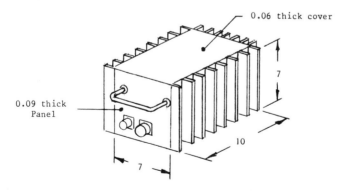

Figure 4.17. Convectively Cooled Finned Enclosure.

finned determine (a) the fin configuration required and (b) the interface surface temperature at an altitude of 30k ft with an ambient of 30°C using this fin configuration.

(a) *Determination of Fin Configuration.*

TOP COVER. The characteristic length from Table 2.2 becomes

$$L_c = \frac{2(7)(10)}{17} = 8.23 \text{ in}$$

From (4.32) the heat transfer coefficient is

$$h_c = 0.00414(0.54) \left(\frac{20}{8.23}\right)^{0.25}$$

$$= 0.00279 \text{ W/in}^2 \text{ °C}$$

The fin impedances from (4.30) are

$$R_s = \frac{1}{0.00279(7)(10)} = 5.12 \text{ °C/W}$$

$$R_{sq} = \frac{1}{(2/1.8)(0.06)} = 15.0 \text{ °C/W}$$

The configuration impedance ratio for Case I of Figure 4.16 is

$$\frac{R_{sq}}{R_s} = \frac{15}{5.12} = 2.9$$

which corresponds to a fin efficiency $\eta = 0.90$. From (4.26)

$$Z_1 = \frac{L_c^{0.25}}{\eta A C} = \frac{8.23^{0.25}}{0.9(70)(0.54)} = 0.049$$

The process of obtaining Z is repeated for each convectively cooled surface.

PANEL.

$$L_c = 7.0 \text{ in};$$

$$h_c = 0.00414(0.59) \left(\frac{20}{7}\right)^{0.25} = 0.00317 \text{ W/in}^2\text{°C};$$

$$R_s = \frac{1}{0.00317(7)(7)} = 6.44 \text{°C/W};$$

$$R_{sq} = \frac{1}{(2/1.8)(0.09)} = 10.0 \text{°C/W}.$$

The configuration impedance ratio for Case II of Figure 4.16 is

$$\frac{R_{sq}}{R_s}\left(\frac{L}{H}\right) = \frac{10}{6.44}(1) = 1.55$$

which corresponds to a fin efficiency $\eta = 0.89$. From (4.26)

$$Z_2 = \frac{7^{0.25}}{0.89(0.59)49} = 0.063$$

SIDES. Since the fin area remains to be determined, (4.26) is modified to

$$Z_i = \left(\frac{L_c^{0.25}}{\eta ACN_s}\right)_i \tag{4.33}$$

where N_s = heat transfer area/surface area.
For the configuration of Figure 4.17

$$Z_3 = Z_4 = \frac{7^{0.25}}{(1)(70)(0.59)\, N_s} = 0.0394/N_s$$

REAR FACE.

$$Z_5 = \frac{7^{0.25}}{(1)49(0.59)\, N_s} = 0.0563/N_s$$

Equation (4.26) after substituting parameters becomes

$$20 = \left[\frac{241.54(50)}{\dfrac{1}{0.049} + \dfrac{1}{0.063} + \dfrac{2N_s}{0.0394} + \dfrac{N_s}{0.0563}}\right]^{0.8}$$

$$= \left(\frac{12077}{36.3 + 68.5N_s}\right)^{0.8}$$

The solution for N_s is

$$N_s = \frac{285.5 - 36.3}{68.5} = 3.64$$

Handling provisions require a fin thickness of at least 0.06 in. to avoid damage. The minimum fin spacing to avoid flow cho king is determined from (4.31) as

$$S = 0.86\left(\frac{7}{20}\right)^{0.25} 1.0087^{0.45(20)} = 0.72 \text{ in}$$

The fin density using these values for spacing and thickness is

$$N = \frac{1}{(S + a)}$$

$$N = \frac{1}{(0.72 + 0.06)} = 1.28 \text{ or } 2 \text{ fins/inch}$$

From fin geometry, the value of N_s becomes

$$N_s = 2LN + 1 \tag{4.34}$$

The fin length, the final parameter for defining the required fin configuration, is found using (4.34) as

$$L = \frac{N_s - 1}{2N}$$

$$L = \frac{3.64 - 1}{2(2)} = 0.66 \text{ in}$$

Now that the fin length is known, the fin effectiveness used in evaluating the Z values for the sides and rear face should be verified. A computed effectiveness of $\eta = 0.99$ substantiates the derived fin design.

(b) *Determination of Surface Temperature at 30k ft Altitude.*

An estimate of the surface-to-ambient temperature difference is obtained from (4.26) using the altitude-to-sea level pressure ratio of 0.2975 from Appendix C as

$$\Delta T = \left[\frac{241.54(50)}{0.2975^{0.5}(36.3 + 68.5 \times 3.64)} \right]^{0.8} = 32.5\,^{\circ}C$$

Lower values of h_c that occur at increasing altitudes cause an increase in the fin efficiency. Since the efficiency of the fin configuration is high, corresponding values of Z for increasing altitude are slowly changing. In this case, the estimated ΔT accurately predicts the temperature difference at 30k ft. The surface temperature at this altitude becomes

$$T_{\text{surface}} = T_{\text{amb}} + \Delta T$$

$$T_{\text{surface}} = 30 + 32.5 = 62.5\,^{\circ}C \qquad \text{Q.E.D.}$$

4.3 FORCED CONVECTION DESIGN REQUIREMENTS

As the density of electronic components increases, the effectiveness of natural convection is reduced due to close spacings and choked flow conditions. Although natural convection and radiation is the simplest to use and should be employed whenever possible, spacial limitations imposed by large fins or increased package volume due to reduced component densities may require the use of forced convection. Either direct or indirect cooling may be required, depending on the nature of environment and the equipment configuration. Direct cooling describes the condition in which the surfaces of heat dissipating components are exposed directly to the coolant. The relatively small size of most heat producing electronic parts develops high heat transfer coefficients, which leads to an order of magnitude increase in heat transfer over natural methods. Airborne contaminants or moisture that can short circuit connections and foul connectors may preclude the use of the direct approach.

An indirect approach to forced cooling prevents fouling and performance loss by confining the coolant to a specially designed duct or cold plate. Generated heat from parts, modules or circuit boards is conducted to the cold plate, which transfers heat to the coolant. Methods of arrangement, placement, mounting, fastening, baffling, ducting and other factors lead to many possible cooling design configurations for any particular package. The cooling system is an integral part of the package design which includes consideration of weight, structural suitability for shock and vibration, accessibility and producibility factors. Optimum use of the indirect method requires tradeoffs concerning part or assembly arrangement accounting for the resulting temperature gradients to obtain the lowest operating temperature possible. Characteristics of the indirect and direct methods are summarized in Table 4.3.

The design process for forced convection involves the application of Procedure 4.1 to establish goals and a viable cooling solution.

Table 4.3. Characteristics of Forced Convection Cooling

Indirect	Direct
1. Coolant flow is separated from electronics.	1. Limitless possibilities of arrangement, placement and ducting.
2. Cold plates provide a high section modulus structure to resist shock, vibration or handling loads.	2. Ability to cool specific units as required.
3. Many varieties of practical flow channel configurations can be selected to remove heat and minimize flow losses.	3. Creates flexibility in packaging.
4. The cold plate conducts heat, concentrated at a region due to an attachment, by spreading it over a relatively large area.	
5. Indirect methods provide a practical solution to cooling sealed units.	

PROCEDURE 4.1

a. Obtain the equipment heat dissipation from the best information available, a technical proposal or experience with similar equipment.

b. Determine the required coolant flow rate to provide an acceptable temperature rise from the inlet to exhaust. An exit air temperature is selected that will provide a suitable convective sink for an additional surface to air temperature rise. Generally parts located near the coolant inlet are exposed to cooler sink temperatures than those near the exhaust. The air temperature increases as it flows through the unit due to the heat it absorbs. The temperature rise of the air is independent of the geometry or dimensions of the unit. The flow rate, a function of the temperature rise of the air and the total power dissipated to the coolant, is given by

$$\dot{m} = \frac{Q}{C_p \Delta T} = \frac{0.237Q}{\Delta T} \text{ lb/min} \qquad (4.35a)$$

In metric units, (4.35a) becomes

$$\dot{m} = \frac{0.996Q}{\Delta T} \text{ gm/s} \qquad (4.35b)$$

where

Q is the total power transferred to the coolant, watts;
ΔT is the temperature rise of the air, °F (4.35a), °C (4.35b).

Equation (4.35) is applicable for altitudes <20k feet (6.1 km). The leading constant can be corrected for applications at greater altitudes.

c. Estimate the pressure drop through the unit as a function of flow rate. A concept of the packaging design with approximate dimensions of the flow cross-section along the coolant path is converted to an equivalent duct for determining related flow losses. Flow losses are due to inlet expansions, exhaust contractions, straight sections, changes in direction, internal constrictions, filters and similar conditions.

d. A fan or blower can be selected which supplies the required flow as a function of inlet air temperature (4.35) and estimated pressure drop. The selection is also based on other system requirements: size, weight, power consumption, acoustical noise, life and cost.

e. A thermal schematic can be constructed which graphically illustrates the air and component temperature profiles along the coolant path. To reflect the local effects in this profile, the power dissipation along the path is treated in increments so that components in each increment can be assigned an estimated local air temperature. The schematic may then be used to

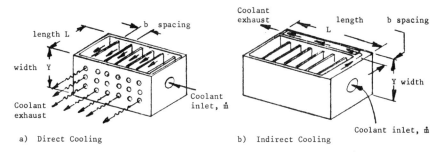

a) Direct Cooling   b) Indirect Cooling

Figure 4.18. Forced Convection Cooling Methods.

determine the need for reducing specific component temperatures. This may be accomplished with direct cooling by using finned surface extenders which attach to the part. For indirectly cooled equipment, this may require an improvement of the interface, attachment or path conductance to the component.

Impingement Cooling Uniformly Heated Assemblies.   The equation for heat transfer between a surface and the coolant in contact with it is given by (2.2). The difficulty with (2.2) is that the heat transfer coefficient h_c is not constant and the coolant temperature varies along the flow path. The value of h_c is also dependent upon whether the flow regime is laminar or turbulent. These complications arise because convection is a combination of heat conduction and fluid flow. Generally, flow parallel to smooth large surfaces is laminar whereas flow over electronic parts is turbulent. An example of direct and indirect cooling is shown in Figure 4.18.

The average heat transfer coefficient evaluated at average coolant temperatures may be determined with the aid of Figure 4.19. For direct cooling, parameter b, Figure 4.19, is the average free flow spacing accounting for components.

EXAMPLE 4.7.   Direct cooling is to be used to cool the 61 watt unit of Figure 4.20, which contains 7 circuit card assemblies, 4 of which dissipate 7 watts each and 3 of which dissipate 11 watts each. The unit operates at sea level with an inlet air temperature of 30°C maximum. Determine the minimum coolant flow rate required to limit component case temperatures to an 80°C worst case.

This design example demonstrates the process of Procedure 4.1(b). The flow can be metered using cutouts on the plenum bulkhead with an area proportional to the rate required by the assembly being cooled. The temperature rise of the coolant flowing along an 11 watt card is determined from (4.35) as

$$\Delta T = \frac{0.237Q}{\dot{m}} \left(\frac{°C}{1.8°F} \right) = \frac{1.448}{\dot{m}} \, °C \qquad (4.36a)$$

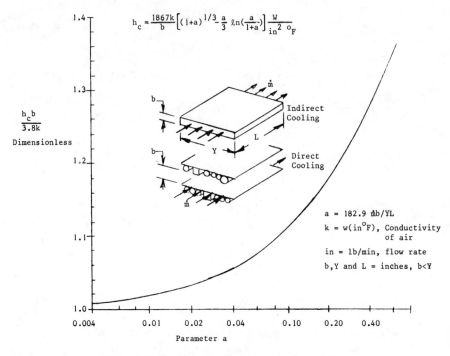

Figure 4.19. Heat Transfer Coefficients for Ducted Flow.

which in metric units is

$$\Delta T = \frac{0.996Q}{\dot{m}} = \frac{10.956}{\dot{m}} \ ^\circ C \tag{4.36b}$$

The coolant temperature rise is usually much greater than the convective temperature difference across the boundary layer. Additionally, the convective temperature difference varies slowly with large changes in flow rate $\dot{m}$ as seen in Figure 4.19. Assuming

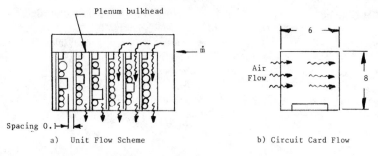

Figure 4.20. Direct Component Cooling.

a negligible air temperature rise in the distribution plenum, an initial flow rate limiting the coolant rise to about 80 percent of the inlet to component case temperature difference using (4.36) becomes

$$\dot{m}_{11} = \frac{1.448}{0.8(80 - 30)} = 0.036 \text{ lb/min}$$

$$= \frac{10.956}{0.8(80 - 30)} = 0.274 \text{ g/s} \tag{4.37}$$

Parameter a of Figure 4.19 is evaluated for this rate as

$$a = 182.9\dot{m}b/YL = \frac{182.9(0.036)(0.1)}{48}$$

$$a = 0.0137 \tag{4.38}$$

The convective film temperature difference for an average coolant temperature of $50°C$ using (2.2) and Figure 4.19 becomes

$$\Delta T = \frac{Q}{h_c A} = \frac{11}{0.0279(6)(8)} = 8.2°C \tag{4.39}$$

With the flow rate of (4.37), the worst case component temperature for the 11 watt assembly is

$$T_{component} = 30 + 0.8(80 - 30) + 8.2 = 78.2°C$$

The flow rate of (4.37) can be minimized by increasing the coolant temperature rise by 1.44°C, or about 80 percent of the residual remaining to a value of 41.44°C. The reduced flow rate of 0.0349 lb/min is used to reevaluate the convective temperature difference which experiences a slight increase to 8.24°C as expected. The resulting maximum component temperature becomes

$$T_{component} = 30 + 41.44 + 8.24 = 79.68°C$$

The flow rate for the 7 watt cards is determined by factoring (4.39) down by the power dissipation ratio of $7/11$. The result is added to the inlet temperature and subtracted from the allowable maximum case temperature to give

$$\Delta T_7 = 80 - [(7/11)(8.24) + 30] = 44.7°C$$

The flow rate from (4.36) becomes

$$\dot{m}_7 = \frac{(7/11)1.448}{44.7} = 0.0206 \text{ lb/min}$$

Using (4.38)

$$a = \frac{182.9(0.0206)(0.1)}{48} = 0.00785$$

In conjunction with Figure 4.19, the heat transfer coefficient for the 7 watt card becomes

$$h_c = 0.0276 \text{ W}/(\text{in}^2 \cdot {}^{\circ}\text{C})$$

The convective temperature difference from (4.39) is

$$\Delta T = \frac{7}{0.0276(6)(8)} = 5.28 {}^{\circ}\text{C}$$

The maximum component temperature for the 7 watt card is

$$T_{\text{component}} = 30 + 44.7 + 5.28 = 79.98 {}^{\circ}\text{C}$$

The minimum flow rate for the unit becomes

$$\dot{m}_{\text{min}} = 4\dot{m}_7 + 3\dot{m}_{11}$$
$$= 4(0.0206) + 3(0.0349) = 0.187 \text{ lb/min} \qquad \text{Q.E.D.}$$

The cooling layout requires planning so that all heat dissipating components are sufficiently cooled. The flow route must not short-circuit itself or choke off needed coolant. Short-circuiting or mixing, Figures 4.21(a) and (b), results if the exhaust is placed too close to the inlet or if inadequate baffling is used. Choking may be the result of an unexpected restriction due to an oversize component or to insufficient duct spacing, Figure 4-21(c). These conditions develop overheating, which reduces the useful life of the equipment. In developing the layout, higher power parts should be placed closest to the exhaust with lower power parts near the inlet. This will minimize the average component temper-

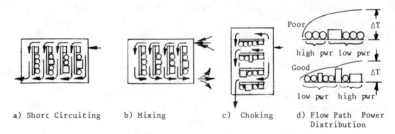

a) Short Circuiting   b) Mixing   c) Choking   d) Flow Path Power Distribution

Figure 4.21. Undesirable Flow Layout Characteristics.

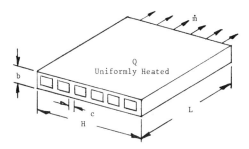

Figure 4.22. Cold Plate Configuration.

ature for the unit, as shown in Figure 4.21(d). Higher temperatures result for all components if higher power parts are placed near the inlet.

Cold Plate Design Considerations for Uniform Heating. Instead of channelling the coolant over the electronic parts which provide the surface area for heat transfer, the indirect method may be used. In this technique, the coolant passes through ducts or cold plates which provide the surface area necessary for heat transfer. The coolant does not contact the component directly. Good conductive paths transfer the generated heat from the part to the coolant via the duct configuration. The heat transfer surface area of these cold plates may be increased by finning. Producibility considerations establish an upper limit to the fin density that can be used. High fin densities also contribute high flow losses due to friction which may limit fan selection.

Cold plates which are integrated into the chassis usually develop a higher strength assembly. The cold plates prevent part and circuit contamination due to moisture and other impurities transported by the coolant. Since the flow path is bounded structurally, the removal of an access cover for test point or electronic monitoring does not alter the cooling effectiveness of the system, which does occur in most implementations of the direct method.

The variables used in defining the cold plate geometry of Figure 4.22 are described in Table 4.4. Only smooth rectangular passages are considered since

Table 4.4. Cold Plate Parameters

| Parameter | Units | | Parameter | Units | |
	English	Metric		English	Metric
$\dot{m}$ Total flow	lb/min	g/s	N Fin density	fins/inch	fins/mm
$Q(x)$ Heat flux	watts	watts	μ Viscosity	lb/min·in	g/s·mm
b Passage height	in	mm	k Thermal cond.	watt/in·°F	watt/mm·°C
			h Heat transfer		
c Fin thickness	in	mm	coef.	watt/in²·°F	watt/mm²·°C
L Length	in	mm	T Temperature	°F	°C
			A Heat transfer		
H Width	in	mm	area	in²	mm²

these are commonly used for most applications. The flow tube aspect ratio of Figure 4.22 is defined as

$$\alpha = b \Big/ \left(\frac{1}{N} - c \right) \tag{4.40}$$

The equivalent diameter for a single passage is

$$D_e = 2b/(1 + \alpha) \tag{4.41}$$

The total heat transfer area assuming 100 percent fin efficiency becomes

$$A = \frac{2bLH(1 + \alpha)}{B(b + c)} \tag{4.42}$$

where B defines the configurations which are heated from one or both sides ($B = 2$, one side heated; $B = 1$, two sides heated).

The cold plate mass velocity becomes

$$G = \dot{m}(b + c)/Hb^2 \tag{4.43}$$

The Reynolds number, a dimensionless parameter used to distinguish between laminar and turbulent flow, is

$$R_e = \frac{2\dot{m}(b + c)}{b(1 + \alpha)\mu H} \tag{4.44}$$

The heat transfer coefficient, wall temperature rise and loss characteristics are dependent upon the nature of flow.

Laminar Flow.

$$h = \frac{2015k(1 + \alpha)}{\left(\dfrac{6}{5\alpha^{2/3}} + 1 \right) b} \tag{4.45}$$

$$T_W(X) = T_a(X) + \frac{0.1227Q(X/L)B(b + c)\left(\dfrac{6}{5\alpha^{2/3}} + 1 \right)}{(1 + \alpha)^2 LHk} \tag{4.46}$$

$$\Delta P = \frac{5.792\dot{m}(1 + \alpha)^2 \mu L(b + c)}{(0.7/\alpha^{2/3} + 1)Hb^4} \text{ inches } H_2O \tag{4.47a}$$

In metric units

$$\Delta P = \frac{1.069(10^6)\dot{m}(1 + \alpha)^2\mu L(b + c)}{(0.7/\alpha^{2/3} + 1)Hb^4} \text{ mm H}_2\text{O} \qquad (4.47b)$$

Turbulent Flow.

$$h = \frac{19208k(1 + \alpha)^{0.2}(\dot{m}/H)^{0.8}}{b} \qquad (4.48a)$$

In metric units

$$h = \frac{50620k(1 + \alpha)^{0.2}(\dot{m}/H)^{0.8}}{b} \qquad (4.48b)$$

$$T_W(X) = T_a(X) + \frac{0.5239Q(X/L)Bb}{(1 + \alpha)^{1.2}(\dot{m}/H)^{0.8}} \qquad (4.49a)$$

In metric units

$$T_W(X) = T_a(X) + \frac{4.349(10^{-3})Q(X/L)Bb}{(1 + \alpha)^{1.2}(\dot{m}/H)^{0.8}} \qquad (4.49b)$$

$$\Delta P = \frac{2.83(10^{-3})L(1 + \alpha)^{1.2}}{b^3}\left(\frac{\dot{m}}{H}\right)^{1.8} \text{ inches H}_2\text{O} \qquad (4.50a)$$

In metric units

$$\Delta P = \frac{410.2L(1 + \alpha)^{1.2}}{b^3}\left(\frac{\dot{m}}{H}\right)^{1.8} \text{ mm H}_2\text{O} \qquad (4.50b)$$

In the above expressions, the temperature rise of the coolant is

$$T_a(X) = T_{in} + \frac{0.237Q(X/L)}{\dot{m}} \qquad (4.51a)$$

which in metric units becomes

$$T_a(X) = T_{in} + \frac{0.996Q(X/L)}{\dot{m}} \qquad (4.51b)$$

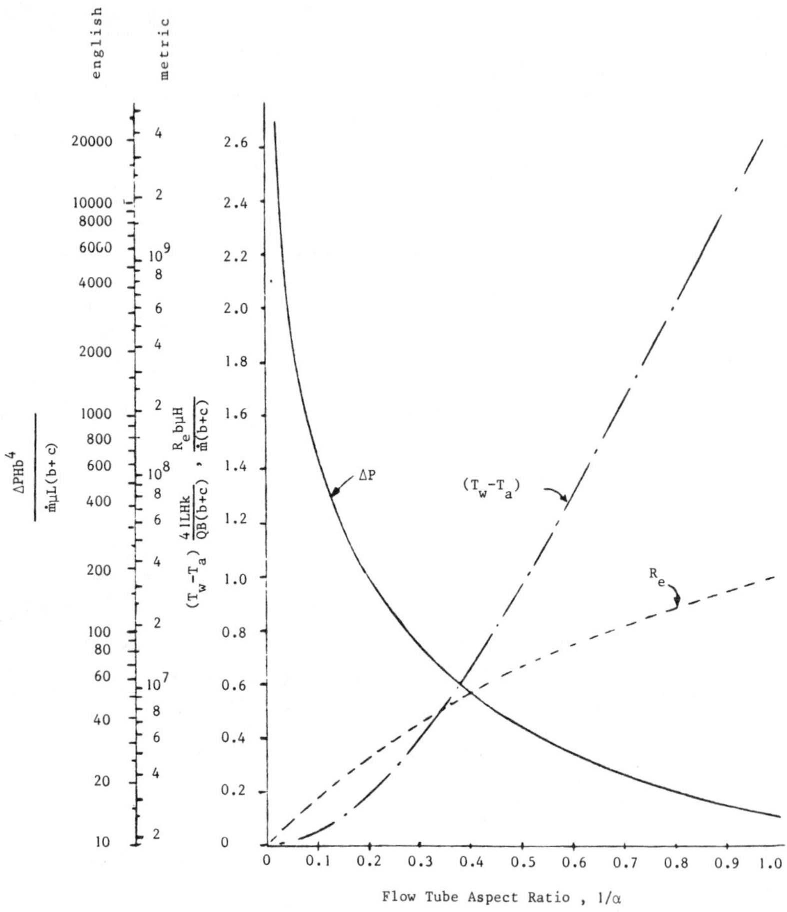

Figure 4.23. Cold Plate Design Characteristics.

Equations (4.44), (4.46) and (4.47) are shown graphically on Figure 4.23 as a function of the flow tube aspect ratio for laminar flow conditions. Figure 4.23 shows that as the flow tube aspect ratio (4.40) is decreased, the wall-to-local air temperature ($T_W - T_a$) increases with an attendant decrease in pressure drop. A design chart suitable for performing design tradeoffs is illustrated in Figure 4.24 for laminar flow and Figure 4.25 for turbulent flow. These charts are based on English units. The use of these charts is illustrated in the following design example:

EXAMPLE 4.8.   A cold plate is to be used to cool the 61 watt unit of Figure 4.20. The configuration, which is similar to Figure 4.18(b), is shown on Figure 4.26. Since

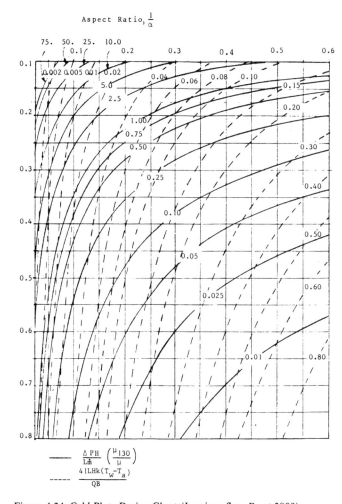

Figure 4.24. Cold Plate Design Chart (Laminar flow $R_e < 2800$).

an indirect cooling approach is used with internal conduction to the cold plate, a higher part density is used requiring only 4.5 × 5.5 inch (114.3 × 139.7 mm) cards which are placed on 0.5 inch (12.7 mm) centers. The resulting configuration is smaller than the comparable direct cooled unit. If the temperature rises for the 7 watt and 11 watt card assemblies are 17°C and 25°C respectively above the cold plate wall temperature, determine the card location, minimum coolant flow rate, and fin density to limit worst case component temperature to 80°C and the cold plate pressure loss to 0.2 inches of water.

The flow rate can be minimized by organizing the card assemblies in the unit to maximize the component temperatures on each. The resulting distribution shown on

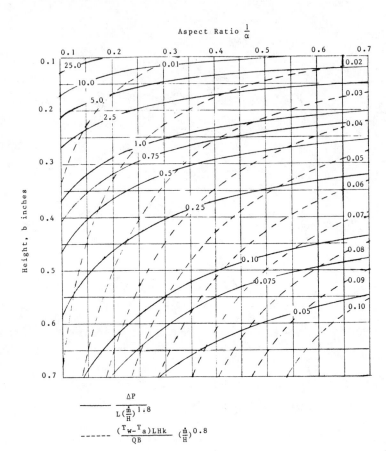

Figure 4.25. Cold Plate Design Chart (Turbulent flow $R_e > 2800$).

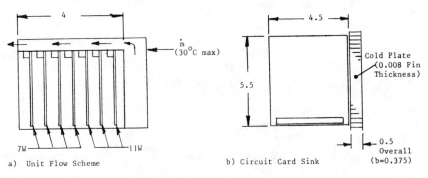

Figure 4.26. Cold Plate Cooled Unit.

Figure 4.26(a) permits the largest coolant temperature rise consistent with maintaining 80°C component temperatures. The allowable temperature rise for the coolant and coolant-to-wall is then limited to

$$\Delta t_{allowable} = 80 - (17 + 30) = 33°C \qquad (4.52)$$

The coolant temperature rise is usually much greater than the convective temperature difference across the boundary layer. Also, the convective temperature difference varies slowly with a change in flow rate.

The temperature rise of the coolant flowing along the cold plate is given by (4.35) as

$$\Delta T = \frac{0.237Q}{\dot{m}}\left(\frac{°C}{1.8°F}\right) = 0.1317\frac{Q}{\dot{m}}°C \qquad (4.53)$$

Considering what is allowable (4.52), a 4°C temperature difference is assumed between the wall and the coolant. Following (4.53), the flow rate is

$$\dot{m} = \frac{0.1317(61)}{(33 - 4)} = 0.277 \text{ lb/min} \qquad (4.54)$$

Referring to values which determine ΔP from Figure 4.24 as $\overline{N}$, the expression for the pressure loss becomes

$$\Delta P = \frac{\overline{N}L\dot{m}}{H} = \frac{\overline{N}(4)(0.277)}{5.5}$$

$$\Delta P = 0.201\overline{N} \qquad (4.55)$$

Since the cold plate pressure loss is limited to 0.2 inches of water, (4.55) gives

$$\overline{N} = 0.2/0.201 \simeq 1.0 \qquad (4.56)$$

Figure 4.26(b) defines a flow passage height of 0.375 inches. The intersection of $\overline{N}$ and $b = 0.375$ is located on Figure 4.24. This point establishes the cold plate aspect ratio and the wall-to-coolant temperature rise of the resulting configuration. Referring to values which determine $(T_W - T_a)$ from Figure 4.24 as N^*, the expression for the wall-to-coolant temperature difference becomes

$$(T_W - T_a) = \frac{N^*QB}{41LHk} = \frac{N^*(61)2}{41(4)(5.5)3.98(10^{-4})}\left(\frac{°C}{1.8°F}\right)$$

$$(T_W - T_a) = 189N^*°C \qquad (4.57)$$

The intersection of b and $\overline{N}$ gives $N^* = 0.015$ so that (4.57) becomes

$$(T_W - T_a) = 189(0.015) = 2.84°C \qquad (4.58)$$

This result is less than the assumed value of 4°C used to establish the flow rate (4.54). Since the temperature difference (4.58) is slowly changing, an improved estimate using 2.8°C for determining the flow rate (4.53) gives

$$\dot{m} = \frac{0.1317(61)}{(33 - 2.8)} = 0.266 \text{ lb/min} \tag{4.59}$$

Repeating the process, (4.55) gives the pressure loss $\Delta P = 0.193\overline{N}$ which determines $\overline{N}$ as

$$\overline{N} = 0.2/0.193 = 1.036 \tag{4.60}$$

The intersection of b and $\overline{N}$ from (4.60) on Figure 4.24 gives $N^* = 0.014$ so that (4.57) becomes

$$(T_W - T_a) = 189(0.014) = 2.6°C \tag{4.61}$$

This value is reasonably close to the improved estimate establishing the flow rate of 0.266 lb/min from (4.59) as a minimum. The corresponding aspect ratio from Figure 4.24 is $\alpha = 11.4$. The fin density is obtained from (4.40) as

$$N = \frac{\alpha}{b + \alpha c} = \frac{11.4}{0.375 + 11.4(0.008)} \tag{4.62}$$

$$N = 24 \text{ fins/inch}$$

The fin density can be reduced by decreasing the allowable pressure drop, which would increase the required flow rate. For an allowable pressure drop of 0.1 inch of water, the minimum flow rate is 0.309 lb/min, which develops a requirement for 17 fins/inch. In this case, the wall-to-coolant temperature difference would become 7°C. Laminar flow conditions exist as can be verified using (4.44).

The worst case component temperature on the 11 watt cards can be determined using (4.53) with (4.58) and (4.59) to evaluate the wall temperature corresponding to the card assembly location. The air temperature in the cold plate at the third 11 watt card from the inlet using (4.51a) modified by (4.53) becomes

$$T_a = 30 + 0.1317 \frac{Q}{\dot{m}}$$

$$= 30 + \frac{0.1317(3)(11)}{0.266}$$

$$T_a = 46°C$$

The wall temperature at this location is found using (4.58) as

$$T_W = T_a + 2.8 = 48.8°C$$

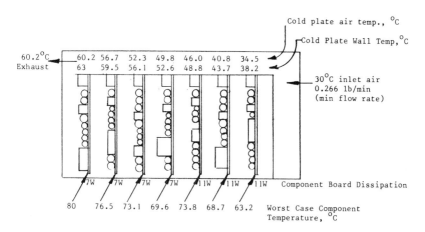

Figure 4.27. Unit Temperature Schematic.

The worst case component on the third 11 watt card is

$$T_{\text{comp}} = T_W + 25 = 73.8°C$$

This confirms that component temperatures are less than 80°C as required. This process can be repeated for each card of Figure 4.26(a). The results shown on Figure 4.27 demonstrate the process of Procedure 4.1(e). Average component temperatures may be estimated and used in reliability analysis from the schematic of Figure 4.27.

Q.E.D.

Equations (4.45) through (4.51) may be used to verify the results obtained from Figures 4.24 or 4.25. Either of these figures may be used to optimize a cold plate configuration by allowing any of the parameters to vary from a trial point. The trend established by the lines of constant pressure loss and wall-to-coolant temperature difference permits rapid assessment of trades which affect the cold plate configuration.

Unit pressure losses include entrance, exit, contraction, expansion and duct losses. Losses due to cold plate flow tube geometry are obtained from Figures 4.24 and 4.25. Cold plate entrance or exit losses for coolant contraction or expansion due to the geometrical transition associated with the flow tubes can be estimated using

$$\Delta P = \frac{0.218}{b^4} \left[\frac{\dot{m}(b + c)}{H} \right]^2 \text{ inches H}_2\text{O} \qquad (4.63a)$$

or in metric units

$$\Delta P = \frac{4028}{b^4} \left[\frac{\dot{m}(b + c)}{H} \right]^2 \text{ mm H}_2\text{O} \qquad (4.63b)$$

where $b \gg c$.

Convective Cooling of Nonuniformly Heated Assemblies.   Components are generally mounted on printed circuit boards because they facilitate test, repair and replacement. Large, heavy, high heat dissipating components may be mounted on removable panels or structural elements which provide the necessary support and heat removal paths. In most cases, these mounting surfaces are rectangular and attach to mating structure along opposite edges.

The heat distribution on these surfaces may be discrete or distributed and may require forced air cooling in addition to edge conduction provided by the attachment interface. A model of a general forced-convection-cooled component mounting surface with any number of concentrated line or distributed heat sources is shown in Figure 4.28.

In this figure, q_1, q_2, q_{a1} and q_{a2} are line heat sources expressed in watts whereas distributed sources $\dot{q}_{a3}$ and $\dot{q}_{a4}$ are expressed in watts/in (W/mm). These sources are due to components or assemblies fastened at the corresponding location ai. Distributed sources may be added or subtracted to develop any desired contour. Subtracting $\dot{q}_{a4}$ from $\dot{q}_{a3}$ in Figure 4.28, for example, leads to the distribution of Figure 4.29.

The differential equations that describe the heat flow in the system of Figure 4.28 which are a result of a heat balance on a differential element become

$$kab \frac{d^2 T_s}{dx^2} - hb(T_s - T_a) = -\dot{Q}(x) \qquad (4.64a)$$

$$\dot{m}cp = \frac{dT_a}{dx} = hb(T_s - T_a) \qquad (4.64b)$$

The boundary conditions from Figure 4.28 are

$$kab \frac{dT_s}{dx}(0) = G_1 \left[T_s(0) - T_1 \right] - q_1 \qquad (4.65a)$$

$$kab \frac{dT_s}{dx}(L) = -G_2 \left[T_s(L) - T_2 \right] + q_2 \qquad (4.65b)$$

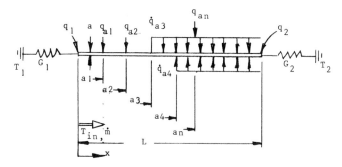

Figure 4.28. Forced Convection Cooled Component Assembly.

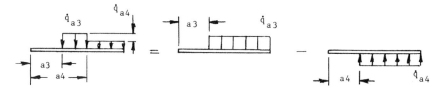

Figure 4.29. Technique for Developing Desired Distributed Heating.

The solutions to the coupled differential equations (4.64) subject to conditions (4.65) are

$$T_a(x) = e^{-\phi x}(B_o \cosh \theta x + B_1 \sinh \theta x)$$

$$+ T_{in} - B_o + \sum_{i}^{n} \langle T_{ci}(x) \rangle_{ai} \quad (4.66a)$$

$$T_s(x) = \frac{1}{2} e^{-\phi x} \left[\left(\frac{\theta}{\phi} B_1 + B_o \right) \cosh \theta x \right.$$

$$\left. + \left(\frac{\theta}{\phi} B_o + B_1 \right) \sinh \theta x \right] + T_{in} - B_o + \sum_{i}^{n} \langle T_{di}(x) \rangle_{ai} \quad (4.66b)$$

In (4.66), the angular brackets are used to describe a function singularity with the meaning

$$\langle f(x) \rangle_{ai} \quad \begin{array}{l} = 0 \qquad x < ai \\ = f(x) \qquad x \geq ai \end{array} \quad (4.67)$$

Temperature functions T_{ci} and T_{di} defined for concentrated and distributed sources are equal to zero at $x = 0$ regardless of the value of ai. The initial temperatures for the coolant and component mounting surface at $x = 0$ become

$$T_a(0) = T_{in} \tag{4.68}$$

$$T_s(0) = T_{in} + \frac{B_1}{2}\left(\frac{\theta}{\phi}\right) - \frac{B_o}{2}$$

The concentrated temperature functions are

$$T_{ci}(x) = \frac{q}{\dot{m}c_p}\left\{1 - e^{-\phi(x-ai)}\left[\frac{\phi}{\theta}\sinh\theta(x-ai) + \cosh\theta(x-ai)\right]\right\} \tag{4.69a}$$

$$T_{di}(x) = \frac{q}{\dot{m}c_p}\left\{1 - e^{-\phi(x-ai)}\left[\frac{1}{2}\left(\frac{\theta}{\phi} + \frac{\phi}{\theta}\right)\sinh\theta(x-ai)\right.\right.$$

$$\left.\left. + \cosh\theta(x-ai)\right]\right\} \tag{4.69b}$$

$$\dot{T}_{di}(x) = -\frac{qN^2}{2\dot{m}c_p}e^{-\phi(x-ai)}\left[\frac{\cosh\theta(x-ai)}{\phi} + \frac{\sinh\theta(x-ai)}{\theta}\right] \tag{4.69c}$$

The distributed temperature functions are

$$T_{ci}(x) = \frac{\dot{q}}{\dot{m}C_p}\left\{\frac{2\phi}{N^2} + (x-ai) - \frac{e^{-\phi(x-ai)}}{N^2}\left[\left(\frac{\phi^2 + \theta^2}{\theta}\right)\sinh\theta(x-ai)\right.\right.$$

$$\left.\left. + 2\phi\cosh\theta(x-ai)\right]\right\} \tag{4.70a}$$

$$T_{di}(x) = \frac{\dot{q}}{\dot{m}C_p}\left\{\left(\frac{\theta^2 + 3\phi^2}{2\phi N^2}\right) + (x-ai)\right.$$

$$\left. - e^{-\phi(x-ai)}\left[\left(\frac{3\theta^2 + \phi^2}{2\theta N^2}\right)\sinh\theta(x-ai) + \left(\frac{\theta^2 + 3\phi^2}{2\phi N^2}\right)\cosh\theta(x-ai)\right]\right\} \tag{4.70b}$$

$$\dot{T}_{di}(x) = \frac{\dot{q}}{\dot{m}C_p}\left\{1 - e^{-\phi(x-ai)}\left[\cosh\theta(x-ai) + \left(\frac{\theta^2 + \phi^2}{2\phi\theta}\right)\sinh\theta(x-ai)\right]\right\} \tag{4.70c}$$

Parameters which are used for determining temperatures become

$$N_s = \frac{A_H}{bL} = \frac{\text{total heat transfer area}}{\text{surface area}}$$

$$\phi = \frac{hN_s b}{2\dot{m}C_p}$$

$$N^2 = \frac{hN_s}{ka}$$

$$\phi = \sqrt{\phi^2 + N^2}$$

$$B_{11} = \frac{G_1\theta/\phi}{2\dot{m}C_p + G_1}$$

$$B_{12} = \frac{2G_1(T_{\text{in}} - T_1) - 2q_1}{2\dot{m}C_p + G_1}$$

(4.71a)

Boundary functions and constants of integration are

$$G_{s1} = \left[\dot{m}C_p + \frac{G_2}{2}\left(\frac{\theta}{\phi}B_{11} + 1\right) \right] \sinh\theta L + \left[\dot{m}C_p B_{11} + \frac{G_2}{2}\left(\frac{\theta}{\phi}\right.\right.$$

$$\left.\left. + B_{11}\right) \right] \cosh\theta L - G_2 B_{11} e^{\phi L}$$

$$G_{s2} = \left(\frac{G_2}{2} + \dot{m}C_p\right)\cosh\theta L + \frac{G_2}{2}\left(\frac{\theta}{\phi}\right)\sinh\theta L$$

(4.71b)

$$Q = q_2 - G_2\left[\sum_i^n T_{di}(L) + T_{\text{in}} - B_{12}\right] - kab\sum_i^n \dot{T}_{di}(L)$$

$$B_1 = \frac{Qe^{\phi L} - B_{12}G_{s2}}{G_{s1}}$$

$$B_0 = B_{11}B_1 + B_{12}$$

For uniform heating over length L in the absence of concentrated sources, q_{ai} simplifies (4.66) to

$$T_a(x) = e^{-\phi x}(B_o \cosh\theta x + B_1 \sinh\theta x) + T_{\text{in}} - B_o + \frac{\dot{q}}{\dot{m}C_p}\left\{\frac{2\phi}{N^2}\right.$$

$$\left. + x - \frac{e^{-\phi x}}{N^2}\left[\left(\frac{\phi^2 + \theta^2}{\theta}\right)\sinh\theta x + 2\phi\cosh\theta x\right]\right\}$$

(4.72a)

$$T_s(x) = \frac{1}{2}e^{-\phi x}\left[\left(\frac{\theta}{\phi}B_1 + B_o\right)\cosh\theta x + \left(\frac{\theta}{\phi}B_o + B_1\right)\sinh\theta x\right]$$

$$+ T_{in} - B_o + \frac{\dot{q}}{\dot{m}C_p}\left\{\left[\left(\frac{\theta^2 + 3\phi^2}{2\phi N^2}\right) + x\right.\right.$$

$$\left.\left. - e^{-\phi x}\left[\left(\frac{3\theta^2 + \phi^2}{2\theta N^2}\right)\sinh\theta x + \left(\frac{\theta^2 + 3\phi^2}{2\phi N^2}\right)\cosh\theta x\right]\right]\right\} \qquad (4.72b)$$

In the special case of uniform heating where $q_1 = q_2 = 0$, $G_1 = G_2 = G$ and $T_1 = T_2 = T$, the maximum surface temperature occurring at $L/2$ following (4.72b) becomes

$$T_{s(max)} = T + \frac{\dot{q}L}{2G} + \frac{\dot{q}L^2}{8kab} \qquad (4.73)$$

EXAMPLE 4.9. The forced convection cooled 7W card of Example 4.7 interfaces to a 35°C plenum wall with a card guide which has a conductance of 0.1 W/°C. Cooling air at 30°C flows over the card at a rate of 0.0207 lb/min as determined by the example. The 0.06 in (1.52 mm) thick card has a conductivity $k = 0.44$ W/in·°C. With components considered, the ratio of heat transfer area to card surface area is 3. Determine (a) the part surface temperature profile considering uniform heating, then (b) the part surface temperature if 2W are concentrated along a line 2 inches from the plenum with the remaining uniformly distributed.

(a) *7W Uniform Heating.* From (4.36a)

$$\dot{m}C_p = \frac{\dot{m}(1.8)}{0.237} = \frac{0.0207(1.8)}{0.237}$$

$$\dot{m}C_p = 0.157W/°C$$

The system parameters from (4.71a) are

$$N_s = 3;$$

$$\phi = \frac{0.0276(3)(8)}{2(0.157)} = 2.109 \text{ in}^{-1};$$

$$N^2 = \frac{0.0276(3)}{0.44(0.06)} = 3.136 \text{ in}^{-2};$$

$$\theta = \sqrt{\phi^2 + N^2} = 2.754 \text{ in}^{-1};$$

$$B_{11} = \frac{0.1(2.754)/2.109}{2(0.157) + 0.1} = 0.3154;$$

$$B_{12} = \frac{2(0.1)(30 - 35)}{2(0.157) + 0.1} = 2.415°C.$$

The distributed temperature function (4.70c) evaluated at $x = L = 6$ is

$$\dot{T}_{di}(6) = \frac{7/6}{0.157}\left\{1 - e^{-2.109(6)}\left[\cosh 2.754(6) + \frac{2.754^2 + 2.109^2}{2(2.754)(2.109)}\sinh 2.754(6)\right]\right\}$$

$$\dot{T}_{di}(6) = -354.67°C/in$$

Boundary functions and constants of integration from (4.71b) are

$$G_{s1} = \dot{m}C_p\,(\sinh\theta L + B_{11}\cosh\theta L) = 0.157[7.503 \times 10^6$$

$$+ (0.3154)7.503 \times 10^6];$$

$$G_{s1} = 1.549(10^6)W/°C;$$

$$G_{s2} = \dot{m}C_p\cosh\theta L = 1.178(10^6)W/°C;$$

$$Q = -kab\dot{T}_{di}(L) = -0.44(0.06)(8)(-354.67);$$

$$Q = 74.906W;$$

$$B_1 = \frac{74.906e^{2.109(6)} + 2.415(1.178 \times 10^6)}{1.549 \times 10^6} = 16.973°C;$$

$$B_o = 0.3154(16.973) - 2.415 = 2.938°C.$$

Air and surface temperatures from (4.72) are

$$T_a(x) = e^{-2.109x}(2.938\cosh 2.754x + 16.973\sinh 2.754x) + 27.062$$

$$+ 7.42\left[1.345 + x - \frac{e^{-2.109x}}{3.136}(4.369\sinh 2.754x\right.$$

$$+ 4.218\cosh 2.754x)\Big]$$

$$T_a(x) = e^{-2.109x}(-7.42\cosh 2.754x + 6.636\sinh 2.754x)$$

$$+ 7.42x + 37.042 \tag{4.74}$$

$$T_s(x) = \frac{1}{2}e^{-2.104x}(25.102\cosh 2.754x + 20.809\sinh 2.754x) + 27.062$$

$$+ 7.42[1.582 + x - e^{-2.109x}(1.575\sinh 2.754x$$

$$+ 1.582\cosh 2.754x)]$$

$$T_s(x) = e^{-2.109x}(0.8125\cosh 2.754x - 1.282\sinh 2.754x) + 7.42x + 38.8$$

Equation (4.74) is graphically depicted in Figure 4.30.

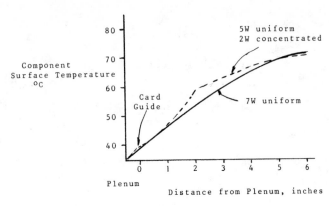

Figure 4.30. 7W Card Component Temperature Profile.

(b) *2W Concentrated with 5W Uniform Heating.* Most of the work performed for (a) applies. The temperature function for the 5W uniform heating can be obtained by factoring the previous result to give

$$\dot{T}_{di} = \frac{5}{7}(-354.67) = -253.33°C/in$$

The temperature function for the concentrated 2W line heat load is obtained from (4.69c) using $(x - ai) = 4$ at $L = 6$ to give

$$\dot{T}_{d2}(6) = -\frac{2(3.136)}{2(0.157)} e^{-2.109(4)} \left[\frac{\cosh 2.754(4)}{2.109} + \frac{\sinh 2.754(4)}{2.754} \right]$$

$$\dot{T}_{d2}(6) = -110.35°C/in$$

New integration constants become

$$Q = -kab\,[\dot{T}_{di}(L) + \dot{T}_{d2}(L)] = -0.44(0.06)(8)(-363.68);$$

$$Q = 76.81W;$$

$$B_1 = \frac{76.81e^{2.109(6)} + 2.415(1.178 \times 10^6)}{1.549 \times 10^6} = 17.358°C;$$

$$B_o = 0.3154(17.358) - 2.415 = 3.06°C.$$

Air and surface temperatures from (4.66) become

$$T_a(x) = e^{-2.109x}(3.06 \cosh 2.754x + 17.358 \sinh 2.754x) + 26.94$$

$$+ 5.31 \left[1.345 + x - \frac{e^{-2.109x}}{3.136} (4.369 \sinh 2.754x + 4.218 \cosh 2.754x) \right]$$

$$+ \langle 12.74 \{1 - e^{-2.109(x-2)} \times [0.766 \sinh 2.754(x - 2)$$
$$+ \cosh 2.754(x - 2)]\}\rangle$$

$$T_a(x) = \frac{1}{2} e^{-2.109x}(25.72 \cosh 2.754x + 21.354 \sinh 2.754x) + 26.94$$

$$+ 5.31 [1.582 + x - e^{-2.109x}(1.575 \sinh 2.754x + 1.582 \cosh 2.754x)]$$
$$+ \langle 12.74 \{1 - e^{-2.109(x-2)} \times [1.036 \sinh 2.754(x - 2)$$
$$+ \cosh 2.754(x - 2)]\}\rangle$$

Component surface temperatures for a configuration with 2W concentrated and 5W uniformly distributed are illustrated in Figure 4.30. The effect of concentrated heating causes an increase to part temperatures over 70 percent of the card surface. Q.E.D.

Pressure Losses Due to Forced Cooling. Flow losses occur in air cooled electronic equipment because of drag and friction caused by movement along surfaces, through openings and across components. The movement of air is due to a difference in pressure between any two points in the flow path. Air flows from a high pressure to a low pressure area.

When a fan is used to move air in an enclosure, the upstream pressure is lower than the downstream pressure along the direction of flow. The upstream pressure is also slightly lower than the external ambient pressure while the downstream pressure is somewhat higher. The pressure change at the fan is due to the energy added to the flow by the fan. Flow losses dissipate the available energy. The size of the fan can be minimized if smooth duct surfaces are used and sharp turns, throttling openings and rapid transitions can be avoided.

When the air is required to turn, contract or expand, the pressure drop through the system increases in proportion to the energy dissipated. The flow system for an electronics unit may resemble a network with numerous flow branches. High losses in any branch will reduce the flow to that branch. If these conditions are not anticipated, overheating and loss of performance may result.

A method analogous to an electrical network is used to represent the flow system network. This approach is suited to parametric studies structured to optimize a particular configuration since various flow resistances can be used as a variable. The basis for this method assumes that the effect of compressibility on atmospheric air can be neglected at velocities less than the speed of sound. Velocities are generally low in order to minimize losses. Velocities approaching the speed of sound through equipment flow passages develop a condition that chokes the flow. The critical velocity is reached when the ratio of the downstream to upstream pressures is less than 0.528. For normal flow, the pressure relationship of (4.75) applies to an enclosure.

$$\frac{P_{ambient}}{P_{ambient} + \Delta P} \geq 0.528 \qquad (4.75)$$

The flow resistance may be expressed in terms of a pressure loss factor R which is derived using the expression for flow rate and head loss

$$\dot{m} = A\rho V \qquad (4.76a)$$

$$V = \sqrt{2gH_L/K} \qquad (4.76b)$$

A relationship for the pressure drop in terms of the flow rate and loss factor developed for commercial standard conditions, Table 4.6, using (4.76) becomes

$$\Delta P = 0.2287(R\dot{m})^2\rho^* \text{ in } H_2O \qquad (4.77a)$$

and in metric units

$$\Delta P = 42534(R\dot{m})^2\rho^* \text{ mm } H_2O \qquad (4.77b)$$

given $\dot{m}$ = flow rate, lb/min (g/s);
$R = \sqrt{K}/A$, flow resistance, in^{-2} (mm^{-2});
K = flow coefficient, dimensionless;
ρ^* = density ratio, $\rho_{commercial}/\rho$.

A cooling flow path system consists of both parallel and series paths. In a parallel system, the pressure drops in each branch are equal. Additionally, the total flow rate equals the sum of the flow rates through each branch. These conditions are expressed as

Parallel System $\begin{cases} \Delta P_T = \Delta P_1 = \Delta P_2 = \cdots \Delta P_n & (4.78a) \\ \dot{m}_T = \dot{m}_1 + \dot{m}_2 + \cdots + \dot{m}_n & (4.78b) \end{cases}$

Based on (4.77) and (4.78), the effective total resistance for parallel branches becomes

$$\frac{1}{R_T} = \frac{1}{R_1} + \frac{1}{R_2} + \cdots + \frac{1}{R_n} \qquad (4.79)$$

In a series system, the flow rates through each branch are equal. Also, the total pressure drop equals the sum of the individual branch losses. These conditions are expressed as

$$\text{Series System} \begin{cases} \Delta P_T = \Delta P_1 + \Delta P_2 + \cdots + \Delta P_n & (4.80a) \\ \dot{m}_T = \dot{m}_1 = \dot{m}_2 = \cdots = \dot{m}_n & (4.80b) \end{cases}$$

Following (4.77) and (4.80), the effective total resistance for series branches becomes

$$R_T^2 = R_1^2 + R_2^2 + \cdots + R_n^2 \qquad (4.81)$$

The flow coefficient, K, in (4.77) assumes that the upstream flow area and the frictional resistance of an actual fluid flowing through the restriction are accounted for. K is essentially unaffected by variations in temperature, pressure and flow. Values for different configurations which were developed empirically are shown in Figure 4.31. The pressure loss factor for fully developed flow in square or rectangular ducts can be evaluated using Figure 4.32.

EXAMPLE 4.10.   Assuming that Figure 4.33 represents the flow resistance network for the unit of Example 4.7, determine for the duct areas of Table 4.5 (a) the plenum discharge area corresponding to R_{7a} for the 7W cards and (b) the electronics unit pressure drop.

(a) *Plenum Discharge Area for 7W Card.* The pressure drop for the 11W cards must be determined in order to evaluate the plenum discharge area for the 7W card.

Air flowing into the card ducts first contracts when traversing the plenum discharge openings then expands into the duct. The contraction area ratio from Table 4.5 is

$$A_3/A_2 = 0.3/3.5 = 194/2258 = 0.086$$

The loss factor from (4.77b), following Figure 4.31 (XI) which gives $K = 0.5$ (interpolated), is

$$R_{11a} = \frac{\sqrt{K}}{A_3} = \frac{\sqrt{0.5}}{0.3} = 2.36 \text{ in}^{-2}$$

$$= \frac{\sqrt{0.5}}{194} = 0.0036 \text{ mm}^{-2}$$
$$(4.82)$$

The expansion loss factor based on Figure 4.31 (X) and Table 4.5 is

$$R_{11b} = \frac{1}{A_3} - \frac{1}{A_4} = \frac{1}{0.3} - \frac{1}{0.8} = 2.08 \text{ in}^{-2}$$

$$= \frac{1}{194} - \frac{1}{516} = 0.0032 \text{ mm}^{-2}$$
$$(4.83)$$

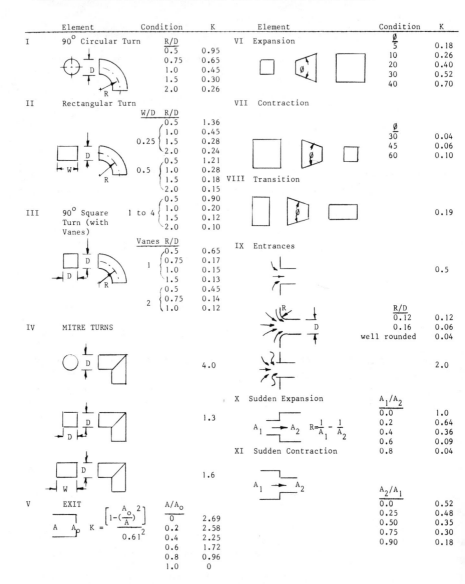

Element	Condition	K		Element	Condition	K
I 90° Circular Turn	R/D			VI Expansion	∅/5	
	0.5	0.95			5	0.18
	0.75	0.65			10	0.26
	1.0	0.45			20	0.40
	1.5	0.30			30	0.52
	2.0	0.26			40	0.70
II Rectangular Turn	W/D R/D			VII Contraction		
	0.25 { 0.5	1.36			∅	
	1.0	0.45			30	0.04
	1.5	0.28			45	0.06
	2.0	0.24			60	0.10
	0.5 { 0.5	1.21		VIII Transition		
	1.0	0.28				
	1.5	0.18				0.19
	2.0	0.15				
III 90° Square Turn (with Vanes) 1 to 4	0.5	0.90		IX Entrances		
	1.0	0.20				0.5
	1.5	0.12				
	2.0	0.10				
	Vanes R/D					
	1 { 0.5	0.65			R/D	
	0.75	0.17			0.12	0.12
	1.0	0.15			0.16	0.06
	1.5	0.13			well rounded	0.04
	2 { 0.5	0.45				
	0.75	0.14				2.0
	1.0	0.12				
IV MITRE TURNS		4.0		X Sudden Expansion	A₁/A₂	
					0.0	1.0
				A₁ → A₂ R = 1/A₁ − 1/A₂	0.2	0.64
		1.3			0.4	0.36
					0.6	0.09
				XI Sudden Contraction	0.8	0.04
		1.6		A₁ → A₂		
					A₂/A₁	
V EXIT K = [1−(Aₒ/A)²]/0.61²	A/Aₒ				0.0	0.52
	0	2.69			0.25	0.48
	0.2	2.58			0.50	0.35
	0.4	2.25			0.75	0.30
	0.6	1.72			0.90	0.18
	0.8	0.96				
	1.0	0				

Figure 4.31. Flow Coefficients for Smooth Duct Elements.

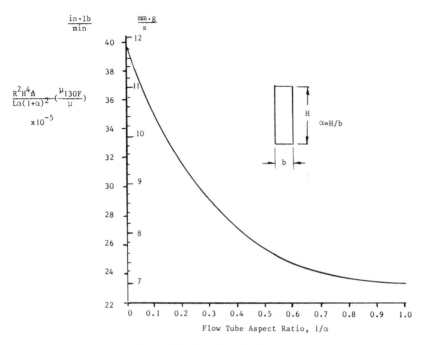

Figure 4.32. Duct Loss Coefficient.

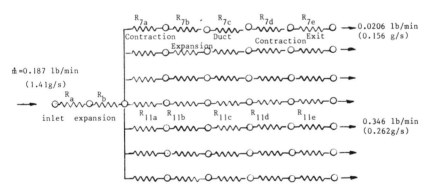

Figure 4.33. Assembly Flow Loss Schematic.

Table 4.5. Duct Characteristics

Flow Element		Cross Sectional Area		Length	
		in^2	mm^2	in	mm
A_1	Inlet	0.78	506	—	—
A_2	Plenum	3.5	2258	—	—
A_3	Flow Opening (11W Card)	0.3	194	—	—
A_4	Card Ducts	0.8	516	6	152
A_5	Exhaust Openings	0.3	194	—	—

The duct loss based on $\alpha = 80$ from Example 4.7 is determined from Figure 4.32 as

$$\frac{R_{11c}^2 H^4 \dot{m}}{L\alpha(1 + \alpha)^2} = 39(10^{-5})$$

Solving for R_{11c} and substituting configuration parameters give

$$R_{11c}^2 = \frac{39(10^{-5})6(80)(81)^2}{8^4 \dot{m}} = \frac{0.3}{\dot{m}} \text{ in}^{-4}$$

$$= \frac{11.62(10^{-5})(152)(80)81^2}{203.2^4 \dot{m}} = \frac{5.44(10^{-6})}{\dot{m}} \text{ mm}^{-4}$$

The air contracts in order to exit the unit through the openings A_5. The contraction area ratio from Table 4.5 is

$$A_5/A_4 = 0.3/0.8 = 194/516 = 0.375$$

The loss factor from (4.77b), following Figure 4.31 (XI) which gives $K = 0.41$ (interpolated), is

$$R_{11d} = \frac{\sqrt{K}}{A_5} = \frac{\sqrt{0.41}}{0.3} = 2.13 \text{ in}^{-2}$$

$$= \frac{\sqrt{0.41}}{194} = 0.0033 \text{ mm}^{-2}$$

Finally, the air expands after leaving the openings A_5. From Figure 4.31 (X), $K = 1$ so that the loss factor becomes

$$R_{11e} = \frac{\sqrt{K}}{A_5} = 3.33 \text{ in}^{-2}$$

$$= 0.0052 \text{ mm}^{-2}$$

From (4.81) the total effective resistance for the 11W card is

$$R_{T(11)}^2 = \sum_1^5 R_i^2 = R_{11a}^2 + R_{11b}^2 + R_{11c}^2 + R_{11d}^2 + R_{11e}^2$$

$$R_{T(11)}^2 = 25.54 + \frac{0.3}{\dot{m}} \text{ in}^{-4} \qquad (4.84)$$

$$= 6.113(10^{-5}) + \frac{5.44(10^{-6})}{\dot{m}} \text{ mm}^{-4}$$

The pressure loss from (4.77) is

$$\Delta P_{11} = 0.2287\dot{m}^2\left(25.54 + \frac{0.3}{\dot{m}}\right) \text{ in } H_2O$$

$$= 42534\dot{m}^2\left[6.113(10^{-5}) + \frac{5.44(10^{-6})}{\dot{m}}\right] \text{ mm } H_2O$$

For the 11W card flow rate of 0.0349 lb/min (0.26 g/s), the pressure drop becomes

$$\Delta P_{11} = 0.0095 \text{ in } H_2O$$

$$= 0.236 \text{ mm } H_2O$$

Except for R_{11a}^2 and R_{11b}^2, (4.84) describes the resistance for the 7W card since the card spacing and exit configuration are the same. With the substitution of R_{7a} and R_{7b} for R_{11a} and R_{11b}, (4.84) for the 7W card becomes

$$R_{T(7)}^2 = R_{7a}^2 + R_{7b}^2 + 15.644 + \frac{0.3}{\dot{m}} \text{ in}^{-4}$$

$$= R_{7a}^2 + R_{7b}^2 + 3.793(10^{-5}) + \frac{5.44(10^{-6})}{\dot{m}} \text{ mm}^{-4} \qquad (4.85)$$

Based on (4.78a), $\Delta P_{11} = \Delta P_7$, which gives

$$0.0095 = 0.2287\dot{m}^2\left(R_{7a}^2 + R_{7b}^2 + 15.644 + \frac{0.3}{\dot{m}}\right) \qquad (4.86a)$$

or in metric units

$$0.236 = 42534\dot{m}^2\left[R_{7a}^2 + R_{7b}^2 + 3.793(10^{-5}) + \frac{5.44(10^{-6})}{\dot{m}}\right] \qquad (4.86b)$$

The solution of (4.86) for $R_{7a}^2 + R_{7b}^2$ using the flow rate for the 7W card gives

$$R_{7a}^2 + R_{7b}^2 = 67.7 \text{ in}^{-4} \tag{4.87a}$$

$$= 1.55(10^{-4}) \text{ mm}^{-4} \tag{4.87b}$$

Using (4.82), (4.83) and (4.87), the value of the plenum discharge opening for the 7W card, A_3 is determined:

$$\left(\frac{\sqrt{0.5}}{A_3}\right)^2 + \left(\frac{1}{A_3} - \frac{1}{0.8}\right)^2 = 67.7 \tag{4.88}$$

The plenum discharge area for the 7W card is obtained from (4.88) as $A_3 = 0.14$ in (3.6 mm) since $K = 0.5$ remains essentially unchanged. Discharge openings should be oblong cutouts to avoid trapping particles, dust and lint. The value developed for the discharge openings should be verified empirically.

(b) *Pressure Drop for the Electronics Unit.* Using the flow coefficient for the squared opening, Figure 4.31 (IX), the entrance loss coefficient is computed:

$$R_a = \frac{\sqrt{0.5}}{A_1} = \frac{\sqrt{0.5}}{0.78} = 0.91 \text{ in}^{-2}$$

$$= \frac{\sqrt{0.5}}{506} = 0.0014 \text{ mm}^{-2}$$

The loss factor for the air expanding out of the inlet opening is found using Figure 4.31(X) as

$$R_b = \frac{1}{A_1} - \frac{1}{A_2} = \frac{1}{0.78} - \frac{1}{3.5} = 0.99 \text{ in}^{-2}$$

$$= \frac{1}{506} - \frac{1}{2258} = 0.0015 \text{ mm}^{-2}$$

From (4.81), the total effective entrance resistance is

$$R_T^2 = 0.91^2 + 0.99^2 = 1.81 \text{ in}^{-4}$$

$$= 0.0014^2 + 0.0015^2 = 4.21(10^{-6}) \text{ mm}^{-4}$$

Using (4.77) for the entrance pressure drop and adding the parallel loss determined in (a) give the equipment flow loss as

$$\Delta P = 0.2287(1.81)(1.87^2) + 0.0095 = 0.024 \text{ in } H_2O$$

$$= 42534(4.21 \times 10^{-6})(1.41^2) + 0.236 = 0.47 \text{ mm } H_2O \qquad \text{Q.E.D.}$$

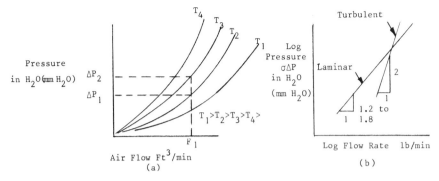

Figure 4.34. Flow Characteristics of an Assembly. (a) Equipment pressure loss; (b) Normalized equipment pressure loss.

Effect of Temperature and Altitude on Pressure Drop. Increasing temperature or altitude, or both, reduces the density of the cooling air flow so that a higher velocity is required to maintain constant mass flow. An increasing velocity also increases the pressure loss through the unit. The effect of temperature and altitude on equipment pressure losses should be assessed to assure that adequate flow is provided to prevent overheating and performance loss. Equipment pressure loss, Figure 4.34(a), can be normalized, Figure 4.34(b), by relating it to the density of air at a standard condition. Standard conditions for military and commercial equipment are given in Table 4.6.

Consider the flow impedance curve of Figure 4.34, developed for an electronics unit for several average cooling air temperatures at a particular altitude.

At an average cooling air temperature T_2, flow F_1 of Figure 4.34 develops a pressure loss ΔP_1. A reduction of the inlet temperature, which could be the result of a different environment, will increase the losses through the unit. Figure 4.34 also reflects the increased flow relative to inlet conditions due to heating along the flow path.

Curves of Figure 4.34 may be constructed at a specific ambient pressure using (4.89a) for constant temperature through the unit or (4.89b) for constant pressure loss based on the weight flow through the unit.

Table 4.6. Standard Air

Application	Temperature °F	°C	Pressure lb/in²	MPa	Density lb/ft³	kg/m³
Commercial	68	20	14.7	0.102	0.0752	1.204
Military	59	15	14.7	0.102	0.0765	1.225

$$\left(\frac{F_1}{F_2}\right)^n = \frac{\Delta P_1}{\Delta P_2} \qquad P, T \text{ const} \tag{4.89a}$$

$$\frac{F_1}{F_2} = \frac{T_1}{T_2} \qquad P, \Delta P \text{ const} \tag{4.89b}$$

where ΔP_i = unit pressure loss;
P = ambient pressure;
T_i = absolute temperature;
F_i = volumetric flow;

$n \begin{cases} 2 = \text{fully turbulent flow} \\ 1.4 = \text{flow between average spaced circuit boards} \\ 1.2 = \text{flow between widely spaced circuit boards.} \end{cases}$

A family of curves similar to Figure 4.34(a) can be constructed for varying temperature or pressure (altitude). If the pressure loss and inlet and exhaust temperatures for a specific altitude are known, the pressure loss can be determined for the same flow rate at different conditions using

$$\Delta P_2 = -P_2 + \sqrt{P_2^2 + 2\overline{P}_1 \, \Delta P_1 \left(\frac{T_2}{\overline{T}_1}\right)} \tag{4.90}$$

given

$\overline{P}_1$ = average pressure loss through unit = $P_1 + \Delta P_1/2$;

$\overline{T}_i$ = average cooling air temperature through unit.

EXAMPLE 4.11. Forced air entering a 120W unit at 70°F (21.1°C) exits at 150°F (65.6°C). A 0.8 in H_2O (20.3 mm H_2O) pressure drop was measured at sea level for these conditions. Determine the pressure loss at 20,000 ft (6.1 km) with an inlet temperature of 50°F (10°C) for (a) the same flow rate, and (b) a flow rate of 0.27 lb/min (2g/s) assuming fully turbulent flow.

(a) *Pressure Loss for an Inlet of 50° F (65.6° C).* The average pressure through the unit at sea level is

$$\overline{P}_1 = \frac{14.7 \text{ lb/in}^2}{0.036 \text{ lb/in}^2 \cdot \text{in } H_2O} + \frac{0.8 \text{ in } H_2O}{2} = 408.73 \text{ in } H_2O$$

which in metric units becomes

$$\overline{P}_1 = \frac{1.013(10^5) \text{ Pa}}{9.806 \text{ Pa/mm H}_2\text{O}} + \frac{20.3}{2} = 10340 \text{ mm H}_2\text{O}$$

The average cooling air temperature through the unit at sea level is

$$\overline{T}_1 = \left(\frac{70 + 150}{2}\right) + 460 = 570\,^\circ\text{R} \ (316.3\,^\circ\text{K})$$

Since the temperature rise at altitude is the same as that at sea level for the same flow rate,

$$\overline{T}_2 = 50 + 80/2 + 460 = 550\,^\circ\text{R} \ (305.2\,^\circ\text{K})$$

Based on the atmosphere pressure ratio from Appendix C the pressure at 20,000 ft (6.1 km) becomes

$$P_2 = \frac{0.4599(14.7)}{0.036} = 187.79 \text{ in H}_2\text{O}$$

$$= \frac{0.4599(1.013)(10^5)}{9.806} = 4751 \text{ mm H}_2\text{O}$$

Substituting into (4.90)

$$\Delta P_2 = -187.79 + \sqrt{187.79^2 + 2(408.73)(0.8)\frac{550}{570}}$$

$$\Delta P_2 = 1.67 \text{ in H}_2\text{O}$$

In metric units

$$\Delta P_2 = -4751 + \sqrt{4751^2 + 2(10340)(20.3)\left(\frac{305.2}{316.3}\right)}$$

$$\Delta P_2 = 42.0 \text{ mm H}_2\text{O}$$

(b) *Pressure Loss for an Inlet of 50°F (65.6°C) and* $\dot{m} = 0.27$ *lb/min (2g/s)*. Based on (5.35), the temperature difference through the unit becomes

$$\Delta T = \frac{0.237(120)}{0.27} = 105.33\,^\circ\text{F}$$

$$= \frac{0.996(120)}{2} = 59.8\,^\circ\text{C}$$

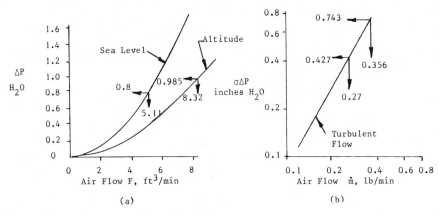

Figure 4.35. Effect of Altitude on Flow Characteristics. (a) Equipment pressure loss; (b) Normalized equipment pressure loss. (Vertical scale = inches H_2O)

Following (4.89a), the pressure loss at 0.27 lb/min (2g/s) for fully turbulent flow at sea level becomes

$$\Delta P = 0.8 \left(\frac{0.27}{0.356} \right)^2 = 0.46 \text{ in } H_2O$$

$$= 20.3 \left(\frac{2}{2.69} \right)^2 = 11.22 \text{ mm } H_2O$$

The average pressure through the unit at sea level becomes

$$\overline{P}_1 = 408.33 + 0.46/2 = 408.46 \text{ in } H_2O$$

$$= 10340 + 11 = 10351 \text{ mm } H^2O$$

Substituting into (4.90)

$$\Delta P_2 = -187.79 + \sqrt{187.79^2 + 2(408.46)0.46 \left(\frac{562.7}{570} \right)}$$

$$\Delta P_2 = 0.985 \text{ in } H_2O$$

Figure 4.35(a) illustrates the unit loss characteristics at sea level and altitude.

Q.E.D.

The volumetric flow rate corresponding to the unit pressure loss at altitude is determined from

$$F = \dot{m}/\rho \tag{4.91}$$

where the density is

$$\rho = \frac{\overline{P}}{R\overline{T}} = 0.0975 \frac{\overline{P}}{\overline{T}} \, \text{lb/ft}^3 \tag{4.92a}$$

$$= 0.342 \frac{\overline{P}}{\overline{T}} \, \text{kg/m}^3 \tag{4.92b}$$

given

$R =$ universal gas constant, 53.3 ft/°R (29.24 m/°K)

$\overline{P} =$ average pressure loss through unit, in H_2O (mm H_2O)

$\overline{T} =$ average temperature through unit, °R (°K)

The density of cooling air flowing through a heat dissipating unit decreases due to its temperature rise. This causes a corresponding increase in volumetric flow rate since the weight flow is constant through the unit. Increasing altitude also results in decreasing air densities. In order to maintain the same weight flow, the volumetric flow rate must increase. Increasing volumetric flow increases pressure losses through the unit per (4.89).

Determination of unit pressure loss can be simplified by normalizing unit curves for different conditions of Figure 4.34(a), to a simple curve of Figure 4.34(b) representing all air density values for either laminar or turbulent flow. The normalization parameter sigma (σ) is defined as

$$\sigma = \frac{\text{air density at any condition}}{\text{air density at standard conditions}} = \frac{\rho}{\rho_{\text{std}}} \tag{4.93}$$

for the standard conditions depicted in Table 4.6. The loss for nonoperating equipment is $\sigma \, \Delta P$ given the air temperature through the unit remains constant at standard conditions for a known weight flow. This loss, $\sigma \, \Delta P$, can be computed by multiplying the loss ΔP at any condition by σ from (4.93). The $\sigma \, \Delta P$ value is plotted on the coordinates of Figure 4.34(b). A straight line with slope n is drawn through $\sigma \, \Delta P$ for laminar or turbulent unit flow characteristics as defined by (4.89). For any flow rate, $\sigma \, \Delta P$ values are determind from (4.89a) as

$$(\sigma \, \Delta P)_i = (\sigma \, \Delta P)_j \left(\frac{\dot{m}_i}{\dot{m}_j} \right)^n \tag{4.94}$$

The single $\sigma \, \Delta P$ curve is a flow-loss characteristic of the forced cooled electronic unit. The pressure loss for any altitude or flow condition may be determined from the $\sigma \, \Delta P$ curve using

$$\Delta P_2 = \frac{(\sigma \, \Delta P)_2}{\sigma_2} \qquad (4.95)$$

In (4.95), $(\sigma \, \Delta P)_2$ is a value from the $\sigma \, \Delta P$ curve for a flow rate $\dot{m}_2$ corresponding to the condition being evaluated. The value σ_2 corresponding to the flow condition may be determined from

$$\sigma_2 = B \left(P_2 + \sqrt{P_2^2 + \frac{(\sigma \, \Delta P)_2}{B}} \, \right) \qquad (4.96)$$

where $B = 0.0487 / \overline{T}_2 \rho_{std}$ (English)

$B = 0.0171 / \overline{T}_2 \rho_{std}$ (metric)

In (4.96), the units of measure are consistent with (4.92) and the units of $(\sigma \, \Delta P)_2$ are inches H_2O (mm H_2).

Equations (4.90) and (4.96) account for average pressure and temperature conditions within the units. This is an important consideration for equipment with large pressure losses or operating at altitude. Cooling system designs based on either inlet or exhaust conditions may provide inadequate heat removal in actual hardware, leading to the development of excessive operating temperatures that can lead to performance loss and premature failure.

EXAMPLE 4.12. Solve Example 4.11 using a normalized $\sigma \, \Delta P$ curve based on the given data for commercial equipment.

(a) *Pressure Loss for an Inlet of 50° F (65.6° C)*. The sea level air density for average conditions through the unit is determined from (4.92) as

$$\rho = 0.0975 \left(\frac{408.73}{570} \right) = 0.0699 \text{ lb/ft}^3 \qquad (4.97a)$$

which in metric units becomes

$$\rho = 0.0342 \left(\frac{10340}{316.3} \right) = 1.118 \text{ kg/m}^3 \qquad (4.97b)$$

From Table 4-6, the value of sigma using (4.93) and (4.97) is

$$\sigma = \frac{0.0699}{0.0752} = \frac{1.118}{1.204} = 0.929$$

The value of $\sigma \, \Delta P$ becomes

$$\sigma \, \Delta P = 0.929(0.8) \quad = 0.743 \text{ in } H_2O \tag{4.98a}$$

$$= 0.929(20.3) = 18.86 \text{ mm } H_2O \tag{4.98b}$$

This point is plotted in Figure 4.35(b) at the corresponding weight flow determined from (4.35). The $\sigma \, \Delta P$ at altitude is given by (4.98) since the flow rate has not changed. The sigma at altitude is determined from (4.96) with

$$B = 0.0487/(550)(0.0752)$$

$$= 0.001177 \tag{4.99a}$$

$$\sigma = 0.001177 \left[187.79 + \sqrt{187.79^2 + \frac{0.743}{0.001177}} \right]$$

$$= 0.444$$

In metric units

$$B = 0.0171/(305.2)(1.204) = 4.65(10^{-5})$$

$$\sigma = 4.65(10^{-5}) \left[4751 + \sqrt{4751^2 + \frac{18.86}{4.65(10^{-5})}} \right] \tag{4.99b}$$

$$= 0.444$$

The pressure loss at altitude is given by (4.95), using (4.98) and (4.99), as

$$\Delta P_2 = \frac{0.743}{0.444} = 1.67 \text{ in } H_2O$$

$$= \frac{18.86}{0.444} = 42.5 \text{ mm } H_2O$$

(b) *Pressure Loss for Air Inlet of 50° F (65.6° C) and $\dot{m} = 0.27 \, lb/min \, (2g/s)$.* The $\sigma \, \Delta P$ corresponding to the given flow rate is found from (4.94) for turbulent conditions, using (4.98), as

$$\sigma \, \Delta P = 0.743 \left(\frac{0.27}{0.356} \right)^2 = 0.427 \text{ in } H_2O \tag{4.100a}$$

$$= 18.86 \left(\frac{2}{2.69} \right)^2 = 10.42 \text{ mm } H_2O \tag{4.100b}$$

This quantity in addition to that from (4.98) defines the unit pressure loss curve for turbulent flow shown on Figure 4.35(b). By means of (4.96) and (4.100), the sigma corresponding to the flow conditions at altitude becomes

$$B = 0.0487/(562.7)(0.0752) = 0.00115$$

$$\sigma = 0.00115 \left[187.79 + \sqrt{187.79^2 + \frac{0.427}{0.00115}} \right] \qquad (4.101a)$$

$$= 0.433$$

In metric units

$$B = 0.0171/(312.6)(1.204) = 4.54(10^{-5})$$

$$\sigma = 4.54(10^{-5}) \left[4751 + \sqrt{4751^2 + \frac{10.42}{4.54(10^{-5})}} \right] \qquad (4.101b)$$

$$= 0.432$$

The pressure loss is given by (4.95), using (4.100) and (4.101), as

$$\Delta P_2 = \frac{0.427}{0.433} = 0.986 \text{ in } H_2O$$

$$= \frac{10.42}{0.432} = 24.12 \text{ mm } H_2O \qquad \text{Q.E.D.}$$

Fan Application Considerations.   Fan performance must match the equipment's flow requirements. The matching condition occurs at point a of Figure 4.36 where the equipment and fan curves intersect. The pressure developed by

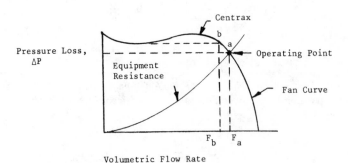

Figure 4.36. Fan and Equipment Performance Matching.

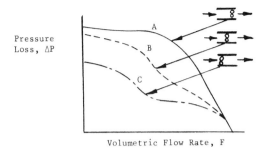

Pressure
Loss, ΔP

Volumetric Flow Rate, F

Figure 4.37. Entrance and Exit Duct Effect on Fan Performance.

the fan at these points exactly matches the equipment resistance and the equipment flow equals the fan capacity. If the flow rate at this operating point does not meet design requirements, either the fan or the loss characteristic of the equipment must be changed.

A fan may be used to blow or draw air through the unit. A blowing configuration maintains a higher than ambient pressure through the unit, producing more turbulence and a greater weight flow. Turbulence increases heat transfer from surfaces with greater pressure losses. The heat dissipated by the fan motor must be added to the equipment dissipation in a blowing system.

A configuration that pulls the air through the equipment reduces the pressure in the equipment below ambient. The lower than ambient pressure can suck dust, dirt and other contaminants into the equipment through unintentional openings that may also short-circuit planned flow routes.

Fan performance is enhanced if the velocity profile is fully developed prior to entering the fan. This is facilitated during equipment design by providing a duct or housing on the upstream end of the fan, curve A of Figure 4.37. If the duct projects on both the up and downstream ends of the fan curve B, the available pressure is lower and some instability could occur. Irregular performance and pronounced instability occur for ducts extended only on the downstream end of the fan, curve C of Figure 4.37. Instability also occurs at flows less than F_b on Figure 4.36. The operating point should always be to the right of F_b to avoid stalling and overheating of equipment and parts.

Multiple application of fans may be used to increase pressure capacity for the same flow rate (series deployment) or increase flow rate for the same pressure capacity (parallel deployment). For a series system, the combined fan curve may be developed by adding the fan pressures at each flow for the two fans, Figure 4.38(a). Although the available capacity is doubled, actual capacity through equipment is not. Fans used in series should be mounted a sufficient distance apart to allow a redevelopment of the velocity profile.

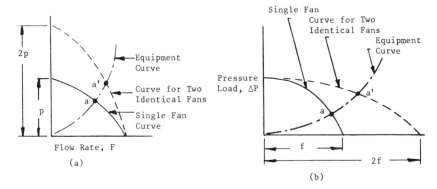

Figure 4.38. Multiple Application of Fans. (a) Operating characteristic for fans in series; (b) Operating characteristics for fans in parallel.

When fans are connected in parallel, the sum of the flow rates through each fan at the same pressure is used to develop the fan curve, Figure 4.38(b). Although the available capacity is doubled, actual capacity through equipment is not. A parallel system is more efficient for low loss equipment since the available flow and pressure capacities are greater. On the other hand, a series system is more efficient for high loss equipment.

Constant speed fans deliver approximately the same volumetric flow at any altitude but with a pressure capacity diminishing in proportion to σ, from (4.93). The temperature rise through the equipment will increase due to a corresponding reduction in the weight flow of air.

EXAMPLE 4.13.   A fan with the characteristics listed in Table 4.7 has been selected to cool the unit of Example 4.7 having the pressure drop characteristics developed in Example 4.10. Determine (a) the sea level weight flow and the average unit exhaust temperature for a blowing configuration with a fan dissipation of 15W, (b) the maximum component case temperature for the 7W cards, and (c) the cooling weight flow at 20 kft (6.1 km) and the required inlet air temperature to obtain the sea level exhaust temperature determined in (a).

Table 4.7. Fan Characteristics

Volume Flow		Pressure	
ft^3/min	cm^3/s	in H$_2$O	mm H$_2$O
0	0	0.13	3.3
2	944	0.10	2.5
4	1888	0.05	1.3
5.4	2549	0.0	0.0

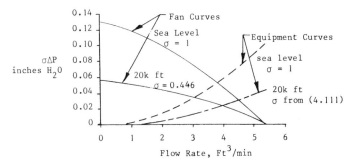

Figure 4.39. Unit Operating Characteristics.

(a) *Sea Level Weight Flow and Average Unit Exhaust Temperature.* The equipment pressure drop computed at commercial standard condition in Example 4.10 may be expressed as

$$\Delta P = 0.598\dot{m}(\dot{m} + 0.0215) \text{ in } H_2O \qquad (4.102a)$$

$$= 0.259\dot{m}(\dot{m} + 0.167) \text{ mm } H_2O \qquad (4.102b)$$

using the card-to-unit weight flow ratio developed in Example 4.7. Equations (4.102) may be expressed in terms of the volumetric flow rate using the characteristics of Table 4.6. The resulting equipment resistance and fan curve are shown graphically in Figure 4.39.

The volumetric flow rate at sea level is given by Figure 4.39 as 3.8 ft^3/min (1793 cm^3/s). Based on standard air conditions of Table 4.6, the weight flow becomes

$$\dot{m} = \rho F = 0.0752(3.8) = 0.29 \text{ lb/min} \qquad (4.103a)$$

$$= 1204 \frac{g}{m^3} \left(\frac{1793 \ m^3}{10^6 \ s} \right) = 2.16 \text{ g/s} \qquad (4.103b)$$

The average exhaust temperature accounting for the fan dissipation, using (4.53), becomes

$$T_{\text{out}} = 30 + \frac{0.1317(15 + 61)}{(0.29)} = 64.5°C \qquad (4.104a)$$

In metric units

$$T_{\text{out}} = 30 + \frac{0.996(15 + 61)}{2.16} = 65.0°C \qquad (4.104b)$$

(b) *Maximum Component Case Temperature for the 7W Cards.* The fraction of the unit weight flow delivered to the 7W cards is determined from the results of Example 4.7 as

$$\dot{m}_7 = 0.11\dot{m}_{unit} \tag{4.105}$$

Based on (4.103) and (4.105), the 7W card weight flow becomes

$$\dot{m}_7 = 0.11(0.29) = 0.0319 \text{ lb/min} \tag{4.106a}$$

$$= 0.11(2.16) = 0.237 \text{ g/s} \tag{4.106b}$$

Parameter a of Figure 4.19 is obtained from (4.38), using (4.106), as

$$a = 0.0137\left(\frac{0.0319}{0.036}\right) = 0.0121$$

The heat transfer coefficient from Figure 4-19 becomes

$$h_c = \frac{1.02(3.8)k}{b} = \frac{1.02(3.8)7.47(10^{-4})}{0.1}$$

$$h_c = 0.0298 \frac{W}{in^2 \cdot {}^\circ C}$$

The convective film temperature difference is determined from (4.39) as

$$\Delta T = \frac{7}{0.0289(6)(8)} = 5.0\,^\circ C \tag{4.107}$$

Using (4.104) and (4.107) the maximum component case temperature becomes

$$T_{component} = 65 + 5 = 70\,^\circ C \text{ (max)}$$

(c) *Altitude Weight Flow and Inlet Air Temperature to Achieve Comparable Maximum Component Temperatures.* Based on the pressure ratio from Appendix C, the density at 20 kft (6.1 km) for an inlet of $30\,^\circ C$ ($86\,^\circ F$) from (4.92) becomes

$$\rho = \frac{0.0975(0.4599)\left(\dfrac{14.7}{0.036}\right)}{(86 + 460)} = 0.0335 \text{ lb/ft}^3 \tag{4.108a}$$

$$= \frac{0.0342(0.4599)(10371)}{(30 + 273)} = 0.538 \text{ kg/m}^3$$

The proportionality constant for fan performance at 20 kft is determined from (4.93), using Table 4.6 and (4.108), as

$$\sigma = \frac{0.0335}{0.0752} = \frac{0.538}{1.204} = 0.446 \tag{4.109}$$

Pressure characteristics for fan performance at sea level, Table 4.7, are multiplied by (4.109) to obtain the delivery capacity at 20 kft (6.1 km). The results are plotted in Figure 4.39.

The density of the air flowing through the unit at 20 kft (6.1 km) accounting for the average pressure and temperature within the unit is determined from (4.92), using (4.35), (4.102) and (4.103), as

$$\rho = 0.0975 \left[\frac{P_2 + \dfrac{0.598}{2} \rho F(\rho F + 0.0215)}{T_{in} + \dfrac{0.237Q}{2\rho F}} \right] \text{lb/ft}^3$$

Solving for ρ by eliminating negligible terms gives

$$\rho = 0.0975 \frac{(P_2 - 1.215Q/F)}{T_{in}} \text{lb/ft}^3 \tag{4.110}$$

The proportionality factor for the equipment pressure loss at altitude is determined using (4.93) and Table 4.6 for commercial standard air as

$$\sigma = \frac{1.297(P_2 - 1.215Q/F)}{T_{in}} \tag{4.111a}$$

which in metric units becomes

$$\sigma = \frac{0.0284(P_2 - 0.0146Q/F)}{T_{in}} \tag{4.111b}$$

given P_2 = ambient pressure, in H_2O (mm H_2O);

Q = unit power dissipation, watts;

F = flow rate, ft^3/min (m^3/s);

T_{in} = inlet temperature, $^\circ R$ ($^\circ K$).

Equation (4.110) applies specifically to the unit considered in this example since the unit pressure loss characteristics of (4.102) is used in their development. The outlined procedure for deriving (4.111) for any application, however, would be the same.

The equipment loss characteristics at altitude are determined from the product of (4.102) and (4.111) for different flow rates. The results are shown graphically in Figure 4.39. A volumetric flow rate of 3.9 ft^3/min through the unit at 20 kft (6.1 km) is determined from the intersection of the equipment and fan curves in Figure 4.39. Based on (4.103) and (4.110), the weight flow becomes

$$\dot{m} = F\rho = \frac{3.9(0.0975)[187.79 - 1.215(61)/3.9]}{(460 + 86)}$$

$$\dot{m} = 0.118 \text{ lb/min}$$

The exhaust temperature at altitude using this weight flow, which corresponds to the sea level inlet temperature of 30°C, follows (4.104) as

$$T_{out} = 30 + \frac{0.1317(15 + 61)}{0.118} = 114.8°C$$

As the inlet air temperature is decreased in order to achieve sea level exhaust values, the density through the unit increases developing a weight flow increase. The inlet air temperature accounting for this variation using (4.110) in the format of (4.104) becomes

$$T_{out} - T_{in} = \frac{0.1317(76)}{F\rho}$$

$$= \frac{0.1317(76)1.8(273 + T_{in})}{F(0.0975)\left[187.79 - \dfrac{1.215(61)}{F}\right]}$$

The solution for T_{in} based on the sea level exhaust $T_{out} = 64.5°C$ is

$$T_{in} = \frac{12112F - 55227}{110.7 + 187.8F} °C$$

The inlet air temperature which develops the sea level exhaust temperature is determined using $F = 3.9$ ft^3/min from Figure 4.39 as

$$T_{in} = -9.5°C$$

The average air density through the unit for this inlet temperature based on (4.110) is

$$\rho = \frac{0.0975\left[187.79 - \dfrac{1.215(61)}{3.9}\right]}{1.8(273 - 9.5)} \tag{4.112}$$

$$= 0.0347 \text{ lb/ft}^3$$

This gives a weight flow through the unit at altitude of

$$\dot{m} = \rho F = 0.0347(3.9) = 0.135 \text{ lb/min}$$

Actually, the proportionality factor based on (4.112) using (4.111) generates a new equipment resistance curve on Figure 4.39 which affects the flow rate at the fan curve intersection. Since the flow, F, indicated by the intersection is slowly changing, the value used and the results obtained are representative of the unit characteristics at altitude.   Q.E.D.

4.4 COMBINED CONVECTION WITH RADIATION

Radiation, which depends on configuration, orientation and surrounding surface temperatures, often occurs simultaneously with the other heat transfer modes. The nature of the configuration and the unknowns under study can rapidly increase the problem complexity and the effort to derive a solution.

The simplest situation involves individual independent contributions which can be determined separately and summed to obtain the result desired. Often the unknown sought is the heat dissipation for known component or surface temperatures.

EXAMPLE 4.14.   Determine the heat dissipated at sea level by the electronics assembly of Example 4.4 if the three dissipating surfaces which are at $60°C$ also radiate to surrounding compartment walls at $40°C$ having an emissivity of 0.8.

The heat dissipated by natural convection at sea level is determined using (4.26) and Table 4.2 which becomes

$$Q_1 = 0.00414 \sum_{1}^{3} \frac{\Delta T^{1.25}}{Z_i}$$

$$= 0.00414 \left(\frac{1}{0.127} + \frac{1}{0.114} + \frac{1}{0.084} \right) (60 - 40)^{1.25} \qquad (4.113)$$

$$Q_1 = 5\text{W}$$

The heat dissipated by radiation for the configuration of Table 4.1(III) using (4.23) and the total surface area of 74 in^2 from Table 4.2 as

$$Q_2 = 3.485(10^{-12})(0.8)(74)[\,(140 + 460)^4 - (104 + 460)^4] \qquad (4.114)$$

$$Q_2 = 5.9\text{W}$$

The heat dissipated by the unit is the total of the individual dissipations, giving

$$Q_1 = Q_1 + Q_2 = 10.9\text{W}$$

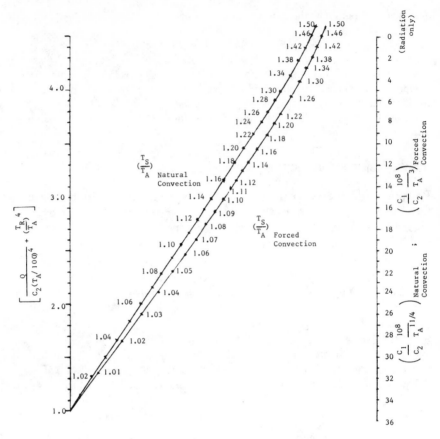

Figure 4.40. Combined Radiation and Convection Heat Transfer.

Table 4-8. Dimensional Relationship for N_1 and N_2

		Dimensional Group			
Equation (4.115) Variable		I	II	III	IV
Q	Dissipated Heat	Btu/hr	W	W	W
L_c	Characteristic Length	ft	in	in	in
A	Area	ft^2	in^2	in^2	mm^2
T	Absolute Temperature	°R	°R	°K	°K
N_1	Natural Convection; $\alpha = 1.25$	0.524	1.986(10^{-3})	4.14(10^{-3})	1.4406(10^{-5})
N_2	Radiation	0.1713	3.485(10^{-4})	36.8(10^{-4})	5.704(10^{-6})

Radiation has increased the power dissipating capacity of the unit 118 percent for the same surface temperatures. At higher altitudes, the contribution due to radiation is even greater. Q.E.D.

Usually the surface temperatures of units, assemblies or components are not generally known. Temperatures of the ambient or surrounding surfaces which are dependent upon the equipment application or deployment are known or have been specified. The problem is to determine the equipment and part temperatures knowing the distribution and magnitude of the dissipated heat. The determination of surface temperatures under these conditions involves coupled nonlinear temperature relationships. For many applications, these relationships can be solved with the aid of the nomogram of Figure 4.40 which applies to the expression

$$Q = C_1(T_s - T_A)^\alpha + C_2\left[\left(\frac{T_s}{100}\right)^4 - \left(\frac{T_R}{100}\right)^4\right] \qquad (4.115)$$

For natural convections, $\alpha = 1.25$ with C_1 determined from (4.26) for parallel systems, (4.27) for series systems and (4.28) for series parallel systems. The radiation coefficient is $C_2 = N_2 A \epsilon F$ where N_2 depends on the system of units used. A similar numeric parameter applies to C_1. Values of N_1 and N_2 are given for different units in Table 4.8.

EXAMPLE 4.15. Determine the surface temperature of the three dissipating surfaces of the electronics assembly of Example 4.4 when cooled by natural convection and radiation at sea level conditions. The surrounding compartment walls, which have an emissivity of 0.8, are at 40°C.

The value of C_1 is easily determined from (4.113) as

$$C_1 = 0.00414\left(\frac{1}{0.127} + \frac{1}{0.114} + \frac{1}{0.084}\right) = 0.118$$

where $N_1 = 0.00414$ corresponding to Case III of Table 4.8. Based on Case III with the same dimensions, $C_2 = N_2 A \epsilon F = 36.8(10^{-4})(74)(0.8) = 0.218$.

Parameters of Figure 4.40 with

$$T_A = \frac{(40 + 60)}{2} = 50°C (323°K)$$

and

$$T_R = 40°C (313°K)$$

become

$$\left(\frac{10^8 C_1}{T_A^{11/4} C_2}\right) = \frac{0.118(10^8)}{(273 + 50)^{11/4} 0.218} = 6.81$$

$$\left[\frac{Q}{C_2(T_A/100)^4} + \left(\frac{T_R}{T_A}\right)^4\right] = \left[\frac{5}{0.218\left(\frac{323}{100}\right)^4} + \left(\frac{313}{323}\right)^4\right] = 1.09$$

From Figure 4.40

$$\frac{T_s}{T_A} = 1.015$$

The surface temperature becomes

$$T_s = 1.015 T_A - 273 = 1.015(323) - 273$$

$$T_s = 54.8°C$$

The process is repeated using an updated value of T_A until the change in T_s becomes negligible. This gives a surface temperature of 53°C, which is 11.7 percent lower than that obtained without radiation.   Q.E.D.

Figure 4.40 also applies when forced convection is combined with radiation. In this case, $\alpha = 1$ and $C_1 = h_c A$ from (2.2); the units used for the heat transfer coefficient h_c are consistent with other parameters of (4.115) per Table 4.8.

5

Thermal Elastic Effects

5.1 BASIC ASPECTS OF THERMAL ELASTIC DESIGN

Temperature variations due to heating and cooling during equipment operation or changes in the environment can cause high mechanical stresses which reduce the useful life of equipment. Similar conditions which can induce temperature related mechanical stress occur during fabrication, testing, shipment and storage. These stresses are developed on assemblies or at attachments where materials exhibiting different thermal expansion coefficients are used. The effects of stress caused by differential expansion usually emerge as cracked solder joints, cracked interface material of bonded or joined surfaces, and fractured or pulled component leads.

Although a variety of environments such as shock, vibration, and so forth, may induce failures of this type, they are primarily caused by thermal fatigue. Heating and cooling cycles, which are a consequence of normal equipment operation or of variations in ambient temperature, cause alternating expansion and contraction processes proportional to the temperature difference relative to the original temperature at assembly. Sometimes only a few cycles are sufficient to cause failure of a solder joint, fastened connection or electrical interconnection. These failures are not always catastrophic but emerge as intermittencies, errors or variations in performance due to increased or erratic resistance changes.

The magnitude of the forces generated and the degree to which these forces are transmitted in an assembly are dependent upon the specific geometry and the physical and mechanical properties of the materials involved. The forces which develop in a solder joint, for example, are due primarily to the differences among the linear coefficients of expansion for the solder alloy, component lead material and printed circuit board. Other factors which influence the magnitude and development of these forces are conformal coatings or encapsulants, component attachments and lead configuration. Proper selection of geometry, materials and component lead strain reliefs will reduce and uncouple the magnitude of the destructive forces. Similar considerations apply to any joined assembly whether it be bonded or bolted.

The evaluation of thermal elastic effects requires a fundamental understanding of materials, structures, heat transfer and the effect of the thermal environment on the system. The temperature fluctuations of an assembly may be reduced by examining and improving upon local and system heat transfer characteristics. Mounting and attachment methods must include the effects of temperature as well as inertial forces due to vibration and shock.

Material Considerations.   Materials exhibit both short and long term effects due to the thermal environment. Short term effects include the variation of material properties with temperature. Since both epoxy-glass printed circuit board material and low temperature eutectic solder (63/37) comprise a major part of electronic packages, each is examined in Figure 5.1.

At relatively moderate operating temperatures, Figure 5.1(a) shows a large reduction in the mechanical strength of solder. Consequently, analyses should use values at approximately the operating temperature of the assembly when considering loads under those conditions. The variation of the coefficient of expansion of expoxy fiberglass normal to the board surface must also be accounted for to avoid failures at operating temperatures. Figure 5.1(b) shows the effect on the coefficient of expansion as the material undergoes a transition from a glassy to a softening rubbery state at temperatures above 240°F (130°C).

Long term effects include stress relaxation, creep and plastic flow. Stress relaxation and creep occur in most plastic and metallic materials and are most obvious in low-melting-point alloys. These effects relieve some of the stresses developed during fabrication. The degree of stress relaxation is dependent upon

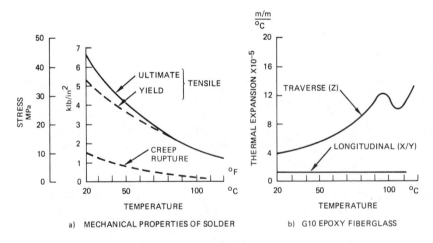

a)   MECHANICAL PROPERTIES OF SOLDER   b)   G10 EPOXY FIBERGLASS

Figure 5.1. Temperature Effect of Two Circuit Board Materials.

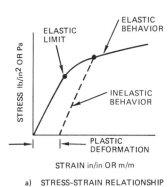

a) STRESS-STRAIN RELATIONSHIP

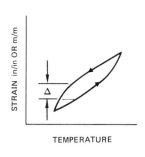

b) DEFLECTION-TEMPERATURE RELATIONSHIPS

Figure 5.2. Load-Deformation Characteristics.

the magnitude of the initial stress and the temperature of the assembly. A soldered joint initially stressed at 1100 lb/in^2 (7.6 MPa), for example, will relax to 1050 lb/in^2 (7.2 MPa) at 80°F (26.7°C) or 880 lb/in^2 (6.1 MPa) at 115°F (46°C) over a 10 hour period. The reduction varies exponentially with time. Given sufficient time, room temperature environments are sufficient to relieve some solder joint stresses. Epoxy fiberglass, on the other hand, does not exhibit any stress reduction at temperatures less than 190°F (87.8°C). With an initial stress of 3300 lb/in^2 (22.7 MPa), 10 days had elapsed before a 20 percent reduction was observed at a soak temperature of 260°F (126.7°C).

Plastic flow is particularly important in thermal cycling since it can lead to fatigue failures. Tensile loads which do not exceed the elastic limit of the material do not incur plastic strain. A load that exceeds the elastic limit, however, can change the original length of a system even after the load is removed. This characteristic, shown in Figure 5.2(a), applies particularly to metallic materials since plastics and notably epoxy do not experience plastic deformation. If any element or lead of an assembly deforms plastically during a thermal cycle but returns to its original length when the assembly is returned to its initial temperature, then the stress-temperature curve has a configuration similar to that of Figure 5.2(b).

The mean number of cycles to failure N_f for the plastic flow cycle of Figure 5.2(b) is proportional to the hysteresis loop width Δ, as follows:

$$N_f \, \alpha \, \frac{1}{\Delta^2} \qquad (5.1)$$

The life of the assembly is strongly dependent upon the magnitude of the cyclic strain. The cyclic strain of component leads in a soldered assembly increases if stress relaxation in the solder joints occurs, since the leads carry a greater portion of the load. Plastic deformations are likely to occur in areas of stress concentrations or geometric irregularities. Temperatures above the glass-transition range of the epoxy should also be avoided to minimize the magnitude of the cyclic strain on the assembly.

Geometrical/Arrangement Considerations. Displacements due to differential expansion develop stresses only when restricted or constrained by structure or other components of an assembly. Encapsulated assemblies and conformally coated circuit boards provide some protection against vibration and airborne contaminants such as moisture, dirt or other conductive matter. Unfortunately, the encapsulants and coatings are somewhat inelastic, exhibiting expansion coefficients greater than the components themselves thus generating higher forces than would otherwise occur. Brackets, bolts, adhesives or other mechanical supports which constrain thermal displacements will also reduce the fatigue life of the assembly.

The leads of many configurations may be used to provide strain relief provided they are not rigid and heavy. Parts should be positioned to avoid bridging by coating, or thin coatings should be used. Common configurations that limit fatigue life are shown in Figure 5.3(a).

Many components use their leads for support. Forces due to differential expansion could crack the solder joints or component seals when any part of a component body contacts the circuit board after the leads are soldered if strain relief is not used. Spacers or heat sinks made of material of equal or greater stiffness than the component are equally as damaging. Figure 5.3(a) illustrates several configurations (A) through (E) that constrain lead deflection. Component body constraints shown in (F) and (G) can cause damage to the component or failure at the attachments. Encapsulants that bridge between the component and board develop forces that lift the component. Lead configurations that restrain movement such as in (D) and the encased leads of (E) can cause joint fracture. Joints on single-side boards (H) have very little capacity to resist forces developed by restrictive attachments. Larger pads or plated-through holes (P) provide improved strength. Double-sided boards without plated-through holes (I) have a short fatigue life due to the extended lead length without added joint strength.

Resilient pads or spacers (J) and (O) absorb the differential displacements by compression. Formed or bent leads (K) and (N) also compensate for thermal displacements. Bolted or secured components should use clearance holes (L) for the leads and jumper wires for connection to the copper circuit trace. Components mounted with clearance above the board (M) and (Q) without intervening materials develop insignificant interface stresses. Strain relief is also provided by inverting the component as in (R).

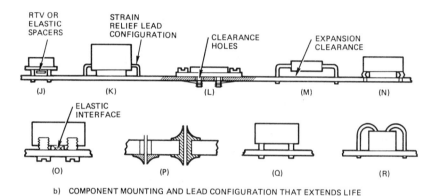

HEAT SINK
OR HARD   ENCAPSULATED
SPACERS   ASSY   EPOXY
                                                   FIBERGLASS   CONFORMAL COATING
                                                                 (ENCAPSULANT)

(A)   (B)   (C)   (D)   (E)

                                                                          BOARD
INSERT   EPOXY   TAPPED
            ASSY   POLE

(F)   (G)   (H)   (I)

a)   COMPONENT MOUNTING AND LEAD CONFIGURATIONS THAT LIMIT LIFE

RTV OR   STRAIN
ELASTIC   RELIEF LEAD
SPACERS   CONFIGURATION   CLEARANCE   EXPANSION
                                     HOLES   CLEARANCE

(J)   (K)   (L)   (M)   (N)

ELASTIC
INTERFACE

(O)   (P)   (Q)   (R)

b)   COMPONENT MOUNTING AND LEAD CONFIGURATION THAT EXTENDS LIFE

Figure 5.3. Component Attachment Configurations That Influence Equipment Lifetime.

Vibration and shock environments must also be considered when developing and evaluating the assembly design. Similar forces at the attachments and lead interfaces developed during vibration may also reduce fatigue life.

5.2 TEMPERATURE RELATED DISPLACEMENTS AND STRESSES

Thermal-stress relationships that are applicable to many configurations found in electronic assemblies are based on an idealized one dimensional symmetrical model. Symmetry simplifies development by assuring that displacements do not produce bending moments in an assembly. A generalized system that represents complexly shaped elements of different materials is shown in Figure 5.4(a).

Imposing a temperature different from the original assembled temperature

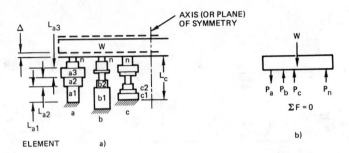

Figure 5.4. Generalized Thermal-Elastic Component Model. (a) Generalized component configuration; (b) Force balance on component.

causes each element of Figure 5.4(a) to elongate or contract. The change in length of element a, for example, due to the temperature difference is

$$\delta_a = [(\alpha L)_{a1} + (\alpha L)_{a2} + \cdots + (\alpha L)_{an}] \, \Delta T \tag{5.2}$$

given α_i material coefficient of expansion (Appendix C).

If the element is free to elongate or contract without resistance or restraint, no force or stress is developed in any part of the element. Constraint may develop, however, by the elongation of a different element of the system expanding to, say, δ_i. In general, each element constrains the other from an incremental change in length due to the forces of constraint. The resultant change in length, Δ, due to the pushing and pulling of each element, is a consequence of the force balance of Figure 5.4(b). If the force W acts to reduce Δ, we obtain

$$\Delta = \frac{\sum (K\delta)_i - W/2}{\sum K_i} \tag{5.3}$$

where K_i is the element stiffness given by

$$\frac{1}{K_i} = \frac{1}{\left(\dfrac{AE}{L}\right)_{i1}} + \frac{1}{\left(\dfrac{AE}{L}\right)_{i2}} + \cdots + \frac{1}{\left(\dfrac{AE}{L}\right)_{in}} \tag{5.4}$$

The internal load developed in any element is due to the difference between the change in free length, δ_i, and the resultant change in length Δ which may be written as

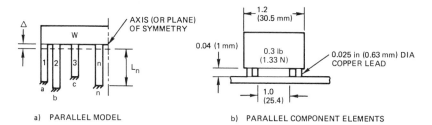

Figure 5.5. Constant Area Element Component Model.

$$F_i = K_i (\Delta - \delta_i) = \left(\frac{AE}{L}\right)_i (\Delta - \delta_i) \qquad (5.5)$$

The stress in any segment is found by dividing (5.5) by the segment area:

$$\sigma_i = \frac{F_i}{A_i} = \left(\frac{E}{L}\right)_i (\Delta - \delta_i) \qquad (5.6)$$

Positive values of F_i and σ_i indicate tension, negative indicates compression. Equation (5.6) may be specialized for configurations similar to those of Figure 5.3 where each element is of constant area. The model of Figure 5.4 may be simplified to a parallel model of Figure 5.5(a).

The element stress for any element of Figure 5.5(a) can be determined from

$$\sigma_i = \left(\frac{E}{L}\right)_i \left[\frac{\Delta T \sum (AE\alpha) - W/2}{\sum \left(\frac{AE}{L}\right)} - (\alpha L \, \Delta T)_i \right] \qquad (5.7)$$

EXAMPLE 5.1.   Determine the stress in the copper leads of Figure 5.5(b) for an epoxy encapsulated component subject to a temperature difference of 90°F (50°C). Material properties from Appendix C are summarized as follows:

	Modulus of Elasticity, E		Coefficient of Expansion, α × 10^{-5}	
	M lb/in^2	GPa	in/in·°F	m/m·°C
Epoxy	2.3	16.3	4.0	7.2
Copper	17.2	119.0	0.93	1.68

The length of the copper leads extends into the encapsulated body and circuit board at least one diameter. The effective length of the lead is therefore 0.075 in (1.9 mm).

Using symmetry, two elements describe the model so (5.7) becomes

$$\sigma_c = \left(\frac{E}{L}\right)_c \left\{ \frac{\Delta T\left[(AE\alpha)_c + \left(\frac{AE\alpha}{2}\right)_e\right] - W/2}{\left(\frac{AE}{L}\right)_c + \left(\frac{AE}{2L}\right)_e} - (\alpha L\,\Delta T)_c \right\} \qquad (5.8)$$

given

$$A_c = 4.908(10^{-4}) \text{ in}^2 \ (0.316 \text{ mm}^2) \qquad \text{copper lead area}$$

$$A_e = \pi r^2 = 0.785 \text{ in}^2 \ (506.7 \text{ mm}^2) \qquad \text{epoxy area}$$

$$\sigma_c = \frac{17.2(10^6)}{0.075}\left[\frac{90(0.0785 + 36.06) - 0.3/2}{1.126(10^5) + 2.254(10^7)} - 6.2775(10^{-5})\right] = 18529 \text{ lb/in}^2$$

In metric units with

$$1.33N\left(10^3 \frac{\text{mm}}{\text{m}}\right)^2 = 1.33(10^{-3}) \text{ GPa}\cdot\text{mm}^2,$$

$$\sigma_c = \frac{119}{1.9}\left[\frac{50(6.32 \times 10^{-4} + 0.297) - 6.65(10^{-4})}{19.79 + 4130}\right.$$

$$\left. - 1.596(10^{-3})\right] = 0.1246 \text{ GPa}$$

The stress in the copper lead is below the elastic limit so plastic deformation will not occur. However, if stress concentrations exist, the local stresses will be greater than calculated and plastic deformation would likely occur. The effect of the resulting tensile force of 9.1 lb (4.1 kg) on solder joint stresses at the elevated temperature must be examined. This magnitude is sufficient to cause creep rupture failure on most joint configurations. Q.E.D.

Configurations (D) and (F) of Figure 5.3 are easily evaluated using (5.6). Consider the equivalent model Figure 5.6(a). The change of length of element a of Figure 5.6(a), based on (5.2) with $\alpha_{a1} = \alpha_{a3}$, becomes

$$\delta_a = [\alpha_{a1}L + (\alpha L)_{a2}]\,\Delta T \qquad (5.9)$$

where $L = L_{a1} + L_{a3}$. Equation (5.9) represents the configuration of Figure 5.6(b). The stiffness of element a of Figure 5.6(b) following (5.4) is

$$\frac{1}{K_a} = \frac{L}{(AE)_{a1}} + \left(\frac{L}{AE}\right)_{a2} \qquad (5.10)$$

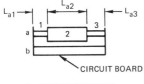 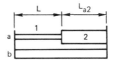

Figure 5.6. Thermal Elastic Model of Components in Figure 5.3(D) and (F).

so that (5.3) with $W = 0$ may be written as

$$\Delta = \frac{\dfrac{\delta_a}{K_b} + \dfrac{\delta_b}{K_a}}{\dfrac{1}{K_a} + \dfrac{1}{K_b}} \qquad (5.11)$$

Substituting (5.11) into (5.6) with $\delta_i = \delta_a$ gives a relationship for the force in element a as

$$F_a = \frac{\delta_b - \delta_a}{\dfrac{1}{K_a} + \dfrac{1}{K_b}} \qquad (5.12)$$

The stress in element $a2$ of Figure 5.6(b) is determined by dividing (5.12) by the area A_{a2}. With the substitution of (5.9) and (5.10) into (5.12), the stress for element $a2$ with $\delta_b = (\alpha L)_b \, \Delta T$ becomes

$$\sigma_{a2} = \frac{\Delta T}{A_{a2}} \left[\frac{(\alpha L)_b - \alpha_{a1} L - (\alpha L)_{a2}}{\left(\dfrac{L}{AE}\right)_{a1} + \left(\dfrac{L}{AE}\right)_{a2} + \left(\dfrac{L}{AE}\right)_b} \right] \qquad (5.13)$$

EXAMPLE 5.2.   Determine the stresses of the conformally coated glass-cased diode with Dumet leads mounted to a 4 inch (101.6 mm) wide epoxy glass fiberboard for the configuration of Figure 5.7 subject to a ΔT of 90°F (50°C).

a) GLASS DIODE MOUNTING   b) CONFORMAL COAT BRIDGING

Figure 5.7. Component Configuration in Example 5.2.

Material properties from Appendix C are summarized as follows:

	Modulus of Elasticity, E		Coefficient of Expansion, $\alpha \times 10^{-5}$	
	M lb/in²	GPa	in/in·°F	m/m·°L
Epoxy (long)	2.3	16.3	0.55	1.00
Glass	10.0	69.0	0.40	0.72
Dumet	26.1	180.0	0.28	0.50
Conformal Coating	0.4	2.8	14.0	25.2

The cross-sectional area of each element becomes

$$A_{a1} \text{ (lead)} = 4.908(10^{-4}) \text{ in}^2 \text{ (0.316 mm}^2)$$

$$A_{a2} \text{ (glass)} = 6.36(10^{-3}) \text{ in}^2 \text{ (4.104 mm}^2)$$

$$A_b \text{ (board)} = 0.24 \text{ in}^2 \text{ (154.8 mm}^2)$$

With the substitution of values into (5.13), the stress in the glass component case for in-plane expansion, Figure 5.7(a), becomes

$$\sigma_{\text{glass}} = \frac{90}{6.36(10^{-3})} \left[\frac{2.75(10^{-6}) - 7(10^{-7}) - (10^{-6})}{1.951(10^{-5}) + 3.93(10^{-6}) + 9.06(10^{-7})} \right]$$

$$= 610 \text{ lb/in}^2$$

The stress in the lead is determined directly from

$$\sigma_{\text{lead}} = \frac{A_{a2}}{A_{a1}} \sigma_{\text{glass}} = \frac{6.36(10^{-3})}{4.908(10^{-4})} (610) = 7908 \text{ lb/in}^2$$

since the force in each element of a is equal. The calculated stresses are well below element ultimate strength (11000 lb/in² for glass).

The lead stress for normal-to-plane expansion is found using (5.7) for an average thickness of 0.004 in (0.1 mm) of conformal coating bridging between the component and board over an area of 0.015 in² (9.67 mm²). Substituting values using one-half the footprint area due to symmetry into (5.7) gives

$$\sigma_{\text{lead}} = \frac{90(26.1 \times 10^6)}{0.045} \left[\frac{0.0358 + 0.42}{2.85(10^5) + 7.5(10^5)} - 1.26(10^{-7}) \right]$$

$$= 16411 \text{ lb/in}^2$$

In metric units

$$\sigma_{\text{lead}} = \frac{50(180)}{1.14} \left[\frac{0.00028 + 0.0034}{49.76 + 135.38} - 5.715(10^{-6}) \right] = 0.1118 \text{ GPa}$$

The resultant tensile lead stress is therefore

$$\sigma_\ell = \sqrt{16411^2 + 7908^2} = 18217 \, \text{lb/in}^2$$

or

$$\sigma_\ell = \sqrt{0.1118^2 + 0.0545^2} = 0.124 \, \text{GPa}$$

The substantial increase in lead stress due to bridging of the conformal coating should be noted. As in Example 5.1, the effect of the resulting tensile force on solder joint stresses at operational temperatures must be examined.   Q.E.D.

Stresses at Soldered and Threaded Attachments.   Solder joint capability is dependent on the selected design configuration. Some configurations use single or double sided circuit boards with "poke thru" joints similar to those of Figure 5.3(H), (I) and (P). Another configuration is represented by planar lead components.

Failure of poke-thru configurations is generally due to excessive creep rupture stresses. This failure mode develops annular cracks in the solder around the lead. When cracks occur, they are usually apparent with a magnification of 30×. The discontinuous nature of the cracks suggests tensile/compressive plastic flow due to temperature cycling. Catastrophic failures occur when the lead is pulled free of the solder or the seal at the component to lead interface fractures.

The shear area for two solder joint configurations is shown in Figure 5.8.

Creep-rupture stress levels may be computed with the aid of Figure 5.8. At operating temperatures of about $176°F$ ($80°C$), stress levels should not exceed $400 \, \text{lb/in}^2$ (2.76 MPa), following Figure 5.1(a). Stresses above creep rupture levels cause failure for temperature cycled equipment. For example, twenty

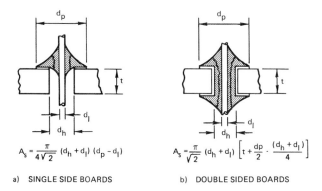

a)   SINGLE SIDE BOARDS   b)   DOUBLE SIDED BOARDS

Figure 5.8. Shear Area of Typical Solder Joints.

Table 5.1. Design Criteria for Molded Inserts

Knurled/Slotted Configuration

Outside Diameter (in)	⅛	3/16	¼	⅜
Minimum Surrounding Wall Thickness (in)	0.094	0.140	0.188	0.281
Length	0.156	0.250	0.375	0.500
Pull-out Load (lb)	90	240	450	750

Helical Configuration

Outside Diameter (in)	⅛	3/16	¼	⅜
Minimum Surrounding Wall Thickness (in)	0.062	0.100	0.100	0.110
Insert Length	0.162	0.247	0.328	0.364
Pull-out Load	80	120	215	260

cycles between $-40°$ to $+160°$F ($-40°$ to $+71°$C) cause cracked joints which exhibit a 5 MΩ increase in resistance.

Forces sufficient to pull or loosen inserts of bolted encapsulated assemblies (Figure 5.3(G)) have occurred. Typical insert failure loads are listed in Table 5.1.

Reflow soldered planar component leads usually fail by a peeling action which fractures the solder joint (Figure 5.9). Based on interface geometry and mechanical properties of the joint materials, peeling occurs when the force normal to the board surface equals or exceeds

$$F_{\text{peel}} = 0.38 \, W \sigma_{\text{ult}} \left[ht^3 \left(\frac{E_\ell}{E_s} \right) \right]^{1/4} \qquad (5.14)$$

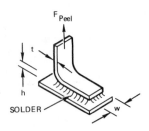

Figure 5.9. Peel Force on a Soldered Planar Lead Interface.

given

E_ℓ, E_s = lead and solder modulus of elasticity respectively;
δ_{ult} = ultimate solder strength, Figure 5.1;
t, W = lead thickness and width respectively;
h = solder thickness at peel zone.

EXAMPLE 5.1 (CONTINUED). A centrally located insert/screw attachment, Figure 5.10(a), is considered to reduce solder joint stresses for the component of Figure 5.5(b). The leads are soldered into 0.035 in (0.88 mm) diameter plated-through holes with 0.075 in (1.9 mm) solder pads. Evaluate lead and solder joint stresses and the potted insert tensile force to determine a suitable modification.

The change in length of the series configuration for the screw and the epoxy board in accordance with (5.2) becomes

$$\delta_{s/b} = [(\alpha L)_s - (\alpha L)_b]\,\Delta t \qquad (5.15)$$

The change in length for the parallel-series combination using (5.3) is

$$\Delta = \frac{\Delta T\left[(AE\alpha)_\ell + \left(\dfrac{AE\alpha}{2}\right)_e\right] + K_{s/b}\,\delta_{s/b} - W/2}{\left(\dfrac{AE}{L}\right)_\ell + \left(\dfrac{AE}{2L}\right)_e + K_{s/b}} \qquad (5.16)$$

given

$$K_{s/b} = \frac{1}{2}\frac{K_b K_s}{K_b + K_s}\,;\; K_b = \left(\frac{AE}{L}\right)_b,\; K_s = \left(\frac{AE}{L}\right)_s$$

Equation (5.16) accounts for the apparent shortening of the screw due to normal to surface expansion of the board relative to the referenced plane in Figure 5.10(a). Stress areas of various size screws are listed in Table 5.2.

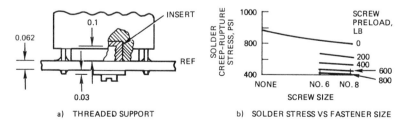

a) THREADED SUPPORT   b) SOLDER STRESS VS FASTENER SIZE

Figure 5.10. Influence of a Threaded Attachment on a Lead Supported Component.

Table 5.2. Dimensional Characteristics of Screws

	UNC/UNF				Metric		
	Min. Dia.		Stress Area, in^2		Major Dia. $\times$ Pitch	Min. Dia., mm	Stress Area, mm^2
Size	Course	Fine	Course	Fine			
4	0.0813	0.0864	0.0060	0.0065	M3.0 $\times$ 0.5	2.459	5.03
6	0.0997	0.1073	0.0090	0.0101	M3.5 $\times$ 0.6	2.850	6.78
8	0.1257	0.1299	0.0139	0.0146	M4.0 $\times$ 0.7	3.242	8.78
10	0.1389	0.1517	0.0174	0.0199	M5.0 $\times$ 0.8	4.134	14.20
¼	0.1887	0.2062	0.0317	0.0362	M6.3 $\times$ 1.0	5.217	22.60
⁵⁄₁₆	0.2443	0.2614	0.0522	0.0579	M8.0 $\times$ 1.25	6.647	36.60

Using a size 6 screw with a bearing area due to the washer of 0.0755 in^2 (48.7 mm^2) in (5.16), neglecting preload, gives

$$\Delta = \frac{90[0.0785 + 36.06] - 7.38 \times 10^5(6.38 \times 10^{-5}) - 0.3/2}{1.126(10^5) + 2.254(10^7) + 7.38(10^5)} = 1.37(10^{-4}) \text{ in}$$

where $K_s/_b = 7.38(10^5)$ lb/in. Substitution of parameters into (5.6) for the lead stress gives

$$\sigma_{\text{lead}} = \left(\frac{E}{L}\right)_\ell (\Delta - \delta_\ell)$$

$$= \frac{17.2(10^6)}{0.075} (1.37 \times 10^{-4} - 6.2775 \times 10^{-5}) = 17029 \text{ lb/in}^2$$

The solder shear area based on Figure 5.8(b) becomes

$$A_s = \frac{\pi}{\sqrt{2}} (0.025 + 0.035) \left(0.062 + \frac{0.075}{2} - \frac{0.06}{4} \right) = 0.0113 \text{ in}^2$$

The solder joint creep-rupture stress is

$$\sigma_{\text{creep-rupture}} = \sigma_{\text{lead}} \left(\frac{A_\ell}{A_s} \right) \tag{5.17}$$

$$= 17029 \left(\frac{4.908 \times 10^{-4}}{0.0113} \right) = 739 \text{ lb/in}^2$$

Substituting (5.15) and (5.16) into (5.5) gives the change in the fastener tensile force as

$$F_s = K_{s/b} (\Delta - \delta_{s/b}) = 7.38(10^5)[1.37(10^{-4}) + 7.09(10^{-7})90] = 148.2 \text{ lb}$$

Even though the screw carries some of the expansion load, the reduction of the creep rupture stress, neglecting preload, is insufficient to obtain positive margins. The preload developed by a threaded fastener is a function of its minimum diameter, D_{min}, and the applied torque, T_q as follows:

$$W_{preload} = \frac{5 T_q}{D_{min}} \qquad (5.18)$$

Results obtained by substituting (5.18) into (5.16) and recomputing (5.6) are plotted on Figure 5.10(b). Approximately 800 lb (3.6 kN) is required to obtain acceptable solder stress levels. The torque required to develop 800 lb on a size 6 screw using (5.18) and D_{min} from Table 5.2 becomes

$$T_q = \frac{W_{preload}(D_{min})}{5} = 800 \frac{(0.0997)}{5} = 15.9 \text{ in} \cdot \text{lb}$$

If the constraints imposed by the use of threaded fasteners are impractical, other lead strain relief configurations similar to those of Figure 5.3(K) and (N) should be examined.   Q.E.D.

5.3 STRAIN RELIEF CONCEPTS

Relationships developed in Section 5.2 are based on the assumption that temperature related forces develop only elongation or compressive actions in the component lead and attachment interface. The stresses that develop by inhibiting or preventing expansion or contraction can be alleviated if configurations that enable bending or flexure can be used.

The lead geometry, size, shape and material of electronic components vary considerably. Other factors that influence the adaptation of strain relief configurations are the availability of circuit board space and cost of assemblies. Automated design, fabrication and assembly of circuit boards have reduced their cost while increasing assembly predictability and reliability. Auto-insertion equipment either preforms leads or installs components with leads configured as in Figure 5.3(B), (E) or (K). The formed leads of Figure 5.3(K) flex during component expansion. Low stresses in the lead and joints will occur provided the lead is sufficiently compliant.

Higher packaging densities in addition to characteristics of the lead and component geometry may limit the size and effectiveness of the strain relief. Strain relief configurations that require more space reduce the number of components that can be mounted on the available surface. Use of elastic interface materials, lead spreaders or clearance between the component and board surface minimize stresses without increasing the required mounting area. These methods, however, require special processing which can impact the assembly cost.

Assemblies which are to be conformally coated for protection from dirt and moisture are particularly prone to both thermally and dynamically induced failures. Coatings that bridge between the component and circuit board or around the formed leads decrease the effectiveness of the strain relief. Most coatings have high coefficients of expansion that increase even more at high operating temperatures. The increased stiffness due to bridging develops greater forces on the joints. When vibration is present high flexural stresses can lead to fatigue failure. Under these conditions the assembly exhibits a greater sensitivity to workmanship flaws and stress concentrations developed by nicks, scratches and configuration abnormalities.

The compliance of the circuit board in contact with the component may provide some strain relief. Consider a configuration analogous to that of Figure 5.3(B). The change in length due to the incremental displacements of the leads, component body and circuit board is

$$\Delta = \frac{\Delta T[(AE\alpha)_\ell + K(\alpha L)_e] - W/2}{\left(\dfrac{AE}{L}\right)_\ell + K} \tag{5.19a}$$

given

$$K = \frac{1}{2} \frac{K_f K_e}{K_f + K_e}$$

where K_f = flexural stiffness of board between leads;
$\quad K_e$ = axial stiffness of component body in board contact.

The lead stress can be determined from (5.6). The board stiffness is in series with the threaded support for the configuration of Figure 5.10(a). In this case, the change in length of (5.16) is modified to

$$\Delta = \frac{\Delta T\left[(AE\alpha)_\ell + \left(\dfrac{AE\alpha}{2}\right)_e\right] + K\delta_{s/b} - W/2}{\left(\dfrac{AE}{L}\right)_\ell + \left(\dfrac{AE}{2L}\right)_e + K} \tag{5.19b}$$

given

$$K = \frac{1}{2} \frac{K_f K_{s/b}}{K_f + K_{s/b}}$$

where K_f = flexural stiffness of board between leads.

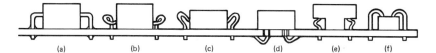

Figure 5.11. Various Formed Component Lead Configurations.

Formed lead configurations can have a variety of shapes and behave as an effective strain relief. Some configurations require more board space and are more convenient to form than others. The geometry and shape are dependent upon lead location on the component relative to the mounting face selected. Typical shapes are shown in Figure 5.11.

The lead exit of Figure 5.11(a) is sufficiently far from the board surface to prevent solder from wicking into the bend radius. In those cases, when the lead exit is close to the board, Figure 5.11(b) and (c), a more exotic shape is required to prevent solder wicking. When solder wicks up into the formed lead, the strain relief effectiveness decreases dramatically. Forms that provide strain relief for bottom-exiting leads are shown in Figure 5.11(d), (e) and (f).

Figure 5.12 illustrates the configuration parameters and stiffness of two configurations. The stiffness of any shape can be determined using the techniques of Chapter 7. The stiffness of Figure 5.12(a) represents a configuration which has a bend radius $\leq 2d$ provided solder does not wick into its curvature. The flexural length ℓ and y of Figure 5.12 should be greater than $2d$ to obtain a marked reduction in joint forces.

EXAMPLE 5.3.   Determine the lead tensile force and solder creep rupture stress for the component of Example 5.1 reconfigured with side-exiting leads as shown in Figure 5.13.

An expression for the change in height based on (5.8) and (5.19a) is

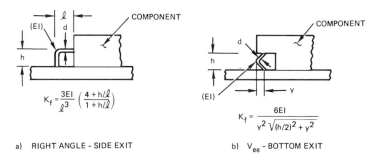

a)   RIGHT ANGLE - SIDE EXIT

$$K_f = \frac{3EI}{\ell^3} \left(\frac{4 + h/\ell}{1 + h/\ell} \right)$$

b) V_{ee} - BOTTOM EXIT

$$K_f = \frac{6EI}{y^2 \sqrt{(h/2)^2 + y^2}}$$

Figure 5.12. Stiffness of Formed Leads in Figure 5.11(a) and (e).

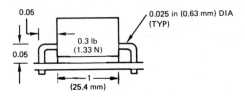

Figure 5.13. Component Lead Configuration Evaluated in Example 5.3.

$$\Delta = \frac{\left(\dfrac{AE\alpha}{2}\right)_e \Delta T + K(\alpha h \, \Delta T)_c - W/2}{K + \left(\dfrac{AE}{2L}\right)_e} \qquad (5.20)$$

given

$$K = \frac{K_f K_\ell}{K_f + K_\ell}$$

where

$$K_f = \frac{3EI}{\ell^3}\left(\frac{4 + h/\ell}{1 + h/\ell}\right) \quad \text{from Figure 5.12(a)}$$

$$K_\ell = \left(\frac{AE}{h}\right)_c$$

Material property and parametric data of Example 5.1 apply. The effective length, ℓ, and height, h, of the lead is 0.075 in (1.9 mm) since one diameter is assumed extended into the component and board. The lead moment of inertia is

$$I = \frac{\pi d^4}{64} = \frac{\pi(0.025)^4}{64} = 1.92(10^{-8}) \text{ in}^4 = \pi \frac{(0.63)^4}{64} = 7.73(10^3) \text{ mm}^4$$

$$K_f = \frac{3EI}{\ell^3}\left(\frac{4 + h/\ell}{1 + h/\ell}\right) = \frac{3(17.2 \times 10^6)(1.92 \times 10^{-8})}{0.075^3}\left(\frac{4 + 1}{1 + 1}\right)$$

$$= 5871 \text{ lb/in} = \frac{3(119 \times 10^9)(7.73 \times 10^{-3})}{1.9^3(1000 \text{ mm/m})}\left(\frac{5}{2}\right) = 10^6 \text{ N/m}$$

$$K_\ell = 1.126(10^5) \text{ lb/in} = 1.972(10^7) \text{ N/m}$$

$$K = \frac{5871(1.126 \times 10^5)}{5871 + 1.126 \times 10^5} = 5580 \text{ lb/in}$$

$$= \frac{10^6(1.972 \times 10^7)}{10^6 + 1.972 \times 10^7} = 9.517(10^5) \text{ N/m} \ (0.9517 \text{ GPa}\cdot\text{mm})$$

The change in height using (5.20) becomes

$$\Delta = \frac{36.06(90) + 5580(6.2775 \times 10^{-5}) - 0.3/2}{5580 + 2.254 \times 10^7} = 1.439(10^{-4}) \text{ in}$$

$$= \frac{0.297(50) + 0.9517(1.596 \times 10^{-3}) - 6.65(10^{-4})}{0.9517 + 4130} = 3.59(10^{-3}) \text{ mm}$$

The axial tensile force in the lead is determined from (5.5) as

$$F = K(\Delta - \delta_\ell) = K(\Delta - \alpha h \, \Delta t)$$

$$= 5580(1.439 \times 10^{-4} - 6.2775 \times 10^{-5}) = 0.45 \text{ lb}$$

$$= \frac{9.517(10^5)(3.59 \times 10^{-3} - 1.596 \times 10^{-3})}{(1000 \text{ mm/m})} = 1.89 \text{ N}$$

The solder shear area was previously determined as 0.0113 in^2 (7.29 mm^2). The creep rupture stress becomes

$$\sigma_{\text{creep rupture}} = \frac{F}{A_s} = \frac{0.45}{0.0133} = 39.8 \text{ lb/in}^2$$

$$= \frac{1.89}{7.29}\left(10^6 \frac{\text{mm}^2}{\text{m}^2}\right) = 0.26 \text{ MPa}$$

The right angle lead strain relief, Figure 5.12(a), is very effective in reducing temperature-induced lead strain and joint stress. The maximum allowable solder joint stress to prevent temperature-induced failure is 400 lb/in^2 (2.75 MPa). The factor of safety, which is defined as

$$FS = \frac{\text{ultimate stress}}{\text{maximum stress caused by the design load}} \tag{5.21}$$

must always be greater than unity. A factor of safety of 10 for this design is indicative of a failure-free design for the specified temperature range.   Q.E.D.

5.4 STRESSES IN BONDED ASSEMBLIES

A variety of bonded joints are used in electronic equipment. Materials are joined by different methods using adhesives and solder among others. Bonded joints are used in the design of printed circuit boards, components and chassis structures.

A change in temperature due to equipment operation or a fabrication process can introduce stresses in joined materials of dissimilar thermal expansion coefficients. Examples of stressed joints include metallic heat sinks and stiffeners

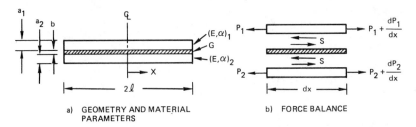

a)   GEOMETRY AND MATERIAL   b)   FORCE BALANCE
     PARAMETERS

Figure 5.14. Configuration and Forces within Bonded Assemblies.

on glass epoxy circuit boards or the structural-heat transfer attachment of electronic devices. Often, elevated cure temperatures are used. Thermal stress, caused by cooling from high bonding temperatures can cause fracture and joint failure.

Continuously Attached Interfaces.   An approach that gives considerable insight regarding the effects of various physical parameters is based on the element strains and equilibrium of forces of Figure 5.14 assuming the joint material is more compliant than the bonded materials.

With P in Figure 5.14(b) defined as the axial force per unit width of the bond material, force equilibrium gives

$$\frac{dP_1}{dx} - S = 0 \tag{5.22a}$$

$$\frac{dP_2}{dx} + S = 0 \tag{5.22b}$$

where S is the joint shear stress. Summing linear strains due to temperature and stress gives

$$\frac{d\delta_1}{dx} = \left(\frac{P}{Ea}\right)_1 + \alpha_1 \, \Delta T \tag{5.23a}$$

$$\frac{d\delta_2}{dx} = \left(\frac{P}{Ea}\right)_2 + \alpha_2 \, \Delta T \tag{5.23b}$$

The stress-strain relationship due to joint shear is

$$\frac{S}{G} = \frac{\delta_1 - \delta_2}{b} \tag{5.24}$$

The shear gradient as determined from (5.23) and (5.24) is

$$\frac{dS}{dx} = \frac{G}{b}\left[\left(\frac{P}{Ea}\right)_1 - \left(\frac{P}{Ea}\right)_2 + \Delta T (\alpha_1 - \alpha_2)\right] \qquad (5.25)$$

Substituting (5.22) into the derivative of (5.25) gives a differential equation for determining the shear stress as

$$\frac{d^2S}{dx^2} - A^2 x = 0 \qquad (5.26)$$

given

$$A^2 = \frac{G}{b}\left[\left(\frac{1}{Ea}\right)_1 + \left(\frac{1}{Ea}\right)_2\right]$$

The solution of (5.26) subject to the boundary conditions

$$S|_{x=0} = 0 \qquad (5.27a)$$

$$P_1|_{x=\ell} = P_2|_{x=\ell} = 0 \qquad (5.27b)$$

$$\frac{dS}{dx}\bigg|_{x=0} = 0 \qquad (5.27c)$$

is determined as

$$S = \frac{G \Delta T (\alpha_1 - \alpha_2) \sinh Ax}{bA \cosh A\ell} \qquad (5.28)$$

Equilibrium for the element of Figure 5.14(b) gives $P_2 = -P_1$. The axial force for width W becomes

$$F = PW = \frac{WG \Delta T (\alpha_1 - \alpha_2)(\cosh A\ell - \cosh Ax)}{bA^2 \cosh A\ell} \qquad (5.29)$$

EXAMPLE 5.4. Develop the joint shear stress profile and the maximum axial stress for a 0.25 × 0.25 inch (6.35 × 6.35 mm) 6061 aluminum stiffener mounted across the 5 inch width of a 6.5 × 5 inch (165 × 127 mm) glass epoxy circuit board of 0.06 inch (1.52 mm) thickness. The bond thickness of 0.001 in (0.0254 mm) is filled with a cement which has a shear modulus $G = 2(10^5)$ lb/in^2 (1.38 GPa). The assembly is subject to temperature variations of 155°F (86°C).

Material properties from Appendix C are summarized as follows:

	Modulus of Elasticity, E		Coefficient of Expansion, $\alpha \times 10^{-5}$	
	Mlb/in^2	GPa	in/in·°F	m/m·°C
Epoxy (x/y)	2.30	16.3	0.55	1.00
Aluminum 6061	9.95	68.6	1.30	2.34

Parameter A from (5.26) becomes

$$A^2 = \frac{2(10^5)}{0.001} \left[\frac{1}{2.3(10^6)(0.06)} + \frac{1}{9.95(10^6)(0.25)} \right] = 1529 \text{ in}^{-2}$$

In metric units

$$A^2 = \frac{1.38}{0.0254} \left[\frac{1}{16.3(1.52)} + \frac{1}{68.6(6.35)} \right] = 2.32 \text{ mm}^{-2}$$

The shear stress in the cement bond as a function of x over the half width $\ell = 2.5$ in (63.5 mm) using (5.28) is

$$S = \frac{2(10^5)155(1.3 - 0.55) \times 10^{-5} \sinh 39.1x}{0.001(39.1) \cosh (2.5 \times 39.1)}$$

$$= 4.19(10^{-39}) \sinh 39.1x \text{ lb/in}^2$$

In metric units

$$S = \frac{1.38(86)(2.34 - 1.0) \times 10^{-5} \sinh 1.52x}{0.0254(1.52) \cosh (1.52 \times 63.5)}$$

$$= 9.95(10^{-44}) \sinh 1.52x \text{ GPa}$$

The bond shear stress distribution is shown in Figure 5.15. The maximum shear stress of 5936 lb/in^2 (40.9 MPa) occurs at $x = \ell$ as indicated by (5.28). At $x = \ell$ (5.28) becomes

$$S_{max} = \frac{G \Delta T (\alpha_1 - \sigma_2) \tanh A\ell}{bA} \qquad (5.30)$$

For many design configurations, $\tanh A\ell \simeq 1$, simplifying (5.30) to

$$S_{max} = \frac{G \Delta T (\alpha_1 - \alpha_2)}{bA} \qquad (5.31)$$

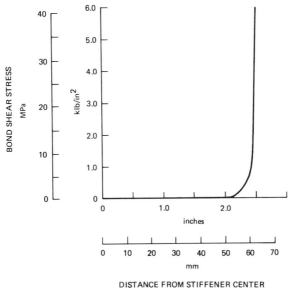

Figure 5.15. Shear Stress Distribution in the Bonded Interface of the Configuration in Example 5.4.

Maximum axial force in the joined members occurs at $x = 0$, simplifying (5.29) to

$$F_{max} = \frac{WG\,\Delta T\,(\alpha_1 - \alpha_2)}{bA^2} \tag{5.32}$$

Maximum axial stress occurs at the center of the aluminum stiffener. The stress magnitude, following (5.32), becomes

$$\sigma_{max} = \frac{F_{max}}{(0.25)^2} = \frac{0.25(2 \times 10^5)155(1.3 - 0.55) \times 10^{-5}}{0.001(1529)(0.25)^2} = 608 \text{ lb/in}^2$$

In metric units

$$\sigma_{max} = \frac{F_{max}}{(6.35)^2} = \frac{6.35(1.38)86(2.34 - 1.0) \times 10^{-5}}{0.0254(2.32)(6.35)^2} = 4.25 \text{ MPa}$$

Q.E.D.

The magnitude of the bond shear stress can be reduced by increasing the joint thickness. Equations (5.26) and (5.28) show that the bond stress varies as $b^{-1/2}$. As seen in Figure 5.15, about 90 percent of the joint material along

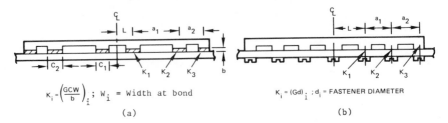

$$K_i = \left(\frac{GCW}{b}\right)_i; \quad W_i = \text{Width at bond}$$

(a)

$$K_i = (Gd)_i; \quad d_i = \text{FASTENER DIAMETER}$$

(b)

Figure 5.16. Discretely Attached Assemblies. (a) Bonded; (b) Bolted or riveted.

the center of the stiffener is ineffective structurally. Only the 0.25 in (6.35 mm) section at the ends of the stiffener resists the thermally induced load.

Fasteners placed near the ends of stiffeners similar to that of Example 5.4 reduce the maximum bond stress significantly. The bond is still required to assure an integral section to resist dynamic bending loads encountered in shock or vibration environments.

The development of joint loads in the preceding discussion has assumed that flexure does not occur. Flexural consideration leads to complicated relationships that mask the contribution of physical parameters. When flexure is allowed, however, significant tensile stress in the joined material can occur with the advantage of reduced bond shear stresses.

Analyses of lamina containing several layers of either rectangular or circular configuration follow an approach similar to that of this section.

Discretely Attached Interfaces. Stiffeners are often added on component assemblies to decrease displacements and stresses. Mounted parts, devices and interconnections on the surfaces of assemblies require the stiffener to be scalloped to avoid interferences and facilitate component replacement. Stiffeners of this type are attached by bonding, rivets or bolts to the mating assembly at discrete locations as shown in Figure 5.16.

Stiffnesses at each discrete attachment are dependent upon the shear modulus, G, of the adhesive or fastener, the fastener-diameter d, or the dimensions of the bonded interface. The evaluation of symmetrical systems with any dis-

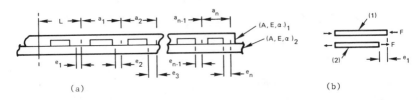

(a)

(b)

Figure 5.17. Discrete Attachment. (a) System model; (b) Free body diagram for first segment.

tribution and quantity of discrete attachments is based on the incremental strain experienced at each interface. Consider the generalized system of Figure 5.17(a).

Neglecting bending or buckling of the system, the displacement at the first interface which is proportional to the load carried, Figure 5.17(b), may be expressed as

$$\delta_1 = (\alpha L \, \Delta T)_1 - \frac{FL}{(AE)_1} \tag{5.33a}$$

$$\delta_2 = (\alpha L \, \Delta T)_2 + \frac{FL}{(AE)_2} + e_1 \tag{5.33b}$$

Since there is no relative displacement at the attachment, $\delta_1 = \delta_2$. The segment strain determined by subtraction becomes

$$F\beta = \epsilon - \frac{e_1}{L} \tag{5.34}$$

given

$$\beta = \left[\frac{1}{(AE)_1} + \frac{1}{(AE)_2} \right]$$

$$\epsilon = (\alpha \, \Delta T)_1 - (\alpha \, \Delta T)_2$$

A similar result is obtained for a free body diagram of the second segment subjected to load F_1 and interface displacement e_2:

$$F_1\beta = \epsilon - \frac{(e_2 - e_1)}{a_1} \tag{5.35}$$

The strain for any segment i may be determined recursively as

$$F_{i-1}\beta = \epsilon - \frac{(e_i - e_{i-1})}{a_{i-1}} \qquad i = 2, 3, \ldots n \tag{5.36}$$

Solving (5.34) and (5.35) for the ratio of displacement e_2/e_1, using $K_1 = (F - F_1)/e_1$ gives

$$\frac{e_2}{e_1} = K_1\beta a_1 + \frac{a_1}{L} + 1 \tag{5.37}$$

Displacement ratios for the system of Figure 5.17(a) are determined by using consecutive expressions of (5.36). The results are

$$\frac{e_3}{e_1} = \frac{a_2}{a_1}\left[\frac{e_2}{e_1}\left(K_2\beta a_1 + \frac{a_1}{a_2} + 1 \right) - 1 \right] \tag{5.38}$$

$$\frac{e_4}{e_1} = \frac{a_3}{a_2}\left[\frac{e_3}{e_1}\left(K_3\beta a_2 + \frac{a_2}{a_3} + 1 \right) - \frac{e_2}{e_1} \right] \tag{5.39}$$

.
.
.

$$\frac{e_i}{e_1} = \frac{a_{i-1}}{a_{i-2}}\left[\frac{e_{i-1}}{e_1}\left(K_{i-1}\beta a_{i-2} + \frac{a_{i-2}}{a_{i-1}} + 1 \right) - \frac{e_{i-2}}{e_1} \right], \, i = 3, 4, \ldots n \tag{5.40}$$

The sum of attachment loads F_i equals the load F on the first segment, which may be expressed as

$$F = \sum_1^n F_i \tag{5.41}$$

Equation (5.41) may be expressed in terms of displacement ratios using the attachment load $F_i = (eK)_i$ as

$$\frac{e_1}{F} = \frac{1}{K_1 \sum_1^n \frac{F_i}{F_1}} \tag{5.42}$$

An expression for attachment load is obtained by expressing (5.34) in terms of e_i/F, giving

$$F_i = \frac{\epsilon K_i(e_i/F)}{\beta + \dfrac{(e_1/F)}{L}} \tag{5.43}$$

Consider a configuration which has only one attachment on each end. Equation (5.43) with $F = F_1$ from (5.41) reduces to

$$F = \frac{\epsilon}{\beta + \dfrac{1}{K_1 L}} \tag{5.44}$$

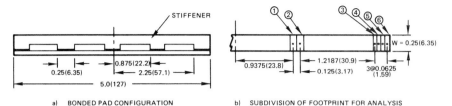

a)   BONDED PAD CONFIGURATION   b)   SUBDIVISION OF FOOTPRINT FOR ANALYSIS

Figure 5.18. Discretely Bonded Assembly.

Maximum stiffener or board load occurs at infinite attachment stiffness, $K_1 = \infty$ in (5.44), which gives

$$F_{max} = \epsilon/\beta \qquad (5.45)$$

Equation (5.45) is comparable to (5.32) for the continuous joint. Fastener clearance reduces the attachment forces. The clearance required for zero attachment force for the attachment configuration of (5.44) with $K_1 = 0$ becomes

$$e_1 = L\epsilon \qquad (5.46)$$

EXAMPLE 5.5.   Determine the joint shear stress profile for the assembly described in Example 5.4 using discrete attachment pads as shown in Figure 5.18(a).

The attachment footprint is subdivided per Figure 5.18(b) to obtain sufficient discrimination of joint stresses. The joint stiffnesses from Figure 5.16(a) become

$$K_1 = K_2 = \frac{GC_1W_1}{b_1} = \frac{2(10^5)(0.125)(0.25)}{0.001} = 6.25(10^6) \text{ lb/in}$$

$$= \frac{1.38(3.17)(6.35)}{0.0254} = 1.093 \text{ GN/m}$$

$$K_3 = K_4 = K_5 = K_6 = \frac{GC_3W_3}{b_3} = \frac{2(10^5)(0.0625)(0.25)}{0.001} = 3.125(10^6) \text{ lb/in}$$

$$= \frac{1.38(1.59)(6.35)}{0.0254} = 548.6 \text{ MN/m}$$

Deflection ratios are calculated successively from (5.37) and (5.40) using constants from (5.34):

$$\beta = \left[\frac{1}{2.3(10^6)(0.06)(0.25)} + \frac{1}{9.95(10^6)(0.25^2)} \right] = 3.06(10^{-5}) \text{ lb}^{-1}$$

$$= \left[\frac{1}{16.3(1.59)(6.35)} + \frac{1}{68.6(6.35^2)} \right] = 6.438(10^{-6}) \text{ N}^{-1}$$

$$\epsilon = 155(1.3 - 0.55)(10^{-5}) = 0.001163$$

$$\frac{e_2}{e_1} = 6.25(10^6)3.06(10^{-5})0.125 + \frac{0.125}{0.9375} + 1 = 25.04$$

$$\frac{e_3}{e_1} = \frac{1.2187}{0.125}\left\{25.04\left[6.25(10^6)3.06(10^{-5})0.125 + \frac{0.125}{1.2187} + 1\right] - 1\right\} = 6095.$$

$$\frac{e_4}{e_1} = \frac{0.0625}{1.2187}\left\{6095\left[3.125(10^6)3.06(10^{-5})1.2187 + \frac{1.2187}{0.0625} + 1\right] - 6095\right\}$$

$$= 42075.$$

$$\frac{e_5}{e_1} = 42075[3.125(10^6)3.06(10^{-5})0.0625 + 1 + 1] - 42075 = 293537$$

$$\frac{e_6}{e_1} = 293537[3.125(10^6)3.06(10^{-5})0.0625 + 1 + 1] - 293537$$

$$= 2.042(10^6)$$

Equation (5.42) gives

$$\frac{e_1}{F} = \frac{1}{K_1\left(1 + \dfrac{F_2}{F_1} + \dfrac{F_3}{F_1} + \cdots + \dfrac{F_6}{F_1}\right)}$$

$$= \frac{1}{6.25(10^6)\{1 + 25.04 + 0.5[6095 + 42075 + 293537 + 2.042(10^6)]\}}$$

$$\frac{e_1}{F} = 1.342(10^{-13}) \text{ in/lb}$$

which in metric units becomes

$$\frac{e_1}{F} = \frac{1}{1.093(10^9)\{1 + 25.04 + 0.5[6095 + 42075 + 293537 + 2.042(10^6)]\}}$$

$$= 7.674(10^{-16}) \text{ m/N}$$

The force on the first joint is determined from (5.43) as

$$F_1 = \frac{\epsilon K_1(e_1/F)}{\beta + \left(\dfrac{e_1/F}{L}\right)} = \frac{0.001163(6.25 \times 10^6)(1.342 \times 10^{-13})}{\left(3.06 \times 10^{-5} + \dfrac{1.342 \times 10^{-13}}{0.9375}\right)} = 3.188(10^{-5}) \text{ lb}$$

$$= \frac{0.001163(1.093 \times 10^9)(7.674 \times 10^{-16})}{\left(6.438 \times 10^{-6} + \dfrac{7.674 \times 10^{-16}}{23.8}\right)} = 1.515(10^{-4}) \text{ N}$$

The forces on the remaining joints may be found consecutively since

$$\frac{e_i}{F} = \frac{e_i}{e_1} \cdot \frac{e_1}{F}$$

therefore,

$$F_i = \frac{K_i}{K_1}\left(\frac{e_i}{e_1}\right) F_1$$

$$F_2 = 25.04(3.188 \times 10^{-5}) = 7.983(10^{-4}) \text{ lb}$$

$$= 25.04(1.515 \times 10^{-4}) = 3.794(10^{-3}) \text{ N} \qquad (5.47)$$

$$F_3 = \tfrac{1}{2}(6095)(3.188 \times 10^{-5}) = 0.097 \text{ lb}$$

$$= \tfrac{1}{2}(6095)(1.515 \times 10^{-4}) = 0.462 \text{ N}$$

Similarly

$$F_4 = 0.670 \text{ lb } (3.184 \text{ N})$$

$$F_5 = 4.678 \text{ lb } (22.232 \text{ N})$$

$$F_6 = 32.539 \text{ lb } (154.645 \text{ N})$$

Average shear stress in the bonded joint segments of Figure 5.18(b) are obtained using

$$\bar{S}_i = (F/CW)_i$$

$$\bar{S}_1 = 3.188 \times 10^{-5}/(0.125 \times 0.25) = 0.001 \text{ lb/in}^2 \qquad (5.48)$$

$$= 1.515 \times 10^{-4}/(3.17 \times 6.35) = 7.53 \text{ Pa}$$

$$\bar{S}_2 = 7.983 \times 10^{-4}/(0.125 \times 0.25) = 0.0255 \text{ lb/in}^2 \ (188.5 \text{ Pa})$$

Similarly

$$\bar{S}_3 = 6.208 \text{ lb/in}^2 \ (45.9 \text{ kPa})$$

$$\bar{S}_4 = 42.88 \text{ lb/in}^2 \ (316.9 \text{ kPa})$$

$$\bar{S}_5 = 299.36 \text{ lb/in}^2 \ (2.21 \text{ MPa})$$

$$\bar{S}_6 = 2083 \text{ lb/in}^2 \ (15.39 \text{ MPa})$$

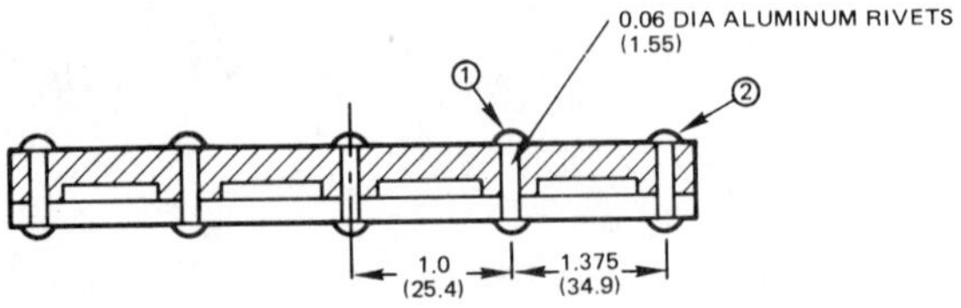

Figure 5.19. Discretely Riveted Assembly.

The maximum stiffener stress using (5.41) becomes

$$\sigma = \frac{F}{A} = \frac{37.98}{0.25^2} = 607.7 \text{ lb/in}^2$$

$$= \frac{180.5}{6.35^2} = 4.48 \text{ MPa} \qquad \text{Q.E.D.}$$

The average shear stresses for the subdivisions used in Example 5.5 follow the trend established by the distribution of peak stresses shown on Figure 5.15. Fractured bonds occur when excessive shear stresses are encountered. This may cause separation of the stiffener, which could short-circuit or damage electronic parts. The riveted assembly examined in Example 5.6 reduces the risk associated with bonding for the configuration of Example 5.4.

EXAMPLE 5.6.   Determine the fastener shear stresses and factor of safety for the riveted configuration of Example 5.5 shown in Figure 5.19.
The joint stiffness from Figure 5.16(b) becomes

$$K_1 = K_2 = Gd = (3.76 \times 10^6)(0.06) = 2.256(10^5) \text{ lb/in}$$

where

$$G = \frac{E}{2(1 + v)} \qquad (5.49)$$

$$= \frac{10^7}{2(1 + 0.33)} = 3.76(10^6) \text{ lb/in}^2 \ (25.9 \text{ GPa})$$

In metric units

$$K_1 = K_2 = (25.9 \text{ GPa})(1.55) = 40.2 \text{ MN/m}$$

The deflection ratio based on constants from Example 5.5 is

$$\frac{e_2}{e_1} = 2.256(10^5)3.06(10^{-5})1.375 + \frac{1.375}{1.0} + 1 = 11.867$$

From (5.42)

$$\frac{e_1}{F} = \frac{1}{2.256(10^5)(1 + 11.867)} = 3.445(10^{-7}) \text{ in/lb}$$

$$= \frac{1}{40.2(10^6)(1 + 11.867)} = 1.933(10^{-9}) \text{ m/N}$$

The force on the first rivet is determined from (5.43) as

$$F_1 = \frac{(0.0011625)(2.256 \times 10^5)(3.445 \times 10^{-7})}{\left(3.06 \times 10^{-5} + \dfrac{3.445 \times 10^{-7}}{1.0}\right)} = 2.92 \text{ lb}$$

$$= \frac{(0.0011625)(40.2 \times 10^6)(1.933 \times 10^{-9})}{\left(6.438 \times 10^{-6} + \dfrac{1.933 \times 10^{-9}}{25.4}\right)} = 14.03 \text{ N}$$

Using (5.47),

$$F_2 = \frac{e_2}{e_1} F_1 = 11.867(2.92) = 34.65 \text{ lb}$$

$$= 11.867(14.03) = 166.5 \text{ N}$$

Average shear stress in each rivet is obtained using

$$\overline{S}_i = 4F_i/(\pi d^2)_i$$

$$\overline{S}_1 = 4(2.92)/(\pi \times 0.06^2) = 1031 \text{ lb/in}^2$$

$$= 4(14.03)/(\pi \times 1.55^2) = 7.43 \text{ MPa} \tag{5.50}$$

$$\overline{S}_2 = 4(34.65)/(\pi \times 0.06^2) = 12254 \text{ lb/in}^2$$

$$= 4(166.5)/(\pi \times 1.55^2) = 88.24 \text{ MPa}$$

The factor of safety for the most highly stressed rivet using (5.21) becomes

$$FS = \frac{30000}{12254} = 2.46 \qquad\qquad \text{Q.E.D.}$$

Asymmetrical configurations may be analyzed by an extension of the approach illustrated.

6

Force Systems in Electronic Equipment

Fractured attachments and component leads, cracked solder joints, distorted component assemblies and mounting structures, abraded wiring and connector contact interfaces are some of the failures that may occur as a result of forces developed by static or dynamic loads and thermally induced differential expansion. Consideration of what could malfunction or fail as a result of motions or forces is fundamental to the design process. Determination of the degree of risk or the design margin necessitates a study of behavior before any equipment is built. This assessment may be obtained from simple physical models when calculable or alternatively from tests on weighted representative configurations. The development of an analytical idealization requires an understanding of the nature of load paths throughout the configuration to accurately simulate the behavior of the actual equipment.

Design characteristics for equipment that experiences temperature dependent forces were addressed in Chapter 5. Methods were presented that enable the design engineer to select mounting and attachment configurations which reduce differential displacements and accompanying stresses in order to minimize the possibility of failure. Related procedures are used to evaluate the effects of static and dynamic loads.

6.1 EQUIPMENT TOPOLOGY

Equipment configurations are uniquely designed in many different forms according to the function provided and the service expected. The assembly may consist of a housing, chassis, modules and printed wiring boards. Each of these subassemblies has a specific function and each is subjected to basically different loads. The structures of these subassemblies may be represented by continuum or lumped mass idealizations. When considered individually, beam and plate elements of a structure may be classified as a continuum. Analytical representations for relatively simple continuum elements have been developed from the governing differential equations. Lumped mass idealizations may be used to represent individual beam and plate elements as well as an assemblage of interconnected discrete members. These idealizations provide the means of quantifying the behavior of the actual structure.

Table 6.1. Common Support Conditions

Case	Symbol	Meaning
I	Fixed	Permits no rotation or translation.
II	Pinned	Permits rotation and no translation.
III	Pin-guided	Permits rotation and longitudinal translation (parallel to the plane).
IV	Hinge	Permits rotation of one member relative to another. Maintains translational continuity.
V	Elastic Hinge	Permits rotation of one member relative to another as a function of the spring constant k. Maintains translational continuity.
VI	k Elastic Support	Permits translation as a function of the spring constant k and also rotation.

Support conditions actually used at each assembly or subassembly level are represented in the idealized model. The particular representation selected is dependent on the reactions that occur and the judgement of the design engineer. Many attachments found in equipment are free to rotate about one local axis. Examples include card guides (Table 3.2A, B and C) and assemblies fastened with rivets, screws or bolts, pins or hinge attachments. Few rigid attachments are available that prevent rotation and translation at the joint.

A welded or brazed attachment develops a rigid condition but precludes disassembly. Any deformation at a rigid interface causes a rotation of the entire structure so that the angles between the connected members at the joint are maintained. A wedge lock retainer, Table 3.2D, exhibits the characteristics of a rigid joint under most conditions. Symbols used to represent various support conditions are shown in Table 6.1. It is not always possible to represent actual support conditions exactly in an idealized model.

A skeletal idealization of virtually any equipment configuration may be constructed. Structural panels, bulkheads and planar component mounting surfaces may be represented by a grid of discrete or finite elements whose properties are based on the characteristics of the surface. Chassis and rack structures may be represented by a framework of elements that consists of rigid and pin-ended members. Rounded configurations may be idealized by folding rectangular elements to approximate the desired shape incrementally. Access covers and other protective panels that experience only in-plane forces may be idealized using members that are pinned at the attachment interface. This sup-

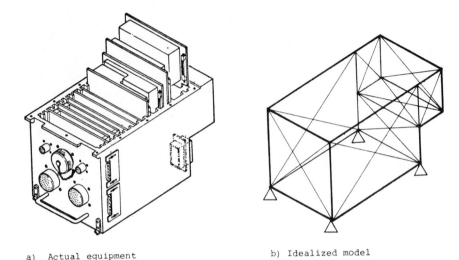

a)   Actual equipment   b)   Idealized model

Figure 6.1. Structural Idealization.

port condition assures that bending loads are not transmitted. These surfaces are often referred to as membranes.

The drawn skeletal idealization, Figure 6.1(b), consists of an assemblage of interconnected lines, each representing a structural element of the equipment. Each line depicts the centroidal axis of the structural interface. Since centroidal axes are represented, the model's dimensions do not necessarily conform to those used to describe the corresponding equipment.

The complexity of a structural model increases dramatically with the addition of each joint. Joints are used to convey information regarding the movements of the structure and the deformations experienced by each member. Each joint of a structure has the potential to translate in each of three Cartesian coordinate directions and to rotate about each of three orthogonal axes that define the coordinate system of the model. Unless some of these motions are constrained, all six displacements may occur. Each of the six joint displacements is independent in that any one does not affect the others. Models containing a large quantity of joints must be evaluated using computer methods since each displacement results in an algebraic equation. Computer methods may be used to rapidly assess the behavior of a large structure and perform parametric investigations. Manual solution techniques outlined in Appendix A provide a basis for preliminary studies on smaller scale models. Manual methods enable the design engineer to achieve a physical understanding of structural behavior, and they provide a means of verifying computer solutions which may be required later in the design cycle.

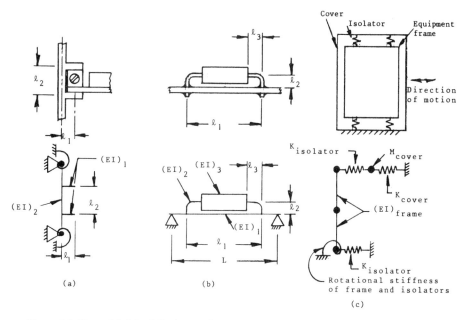

Figure 6.2. Planar Models of Equipment Structures. (a) Wedge lock interface; (b) Components; (c) Suspended equipment.

It is usually not convenient to scale down a large three dimensional model for manual evaluation. Models that develop good insight into equipment behavior may be developed by asking: What moves; how does it move; what forces are acting? The fundamental translational behavior of most equipment configurations is usually uncoupled in each of the three Cartesian coordinate directions. This criterion forms the basis for developing planar models that describe forces and accompanying translations in selected planes. Chassis motion in each coordinate direction, for example, may be sufficiently described using two planar models. Many structural and equipment assemblies can be studied using a single planar model. Figure 6.2 illustrates planar models of common configurations.

An idealized structural framework must be provided with sufficient support to insure geometrical stability under all possible loading conditions. Support conditions defined in Table 6.1 are selected to simulate the load transmission at the model interfaces. This is illustrated in Figure 6.2(a), where elastic hinges are used to account for the rotational stiffness of the attachment surface. For any model, the applied loads and external reactions must be in equilibrium under static conditions and must satisfy the laws of dynamics when exposed to time-dependent loads.

Methods of evaluating the behavior of an idealized framework essentially involve the replacement of the theory of elasticity by a simplified integrated element. In this respect, some simplifying assumptions are implied that reduce the computational effort without seriously compromising accuracy. One assumption requires the structure to exhibit both linear and elastic behavior. This requires that movement of a point obey straight line relationships. Furthermore, this assumption requires the use of an elastic material which has the ability to return to its original position when the load is removed and behaves linearly during loading and unloading. Another assumption limits external displacements to values less than the dimensions of the structural section. This has also been referred to as the *small deflection theory*. Under these conditions, rotations are sufficiently small that small angle approximations apply for sine and cosine functions. This assumption implies that angular rotations cause a point on a structural member to move in a direction perpendicular to the original member axis.

6.2 LOADS, FORCES AND STRESSES

A distinction between *loads* and *forces* is required in order to describe the methods of evaluating a structure. In a strict sense, forces are developed within the elements of a structure as a result of the external application of loads. An element can experience four different internal forces which are categorized as: axial force, shear forces, flexural moments and a torsional moment. Forces are evaluated at the joints of an element. Loads are externally applied torques, concentrated and distributed loads that act along the member and at its joints. Loads include reactions that are necessary to assure geometric stability of the structure.

A free body diagram of any structural member enables the design engineer to isolate and visualize the force and load quantities acting at a particular location. As an example, consider the cantilevered component mounting board of Figure 6.3(a) and its structural idealization in Figure 6.3(b). Although a uniform (constant EI) structure is represented, three elements are used to obtain both force and displacement quantities at intermediate points. The free body diagrams are shown in Figure 6.3(c). Directions of the internal forces at the end of each member are unknown and assumed to act as shown. The directions of internal forces are drawn consistently for each associated free body diagram. These are basic characteristics for the construction of free body diagrams of any structure. The actual directions of the forces are determined by computations. A negative quantity would indicate that the force acts opposite to the assumed direction. Reactions R_1 through R_3 provide support insuring geometric stability. Internal forces F, S and M are in equilibrium with the applied load W in each free body, satisfying

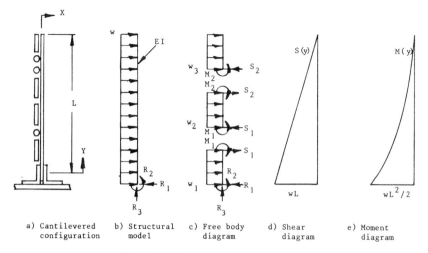

| a) Cantilevered configuration | b) Structural model | c) Free body diagram | d) Shear diagram | e) Moment diagram |

Figure 6.3. Component Board Idealization.

$$\sum \text{forces} = 0$$

$$\sum \text{moments} = 0 \qquad (6.1)$$

Forces along the member vary as a function of length for any applied loading. Considering a differential element (Figure 6.4) of the member of Figure 6.3, force quantities may be considered positive when

1. A moment produces compression in the fibers on the unloaded side and tension in the fibers on the loaded side.
2. The axial force is compressive.
3. The shear force acts to the right on the top face and to the left on the bottom face.
4. The load acts to the right.

Since the differential element is sufficiently small, the load variation $w(y)$ may be considered constant. Equations that relate the applied load to member forces are developed by applying (6.1) to the quantities of the free body diagram shown in Figure 6.4. Eliminating second order terms and simplifying gives these equations as

$$\frac{dS(y)}{dy} = -w(y)$$

$$\frac{dM(y)}{dy} = -S(y) \qquad (6.2)$$

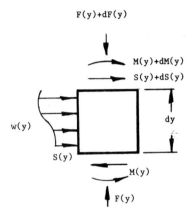

Figure 6.4. Loaded Differential Member Element.

Equation (6.2) provides the basis of obtaining member forces and displacements by direct integration. The axial force shown in Figure 6.4 is not included since it acts along the centroidal axis of the element.

EXAMPLE 6.1   Using direct integration, determine for the cantilevered component board of Figure 6.3 (a) the shear and moment distribution, and (b) the free end rotation and displacement.

(a) *Shear and Moment Distribution.* Using (6.2) in sequence for a uniformly distributed load $w(y)$ = constant gives

$$S(y) = -w \int dy = -wy + C_1 \tag{6.3}$$

$$M(y) = -\int (-wy + C_1)\, dy = \frac{wy^2}{2} - C_1 y + C_2$$

Constants of integration, C_1 and C_2 are determined from known equilibrium conditions $S(L) = M(L) = 0$. Consequently

$$S(y) = w(L - y)$$

$$M(y) = \frac{w(L^2 + y^2)}{2} - wLy$$

These functions are illustrated graphically in Figure 6.3(d) and (e).

(b) *Free End Rotation and Displacement.* The free end rotation and displacement are determined from the moment equation (6.3) by additional integrations. The angular variation as a function of length becomes

$$\theta(y) = \frac{1}{EI} \int M(y)\ dy = \frac{wy(3L^2 + y^2)}{6EI} - \frac{wLy^2}{2EI} + C_3$$

The displacement may be expressed in terms of rotation as

$$X(y) = \int \theta(y)\ dy = \frac{wy^2(6L^2 + y^2)}{24EI} - \frac{wLy^3}{6EI} + C_3 y + C_4$$

Constants of integration, C_3 and C_4 are determined from known geometrical conditions $\theta(0) = X(0) = 0$. Consequently $C_3 = C_4 = 0$, giving the end conditions as

$$\theta(L) = \frac{wL^3}{6EI}$$

$$X(L) = \frac{wL^4}{8EI}$$

where E is the elastic modulus of the material used and I is the moment inertia of the cross-section.   Q.E.D.

The differential equations (6.2) permit an evaluation of shear and flexure on a member subjected to transverse loads. These mathematical techniques, however, are efficient only if the loads are simple functions of length. The complexity and effort necessary to obtain forces and displacements increase rapidly for beams with overhangs, internal hinges and complicated loadings.

Forces and Displacements in Assemblies Idealized as Beams.   Beams can be used to represent an entire unit, assemblies and printed wiring boards. As such, it is desirable to develop direct alternative methods of obtaining behavioral characteristics of these load bearing members. Loads cause the beam to deflect. An accurate assessment of beam displacements is required to prevent misalignment, interference with adjacent assemblies and fracture of solder joints and component lead interfaces. Tabulated expressions for the displacements of beams with simple loading conditions are available in various handbooks. Deflections of more complicated loading conditions may be determined using the principle of superposition. The deflection of point A for the beam of Figure 6.5 is given by the algebraic sum of the deflections at the corresponding location for each loading condition considered separately. The displacement as a function of length for individual loadings can be found in handbooks.

An efficient alternative to the use of superposition is provided by singularity functions which can be used to evaluate the effect of complicated loadings collectively. Figure 6.6 illustrates a beam with a variety of different loads that may be used to represent the effects of components or attachments. The

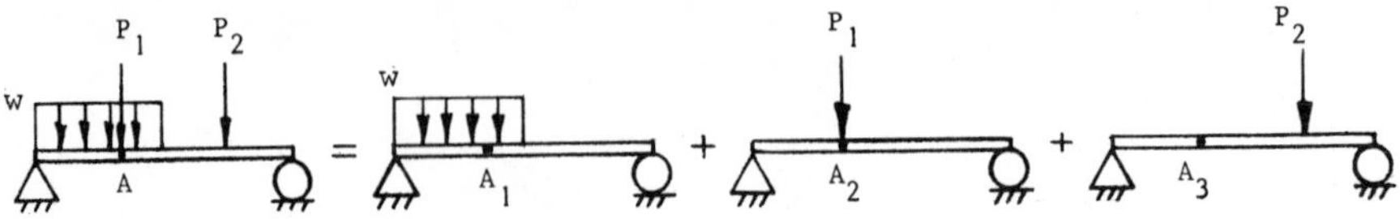

$$\Delta_A = \Delta_{A_1} + \Delta_{A_2} + \Delta_{A_3}$$

Figure 6.5. Method of Superposition for Evaluating Displacements for Complicated Load Configurations.

moment for each type of loading may be determined using (6.2) presuming the right end of the member is constrained. On this basis, individual functions of $M(x)$ can be combined into one expression that is applicable over the entire beam length. The moment representing the five loads of Figure 6.6 becomes

$$M(x) = \sum_{1}^{5} \langle M_i(x) \rangle_{ai} \tag{6.4}$$

where the moment functions $M_i(x)$ are expressed for each load as

Concentrated load: $P_{ai}(x - ai)$ (6.5a)

Distributed load: $\dfrac{w_{ai}(x - ai)^2}{2}$ (6.5b)

Linearly varying load: $\dfrac{w_{ai}(x - ai)^3}{6ai}$ (6.5c)

Concentrated moment: $M_{ai}(x - ai)^0$ (6.5d)

Hinge: $\beta_{ai}EI(x - ai)_*^{-1}$ (6.5e)

In Figure 6.6, M_{a2} and β_{a3} represent a concentrated moment and hinge respectively located ai from the left end of the beam. Distributed loads may be added or subtracted to develop a specific loading. Subtracting w_{a2} from w_{a1} in Figure 6.7 results in the desired load profile.

In (6.4), the angular brackets are used to describe a function singularity with the meaning

$$\begin{aligned} \langle f(x) \rangle_{ai} &= 0 \qquad x < ai \\ &= f(x) \qquad x \geq ai \end{aligned} \tag{6.6}$$

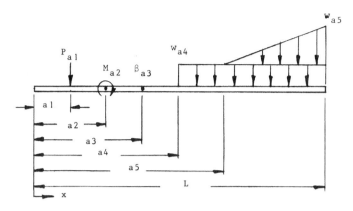

Figure 6.6. Loaded Beam (Positive Acting Loads).

The function $(x - ai)_*^{-1}$ in (6.5e) subject to (6.6) becomes infinite at $x = ai$ and is zero everywhere else. The coefficient $\beta_{ai}EI$ is known as the strength of the singularity. The function $(x - ai)^0$ in (6.5d) subject to (6.6) equals unity for $x \geq ai$.

Reactions are also included when formulating (6.4). An appropriate sign for the direction of the reaction is assumed and included algebraically. Reaction quantities may be determined in terms of the applied loading when the configuration is statically determinate following (6.1). To simplify the computation of reactions, the distributed load is replaced by a concentrated load acting at the centroid. The magnitude of the concentrated load is equal to the total distributed loading. When inclined loads are imposed on the system, it is best to replace its magnitude with orthogonal components aligned to the axis of the structure.

When the geometry contains extra (redundant) reactive loads and/or moments, the structural configuration is statically indeterminate. The unknown extra reactive loads may be evaluated from equations that describe the slope

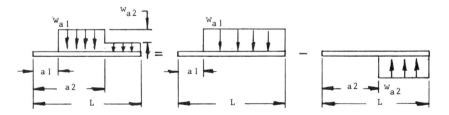

Figure 6.7. Development of a Load Profile.

and deflection of the beam. The slope and deflection of a beam resisting the moment defined by (6.4) become

$$EI\theta(x) = \int_0^x M(x)\,dx + C_3 \qquad (6.7a)$$

$$EIY(x) = \int_0^x dx \int_0^x M(x)\,dx + C_3 x + C_4 \qquad (6.7b)$$

Constants of integration C_3 and C_4 are determined from known geometrical conditions. Rules for integrating moment functions included in (6.4) are

$$\int_0^x \beta_{ai} EI\langle x - ai\rangle_*^{-1}\,dx = \beta_{ai} EI\langle x - ai\rangle^0 \qquad (6.8a)$$

$$\int_0^x Q_i\langle x - ai\rangle^n\,dx = \frac{Q_i\langle x - ai\rangle^{n+1}}{n+1} \qquad n \geq 0 \qquad (6.8b)$$

where Q_i represents a loading defined by (6.5).

The hinge represented by (6.5e) is used to transmit force without resisting rotation. The coefficient β_{ai} in (6.5e) represents the angle between the tangents to the elastic curve on each side of the hinge as shown in Figure 6.8.

EXAMPLE 6.2.   One half of a constant EI mounting plate supports uniformly distributed components weighing 0.75 lb (0.34 kg) and several heavier components that can be represented by the three concentrated loads shown in Figure 6.9. Determine the center displacement.

The reactions are determined using (6.1). Summing moments about B ($\Sigma M_B = 0$) in Figure 6.9 gives R_1 as

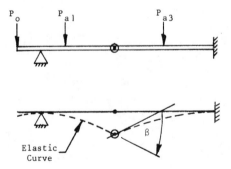

Figure 6.8. Elastic Curve at a Hinge.

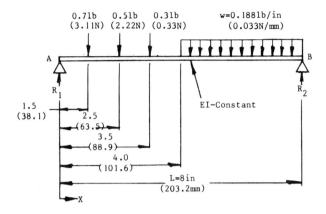

Figure 6.9. Component Loading of Example 6.2.

$$R_1 = \frac{0.7(6.5) + 0.5(5.5) + 0.3(4.5) + 0.75(2)}{8} = 1.27 \text{ lb} \quad \text{(English)}$$

$$R_1 = \frac{3.11(165) + 2.22(140) + 1.33(114) + 3.34(50.8)}{203.2} = 5.64 \text{ N} \quad \text{(metric)}$$

Even though reaction R_2 at B is not needed in the singularity expression for $M(x)$, it may be determined by summing transverse loads using (6.1) or $\Sigma F_v = 0$. The moment as a function of the distance from A in the form of (6.4) following (6.5) becomes

$$M(x) = -1.27x + 0.7\langle x - 1.5 \rangle + 0.5\langle x - 2.5 \rangle$$

$$+ 0.3\langle x - 3.5 \rangle + \frac{0.188}{2} \langle x - 4 \rangle^2 \text{ lb·in} \quad (6.9a)$$

In metric units the moment is

$$M(x) = -5.64x + 3.11\langle x - 38.1 \rangle + 2.22\langle x - 63.5 \rangle$$

$$+ 1.33\langle x - 88.9 \rangle + \frac{0.0329}{2} \langle x - 101.6 \rangle \text{ N·mm} \quad (6.9b)$$

The displacement at any location may be determined from (6.9) using (6.7b) as

$$EIY(x) = \frac{-1.27x^3 + 0.7\langle x - 1.5 \rangle^3 + 0.5\langle x - 2.5 \rangle^3 + 0.3\langle x - 3.5 \rangle^3}{6}$$

$$+ \frac{0.188\langle x - 4 \rangle^4}{24} + C_3 x + C_4 \quad (6.10)$$

Constants of integration C_3 and C_4 are determined from known geometrical conditions for the configuration in Figure 6.9. In this case, the displacement at A and B is zero. This may be stated as $y(0) = y(L) = 0$. The displacement given by (6.10) satisfies the condition at A provided $C_4 = 0$. Constant C_3 may be evaluated for the condition $y(L) = 0$ by setting $x = L$ in (6.10), giving

$$C_3 = \frac{1.27(8^3) - 0.7(6.5^3) - 0.5(5.5^3) - 0.3(4.5^3)}{6(8)} - \frac{0.188(4^4)}{24(8)} = 6.99 \text{ lb} \cdot \text{in}^2$$

In metric units

$$C_3 = \frac{5.64(203.2^3) - 3.11(165.1^3) - 2.22(139.7^3) - 1.33(114.3^3)}{6(203.2)}$$

$$- \frac{0.0329(101.6^4)}{24(203.2)} = 20 \text{ kN} \cdot \text{mm}^2$$

The center deflection at $x = L/2$ may be determined from (6.10) as

$$y(4) = \frac{1}{EI} \left[\frac{-1.27(4^3) + 0.7(2.5^3) + 0.5(1.5^3) + 0.3(0.5^3)}{6} \right.$$

$$\left. + 6.99(4) \right] = \frac{16.5}{EI} \text{ in}$$

$$y(101.6) = \frac{47.3(10^3)}{EI} \text{ mm}$$

where $L/2 = 4$ in (101.6 mm).   Q.E.D.

The versatility of the singularity approach for complicated loadings is illustrated by Example 6.3.

EXAMPLE 6.3. Constraints imposed by an equipment configuration necessitates the use of a mounting surface which is composed of two different structural cross-sections as shown in Figure 6.10. One section supports a linear varying weight distribution of components and is rigidly fastened to a stiff chassis wall. Several components modeled as concentrated loads are supported by a second member which is fastened to the first and a cross-member with screws. Determine the deflection of the free end of the second member at A and at the hinge attachment between the two members at B.

The moment relationship is developed in the form of (6.4) using (6.5). The moment for segment AB becomes

$$M(x) = 0.7x + 0.3\langle x - 1.5 \rangle - R_1 \langle x - 2.5 \rangle$$

$$+ 0.5\langle x - 3.5 \rangle + S_B \langle x - 4 \rangle \text{ lb} \cdot \text{in}   \quad (6.11a)$$

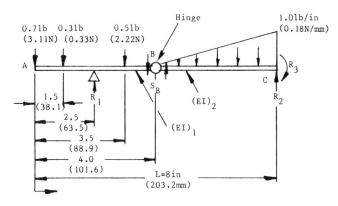

Figure 6.10. Component Loading of Example 6.3.

which in metric units is

$$M(x) = 3.11x + 1.33\langle x - 38.1\rangle - R_1\langle x - 63.5\rangle$$
$$+ 2.22\langle x - 88.9\rangle + S_B\langle x - 101.6\rangle \ \text{N·mm} \quad (6.11\text{b})$$

The reaction R_1 may be determined from (6.11) since the moment at A ($x = 0$) and at the hinge B ($x = 4$) is zero. Based on the condition $M(4) = 0$, R_1 is

$$R_1 = \frac{0.7(4) + 0.3(2.5) + 0.5(0.5)}{1.5} = 2.53 \ \text{lb} \quad \text{(English)}$$

$$R_1 = \frac{3.11(101.6) + 1.33(63.5) + 2.22(12.7)}{38.1} = 11.25 \ \text{N} \quad \text{(metric)}$$

The vertical shear force at the hinge S_B may be determined from a summation of vertical forces (6.1) on segment AB as

$$S_B = 2.53 - 0.7 - 0.3 - 0.5 = 1.03 \ \text{lb} \quad \text{(English)}$$

$$S_B = 11.25 - 3.11 - 1.33 - 2.22 = 4.59 \ \text{N} \quad \text{(metric)}$$

The slope along segment AB may be determined from (6.11a) using (6.7a) as

$$(EI)_1\theta(x)$$

$$= \frac{0.7x^2 + 0.3\langle x - 1.5\rangle^2 - 2.53\langle x - 2.5\rangle^2 + 0.5\langle x - 3.5\rangle^2 + 1.03\langle x - 4\rangle^2}{2}$$

$$+ C_3 \quad (6.12)$$

Relationships for segment BC of Figure 6.10 are similarly developed. The moment becomes

$$M(x) = \beta(EI)_2\langle x - 4\rangle^{-1}_* - S_B\langle x - 4\rangle + \frac{\langle x - 4\rangle^3}{24} \text{ lb·in} \quad \text{(English)} \quad (6.13a)$$

$$M(x) = \beta(EI)_2\langle x - 101.6\rangle^{-1}_* - S_B\langle x - 101.6\rangle \quad (6.13b)$$

$$+ \frac{0.175\langle x - 101.6\rangle^3}{609.6} \text{ N·mm} \quad \text{(metric)}$$

The slope along segment BC may be determined from (6.13a) using (6.7a) as

$$(EI)_2\theta(x) = \beta(EI)_2\langle x - 4\rangle^0 - \frac{1.03\langle x - 4\rangle^2}{2} + \frac{\langle x - 4\rangle^4}{96} \quad (6.14)$$

By virtue of continuity, the slope and displacement at B must be the same for both adjoining segments. The slope at B along segment AB from (6.12) becomes the initial value for (6.14). This slope would remain constant for segment BC if no other loads were imposed. Under these conditions, the relationships defined by (6.12) and (6.14) are additive, giving a continuous function of $\theta(x)$ for the entire beam as

$$\theta(x) = \frac{0.7x^2 + 0.3\langle x - 1.5\rangle^2 - 2.53\langle x - 2.5\rangle^2 + 0.5\langle x - 3.5\rangle^2 + 1.03\langle x - 4\rangle^2}{2(EI)_1}$$

$$+ \beta\langle x - 4\rangle^0 + \frac{\langle x - 4\rangle^4}{96(EI)_2} - \frac{1.03\langle x - 4\rangle^2}{2(EI)_2} + C_3 \quad (6.15a)$$

The slope for the beam expressed in metric units is

$$\theta(x) = \frac{3.11x^2 + 1.33\langle x - 38.1\rangle^2 - 11.25\langle x - 63.5\rangle^2 + 2.22\langle x - 88.9\rangle^2 + 4.59\langle x - 101.6\rangle^2}{2(EI)_1}$$

$$+ \beta\langle x - 101.6\rangle^0 + \frac{0.175\langle x - 101.6\rangle^4}{2438(EI)_2} - \frac{4.59\langle x - 101.6\rangle^2}{2(EI)_2} + C_3 \quad (6.15b)$$

The deflection at any location along the beam may be determined from (6.7b) using (6.15a) as

$$y(x) = \frac{0.7x^3 + 0.3\langle x - 1.5\rangle^3 - 2.53\langle x - 2.5\rangle^3 + 0.5\langle x - 3.5\rangle^2 + 1.03\langle x - 4\rangle^3}{6(EI)_1}$$

$$+ \beta\langle x - 4\rangle + \frac{\langle x - 4\rangle^5}{480(EI)_2} - \frac{1.03\langle x - 4\rangle^3}{6(EI)_2} + C_3x + C_4 \quad (6.16)$$

Constants of integration C_3 and C_4 in addition to the concentrated angle change β are determined from known geometrical conditions for the configuration shown in Figure 6.10. Conditions of slope and deflection at points of constraint becomes $\theta(8) = y(8) = y(2.5) = 0$. Substituting the location for each condition into (6.15a) or (6.16) gives three simultaneous equations for determining C_3, C_4 and β. These may be expressed as

$$
\begin{bmatrix}
1 & 1 & 0 \\
0 & 2.5 & 1 \\
4 & 8 & 1
\end{bmatrix}
\begin{Bmatrix}
\beta \\
C_3 \\
C_4
\end{Bmatrix}
=
\begin{Bmatrix}
-\dfrac{3.774}{(EI)_1} + \dfrac{5.573}{(EI)_2} \\[2mm]
-\dfrac{1.873}{(EI)_1} \\[2mm]
-\dfrac{21.889}{(EI)_1} + \dfrac{8.854}{(EI)_2}
\end{Bmatrix}
$$

The unknown quantities may be determined using the methods of Appendix A as

$$
\beta = -\frac{0.486}{(EI)_1} + \frac{14.53}{(EI)_2} \text{ rad}
$$

$$
C_3 = -\frac{3.282}{(EI)_1} - \frac{8.959}{(EI)_2} \text{ rad}
$$

$$
C_4 = \frac{6.331}{(EI)_1} + \frac{22.39}{(EI)_2} \text{ in}
$$

The deflection at A, $(x = 0)$ is given by (6.16) as C_4, or

$$
y(0) = \frac{6.331}{(EI)_1} + \frac{22.39}{(EI)_2} \text{ in} \quad \text{(English)}
$$

$$
y(0) = \frac{4.6(10^5)}{(EI)_1} + \frac{1.63(10^6)}{(EI)_2} \text{ mm} \quad \text{(metric)}
$$

Positive signs indicate a downward deflection. The deflection at B becomes

$$
y(4) = -\frac{13.4}{(EI)_2} \text{ in} \quad \text{(English)}
$$

$$
y(101.6) = -\frac{9.76(10^5)}{(EI)_2} \text{ mm} \quad \text{(metric)}
$$

Negative signs indicate an upward deflection at point B.   Q.E.D.

Moment functions similar to (6.5) may be derived which include loads due to longitudinal tension or compression and those developed by an elastic foundation constraint. These loads introduce trigonometric relationships that increase the complexity and the effort required to obtain behavioral characteristics of the structure. The effect of these loads can be included in a matrix approach described in Chapter 7.

Influence of Materials and Section Properties.   Values of shear and moment along a member provide the basis of selecting a structural cross-section that would avoid overstress, permanent deformation and fracture at structural attachments and component interfaces. This selection is facilitated by a comparison of calculated member stresses with limiting material stresses. Material stresses are affected by the processing and heat treatment used in developing the raw material. Corrosion, temperature and the nature of loading can alter the limiting stress of materials. A force that ruptures a specimen of known material and configuration establishes its ultimate strength (stress). The selection of structural sections is based on an allowable design stress which is considerably less than the ultimate material strength due to variations in properties and unknown force distributions across the loaded cross-section.

The capabilities of a design depend on the relationship between material stress and member stress due to static, transient, oscillatory or randomly fluctuating loads. Ultimate material stresses are used as limiting quantities to assure that catastrophic failure (rupture) does not occur.

Even though rupture is prevented, distortion and loss of performance may still occur. Allowable design stress levels which are tailored to the anticipated loading will assure that performance and mechanical alignments are preserved. Fatigue data is employed to select materials and structural sections that will resist oscillatory and randomly fluctuating loads.

Materials can withstand an increasing number of cycles when the applied oscillating load develops low stresses. Some materials can withstand an infinite number of cycles at low stress levels. The stress level which limits the number of fatigue cycles defines the endurance limit of the material.

If the applied loading consists of a flexural moment in addition to longitudinal and shear forces, complex stresses will result. For pure flexure, the stress in a member is related to the applied moment by

$$\sigma = \frac{Mr}{I} \tag{6.17a}$$

where r represents the perpendicular distance from the neutral axis to the point of consideration on a member cross-section. The stress and the strain vary linearly from the neutral axis. If c is used to designate $|r|_{max}$ for a particular material or interface, then (6.17a) becomes

$$\sigma_{max} = \frac{Mc}{I} \tag{6.17b}$$

In (6.17), I is the moment of inertia of the whole cross-sectional area of a member about its neutral or centroidal axis. Its value is defined by the integral

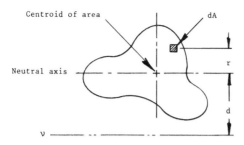

Figure 6.11. Moment of Inertia about a Parallel Axis.

of $r^2\,dA$ over the entire area about the neutral axis. Values of moments of inertia for simple shapes can be found in Appendix C and in most civil or mechanical engineering handbooks. The cross-sections of equipment and most structural members may be partitioned into a combination of these simple shapes to facilitate an evaluation of I. When the structural section of an assembly is composed of different materials, the simultaneous evaluation of EI is required. A compact relationship for determining EI about the centroidal axis of a section may be developed from the parallel axis approach for evaluating I. The moment of inertia about the centroidal axis of the section shown in Figure 6.11 is given by

$$I_{cg} = \int_A r^2\,dA \qquad (6.18)$$

The moment of inertia about any parallel axis ν located a distance d from the centroidal axis of an area is obtained by expressing (6.18) as

$$I_\nu = \int_A (d + r)^2\,dA \qquad (6.19a)$$

Equation (6.19a) may be evaluated by squaring the quantity $(d + r)$, giving

$$I_\nu = d^2 \int_A dA + 2d \int_A r\,dA + \int_A r^2\,dA$$
$$I_\nu = Ad^2 + I_{cg} \qquad (6.19b)$$

where d is constant and $\int_A r\,dA = 0$ about the centroidal axis. The moment of inertia of integral sections and equipment sections assembled using one material may be obtained from expanding (6.19b) as

$$I_{cg} = \sum I_{cg} + \sum Ar^2 - \frac{(\sum Ar)^2}{\sum A} \qquad (6.20a)$$

where $\sum I_{cg}$ represents the sum of each partitioned area of a section about their centroidal axis. Equation (6.20) is easily applied to complicated sections which are partitioned into simple shapes. Additions or modifications to any portion of a section are readily accommodated by (6.20) since only known quantities are used. For a section composed of several materials, (6.20a) becomes

$$(EI)_{cg} = \sum (EI)_{cg} + \sum AEr^2 - \frac{(\sum AEr)^2}{\sum AE} \qquad (6.20b)$$

The location of the neutral axis relative to a selected reference axis is determined from

$$\bar{r} = \frac{\sum AEr}{\sum AE} \qquad (6.21a)$$

For sections composed of one material, (6.21a) becomes

$$\bar{r} = \frac{\sum Ar}{\sum A} \qquad (6.21b)$$

EXAMPLE 6.4. Determine the moment of inertia of the integral section shown in Figure 6.12(a) about its neutral axis and the location of the neutral axis relative to the bottom of the section. Evaluate the effect of equal length ribs on the calculated properties of the sections above.

Figure 6.12(b) illustrates the partitioning of the section into simple shapes of known properties. The summations represented by (6.20a) are determined systematically using the tabular development of Table 6.2. The moment of inertia and location of the

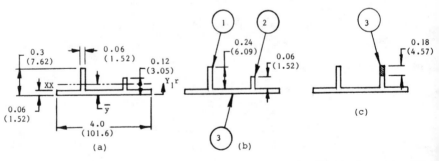

Figure 6.12. Centroid and Moment of Inertia of an Integral Member. (a) Member section; (b) Partitioned section; (c) Design modification.

Table 6.2. Development of Quantities from Eq. 6.20a

Section	Area A in^2 (mm^2)	r in (mm)	Ar in^3 (mm^3)	Ar^2 in^4 (mm^4)	$I_{cg} = bh^3/12$ in^4 (mm^4)
1	0.0144 (9.27)	0.18 (4.57)	0.00259 (42.36)	0.466×10^{-3} (193.6)	6.91×10^{-5} (56.04)
2	0.0036 (2.31)	0.09 (2.28)	0.00032 (5.27)	0.029×10^{-3} (12.0)	0.11×10^{-5} (0.44)
3	0.24 (154.43)	0.03 (0.76)	0.00720 (117.37)	0.216×10^{-3} (89.2)	7.20×10^{-5} (29.73)
	0.258 (166.01)		0.01011 (165)	0.711×10^{-3} (294.8)	14.22×10^{-5} (86.21)

section centroid are determined using (6.20a) with summations from Table 6.2 as

$$I_{xx} = 14.22(10^{-5}) + 0.711(10^{-3}) - \frac{0.01011^2}{0.258} = 4.57(10^{-4}) \text{ in}^4$$

$$\bar{y} = \frac{0.01011}{0.258} = 0.039 \text{ in}$$

English

$$I_{xx} = 86.21 + 294.8 - \frac{165^2}{166.01} = 216.9 \text{ mm}^4$$

$$\bar{y} = \frac{165}{166.01} = 0.99 \text{ mm}$$

metric

Equation (6.20) simplifies the redetermination of section properties as a result of configuration modifications due to changes that evolve during the course of the design process. The effect of extending the right rib in Figure 6.12 is determined using the results obtained in Table 6.2. The updated quantities for the configuration in Figure 6.12(c) become

Section	Area A in^2 (mm^2)	r in (mm)	Ar in^3 (mm^3)	Ar^2 in^4 (mm^4)	$I_{cg} = bh^3/12$ in^4 (mm^4)
Σ 1 + 2 + 3	0.258 (166.01)		0.01011 (165)	0.711×10^{-3} (294.8)	14.22×10^{-5} (86.21)
4	0.0108 (6.95)	0.21 (5.33)	0.00227 (37.04)	0.476×10^{-3} (197.4)	2.91×10^{-5} (12.09)
	0.269 (172.96)		0.01238 (202.04)	1.187×10^{-3} (492.2)	17.13×10^{-5} (98.3)

The moment of inertia and centroid of the modified section are determined from (6.20a) as

$$I_{xx} = 17.13(10^{-5}) + 1.187(10^{-3}) - \frac{0.01238^2}{0.269} = 7.88(10^{-4}) \text{ in}^4$$

$$\bar{y} = \frac{0.01238}{0.269} = 0.046 \text{ in}$$

English

$$I_{xx} = 98.3 + 492.2 - \frac{202.04^2}{172.96} = 354.5 \text{ mm}^4$$

$$\bar{y} = \frac{202.04}{172.96} = 1.168 \text{ mm}$$

metric

Q.E.D.

The moment of inertia of a section about an axis of symmetry can be obtained directly using

$$(EI)_{cg} = \sum_{}^{n} (EI)_i \tag{6.22a}$$

When the assembly is composed of a common material (6.22a) becomes

$$I_{cg} = \sum_{}^{n} I_i \tag{6.22b}$$

EXAMPLE 6.5.   Determine the EI of a composite circuit board assembly which is fabricated by bonding equal thicknesses of epoxy fiberglass to an aluminum plate as show in Figure (6.13).

Elastic moduli for the materials in the section are given in Appendix C. Since the section is composed of two symmetrical components, (6.22a) becomes

$$(EI)_{xx} = (EI)_{epoxy} + (EI)_{alum}$$

$$(EI)_{xx} = 4 \left[\frac{2(10^6)(0.1^3 - 0.04^3)}{12} + \frac{(10^7)(0.04^3)}{12} \right] = 837.31 \text{ lb} \cdot \text{in}^2$$

which in metric units is

$$(EI)_{xx} = 101.6 \left[\frac{1.38(10^4)(2.54^3 - 1.02^3)}{12} \right.$$

$$\left. + \frac{6.89(10^4)(1.02^3)}{12} \right] = 2.41(10^6) \text{ N} \cdot \text{mm}^2 \quad \text{Q.E.D.}$$

Influence of Joining and Fastening Methods.   Electronic equipment assemblies are usually composed of a structural frame or chassis, panels, interconnection boards, circuit boards and component modules. These pieces, parts and components are organized and fastened together into a unique configuration that provides the intended functions and performance. The placement and design of each element should account for the mode of heat transfer used and

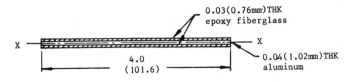

Figure 6.13. Symmetrical Composite Assembly.

Table 6.3. Structural Efficiencies of Common
Attachments

Interface Configuration	Structural Efficiency, %
Brazed, Welded or Bonded	100
Shear Pins and Screws	80
Distributed Cap Head Screws	50
Distributed Pan or Phillips Head Screws	25
Distributed ¼-turn Fasteners	10

the transfer of loads to the equipment mounting interface to assure that the design objectives are met in the service environment.

Methods of attachment can influence the cooling and structural behavior of assembled elements. The amount of heat that can be transferred and the deflection of an assembly depends on whether the attachment interface is bonded or bolted. Bonding minimizes the independence of and motion between adjoining assemblies thereby developing a greater structural stiffness than can be obtained by bolted connections. Bolted attachments do not constrain displacements within the span between fasteners. The structural efficiency of the interface, however, improves with an increasing number of fasteners. The load transfer ability of an attachment also depends on the type of fastener that is used. A quarter-turn fastener, for example, is not as structurally effective as a cap screw. Table 6.3 lists the relative ranking of various attachments based on considering a welded, brazed or bonded conection as being 100 percent structurally efficient.

Many electronic assemblies or modules are designed to be easily removed and replaced in order to facilitate test and repair. The attachment interface for these assemblies must be suitable for transferring the developed loads and the generated heat. In many cases, the objective of minimizing the thermal impedance of an interface also increases the efficiency of the connection. In general, the efficiency of an attachment increases with the level of difficulty and time required for removal or disassembly.

Structural efficiencies are incorporated in (6.20) and (6.22) by derating the section area and moment of inertia of the attached element. The section area 0.5 in^2 (322.6 mm^2) and inertia 0.4 lb$\cdot$in^2 (1148 N$\cdot$mm^2) of an access cover retained with pan head screws are reduced to 0.125 in^2 (80.6 mm^2) and 0.1 lb$\cdot$in^2 (287 N$\cdot$mm^2) respectively before performing the operations defined by (6.20) or (6.22).

EXAMPLE 6.6.   Determine $(EI)_{xx}$ and $\bar{y}$ for the section in Figure 6.14.

The section of Figure 6.14 has been partitioned into six simple shapes of known properties. Assuming section symmetry, the summations represented by (6.20b) are

Table 6.4. Development of Quantities from Eq. 6.20b

Section	Effective Area, A		E		r		AEr		AEr^2		$(EI)_{cg} = Ebh^3/12$	
	in^2	mm^2	lb/in^2	N/mm^2	in	mm	$lb \cdot in$	$N \cdot mm$	$lb \cdot in^2$	$N \cdot mm^2$	$lb \cdot in^2$	$N \cdot mm^2$
1	0.099	63.7	10^7	6.89×10^4	0.03	0.76	2.97×10^4	2.33×10^6	891.	2.53×10^6	297.	8.45×10^5
2	0.099	63.7	10^7	6.89×10^4	7.17	182.1	7.10×10^6	7.99×10^8	5.1×10^7	1.46×10^{11}	297.	8.45×10^5
3	0.340	220.5	10^7	6.89×10^4	0.20	5.1	6.80×10^5	7.75×10^7	1.36×10^5	3.95×10^8	3.27×10^4	9.53×10^7
4	0.340	220.5	10^7	6.89×10^4	7.00	177.8	2.38×10^7	2.70×10^9	1.67×10^8	4.80×10^{11}	3.27×10^4	9.53×10^7
5	1.024	663.0	10^7	6.89×10^4	3.60	91.4	3.68×10^7	4.17×10^9	1.32×10^8	3.81×10^{11}	3.49×10^7	1.01×10^{11}
6	0.126	81.4	2×10^6	1.38×10^4	1.61	40.9	4.06×10^5	4.59×10^7	6.53×10^5	1.88×10^9	170.	4.91×10^5
							6.88×10^7	7.79×10^9	3.51×10^8	1.01×10^{12}	3.49×10^7	1.01×10^{11}

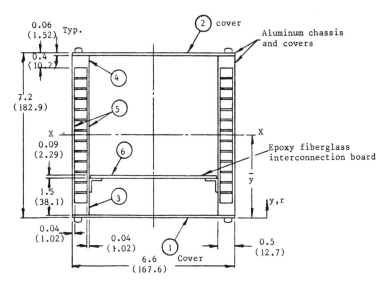

Figure 6.14. Section of a Forced Cooled Assembly.

determined systematically using the tabular development of Table 6.4. The inertia and section area of both pan-head fastened covers and the interconnect board have been reduced 75 percent to account for their structural efficiencies as given in Table 6.3.

$$\Sigma AE = 1.927 \times 10^7 \text{ lb} (8.59 \times 10^7 \text{ N})$$

Following (6.20)b and Table 6.4, $(EI)_{xx}$ and $\bar{y}$ are determined as

$$(EI)_{xx} = 3.49 \times 10^7 + 3.51 \times 10^8 - \frac{(6.88 \times 10^7)^2}{1.927 \times 10^7}$$

$$= 1.4 \times 10^8 \text{ lb} \cdot \text{in}^2$$

$$\bar{y} = \frac{6.88 \times 10^7}{1.927 \times 10^7} = 3.57 \text{ in}$$

⎫
⎬ English
⎭

$$(EI)_{xx} = 1.01 \times 10^{11} + 1.01 \times 10^{12} - \frac{(7.79 \times 10^9)^2}{8.59 \times 10^7}$$

$$= 4.04 \times 10^{11} \text{ N} \cdot \text{mm}^2$$

$$\bar{y} = \frac{7.79 \times 10^9}{8.59 \times 10^7} = 90.7 \text{ mm}$$

⎫
⎬ metric
⎭

Q.E.D.

Table 6.5. Characteristics of Idealized Configurations

Case	Configuration	D_o	M_{max}	F_o	S_r	S_Δ
I		$\dfrac{WL}{48}$	$\dfrac{WL}{4}$	3.13 (15.8)	117.4 (2982)	12.00
II		$\dfrac{3WL}{322}$	$\dfrac{3WL}{16}$	3.16 (15.9)	201.0 (5105)	20.12
III		$\dfrac{WL}{192}$	$\dfrac{WL}{8}$	3.13 (15.8)	234.9 (5966)	24.00
IV		$\dfrac{WL}{3}$	WL	3.13 (15.8)	29.36 (746)	3.00
V		$\dfrac{5WL}{384}$	$\dfrac{WL}{8}$	3.52 (17.7)	119.2 (3028)	9.60
VI		$\dfrac{WL}{185}$	$\dfrac{WL}{8}$	3.54 (17.8)	289.6 (7356)	23.06
VII		$\dfrac{WL}{384}$	$\dfrac{WL}{12}$	3.57 (18.0)	408.7 (10381)	32.00
VIII		$\dfrac{WL}{8}$	$\dfrac{WL}{12}$	3.89 (19.6)	60.6 (1539)	4.00
IX		$\dfrac{M}{8}$	M			8.00
X		$\dfrac{M}{9\sqrt{3}}$	M			15.57

Max Static Displacement	Max Static Moment	Fundamental Frequency, Hz	Flexural Stress, σ	
			$F(f_n)$	$F(\Delta_r)$
$\delta_{st} = \dfrac{D_o L^2}{EI}$	M_{max}	$f_n = \dfrac{F_o}{\sqrt{\delta_{st}}}$	$\dfrac{S_f EcGQ}{(f_n L)^2}$	$\dfrac{S_\Delta Ec\Delta_r}{L^2}$

G denotes gravity units of the applied loading, Δ_{st} denotes static displacement, in (mm), Δ_r denotes the dynamic response displacement ($\Delta_r = QG\delta_{st}$) when for static loads $\Delta_r = \delta_{st}$.

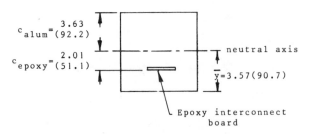

Figure 6.15. Location of Maximum Stresses for Example 6.7.

Flexural Stress and Natural Frequency of Idealized Assemblies. The structural capability of a section is dependent upon the relationship between the flexural stresses due to the applied load and the materials used. Deflections which may be calculated or measured are often used to assess the suitability of an assembly or member. When dynamic loads are imposed the natural frequency provides the best data for determining what displacements and stresses are to be expected.

Expressions that relate stresses to either displacement or frequency may be developed for any member/load configuration. Many equipment configurations can be idealized to obtain insight rapidly in order to determine whether modifications are required to minimize flexure and the accompanying stresses. Common idealized configurations and the relationship that interrelate stresses with deflection or frequency are listed in Table 6.5.

EXAMPLE 6.7. A simply supported 12 in (305 mm) long electronics assembly with section characteristics described in Example 6.6 is designed to withstand an applied acceleration of 10G. Determine the maximum flexural stresses in the aluminum structure, epoxy fiberglass interconnection board and copper trace interconnect for a calculated maximum static displacement of 0.001 to (0.025 mm) for the assembly.

The maximum flexural stresses may be determined from the relationships given by case V of Table 6.5, using distances to the most remote fibers shown in Figure 6.15. Maximum flexural stresses for the aluminum structure and epoxy interconnect board for a dynamic displacement of 0.01 in (0.25 mm) with $Q = 1$ become

$$\left.\begin{array}{l} \sigma_{\text{aluminum}} = \dfrac{9.6(10^7)(3.63)(0.01)}{12^2} = 24200 \text{ lb/in}^2 \\[4mm] \sigma_{\text{epoxy}} = \dfrac{9.6(2 \times 10^6)(2.01)(0.01)}{12^2} = 2680 \text{ lb/in}^2 \end{array}\right\} \text{ English}$$

$$\left.\begin{array}{l} \sigma_{\text{aluminum}} = \dfrac{9.6(68944)(92.2)(0.25)}{305^2} = 164 \text{ N/mm}^2 \\[4mm] \sigma_{\text{epoxy}} = \dfrac{9.6(13788)(51.1))0.25}{305^2} = 18.2 \text{ N/mm}^2 \end{array}\right\} \text{ metric}$$

Maximum stresses in the copper circuit trace may be determined from

$$\sigma_{\text{copper}} = \frac{E_{\text{copper}}}{E_{\text{epoxy}}} \sigma_{\text{epoxy}} \tag{6.23}$$

which gives

$$\sigma_{\text{copper}} = \frac{(17.6 \times 10^6)(2680)}{(2.0 \times 10^6)} = 23584 \text{ lb/in}^2 \quad \text{(English)}$$

$$\sigma_{\text{copper}} = \frac{(1.21 \times 10^5)(18.2)}{(13788)} = 158 \text{ N/mm}^2 \quad \text{(metric)} \qquad \text{Q.E.D.}$$

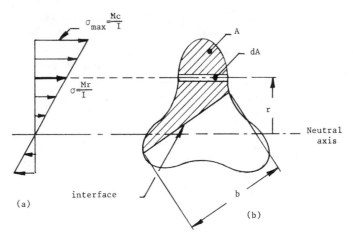

Figure 6.16. Axial Force over a Portion of a Structural Member. (a) Flexural stress distribution; (b) Member cross section.

Equipment structures are often built up using many individual structural sections which contribute collectively to the moment of inertia of the assembly. The transfer of load between connected structural sections is a function of the relative motion or slipping that can occur at the interface. Structural efficiencies listed in Table 6.3 reflect the load transfer ability and integrity of the interface for various attachment methods.

Stresses at structural interfaces and the accompanying attachment requirements to limit slipping are dependent on the flexural moment and temperature gradient experienced by the member. The influence of temperature on structural interfaces and attachments have been addressed in Chapter 5. The degree of relative motion that may occur at an interface of a composite structural member depends on the suitability of the attachment method used to resist the shear forces developed by the applied moment.

Shear Stress in Bonded or Bolted Assemblies Subjected to Flexure. The shear force along the boundary of connected elements in a composite structural section is dependent upon variability of the applied moment over the length of the member. Shear actions are due to the difference between axial forces developed by flexural stresses acting along the member. The axial force on an infinitesimal area dA due to the flexural stress distribution shown in Figure 6.16 may be expressed in terms of (6.17a) as

$$dF = -\left(\frac{Mr}{I}\right) dA \qquad (6.24)$$

The total axial force acting on an element of area A is obtained by integrating (6.24) over the element cross-section, giving

$$F = -\frac{M}{I} \int_A r \, dA = -\frac{M A \bar{r}}{I} \tag{6.25}$$

where $\bar{r}$ is the distance from the neutral axis of the section to the centroid of area A. The shear force at the interface boundary of area A in terms of the change in flexural moment as a function of member length is obtained from (6.25) as

$$\frac{dF}{dx} = -\frac{A\bar{r}}{I}\left(\frac{dM}{dx}\right) \tag{6.26}$$

A convenient expression for determining the axial shear flow in terms of the applied shear load is obtained from (6.2) and (6.26) as

$$q = \frac{S A \bar{r}}{I} \tag{6.27}$$

where $q = dF/dx$ represents the shear force per unit member length. The shear stress in an integrally connected interface may be expressed in terms of (6.27) and the boundary width b in Figure 6.16 as

$$\sigma_\tau = \frac{S A \bar{r}}{I b} \tag{6.28}$$

Integral attachments are developed by brazing, continuous welding and bonding. Fracture or separation at the interface is avoided by assuring that σ_τ is less than the allowable shear stresses of the joining material.

Rivets, screws and spot welds provide attachment at discrete locations along the member. If F_s is the known allowable shear load capability of one attachment, the quantity of attachments required per inch of length is determined from (6.27) as

$$N = q/F_s \tag{6.29}$$

EXAMPLE 6.8.   Aluminum ribs are used to reduce the deflection of a 1.5 lb (0.681 kg) simply supported glass epoxy component board assembly, Figure 6.17, which may be exposed to dynamic loads up to 20g. Both bonding or screw rib attachment methods are being considered along the 8 in (203.2 mm) length of the assembly. For each

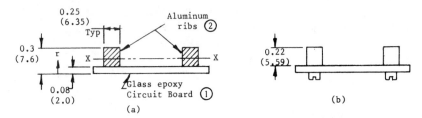

Figure 6.17. Adhesive and Screw Requirements. (a) Bonded configuration; (b) Bolted configuration.

method of attachment determine the epoxy board flexural stress in addition to (a) the maximum adhesive shear stress for the bonded configuration and (b) the quantity of screws required for the bolted configuration if each screw resists 300 lb (1334 N) shear.

(a) *Bonded Configuration.* The summations represented by (6.20b) for the partitioned configuration of Figure 6.17 are determined systematically as

Section	Area A in^2	E lb/in^2	AE, lb	r, in	AEr	AEr^2	$I_{cg} = bh^3/12$ in^4	$(EI)_{cg}$
1	0.48	$2(10^6)$	$0.96(10^6)$	0.04	38400	1536	$2.56(10^{-4})$	512.
2	0.11	10^7	$1.10(10^6)$	0.19	209000	39710	$4.44(10^{-4})$	4440.
			$2.06(10^6)$		247400	41246		4952.

The moment of inertia and centroidal location are given by (6.20b) and (6.21a) as

$$(EI)_{xx} = 4952 + 41246 - \frac{247400^2}{2.06(10^6)} = 16486 \text{ lb} \cdot \text{in}^2 \ (47 \text{ MN} \cdot \text{mm}^2)$$

$$\bar{y} = \frac{247400}{2.06(10^6)} = 0.12 \text{ in} \ (3.05 \text{ mm})$$

Assuming the weight distribution of the assembly is uniformly distributed, the maximum applied shear located at the supports becomes

$$S = 20g(1.5 \text{ lb}/2) = 15 \text{ lb} \quad \text{(English)}$$

$$S = 20g(6.67 \text{ N}/2) = 66.7 \text{ N} \quad \text{(metric)}$$

The first statical moment of area $A\bar{r}$ about the neutral axis of either the aluminum ribs or the epoxy board may be used in (6.28). Considering the ribs with a total area of 0.11 in^2 (70.9 mm^2)

$$A\bar{r} = 0.11(0.3 - 0.11 - \bar{y}) = 0.0077 \text{ in}^3 \quad \text{(English)}$$

$$A\bar{r} = 70.9(7.6 - 2.8 - \bar{y}) = 126.2 \text{ mm}^3 \quad \text{(metric)}$$

The moment of inertia with respect to aluminum becomes

$$I_{xx} \text{ (aluminum)} = \frac{16486}{10^7} = 1.65(10^{-3})\text{in}^4 \ (686.2 \text{ mm}^4)$$

Maximum adhesive shear stress is determined from (6.28) for a boundary width of 0.5 in (12.7 mm) as

$$\sigma_r = \frac{15(0.0077)}{(1.65 \times 10^{-3})(0.5)} = 140 \text{ lb/in}^2 \quad \text{(English)}$$

$$\sigma_r = \frac{66.7(126.2)}{(686.2)(12.7)} = 0.96 \text{ N/mm}^2 \quad \text{(metric)}$$

Maximum epoxy board flexural stress may be determined from case V of Table 6.5. The dynamic displacement for a loading of 20g becomes

$$\Delta_{sr} = G\delta_{s_r} = \frac{20(5)(1.5)(8^3)}{384(16486)} = 0.012 \text{ in} \quad \text{(English)}$$

$$\Delta_r = \frac{20(5)(6.67)(203.2^3)}{384(47 \times 10^6)} = 0.31 \text{ mm} \quad \text{(metric)}$$

Maximum epoxy board flexural stress for the bonded rib configuration is

$$\sigma = \frac{9.6(2 \times 10^6)(0.12)(0.012)}{8^2} = 432 \text{ lb/in}^2 \quad \text{(English)}$$

$$\sigma = \frac{9.6(1.38 \times 10^4)(3.05)(0.31)}{203.2^2} = 3.0 \text{ N/mm}^2 \quad \text{(metric)}$$

(b) *Bolted Configuration.* The reduced structural efficiency of a bolted interface is accounted for by applying the factor for pan head screws (Table 6.3) to the area and moment of inertia of the aluminum ribs. The summation quantities of (6.20b) are redetermined as

Section	Area A in^2	E lb/in^2	AE, lb	r, in	AEr	AEr^2	$I_{cg} = bh^3/12$ in^4	EI_{cg}
1	0.480	2(10^6)	0.96(10^6)	0.04	38400	1536	2.56(10^{-4})	512
2	0.028	10^7	0.28(10^6)	0.19	53200	10108	1.11(10^{-4})	1110
			1.24(10^6)		91600	11644		1622

The moment of inertia and centroidal location are given by (6.20b) and (6.21a) as

$$(EI)_{xx} = 1622 + 11644 - \frac{91600^2}{1.24(10^6)} = 6499 \text{ lb·in}^2 \ (18.6 \text{ MN·mm}^2)$$

$$\bar{y} = \frac{91600}{1.24(10^6)} = 0.074 \text{ in } (1.87 \text{ mm})$$

The first statical moment for the aluminum ribs becomes

$$\bar{Ar} = 0.11(0.3 - 0.11 - 0.074) = 0.0127 \text{ in}^3 \quad \text{(English)}$$

$$\bar{Ar} = 70.9(7.6 - 2.8 - 1.87) = 207.7 \text{ mm}^3 \quad \text{(metric)}$$

The moment of inertia with respect to aluminum becomes

$$I_{xx} \text{ (aluminum)} = \frac{6499}{10^7} = 6.499(10^{-4}) \text{ in}^4 \ (270.5 \text{ mm}^4)$$

Maximum shear flow due to the axial force in the aluminum ribs is determined from (6.27) as

$$q = \frac{15(0.0127)}{(6.449 \times 10^{-4})} = 293 \text{ lb/in} \quad \text{(English)}$$

$$q = \frac{66.7(207.7)}{(270.5)} = 51.2 \text{ N/mm} \quad \text{(metric)}$$

Assuming a uniform screw spacing for both ribs, the number of screws and the spacing required for each rib is determined from (6.29) for a fastener capability of 300 lb (1334 N) as

$$n = NL + 1 = \frac{(293)8}{(2)(300)} + 1 \simeq 5 \text{ screws} \quad \text{(English)}$$

$$n = \frac{51.2(203.2)}{2(1334)} + 1 \simeq 5 \text{ screws} \quad \text{(metric)}$$

$$\text{spacing} < \frac{1}{N} = \frac{300(2)}{293} < 2 \text{ in} \quad \text{(English)}$$

$$\text{spacing} < \frac{1334(2)}{51.2} < 52 \text{ mm} \quad \text{(metric)}$$

The dynamic displacement for a loading of 20g following case V of table 6.5 becomes

$$\Delta_r = \frac{20(5)(1.5)(8^3)}{384(6499)} = 0.03 \text{ in} \quad \text{(English)}$$

$$\Delta_r = \frac{20(5)(6.67)(203.2^3)}{384(18.6 \times 10^6)} = 0.78 \text{ mm} \quad \text{(metric)}$$

Maximum epoxy board flexural stress for a bolted rib configuration is

$$\sigma = \frac{9.6(2 \times 10^6)(0.074)(0.03)}{8^2} = 666 \text{ lb/in}^2 \quad \text{(English)}$$

$$\sigma = \frac{9.6(1.38 \times 10^4)(1.87)(0.78)}{203.2^2} = 4.68 \text{ N/mm}^2 \quad \text{(metric)}$$

Q.E.D.

Degree of Indeterminacy and Symmetry of Structures. Indeterminate structures contain extra reactive loads and internal forces that must be evaluated in terms of the imposed loading in order to determine desired deflections and member stresses. The number of extra reactive loads and internal forces is referred to as *the degree of indeterminacy* (or *redundancy*) of the structure. One method of assessing the degree of redundancy is to add or remove enough constraints to make the idealized structure determinate without compromising stability. When this process is applied to beams, a cantilevered configuration results. A tree-like configuration fixed at a single point results for planar or space frames. The degree of determinacy is equal to the sum of added and deleted constraints.

Up to two reactive loads can be removed at a node (or joint) on a beam. Up to three constraints can be removed from planar frames and up to six from space frames. Constraints may also be added to pin joints and hinges to develop a cantilever or tree-like structure and to assure stability. The quantity of constraints at a node (or joint) due to any additions cannot exceed the number that can be removed. The degree of indeterminacy for a variety of idealized structures is illustrated in Figure 6.18.

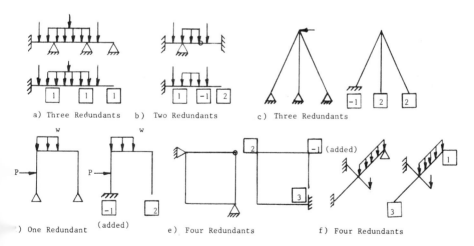

Figure 6.18. Degree of Indeterminacy of Various Configurations.

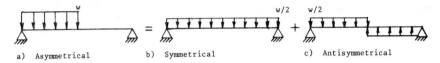

Figure 6.19. Equivalent Loading Using Superpositioning.

The degree of indeterminacy can be reduced if the structure has geometric symmetry. Any reduction in these unknown loads also reduces the effort required to determine the behavior of the structure. The center displacement of the asymmetrically loaded idealized structure of Figure 6.19(a) may be determined using direct integration of the moment equation (6.7b) or by employing superpositioning. Symmetric structures with asymmetrical loadings can be decomposed into the sum of a symmetric and an antisymmetric loading as shown in Figure 6.19. The desired center displacement of the system in Figure 6.19(a) is obtained from the symmetric configuration of Figure 6.19(b) simply by applying case V of Table 6.5 with $W = wL/2$.

Symmetry and antisymmetry may be used to develop a simplified idealized configuration with a minimum of joints and members without affecting the nature of the deflections or stresses. The side sway of the indeterminate structure in Figure 6.18(d) may be determined from the simple determinate configuration of Figure 6.20(d). The determination of member stresses, however, requires consideration of both symmetric and antisymmetric loadings. Simplified configurations of Figure 6.20(c) and (d) are a consequence of member actions in the corresponding structures of Figure 6.20(a) and (b) respectively.

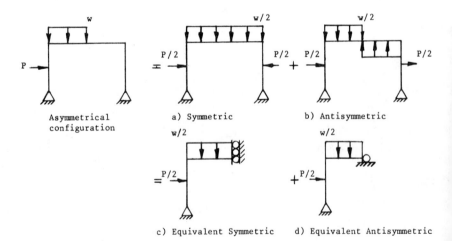

Figure 6.20. Structural Simplifications Using Symmetry.

The only possible displacements along an axis of symmetry of a planar structure as in Figure 6.20(a) are translations. A guided rigid joint along the axis of symmetry in Figure 6.20(c) represents the actions developed in Figure 6.20(a). Translation and rotation of the roller in Figure 6.20(d) simulates displacements of the antisymmetrically loaded configuration of Figure 6.20(b).

Boundary constraints along the axis of symmetry for symmetrical and antisymmetrical loadings are based on joint or member displacements that occur there. Once the idealized configuration has been decomposed into a symmetrical and antisymmetrical loading, the displacements along the axis of symmetry may be visualized by developing the deflected shape due to the imposed loading. Construction of the deflected shape is facilitated by joint rotations and translations that correspond to the loading and support reactions. Deflected shapes assume that axial effects or changes in member length may be ignored.

The deflected shape of the asymmetrical configuration in Figure 6.20 is composed of symmetrical and antisymmetrical displacements as shown in Figure 6.21. The representation of the deflected shapes is facilitated by equal joint rotations developed by the corresponding loading in Figure 6.20(a) and (b). The equivalent structural configurations in Figure 6.20(c) and (d) are based on the actions that occur along the axis of symmetry illustrated by the deflected shape in Figure 6.21(b) and (c).

The determination of the behavior of an idealized structure also quantifies the motion and stresses expected in the corresponding equipment configuration. If the structure is viewed as a whole, its behavior reflects the combined interaction of all the members. The determination of displacements and stresses at a location of interest follows an evaluation of the behavior of the entire structure. The location of particular points can be specified by coordinates of the structure.

The geometry of a structure is normally described in terms of orthogonal Cartesian coordinates such as shown in Figure 6.21(a). Positions of the joints

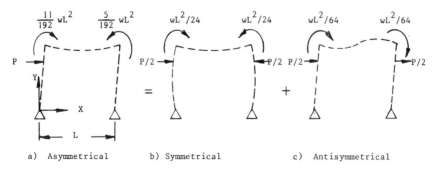

a) Asymmetrical   b) Symmetrical   c) Antisymmetrical

Figure 6.21. Equivalent Displacements Using Superpositioning.

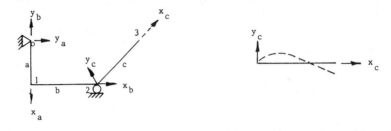

a) Member coordinate assignment   b) Coordinates for member 2-3

Figure 6.22. Member Coordinates.

are specified by x–y coordinates and motion of any point in the structure is described by its two components of motion in the x–y directions. This structural coordinate system uses a common origin and coordinate axes for all members. It is also possible to specify the displacements of a point on any member in terms of an alternative x–y coordinate system. In this case, the x axis aligns with the member and originates at one of the ends. The y axis is orthogonal to the member axis. Each member will then have its own x–y coordinate system, shown in Figure 6.22(a). This member coordinate system facilitates a description of the deformation as illustrated in Figure 6.22(b).

External loads can be applied at any point in the structure in any direction relative to the structure's coordinates. Movements that result are usually referenced to the structure's coordinate axes. Internal deformations due to member movement and the forces that result are usually expressed relative to the member axes.

6.3 LUMPED MASS AND ELASTIC ELEMENT IDEALIZATIONS

Many hardware configurations can be modeled as an equivalent single degree of freedom system using a lumped mass and elastic element to simulate its fundamental behavior with reasonable accuracy. Configurations of various shapes and boundary constraints can be represented by the idealizations shown in Table B.1-1. The development of a substitute structure for distributed mass configurations requires a definition of the deflected shape. A relationship that defines the deflected shape describes the relative deflection of all points, permitting motion of the member to be specified by any point on the element. Shape functions that describe the fundamental behavior of several configurations are described in Table 6.6. These functions are used to determine in Table B.1-1 the percentage of the distributed loading represented by a single lumped quantity that develops comparable behavior.

The results of Table B.1-1 show the effect of end restraints on the value of a lumped mass or concentrated load. The lumped approach accounts for the

Table 6.6. Beam Shape Functions

Case	Configuration	Function, F_s ($\zeta = x/L$)
I		$\zeta - 2\zeta^3 + \zeta^4$
II		$\zeta - 3\zeta^3 + 2\zeta^4$
III		$\zeta^2 - 2\zeta^3 + \zeta^4$
IV		$3 - 4\zeta + \zeta^4$

portion of the distributed mass which is supported by the restraint. A fixed restraint supports a greater portion of the distributed weight than does a hinged restraint. This is illustrated by a comparison of case III and IV of Table B.1-1.

By normalizing a shape function of Table 6.6 to the location of the concentrated load or lumped mass, an equivalent lumped value may be obtained from the kinetic or strain energy of the lumped and distributed systems, expressed as

$$\tfrac{1}{2} M_{eq} V^2 = \tfrac{1}{2} \int_0^1 M(\zeta) V^2(\zeta) \, d\zeta \tag{6.30a}$$

$$\tfrac{1}{2} P_{eq}\delta = \tfrac{1}{2} \int_0^1 P(\zeta)\delta(\zeta) \, d\zeta \tag{6.30b}$$

EXAMPLE 6.9.   An idealized lumped model of a uniformly loaded cantilevered assembly, Figure 6.23, is required in order to simplify the development of a more extensive chassis structural system. Determine (a) an equivalent load acting through the center of gravity (CG) shown on Figure 6.23(b) that produces a displacement comparable to that obtained by a distributed loading, and (b) an equivalent mass that produces a comparable fundamental frequency.

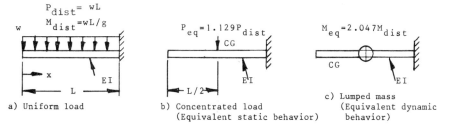

a) Uniform load

b) Concentrated load (Equivalent static behavior)

c) Lumped mass (Equivalent dynamic behavior)

Figure 6.23. Cantilevered Assembly Idealization.

(a) *Equivalent Lumped CG Load.* The shape function from Table 6.6 case IV normalized to unity at $x = L/2$ ($\zeta = \frac{1}{2}$) may be expressed in terms of the center displacement δ_c as

$$\delta(\zeta) = \frac{16}{17}\delta_c(3 - 4\zeta + \zeta^4) \tag{6.31}$$

Substituting (6.31) into (6.30b) for concentrated loading $P(\zeta) = P_{\text{dist}}$ gives

$$P_{\text{eq}}\delta_c = \frac{16}{17}P_{\text{dist}}\delta_c \int_0^1 (3 - 4\zeta + \zeta^4) \, d\zeta$$

$$P_{\text{eq}} = \frac{16}{17}\left(\frac{6}{5}\right)P_{\text{dist}} = 1.129 P_{\text{dist}}$$

The equivalent load also develops a comparable moment (stress) at the built-in assembly interface.

(b) *Equivalent CG Mass.* The velocity $V(\zeta)$ due to assembly motion is assumed to have the same shape as defined for displacement (6.31), giving

$$V(\zeta) = \frac{16}{7}V_c(3 - 4\zeta + \zeta^4) \tag{6.32}$$

Substituting (6.32) into (6.30a) for concentrated mass $M(\zeta) = M_{\text{dist}}$ determines the equivalent mass as

$$M_{\text{eq}} V_c^2 = \left(\frac{16}{17}V_c\right)^2 M_{\text{dist}} \int_0^1 (3 - 4\zeta + \zeta^4)^2 \, d\zeta$$

$$M_{\text{eq}} = \left(\frac{16}{17}\right)^2\left(\frac{104}{45}\right)M_{\text{dist}} = 2.047 M_{\text{dist}}$$

These results are illustrated in Figure 6.23 (b) and (c).   Q.E.D.

Lumped idealizations generally lead to conservative value of displacement, stress and frequency. The conservative nature of this approach tends to compensate for nonidealistic end conditions which are conveniently assumed to be fixed, hinged or free.

The single lump idealizations of Table B.1-1 and Example 6.9 accurately describe the behavior of the distributed system. For most applications, this idealization provides an accurate estimate of the loads and stresses experienced by the actual system and its attachments. When the response to shock, blast or random vibration is being studied, however, additional lumped masses

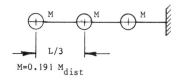

Figure 6.24. Multiple Mass Idealization of a Cantilevered Assembly.

improve the accuracy of the prediction because of the larger quantity of resonant modes available.

Additional nodes representing lumped masses of the structure can be obtained by refining the partitioning of the actual structure. Nodal masses are equivalent to the material volume allocated to each node. If a system supported at both ends is divided into four equal increments, the equivalent mass of each node will be equal to $M_{dist}/4$. A comparable approach applied to a cantilevered assembly supporting a distributed load would give an inaccurate result.

A general method of determining the load or mass on idealized systems represented by multiple equally spaced nodes may be obtained from (6.30). In terms of a single lumped quantity, (6.30) can be expressed as

$$\tfrac{1}{2} \sum_{}^{n} M_i V_i^2 = \tfrac{1}{2} M_{eq} V^2 \tag{6.33a}$$

$$\tfrac{1}{2} \sum_{}^{n} P_i \delta_i = \tfrac{1}{2} P_{eq} \delta \tag{6.33b}$$

EXAMPLE 6.10.   Three modes of vibration are required to accurately assess the loads and stresses of the cantilevered assembly of Example 6.9 exposed to a shock environment. Determine the nodal mass for the idealization shown in Figure 6.24.

Following the nodal geometry of Figure 6.24 and Eq. (6.32) in (6.33a), the mass of each node in terms of the result of Example 6.9(b) becomes

$$\tfrac{1}{2} M [V^2(0) + V^2(1/3) + V^2(2/3)] = \tfrac{1}{2} M_{eq} V_c^2$$

$$(7.972 + 2.497 + 0.249) M V_c^2 = 2.047 M_{dist} V_c^2$$

$$M = \frac{2.047}{10.718} M_{dist} = 0.191 M_{dist}$$

This result accurately represents the modal behavior of the first three resonances of a uniformly distributed cantilevered assembly.   Q.E.D.

The effort required to evaluate structural quantities and behavior of an assembly supporting many discrete irregularly spaced masses can be minimized using (6.33). In this case, a single equivalent load or mass may be deter-

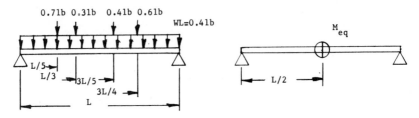

Figure 6.25. Equivalent Lumped Mass Idealization.

mined that represents the system. This equivalent may be located at any point by normalizing the shape function, Table 6.6, to that location. Additional modes of vibration can be defined by increasing the number of discrete masses as illustrated in Example 6.10.

EXAMPLE 6.11. The fundamental frequency of a simply supported power supply assembly carrying distributed as well as discrete masses represented by the model of Figure 6.25(a) is required. Determine the equivalent mass for the idealization of Figure 6.25(b) that develops comparable behavior.

The shape function, Table 6.6, case I, normalized to $x = L/2$ becomes

$$V(\zeta) = \frac{16}{5}V_c(\zeta - 2\zeta^3 + \zeta^4) \tag{6.34}$$

The equivalent mass representing the discrete masses of Figure 6.25(a) is found from (6.33a) using (6.34) as

$$\frac{1}{2g}\left[0.7V^2\left(\frac{1}{5}\right) + 0.3V^2\left(\frac{1}{3}\right) + 0.4V^2\left(\frac{3}{5}\right) + 0.6V^2\left(\frac{3}{4}\right)\right] = \tfrac{1}{2}(M_{\text{eq}})_{\text{conc}}V_c^2$$

$$\frac{V_c^2}{g}[0.7(0.594)^2 + 0.3(0.869)^2 + 0.4(0.952)^2 + 0.6(0.713)^2] = (M_{\text{eq}})_{\text{conc}}V_c^2$$

$$(M_{\text{eq}})_{\text{conc}} = 1.141/g$$

The equivalent mass representing the power supply assembly is determined from Table B.1-1 case III as

$$M_{\text{eq}} = M_{\text{conc}} + 0.504M_{\text{dist}} = [1.141 + 0.504(0.4)]/g \qquad \text{Q.E.D.}$$

$$M_{\text{eq}} = 1.343/g$$

Spring Mass Idealizations. The single lumped mass idealizations of Table B.1-1 and Examples 6.9 and 6.11 lend themselves to spring mass representations of complex assemblies consisting of many circuit cards and modules supported by a chassis or structural framework. This form of idealization, illus-

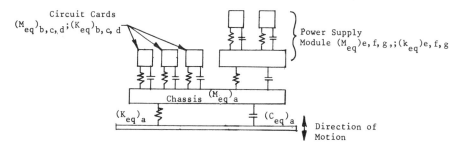

Figure 6.26. Spring Mass Idealization of an Assembly.

trated in Figure 6.26 which is developed for a particular axis of excitation, simplifies the determination of system resonant frequencies. Each mass in this representation is rigid and the springs are usually massless. The effect of this idealization is to discard harmonics of fundamental frequencies due to the distribution of mass and stiffness in the actual hardware. The absence of inertial coupling results in a diagonal mass matrix.

The displacements as a result of vibration at high frequency harmonic modes found in electronic equipment are usually negligible compared to those at the fundamental frequencies. Depending on the nature of suddenly imposed loads, however, stresses due to the contribution of these higher modes may be considerable. The complexity of the structure and inherent damping at the many attachments and interfaces in electronic hardware enable an accurate determination of maximum stresses based on fundamental resonant characteristics.

The chassis rigid body mass $(M_{eq})_a$ idealized in Figure 6.26 is numerically equal to the mass of the assembly after removal of the masses of the three circuit cards and power supply module. The mass of items of the assembly not represented by additional spring-mass configurations, because of similarity to those already shown, are appropriately lumped into the mass configuration at the level of assembly represented. The spring constant $(k_{eq})_a$ in Figure 6.26 represents the sum of the chassis stiffness and the stiffness of the attachment or mounting to the vehicle supporting the assembly. Equivalent mass and stiffness of several common configurations are given in Table B.1-1.

Those relationships exclude the contribution due to support or attachment compliance. Methods for evaluating the behavior of lumped idealizations are delineated in Appendix B. Examples illustrating the development of the transient system response to imposed motion or loads are included in Appendix B.5. Additional examples are used to describe a simplified approach for determining the system frequency response (transmissibility) in Appendix B.6.

Resonant Frequencies of Lumped Mass Idealizations. Methods delineated in Appendix B are applicable to any lumped mass system. This includes those idealizations using discrete spring or elastic element idealizations of the struc-

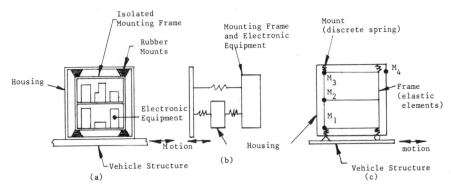

Figure 6.27. Lumped Mass Idealization. (a) Equipment; (b) Spring mass model; (c) Mixed element model.

ture. The isolated mounting frame within a supporting housing in Figure 6.27(a) may be idealized as the spring-mass system of Figure 6.27(b) or as the elastic element model of Figure 6.27(c) incorporating discrete springs representing rubber isolators. The spring-mass model may be used to establish coarse design characteristics for the housing frame and rubber suspension to avoid resonant coupling through the separation of fundamental frequencies.

When vibration or shock isolation is used, the stiffness of the suspended structural system must be considerably higher than the mount stiffness to assure that the modes are uncoupled. This condition can be obtained by separating resonant frequencies by a factor of more than 2.

The number of modes of vibration that a model represents depends on the nature of the idealization used. Usually, the number of masses establishes the lower bound or minimum quantity of modes. The model of Figure 6.27(b) represents two frequencies (modes) of the actual equipment whereas the model of Figure 6.27(c) represents four.

Additional modes that can occur are due to rotations. Loads imposed on a system are applied at its center of mass. In an idealized configuration, loads are imposed on each lumped mass. Rotations are the result of an internal torque due to a lack of coincidence between the elastic center and mass center of a system.

A mass oriented in space can rotate around or translate along the coordinate axes defined for the structure. Up to six distinct movements are possible. Each movement describes a mode of vibration. Constraints imposed by attachments restrict motion and the modes that the mass can ascribe. In a planar system, a mass can translate along the two planar coordinates and rotate about the axis normal to the plane. A planar mass can exhibit up to three modes.

Consider the symmetrically supported body of Figure 6.28(a). The body is

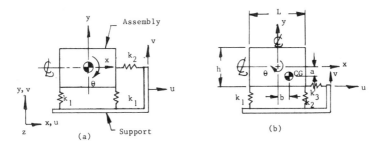

Figure 6.28. Coordinates That Couple or Uncouple Motion. (a) Uncoupled motion; (b) Coupled motion.

free to translate along the x and y axes and rotate about the z axis. Excitation of the support along the x axis produces an x axis response in the body even though movement is unrestricted in the two remaining modes. A single spring-mass model may be used to represent this motion. An analogous approach was used to develop the models of Figures 6.26 and 6.27(b).

When constraints are arranged unsymmetrically with respect to the center of mass of a body, translational and rotational modes may become coupled. Coupled modes consist of components of motion from any of the modes possible. If modes are coupled, vibration along one axis affects the response motion in the other axes. This is illustrated by the configuration shown in Figure 6.28(b). The degree of coupling and the equations of motion can be ascertained by displacing the center of mass of the body in a specific direction. If the body moves without developing a rotational torque or translational forces in the other directions, the mode is uncoupled from the other modes. The rotational mode is uncoupled if a rotation about an axis through the center of mass does not develop torques about the other axes or translational forces along any axes.

It is customary to select axes through the center of gravity of equipment. These axes are the principal axes of the equipment. This choice uncouples the inertia matrix if the elastic and principal axis coincide. If not, inertial coupling occurs through products of inertia which are difficult to determine. Polar moments of inertia I_{xx}, I_{yy}, and I_{zz} may be determined accurately and easily using a wire suspension. One method shown in Figure 6.29(a) uses three equal-length wires to develop an oscillation about the z axis. A second approach, Figure 6.29(b), uses the principle of the pendulum to determine the moment of inertia about the y' axis. The moment of inertia can then be transferred to the y axis through the equipment.

The weight, location of the center of gravity and moment of inertia is required in order to evaluate the behavior of equipment. The dynamic equations for the uncoupled system of Figure 6.28(a) subject to foundation acceleration are

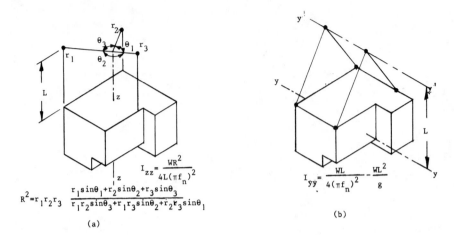

$$I_{zz} = \frac{WR^2}{4L(\pi f_n)^2}$$

$$R^2 = r_1 r_2 r_3 \quad \frac{r_1 \sin\theta_1 + r_2 \sin\theta_2 + r_3 \sin\theta_3}{r_1 r_2 \sin\theta_3 + r_1 r_3 \sin\theta_2 + r_2 r_3 \sin\theta_1}$$

(a)

$$I_{yy} = \frac{WL}{4(\pi f_n)^2} - \frac{WL^2}{g}$$

(b)

Figure 6.29. Empirical Methods for Determining Moments of Inertia. (a) Rotational pendulum; (b) Swinging pendulum.

$$\begin{bmatrix} M & 0 & 0 \\ 0 & M & 0 \\ 0 & 0 & I_{cg} \end{bmatrix} \begin{Bmatrix} \ddot{x} \\ \ddot{y} \\ \ddot{\theta} \end{Bmatrix}$$

$$+ \begin{bmatrix} k_2 & 0 & 0 \\ 0 & 2k_1 & 0 \\ 0 & 0 & k_1 L^2/2 \end{bmatrix} \begin{Bmatrix} x \\ y \\ \theta \end{Bmatrix} = \begin{Bmatrix} -M\ddot{u} \\ -M\ddot{v} \\ 0 \end{Bmatrix} \qquad (6.35)$$

The behavior of the uncoupled system of (6.35) may be represented by three individual spring mass systems. Resonant frequencies are readily determined as

$$f_{xx} = \frac{1}{2\pi}\sqrt{\frac{k_2}{M}}; f_{yy} = \frac{1}{2\pi}\sqrt{\frac{2k_1}{M}} \text{ and } f_\theta = \frac{1}{2\pi}\sqrt{\frac{k_1}{2I_{cg}}}$$

These relationships may be used to select support characteristics that avoid potentially damaging frequencies developed by the foundation.

If the location of the center of gravity is not known accurately or has changed due to modifications that have occurred during the design process, rotations due to inertial coupling can occur. For a small offset between the center of gravity and elastic axes, Equation (6.35) becomes

$$
\begin{bmatrix}
M & 0 & Ma \\
0 & M & Mb \\
Ma & Mb & I_{cg} + M(a^2 + b^2)
\end{bmatrix}
\begin{Bmatrix}
\ddot{x} \\
\ddot{y} \\
\ddot{\theta}
\end{Bmatrix}
$$

$$
+ \begin{bmatrix}
k_2 & 0 & 0 \\
0 & 2k_1 & 0 \\
0 & 0 & k_1 L^2/2
\end{bmatrix}
\begin{Bmatrix}
x \\
y \\
\theta
\end{Bmatrix}
=
\begin{Bmatrix}
-M\ddot{u} \\
-M\ddot{v} \\
0
\end{Bmatrix}
\tag{6.36}
$$

Methods of Appendix B may be used to evaluate the resonant frequencies and response characteristics of (6.36). Balancing using counter weights may be required if response displacements or accelerations exceed equipment allowables.

Spatial, geometrical and weight considerations often make it difficult to obtain an uncoupled configuration. When the motion is coupled as in Figure 6.28(b), the response to a disturbance consists of components from each mode. The dynamic equations of the coupled system shown in Figure 6.28(b) are

$$
\begin{bmatrix}
M & 0 & Ma \\
0 & M & Mb \\
Ma & Mb & I_{cg} + M(a^2 + b^2)
\end{bmatrix}
\begin{Bmatrix}
\ddot{x} \\
\ddot{y} \\
\ddot{\theta}
\end{Bmatrix}
$$

$$
+ \begin{bmatrix}
k_3 & 0 & -k_3 h/2 \\
0 & k_1 + k_2 & (k_1 - k_2)L/2 \\
-k_3 h/2 & (k_1 - k_2)L/2 & (k_1 + k_2)(L/2)^2 + k_3(h/2)^2
\end{bmatrix}
\begin{Bmatrix}
x \\
y \\
\theta
\end{Bmatrix}
$$

$$
=
\begin{Bmatrix}
-M\ddot{u} \\
-M\ddot{v} \\
0
\end{Bmatrix}
\tag{6.37}
$$

The $y - y$ elastic axis may be made to coincide with the center of gravity if the support stiffnesses k_1 and k_2 are selected proportional to the weight distribution using

$$
\frac{k_1}{k_2} = \frac{L/2 - b}{L/2 + b}
\tag{6.38}
$$

Applying (6.38) to (6.37) uncouples the y motion, giving

$$
\begin{bmatrix}
M & 0 & Ma \\
0 & M & 0 \\
Ma & 0 & I_{cg} + Ma^2
\end{bmatrix}
\begin{Bmatrix}
\ddot{x} \\
\ddot{y} \\
\ddot{\theta}
\end{Bmatrix}
$$

$$
+
\begin{bmatrix}
k_3 & 0 & -k_3 h/2 \\
0 & k_2 L/(L/2 + b) & 0 \\
-k_3 h/2 & 0 & k_2 L(L/2 - b) + k_3(h/2)^2
\end{bmatrix}
\begin{Bmatrix}
x \\
y \\
\theta
\end{Bmatrix}
$$

$$
=
\begin{Bmatrix}
-M\ddot{u} \\
-M\ddot{v} \\
0
\end{Bmatrix}
\qquad (6.39)
$$

Even though the y motion is uncoupled from the x and θ motions, the freedom to vibrate in all modes simultaneously is not restricted. The y motion, however, will not influence the other modes.

Rigid body idealizations may be used to evaluate equipment behavior when the stiffnesses throughout the assembly are more than 4 times greater than the elastic support elements. The dynamic equations are developed by examining forces and moments along each of the system coordinates which result from a displacement along one of the coordinates.

Static and Dynamic Relationships for an Idealization. Transformations are used to conveniently express force and displacement quantities in the member coordinates in terms of the system coordinates. Transformation of quantities in the xy plane for a planar system is the result of a rotation about the z axis. Components in the z direction are not affected. Consider the member shown in Figure 6.30(a) subjected to in plane forces. Components of member forces F_x and F_y in the xy system reference are obtained by projecting forces in the member coordinates on to the x and y axes, giving

$$
F_x = F_x' \sin \alpha - F_y' \cos \alpha \qquad (6.40a)
$$

$$
F_y = F_x' \cos \alpha + F_y' \sin \alpha
$$

Equations (6.40a) may be expressed succinctly in matrix form as

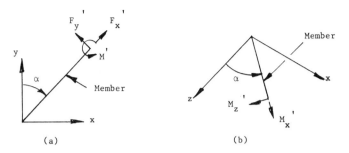

Figure 6.30. Planar Force Coordinates. (a) In plane forces; (b) Normal to plane forces.

$$\begin{Bmatrix} F_x \\ F_y \end{Bmatrix} = \begin{bmatrix} \sin \alpha & -\cos \alpha \\ \cos \alpha & \sin \alpha \end{bmatrix} \begin{Bmatrix} F'_x \\ F'_y \end{Bmatrix} = [A] \begin{Bmatrix} F'_x \\ F'_y \end{Bmatrix} \qquad (6.40b)$$

Moments that are the result of forces normal to the xz plane, Figure 6.30(b), may be transformed similarly, giving

$$\begin{Bmatrix} M_x \\ M_z \end{Bmatrix} = [A] \begin{Bmatrix} M'_x \\ M'_z \end{Bmatrix} \qquad (6.41)$$

Displacements in the system coordinates may be transformed to member coordinates using the matrix transpose

$$\begin{Bmatrix} \delta'_x \\ \delta'_y \end{Bmatrix} = \begin{bmatrix} \sin \alpha & \cos \alpha \\ -\cos \alpha & \sin \alpha \end{bmatrix} \begin{Bmatrix} \delta_x \\ \delta_y \end{Bmatrix} = [A]^T \begin{Bmatrix} \delta_x \\ \delta_y \end{Bmatrix} \qquad (6.42)$$

EXAMPLE 6.12.   Determine the force displacement characteristics in the xy system reference frame for the member of Figure 6.30(a) with characteristics in the member coordinates expressed as

$$\begin{Bmatrix} F'_x \\ F'_y \end{Bmatrix} = \begin{bmatrix} k'_{11} & k'_{12} \\ k'_{12} & k'_{22} \end{bmatrix} \begin{Bmatrix} \delta'_x \\ \delta'_y \end{Bmatrix} = [K'] \begin{Bmatrix} \delta'_x \\ \delta'_y \end{Bmatrix} \qquad (6.43)$$

Premultiplying (6.43) by the transformation matrix $[A]$ gives

$$[A] \begin{Bmatrix} F'_x \\ F'_y \end{Bmatrix} = [A][K'] \begin{Bmatrix} \delta'_x \\ \delta'_y \end{Bmatrix} \qquad (6.44)$$

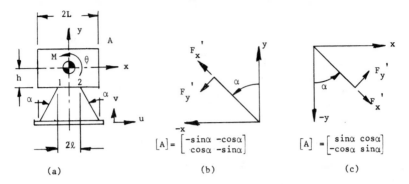

Figure 6.31. Idealization of Compliant Equipment Supports. (a) Equipment and support configuration; (b) Alternative member; (c) 2 coordinates.

Forces in terms of displacements expressed in the system reference frame are determined by substituting (6.40b) and (6.42) into (6.44), giving

$$\begin{Bmatrix} F_x \\ F_y \end{Bmatrix} = [A][K'][A]^T \begin{Bmatrix} \delta_x \\ \delta_y \end{Bmatrix} \tag{6.45}$$

which may be expanded by matrix multiplication described in Appendix A to

$$\begin{Bmatrix} F_x \\ F_y \end{Bmatrix}$$

$$= \begin{bmatrix} k'_{11}S^2 - 2k'_{21}SC + k'_{22}C^2 & (k'_{11} - k'_{22})SC + k'_{21}(S^2 - C^2) \\ (k'_{11} - k'_{22})SC + k'_{21}(S^2 - C^2) & k'_{11}C^2 + 2k'_{21}SC + k'_{22}S^2 \end{bmatrix} \begin{Bmatrix} \delta_x \\ \delta_y \end{Bmatrix}$$

$$\tag{6.46}$$

where $k'_{21} = k'_{12}$, $C = \cos \alpha$ and $S = \sin \alpha$.   Q.E.D.

EXAMPLE 6.13.   Two members with uncoupled stiffnesses in the member axes support the rigid mass shown in Figure 6.31(a). Determine the force displacement characteristics at the center of gravity and the dynamic equation for a foundation
acceleration.

Uncoupled member characteristics may be expressed using (6.43) as

$$\begin{Bmatrix} F'_x \\ F'_y \end{Bmatrix} = \begin{bmatrix} k'_x & 0 \\ 0 & k'_y \end{bmatrix} \begin{Bmatrix} \delta'_x \\ \delta'_y \end{Bmatrix} \tag{6.47}$$

Equation (6.46) describes the force-displacement behavior of member 1 at the point of attachment to the rigid mass in terms of system coordinates as

$$
\begin{Bmatrix} F_x \\ F_y \end{Bmatrix}_1 = \begin{bmatrix} k'_x S^2 + k'_y C^2 & (k'_x - k'_y)SC \\ (k'_x - k'_y)SC & k'_x C^2 + k'_y S^2 \end{bmatrix} \begin{Bmatrix} \delta_x \\ \delta_y \end{Bmatrix} = [\mathbf{K}_1] \begin{Bmatrix} \delta_x \\ \delta_y \end{Bmatrix} \qquad (6.48)
$$

where $k'_{21} = 0$.

The moment about the center of gravity in terms of positive forces at the point of attachment of member 1 is

$$
\{M_z\}_1 = [h \quad -\ell] \begin{Bmatrix} F_x \\ F_y \end{Bmatrix}_1 = [\mathbf{R}_1] \begin{Bmatrix} F_x \\ F_y \end{Bmatrix}_1 \qquad (6.49)
$$

The moment due to displacements of the center of gravity along system axes is determined by substituting (6.48) in (6.49), giving

$$
\{M_{z\delta}\}_1 = [\mathbf{R}_1][\mathbf{K}_1] \begin{Bmatrix} \delta_x \\ \delta_y \end{Bmatrix} \qquad (6.50)
$$

Displacements as a result of a positive rotation, θ, about the center of gravity may be expressed using the matrix transpose $[\mathbf{R}_1]^{\mathrm{T}}$ as

$$
\begin{Bmatrix} \delta_x \\ \delta_y \end{Bmatrix} = \begin{bmatrix} h \\ -\ell \end{bmatrix} \{\theta\} = [\mathbf{R}_1]^T\{\theta\} \qquad (6.51)
$$

The moment due to a rotation about the center of gravity is determined from (6.50) and (6.51) as

$$
\{M_{z\theta}\} = [\mathbf{R}_1][\mathbf{K}_1][\mathbf{R}_1]^T\{\theta\} \qquad (6.52)
$$

An analogous procedure is used to determine the force-displacement behavior at the center of gravity due to member 2. Once suitable member coordinates have been selected, the transformation matrix $[\mathbf{A}]$ is obtained by projecting member forces on to the x and y axes. The transformation matrix for two possible member coordinates is shown in Figure 6.31(b). Using (6.47) and the transformation matrix from Figure 6.31(b) in (6.45) gives

$$
\begin{Bmatrix} F_x \\ F_y \end{Bmatrix}_2 = \begin{bmatrix} k'_x S^2 + k'_y C_2 & (k'_y - k'_x)SC \\ (k'_y - k'_x)SC & k'_x C^2 + k'_y S^2 \end{bmatrix} \begin{Bmatrix} \delta_x \\ \delta_y \end{Bmatrix} = [\mathbf{K}_2] \begin{Bmatrix} \delta_x \\ \delta_y \end{Bmatrix} \qquad (6.53)
$$

The moment about the center of gravity in terms of positive forces at the point of attachment of member 2 and displacements at the center of gravity becomes

$$\{M_z\}_2 = [h \quad \ell] \begin{Bmatrix} F_x \\ F_y \end{Bmatrix}_2 = [R_2] \begin{Bmatrix} F_x \\ F_y \end{Bmatrix}_2$$

(6.54)

$$\{M_{z\delta}\} = [R_2][K_2] \begin{Bmatrix} \delta_x \\ \delta_y \end{Bmatrix}$$

The moment due to a rotation about the center of gravity is determined using the matrix transpose $[R_2]^T$ to relate the rotation to displacements at the point of attachment, giving

$$\{M_{z\theta}\} = [R_2][K_2][R_2]^T\{\theta\}$$

(6.55)

The force-displacement behavior of the configuration in Figure 6.31(a) is obtained by adding relationships for each member, expressed generally as

$$\begin{Bmatrix} F \\ M \end{Bmatrix} = \begin{bmatrix} \sum_{}^{n} [K_i] & \sum_{}^{n} [K_i][R_i]^T \\ \sum_{}^{n} [R_i][K_i] & \sum_{}^{n} [R_i][K_i][R_i]^T \end{bmatrix} \begin{Bmatrix} \delta \\ \theta \end{Bmatrix}$$

(6.56)

Substituting (6.48), (6.50), (6.52), (6.53), (6.54) and (6.55) into (6.56) gives the force displacement behavior at the center of gravity as

$$\begin{Bmatrix} F_x \\ F_y \\ M_z \end{Bmatrix} = \begin{bmatrix} k_x'S^2 + k_y'C^2 & 0 & (k_x'S^2 + k_y'C^2)h \\ 0 & k_x'C^2 + k_y'S^2 & 0 \\ (k_x'S^2 + k_y'C^2)h & 0 & k_{33} \end{bmatrix} \begin{Bmatrix} \delta_x \\ \delta_y \\ \theta \end{Bmatrix}$$

(6.57)

where

$$k_{33} = h^2(k_x'S^2 + k_y'C^2) + \ell^2(k_x'C^2 + k_y'S^2) - 2h\ell(k_x' - k_y')SC$$

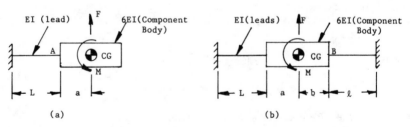

Figure 6.32. Rigid Body Component Idealization. (a) Cantilevered; (b) Fixed support.

The inertia matrix is generated for the dynamic equation consistent with the order of the forces in (6.57). The inertia matrix is also uncoupled since principal axes are used. For a foundation acceleration, the dynamic equation using (6.57) becomes

$$
\begin{bmatrix} M & 0 & 0 \\ 0 & M & 0 \\ 0 & 0 & I_{cg} \end{bmatrix} \begin{Bmatrix} \ddot{x} \\ \ddot{y} \\ \ddot{\theta} \end{Bmatrix}
$$

$$
+ \begin{bmatrix} k_x'S^2 + k_yC^2 & 0 & (k_x'S^2 + k_y'C^2)h \\ 0 & K_x'S^2 + k_y'C^2 & 0 \\ (k_x'S^2 + k_y'C^2)h & 0 & k_{33} \end{bmatrix} \begin{Bmatrix} x \\ y \\ \theta \end{Bmatrix}
$$

$$
= \begin{Bmatrix} -M\ddot{u} \\ -M\ddot{v} \\ 0 \end{Bmatrix} \qquad (6.58)
$$

The y motion is uncoupled due to symmetry about the y axis in Figure 6.31(a). The motion at any point on the body may be determined from components of the displacements of (6.58). At point A on Figure 6.31(a) for example, the vertical and horizontal displacements are given by

$$
y_A = y + L\theta; \ x_A = x - h\theta \qquad (6.59)
$$

Q.E.D.

EXAMPLE 6.14.   Determine the force displacement characteristics at the center of gravity of a lead supported component for (a) the cantilevered configuration of Figure 6.32(a), and (b) and asymmetrical fixed configuration of Figure 6.32(b).

(a) *Cantilevered Component Force Displacement Behavior*. A rigid body idealization can be used since the component body stiffness is sufficiently greater than the leads. The force displacement behavior at point A on the cantilevered leads is given as

$$
\begin{Bmatrix} F \\ M \end{Bmatrix}_A = \frac{EI}{L^3} \begin{bmatrix} 12 & -6L \\ -6L & 4L^2 \end{bmatrix} \begin{Bmatrix} \delta \\ \theta \end{Bmatrix}_A = [\mathbf{K}_A] \begin{Bmatrix} \delta \\ \theta \end{Bmatrix}_A \qquad (6.60)
$$

Forces at the center of gravity of Figure 6.32(a) due to forces acting at point A are related by

$$
\begin{Bmatrix} F \\ M \end{Bmatrix}_{cg} = \begin{bmatrix} 1 & 0 \\ -a & 1 \end{bmatrix} \begin{Bmatrix} F \\ M \end{Bmatrix}_A = [\mathbf{R}_A] \begin{Bmatrix} F \\ M \end{Bmatrix}_A \qquad (6.61)
$$

Displacements at point A as a result of displacements at the center of gravity are determined from the matrix transpose $[\mathbf{R}]^T$ analogous to (6.51), giving

$$\left\{ \begin{array}{c} \delta \\ \theta \end{array} \right\}_A = [\mathbf{R}_A]^T \left\{ \begin{array}{c} \delta \\ \theta \end{array} \right\}_{cg} \tag{6.62}$$

Forces in terms of displacements occurring at the center of gravity are determined by premultiplying (6.60) by $[\mathbf{R}_A]$ analogous to (6.44), then substituting (6.61) and (6.62) to arrive at

$$\left\{ \begin{array}{c} F \\ M \end{array} \right\}_{cg} = [\mathbf{R}_A][\mathbf{K}_A][\mathbf{R}_A]^T \left\{ \begin{array}{c} \delta \\ \theta \end{array} \right\}_{cg} \tag{6.63a}$$

or

$$\left\{ \begin{array}{c} F \\ M \end{array} \right\}_{cg} = \frac{EI}{L^3} \left[\begin{array}{cc} 12 & -(12a + 6L) \\ -(12a + 6L) & 12a^2 + 12aL + 4L^2 \end{array} \right] \left\{ \begin{array}{c} \delta \\ \theta \end{array} \right\}_{cg} \tag{6.63b}$$

(b) *Asymmetrical Fixed Component Force Displacement Behavior.* The force displacement behavior at the center of gravity of the component body in Figure 6.32(b) is obtained by adding the results of this example's part (a), (6.63b), to those with lead ℓ acting alone ($L = 0$).

The force displacement behavior of a cantilevered configuration ($L = 0$) at point B of Figure 6.32(b) is given as

$$\left\{ \begin{array}{c} F \\ M \end{array} \right\}_B = \frac{EI}{\ell^3} \left[\begin{array}{cc} 12 & 6\ell \\ 6\ell & 4\ell^2 \end{array} \right] \left\{ \begin{array}{c} \delta \\ \theta \end{array} \right\}_B = [\mathbf{K}_B] \left\{ \begin{array}{c} \delta \\ \theta \end{array} \right\}_B \tag{6.64}$$

Forces at the component center of gravity due to those acting at point B are related by

$$\left\{ \begin{array}{c} F \\ M \end{array} \right\}_{cg} = \left[\begin{array}{cc} 1 & 0 \\ b & 1 \end{array} \right] \left\{ \begin{array}{c} F \\ M \end{array} \right\}_B = [\mathbf{R}_B] \left\{ \begin{array}{c} F \\ M \end{array} \right\}_B \tag{6.65}$$

Forces in terms of displacements at the center of gravity are determined analogous to (6.63a), giving

$$\left\{ \begin{array}{c} F \\ M \end{array} \right\}_{cg} = \frac{EI}{\ell^3} \left[\begin{array}{cc} 12 & 12b + 6\ell \\ 12b + 6\ell & 12b^2 + 12b\ell + 4\ell^2 \end{array} \right] \left\{ \begin{array}{c} \delta \\ \theta \end{array} \right\}_{cg} \tag{6.66}$$

The force displacement behavior at the center of gravity of the asymmetrical component with fixed lead supports is a result of the contributions from (6.63b) and (6.66) giving

$$\left\{ \begin{array}{c} F \\ M \end{array} \right\}_{cg} = EI \left[\begin{array}{cc} 12\left(\dfrac{1}{L^3} + \dfrac{1}{\ell^3}\right) & 12\left(\dfrac{b}{\ell^3} - \dfrac{a}{L^3}\right) + 6\left(\dfrac{1}{\ell^2} - \dfrac{1}{L^2}\right) \\[3mm] 12\left(\dfrac{b}{\ell^3} - \dfrac{a}{L^3}\right) + 6\left(\dfrac{1}{\ell^2} - \dfrac{1}{L^2}\right) & 12\left(\dfrac{b^2}{\ell^3} + \dfrac{a^2}{L^3}\right) + 12\left(\dfrac{b}{\ell^2} + \dfrac{a}{L^2}\right) + 4\left(\dfrac{1}{\ell} + \dfrac{1}{L}\right) \end{array} \right] \left\{ \begin{array}{c} \delta \\ \theta \end{array} \right\}_{cg} \quad (6.67)$$

Q.E.D.

The component response to vibration or shock may be investigated by augmenting (6.63b) or (6.67) with inertia characteristics analogous to the development of (6.58). The augmented equation describes the dynamic behavior whereas equations similar to (6.67) describe the static behavior of the configuration. The determination of equipment static and dynamic behavior is examined in Chapters 7 and 8 respectively.

The dynamic equation of the cantilevered component of Example 6.14(a) could have been written using (6.60). A coupled inertia matrix results (6.68), since the center of gravity is not located at the end of the lead support.

$$\begin{bmatrix} M & Ma \\ Ma & I_{cg} + Ma^2 \end{bmatrix} \left\{ \begin{array}{c} y \\ \ddot{\theta} \end{array} \right\}$$
$$+ \frac{EI}{L^3} \begin{bmatrix} 12 & -6L \\ -6L & 4L^2 \end{bmatrix} \left\{ \begin{array}{c} y \\ \theta \end{array} \right\} = \left\{ \begin{array}{c} -M\ddot{u} \\ 0 \end{array} \right\} \quad (6.68)$$

The development of (6.63b) using the transformation results in a form which permits a parametric evaluation of the displacement sensitivity as a result of relocating the center of gravity. The use of transformations also uncouples the inertia matrix, converting (6.68) to

$$\begin{bmatrix} M & 0 \\ 0 & I_{cg} \end{bmatrix} \left\{ \begin{array}{c} \ddot{y} \\ \ddot{\theta} \end{array} \right\}$$
$$+ \frac{EI}{L^3} \begin{bmatrix} 12 & -(12a + 6L) \\ -(12a + 6L) & 12a^2 + 12aL + 4L^2 \end{bmatrix} \left\{ \begin{array}{c} y \\ \theta \end{array} \right\} = \left\{ \begin{array}{c} -m\ddot{u} \\ 0 \end{array} \right\} \quad (6.69)$$

6.4 TORSIONAL CONSIDERATIONS

Torsional loads on equipment and assemblies can develop when the center of gravity does not lie on a plane through the attachment interface. The magnitude of the applied torque is the product of the imposed load and the perpendicular distance to an imaginary plane through the attachments.

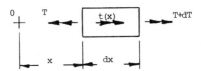

Figure 6.33. Torsional Forces on a Differential Element.

Torsional displacements can increase the required clearance surrounding an assembly. Larger displacements are due to torsional and translational coupling which decreases the apparent fundamental resonant frequency of the equipment. Torsion develops shear forces in the walls, bulkheads and covers. Insufficient shear strength in these surfaces can result in buckling failure and excessive twisting of the unit.

Enclosures, chassis and access covers are generally fabricated using thin materials. Joints may be welded or brazed, or employ fasteners such as rivets or screws. Screws permit relative motion when the structure deflects, reducing the joint effectiveness for transferring forces within the assembly. The effectiveness of different attachments is discussed in Section 6.2.

Torsional loads are resisted by thin external panels and load bearing members that comprise the primary structure of the unit. Bolted panels and access covers resist a lower proportion of the load due to the reduced effectiveness of the attachments. The reduced effectiveness can be accommodated by decreasing the thickness of the attached member.

Torsional Displacements.   An expression relating the angle of twist, θ, experienced by a constant property section to the applied torque, $t(x)$, may be developed by examining the torsional forces acting on the differential element of Figure 6.33. Summing forces on the differential element expresses equilibrium as

$$t(x)\ dx + dT = 0$$

or

$$\frac{dT}{dx} = -t(x) \tag{6.70}$$

The relative angle of twist of two adjoining sections a distance dx apart can be expressed as

$$JG\frac{d\theta}{dx} = T \tag{6.71}$$

where J and G represent the polar moment of inertia of the section and shear modulus of the material respectively. Differentiating (6.71) with respect to x and substituting (6.70) gives

$$JG \frac{d^2\theta}{dx^2} = -t(x) \tag{6.72}$$

Singularity functions similar to (6.4) may be used to evaluate the effect of complicated loadings. Torque functions $t_i(x)$ analogous to (6.5) are

$$\text{Concentrated torque: } T_{ai}(x - ai)_*^{-1} \tag{6.73a}$$

$$\text{Distributed torque: } \quad T_{ai}(x - ai)^0 \tag{6.73b}$$

Torque functions (6.73) representing multiple loadings may be combined into one expression that is applicable over the length of the assembly. The applied torque for n loads may be expressed as

$$t(x) = \sum^{n} \langle T_i(x) \rangle_{ai} \tag{6.74}$$

where the definition of (6.6) applies for the angular brackets. Distributed torques may be added or subtracted to develop a specific loading.

The torque and rate of twist at any location can be determined from (6.72) as

$$T(x) = JG \frac{d\theta}{dx} = -\int_0^x t(x)\,dx + c_0 \tag{6.75a}$$

$$JG\theta = -\int_0^x dx \int_0^x t(x)\,dx + c_0 x + c_1 \tag{6.75b}$$

where the rules of integrating torque functions are analogous to (6.8) and c_i are constants of integration.

EXAMPLE 6.15. Determine the angle of twist and torque at the attachments for the uniform section (JG = constant) assembly shown in Figure 6.34.

Under the action of the applied torque $T_a = Pe$, the unit will twist causing a rotation of one cross-section relative to another. The angle of twist may be described in terms of the applied torque using (6.72) and (6.73a) as

$$(JG) \frac{d^2\theta}{dx^2} = -T_a \langle x - a \rangle_*^{-1}$$

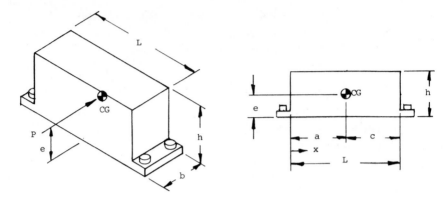

Figure 6.34. Concentrated Torque Idealization for an Assembly.

Performing the integrations indicated in (6.75) gives

$$(JG)\frac{d\theta}{dx} = -T_a\langle x - a\rangle^0 + c_0 \tag{6.76a}$$

$$(JG)\theta = -T_a\langle x - a\rangle + c_0 x + c_1 \tag{6.76b}$$

The angle of twist or torque at the attachments is used to evaluate the quantities c_0 and c_1. In this example, $\theta(0) = \theta(L) = 0$ at the attachments. Substituting these conditions into (6.76b) gives $c_1 = 0$ and $c_0 = T_a c/L$. Using (6.76b), the angle of twist at any location is

$$\theta(x) = \frac{T_a c}{JGL}x - \frac{T_a}{JG}\langle x - a\rangle \tag{6.77}$$

From (6.77), maximum twist occurs at $x = a$ giving

$$\theta_{max} = \frac{T_a c a}{JGL} \tag{6.78}$$

Torque at the attachments may be determined from (6.76a). At $x = 0$, (6.76a) gives

$$T_0 = (JG)\frac{d\theta}{dx}(0) = c_0$$

or $\hspace{8cm}$ (6.79a)

$$T_0 = T_a c/L$$

At $x = L$, (6.76a) gives

$$T_L = (JG) \frac{d\theta}{dx}(L) = -T_a + c_0$$

or   (6.79b)

$$T_L = -\frac{T_a a}{L}$$

Equation (6.79) indicates that a greater proportion of the load is resisted by the support closest to the applied torque. In other words, torque follows the stiffest path.

Q.E.D.

Equipment with nonuniform sections subjected to complicated torsional loads may also be evaluated using singularity functions. An approach similar to that of Example 6.3 is used.

EXAMPLE 6.16.   Determine the angle of twist and torque at the attachments for an assembly with the nonuniform section shown in Figure 6.35.

The assembly is separated at the point of discontinuity into two free body segments shown in Figure 6.35(b). A self-equilibrating torque, T_R, is used to maintain equilibrium between segments. The solution for each free body is initiated independently using a common origin at A. Applying (6.76) to segment AB gives

$$\frac{d\theta}{dx} = \frac{T_R}{(JG)_a} - \frac{T_R}{(JG)_a}\langle x - a \rangle^0$$

(6.80)

$$\theta = \frac{T_R}{(JG)_a} x - \frac{T_R}{(JG)_a}\langle x - a \rangle$$

The angle of twist at B of segment AB is used as the initial angle for segment BC. This quantity is constant for segment BC. Equation (6.80) can therefore be added to the twist relationship for segment BC to obtain a continuous expression for the assem-

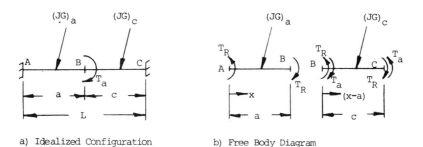

a) Idealized Configuration   b) Free Body Diagram

Figure 6.35. Torque Development of a Non-Uniform Assembly.

bly. Applying (6.76) to segment BC gives

$$\frac{d\theta}{dx} = \frac{(T_R - T_a)}{(JG)_c} \langle x - a \rangle^0$$

$$\theta = \frac{(T_R - T_a)}{(JG)_c} \langle x - a \rangle \qquad (6.81)$$

Adding (6.80) and (6.81) gives an expression for the angle of twist for the assembly as

$$\theta(x) = \frac{T_R}{(JG)_a} x + T_R \left[\frac{1}{(JG)_c} - \frac{1}{(JG)_a} \right] \langle x - a \rangle - \frac{T_a}{(JG)_c} \langle x - a \rangle \qquad (6.82)$$

The angle of twist at $x = 0$ was satisfied by the development of (6.80). At $x = L$, T_R must be selected to assure $\theta(L) = 0$. Solving (6.82) with this condition gives

$$T_R = \frac{T_a c/(JG)_c}{a/(JG)_a + c/(JG)_c} = \frac{T_a k_a}{k_a + k_c} \qquad (6.83)$$

where $k_i = (JG)_i/i$. A simplified expression for the angle of twist at any location based on (6.83) becomes

$$\theta(x) = \left(\frac{T_a}{k_a + k_c} \right) \frac{x}{a} - \left(\frac{T_a L/ac}{k_a + k_c} \right) \langle x - a \rangle \qquad (6.84)$$

Maximum twist occurs at the location of applied torque as

$$\theta_{max} = \frac{T_a}{k_a + k_c} \qquad (6.85)$$

The torque on the attachment at $x = 0$ is given by (6.80) as

$$T_0 = T_R$$

At $x = L$, the torque on the attachment is given by (6.81) as

$$T_L = -\frac{k_c}{k_a} T_R = -\frac{T_a k_c}{k_a + k_c}$$

Q.E.D.

Torsional Stiffness and the Polar Moment of Inertia.   The torsional stiffness of a solid or tubular member increases as the shape of the section approaches a circle. Longitudinal ribs or gussets are not as effective as an increase in thickness of a thin-walled section to improve torsional rigidity. Mounting interfaces and attachments that constrain rotation increases torsional rigidity by preventing shear displacements and warping at those locations. Near these joints the twisting moment is resisted by the torsional stiffness of the whole section and by the bending stiffness of some structural members. These effects, which diminish at locations away from points of constraint, increase the torsional rigidity of the section. It is customary and conservative when developing torsional section properties to use a loading appropriate to the weight distribution of the actual assembly while ignoring the contribution due to constraints. Since this approach assures sufficient design margin, idealizations which further increase conservatism should be avoided.

As an example, the torsional load on a uniformly distributed assembly should also be distributed since the use of concentrated torque idealization would give a torsional stiffness $k_{conc} = k_{dist}/2$. The polar moment of inertia for torque resisting structures that are open sections such as a tee, channel or I may be determined using

$$J = \tfrac{1}{3} \sum_{}^{n} s_i t_i^3 \tag{6.86}$$

where s_i is the length of a segment of constant thickness t_i and n is the number of segments composing the section. The polar moment of inertia of any closed thin wall sections (tubular) may be determined using

$$J = \frac{4A^2}{\oint \dfrac{ds}{t}} \tag{6.87}$$

where A represents the cross-section within the mean boundary and $\oint$ represents $\overset{n}{\Sigma} (s/t)$ for constant thickness segments around the perimeter. The polar moment of inertia of closed sections with thick walls can be obtained by multiplying (6.87) by a factor from Figure 6.36. This factor accounts for the Poisson contribution, which becomes more pronounced as wall thickness is increased. The torsional stiffness of a thick-walled section may be written as

$$k_{TW} = F_\sigma JG/\ell \tag{6.88}$$

EXAMPLE 6.17.   An aluminum traverse tubular member of a cabinet is used to support an assembly weighing 7 lb (3.2 kg) located as shown in Figure 6.37. The center of gravity of the assembly extends 4 in (101.6 mm) from the center of the 1.5 × 2.5 in (38.1 × 63.5 mm) rectangular section mounting structure. Without exceeding the given dimensions of the section, determine a uniform wall thickness to assure

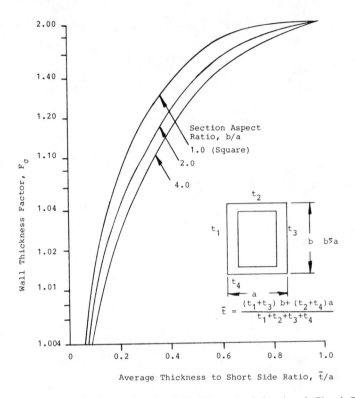

Figure 6.36. Thick Wall Factor for the Polar Moment of Inertia of Closed Sections $(J_{\text{THK WALL}} = F_\sigma J)$.

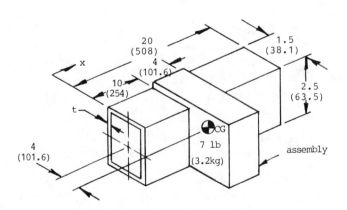

Figure 6.37. Eccentrically Mounted Assembly.

that the torsional resonance is greater than 300 Hertz. The equation for rotation is given by (6.72) and (6.73b) as

$$JG \frac{d^2\theta}{dx^2} = -T_1 \langle x - 10 \rangle^0 + T_1 \langle x - 14 \rangle^0 \quad \text{(English)}$$

$$JG \frac{d^2\theta}{dx^2} = -T_1 \langle x - 254 \rangle^0 + T_1 \langle x - 355.6 \rangle^0 \quad \text{(metric)}$$

which by applying (6.75) become

$$JG\theta = -\frac{T_1}{2} \langle x - 10 \rangle^2 + \frac{T_1}{2} \langle x - 14 \rangle^2 + c_0 x + c_1 \quad \text{(English)}$$

$$JG\theta = -\frac{T_1}{2} \langle x - 254 \rangle^2 + \frac{T_1}{2} \langle x - 355.6 \rangle^2 + c_0 x + c_1 \quad \text{(metric)}$$

$$(6.89)$$

The ends of the tubular member are assumed constrained giving the conditions $\theta(0) = \theta(L) = 0$. Substituting these conditions into (6.89) gives

$$c_1 = 0$$

$$c_0 = \frac{T_1}{2(20)} (10^2 - 6^2) = 1.6 T_1 \text{ lb} \cdot \text{in} \quad \text{(English)} \qquad (6.90)$$

$$c_0 = \frac{T_1}{2(508)} (254^2 - 152.4^2) = 40.64 T_1 \text{ N} \cdot \text{m} \quad \text{(metric)}$$

where $T_1 = 7$ lb·in/in (0.031 N·m/mm). Maximum twist occurs at the location for which $d\theta/dx = 0$ giving $x = 11.6$ in (294.6 mm). Since the total torque $T = 28$ lb·in (3163 N·mm), the maximum angle of twist from (6.89) is

$$\theta_{\text{max}} = -\frac{T(11.6 - 10)^2}{4(2) JG} + \frac{1.6 T(11.6)}{4 JG} = \frac{4.32 T}{JG} \text{ rad} \quad \text{(English)}$$

$$\theta_{\text{max}} = \frac{T(294.6 - 254)^2}{101.6(2) JG} + \frac{40.64 T(294.6)}{101.6 JG} = \frac{109.7 T}{JG} \text{ rad} \quad \text{(metric)}$$

$$(6.91)$$

with JG given in lb·in^2 (N·mm^2). The polar moment of inertia in terms of varying thickness t becomes

$$J = \frac{4[(2.5 - t)(1.5 - t)]^2}{2 \left[\dfrac{1.5 - 2t}{t} - \dfrac{2.5}{t} \right]} = \frac{t[(2.5 - t)(1.5 - t)]^2}{(2 - t)} \text{ in}^4 \quad \text{(English)}$$

$$J = \frac{4[(63.5 - t)(38.1 - t)]^2}{2 \left[\dfrac{38.1 - 2t}{t} + \dfrac{63.5}{t} \right]} = \frac{2t[(63.5 - t)(38.1 - t)]^2}{(101.6 - 2t)} \text{ mm}^4 \quad \text{(metric)}$$

$$(6.92)$$

The torsional stiffness of the configuration may be determined from (6.91) as

$$k_\theta = T/\theta = JG/4.32 \text{ lb·in/rad} \tag{6.93}$$

$$k_\theta = JG/109.7 \text{ N·mm/rad} = 1000JG/109.7 \frac{\text{kg·mm}^2}{\text{s}^2\text{·rad}}$$

The mass moment of inertia with respect to the center of the tabular section is

$$I_\theta = \frac{M_{\text{beam}}}{12}(h^2 + b^2) + M_{\text{assy}}r^2$$

$$I_\theta = \frac{8t(2 - t)}{386(12)}(1.5^2 + 2.5^2) + \frac{7}{386}(4)^2 = 0.0147t(2 - t) + 0.29 \text{ lb·in·s}^2 \tag{6.94}$$

$$I_\theta = 1.285t(101.6 - 2t) + 32764 \text{ kg·mm}^2$$

The torsional frequency accounting for thick wall effects may be expressed as

$$f_\theta = \frac{1}{2\pi} \sqrt{\frac{F_\sigma k_\theta}{I_\theta}} \tag{6.95}$$

Setting $F_\sigma = 1$ initially and substituting (6.92) through (6.94) into (6.95) gives

$$f'_\theta = 149.3(2.5 - t)(1.5 - t)$$

$$\times \left\{ \frac{t}{(2 - t)[0.0147t(2 - t) + 0.29]} \right\}^{1/2} \text{ Hz (English)} \tag{6.96}$$

$$f'_\theta = 110(63.5 - t)(38.1 - t)$$

$$\times \left\{ \frac{t}{(101.6 - 2t)[1.285t(101.6 - 2t) + 32764]} \right\}^{1/2} \text{ Hz (metric)}$$

where $G = 3.8 \times 10^6$ lb/in^2 (2.62×10^4 N/mm^2). The torsional frequency as a function of wall thickness t using (6.96) is shown graphically in Figure 6.38. The quantity F_σ may be determined from Figure 6.36 for the tube aspect ratio of 1.7 for short side ratios t/a using several thicknesses from the ordinate of Figure 6.38. The torsional frequency corresponding to each t may be determined from (6.95) as

$$f_\theta = f'_\theta \sqrt{F_\sigma} \tag{6.97}$$

The torsional frequency from (6.97) is shown in Figure 6.38. A resonance greater than 300 Hz can be assured using a wall thickness greater than 0.27 in (7 mm).   Q.E.D.

Electronic chassis are often divided by walls, bulkheads or interconnection boards in order to provide electrical isolation, functional partitioning or

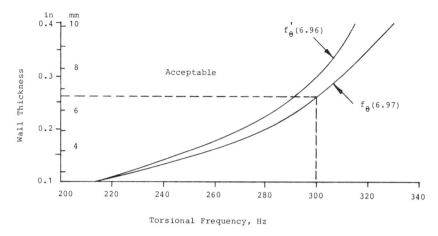

Figure 6.38. Parametric Evaluation of Wall Thickness.

increased structural stiffness. The product of the shear modulus and polar moment of inertia JG for compartmentized configurations fabricated with different materials may be derived from (6.87) as

$$JG = 4\{A\}^T[R]^{-1}\{A\} \qquad (6.98a)$$

where for n compartments

$$\{A\} = \begin{Bmatrix} A_1 \\ A_2 \\ \cdot \\ \cdot \\ \cdot \\ A_n \end{Bmatrix} ; [R] = \begin{bmatrix} r_{11} & -r_{12} & \cdots & -r_{1n} \\ -r_{21} & r_{22} & \cdots & -r_{2n} \\ \cdot & \cdot & & \cdot \\ \cdot & \cdot & & \cdot \\ \cdot & \cdot & & \cdot \\ -r_{n1} & -r_{n2} & \cdots & r_{nn} \end{bmatrix}$$

given $r = \Sigma(s/Gt)$. Subscripts ii refer to compartment i whereas subscripts ij where $ij = ji$ refer to the common partition between compartments i and j. When the section is composed of a single material, (6.98) may be expressed as

$$JG = 4G\{A\}^T[R]^{-1}\{A\} \qquad (6.98b)$$

where $r = \Sigma(s/t)$. The loss in structural efficiency due to the method of joining or attaching surfaces of the section together is accommodated by decreasing the thickness t of the appropriate surface(s). Structural efficiencies in torsion are about 40 percent greater than those given in Table 6.3 for various interface

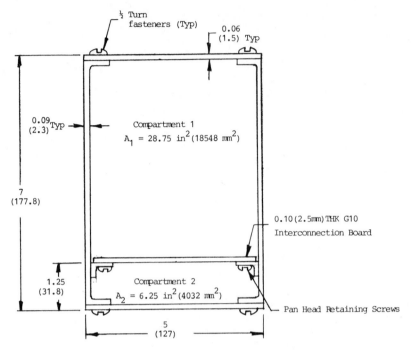

Figure 6.39. Aluminum Chassis Torsion Resisting Section.

configurations. This is due to a combined resistance of the attached surface to both shear and bending. The torsional structural efficiency of distributed pan head screws retaining an access panel may be determined from Table 6.3 as $(1 - 0.75/1.4)100 = 46\%$.

EXAMPLE 6.18.   Determine JG for the section of a chassis shown in Figure 6.39.

The torsional structural efficiency of the top and bottom access covers may be determined from Table 6.3 as $(1 - 0.9/1.4)100 = 36\%$. Following Appendix C for the material modulus, the coefficients of $[\mathbf{R}]$ of (6.98a) become

$$r_{11} = \frac{2(7 - 1.25)}{(3.8 \times 10^6)(0.09)} + \frac{5}{(3.8 \times 10^6)(0.36 \times 0.06)}$$

$$+ \frac{5}{(1.05 \times 10^6)(0.46 \times 0.10)} = 1.98 \times 10^{-4} \frac{\text{in}^2}{\text{lb}}$$

$$r_{11} = \frac{2(177.8 - 31.8)}{(26200)(2.3)} + \frac{127}{(26200)(0.36 \times 1.5)} + \frac{127}{(7239)(0.46 \times 2.5)}$$

$$= 0.029 \frac{\text{mm}^2}{\text{N}}$$

$$r_{12} = \frac{5}{(1.05 \times 10^6)(0.46 \times 0.10)} = 1.035 \times 10^{-4} \frac{in^2}{lb}$$

$$r_{12} = \frac{127}{(7239)(0.46 \times 2.5)} = 0.015 \frac{mm^2}{N}$$

$$r_{22} = \frac{2(1.25)}{(3.8 \times 10^6)(0.09)} + \frac{5}{(3.8 \times 10^6)(0.36 \times 0.06)}$$

$$+ \frac{5}{(1.05 \times 10^6)(0.46 \times 0.10)} = 1.72 \times 10^{-4} \frac{in^2}{lb}$$

$$r_{22} = \frac{2(31.8)}{(26200)(2.3)} + \frac{127}{(26200)(0.36 \times 1.5)} + \frac{127}{(7239)(0.46 \times 2.5)} = 0.025 \frac{mm^2}{N}$$

Matrix **[R]** can be expressed as

$$[\mathbf{R}] = 10^{-4} \begin{bmatrix} 1.98 & -1.035 \\ -1.035 & 1.72 \end{bmatrix} \text{(English)}$$

$$= \begin{bmatrix} 0.029 & -0.015 \\ -0.015 & 0.025 \end{bmatrix} \text{(metric)} \quad (6.99)$$

Inverting (6.99) using methods of Appendix A and substituting in (6.98a) gives

$$JG = 4(10^4) \{28.75 \quad 6.25\} \begin{bmatrix} 0.737 & 0.443 \\ 0.443 & 0.848 \end{bmatrix} \begin{Bmatrix} 28.75 \\ 6.25 \end{Bmatrix} = 3.2 \times 10^7 \text{ lb} \cdot in^2$$

$$JG = 4\{18548 \quad 4032\} \begin{bmatrix} 50 & 30 \\ 30 & 58 \end{bmatrix} \begin{Bmatrix} 18548 \\ 4032 \end{Bmatrix} = 9.05 \times 10^{10} \text{ N} \cdot mm^2$$

Q.E.D.

Fundamental Frequencies of Assemblies That Twist. An equivalent spring mass configuration of a system subjected to dynamic torsional loading may be defined using the parameters of Table 6.7. These parameters and those of Table B.1-1 involve the principle of dynamic similarity, which requires that the work, strain energy and kinetic energy of the equivalent system be identical with those of the actual structure. Shape functions in Table 6.7 may be treated analogously to those of Table 6.6.

EXAMPLE 6.19. Determine the torsional resonance of a 19 in (483 mm) long chassis supported as shown in Figure 6.34 having the section of Figure 6.39 and a weight of 20 lb (9.08 kg).

Table 6.7. Characteristic Equivalents and Shape Functions of Distributed Torsional Systems

Case	Configuration	Twist Shape Function	Mass, M_{eq}	Stiffness, K_{eq}	Dynamic Torque, T_{eq}
I		$\zeta(1 - \zeta/2)$	$M_{\text{conc}} + \%_{15}M_{\text{dist}}$	JG/L	$T_{\text{conc}} + \%T_{\text{dist}}$
II		$\zeta(1 - \zeta)$	$M_{\text{conc}} + \%_{15}M_{\text{dist}}$	$4JG/L$	$T_{\text{conc}} + \%T_{\text{dist}}$

Dynamic behavior at resonance results in a distributed torsional loading on the chassis. The equivalent mass moment of inertia from Table 6.7 for the mounting arrangement shown in Figure 6.34 becomes

$$I_{eq} = \frac{M_{eq}}{12}(4h^2 + b^2) = \frac{8M_{\text{dist}}}{12(15)}(4h^2 + b^2)$$

$$I_{eq} = \frac{8(20)}{12(15)386}(4 \times 7^2 + 5^2) = 0.509 \text{ lb·in·s}^2$$

$$I_{eq} = \frac{8(9.08)}{12(15)}(4 \times 177.8^2 + 127^2) = 5.75 \times 10^4 \text{ kg·mm}^2$$

The torsional resonance for the configuration of case II of Table 6.7, using the results of Example 6.18, is determined as

$$f_\theta = \frac{1}{2\pi}\sqrt{\frac{k_{eq}}{I_{eq}}} = \frac{1}{2\pi}\left[\frac{4(3.2 \times 10^7)}{20(0.509)}\right]^{1/2} = 564 \text{ Hz} \quad \text{(English)}$$

$$f_\theta = \frac{1}{2\pi}\left[\frac{4(9.05 \times 10^{10})1000}{483(5.754 \times 10^4)}\right]^{1/2} = 574 \text{ Hz} \quad \text{(metric)}$$

Q.E.D.

7
Displacements and Stresses in Equipment

The structure of an assembly responds to imposed loads in a manner consistent with basic laws of physics. In essence, work and energy are conserved. Energy losses are insignificant provided the material used behaves elastically. Usually, linearly elastic behavior is assumed when developing a structural idealization of an assembly. In this respect, there will be a linear relationship between the loads that are imposed and the displacements and stresses that result.

A knowledge of the distortion and the corresponding stresses is necessary in order to assess the design integrity of an assembly in its service environment. Displacements and stresses of assemblies that could be idealized as beams or as an elastically supported rigid body have been addressed in Chapter 6. The evaluation of structural configurations that typify those of components and assemblies will be examined in this chapter. These structures may be idealized as a system composed of an array of interconnected flexural members that may provide flexural as well as axial strength and stiffness. In this chapter, the basic concepts used to evaluate beam behavior are extended to include structures that can be idealized as frames.

The mathematical determination of beam displacements described in Chapter 6 was shown to be possible provided an expression for the moment was developed. For design applications with complex loadings, the development of closed-form solutions may be impractical.

Often, a knowledge of the overall deflected shape is not needed. The design engineer usually requires the deflection or stresses at particular locations which may be an interface of a component or attachment points of a substructure. These behavioral characteristics may be obtained by alternative approaches that include moment area, flexibility and stiffness methods. Stiffness methods are equally applicable to the study of circuit board displacements or the behavior of any loaded planar system. These techniques are discussed later in this chapter.

7.1 NATURE OF EQUIPMENT DISPLACEMENTS

Selection of structural members that can withstand imposed loads without exhibiting excessive displacements, catastrophic failure or fatigue requires knowledge of developed moments and shear forces. A moment diagram of an idealized structure also provides the basis of determining displacements by both moment area and flexibility methods. Characteristics that serve to expedite the construction of a moment diagram may be useful.

The moment diagram of a statically determinate structural idealization may be determined from statics for a particular loading. A positive moment which develops a positive curvature is drawn on the tensile side of the element. In this sense, the element curves away from the moment. Any portion of an element that is not subjected to a moment remains in its original shape. Using these conditions, the displaced shape of the structure can be approximated from its M/EI diagram.

The moment diagram of a determinate configuration is constructed by drawing the moment developed by each load. Indeterminate structures are made determinate by replacing supports with appropriate reaction loads or by using geometric symmetry as described in Section 6.2 (Figures 6.18 and 6.20). Once the reaction loads are determined in terms of the applied loading, the moment at any location can be found by algebraically adding the contributing moment quantities at that point using the diagram. The propped cantilever beam shown in Figure 7.1(a), which contains one redundant reaction, illustrates this approach. A moment diagram can be constructed in terms of the unknown reaction R_C or M_A as shown in Figure 7.1(b) or (c) respectively. The moment at point B in Figure 7.1(a) may be expressed in terms of R_C using Figure 7.1(b) as $M_B = R_C b$, or in terms of M_A from Figure 7.1(c) giving $M_B = b(Pa - M_A)/L$.

The moment along any element of the planar frames shown in Figures 7.2 and 7.3 may be determined similarly. The moment distribution of indeterminate idealized structures may be determined from the sum of the moments developed by two statically determinate configurations. One of these determinate configurations represents the moments developed by the applied loads and the other represents the moments developed by the unknown reactions. The

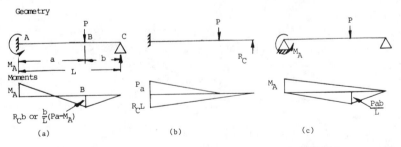

Figure 7.1. Moment Diagrams of a Propped Cantilever. (a) Idealization; (b) Unknown reaction, R_C; (c) Unknown reaction, M_A.

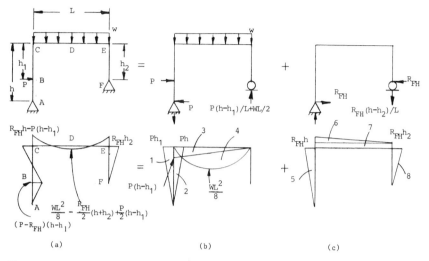

Figure 7.2. Moment Distribution of a Structure With One Redundant Reaction. (a) Idealization; (b) Reactions and moments due to applied loads; (c) Reactions and moments due to unknown reaction, R_{FH}.

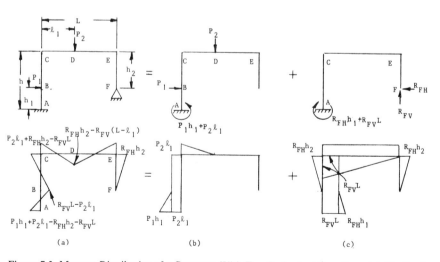

Figure 7.3. Moment Distribution of a Structure With Two Redundant Reactions. (a) Idealized structure; (b) Reaction and moments due to applied loads; (c) Reactions and moments due to unknown reactions, R_{FH} and R_{FV}.

moment distribution for each of these determinate structures is developed by first determining reactions due to the loading and then determining end forces of each member using a free body diagram and (6.1). This process becomes clear if the steps in developing the moment diagrams of Figures 7.2(b) and (c) and Figures 7.3(b) and (c) are retraced.

A qualitative sketch of the deformed structure can be developed once displacements (rotations and translations) at particular locations or joints have been determined. Displacements and redundant reactions may be determined using moment area methods. Applications of this method enable the determination of the total angle change between two points from the area under the M/EI diagram between these two points. The physical nature of the moment acting on the structure can be easily visualized using this approach.

The effort required in determining the area under portions of the M/EI diagrams can be reduced considerably if the components of moment due to individual loads, e.g. Figure 7.2(b) and (c), are used instead of the composite resultant moment diagram, e.g. Figure 7.2(a). Use of components of moment avoids the tedious time consuming determination of points of intersection that occur

Table 7.1 Areas and Centroids

Case	Geometry	Distribution	Area, A	Centroid, $\bar{s}$
I	Triangular		$hL/2$	$L/3$
II	Triangular		$hL/2$	$(\ell_1 + L)/3$
III	Power Law ($n = 2$ parabolic)		$hL/(n + 1)$	$L/(n + 2)$
IV	Parabolic		$2hL/3$	$3L/8$
V	Parabolic		$2hL/3$	$L/2$

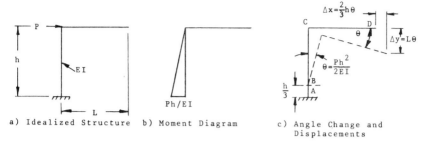

a) Idealized Structure   b) Moment Diagram   c) Angle Change and Displacements

Figure 7.4. Displacements by Moment Area Methods.

in the composite diagram. This approach also simplifies the determination of centroids of area at which point the concentrated angle change occurs. Area and centroidal location of commonly encountered geometrical distributions may be found in Table 7.1.

Discrete Rotation Method for Evaluating Structural Displacements. The load imposed on the structure shown in Figure 7.4(a) results in the moment diagram of Figure 7.4(b). Moments are drawn on the tensile side of the element. For this construction, the element curves away from the moment. This deformation may be represented by a concentrated angle change θ in Figure 7.4(c) which occurs at the centroid of the moment area. The rotation is equal to the moment area. The effect of a rotation at point B in Figure 7.4(c) causes the segment BCD to experience a rigid body rotation. Segment AB is undisturbed since it is unaffected by the rotation θ at point B. The displacements at any point on the structure may be determined using

$$\Delta x = y\theta$$ (7.1)

$$\Delta y = x\theta$$

The horizontal displacement of point D of Figure 7.4(c) using (7.1) becomes

$$\Delta x = \frac{2}{3} h\theta = \frac{ph^3}{3EI}$$

which is equal to the product of the rotation θ and the vertical distance between the point of rotation and the point of interest. The vertical displacement of point D of Figure 7.4(c) based on (7.1) becomes

$$\Delta y = L\theta = \frac{Ph^2L}{2EI}$$

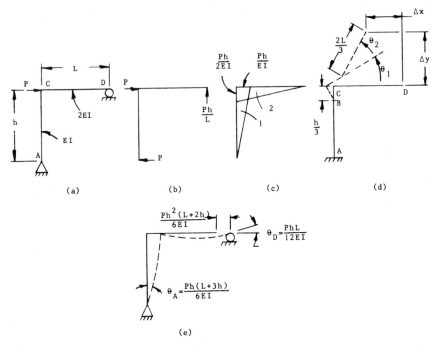

Figure 7.5. Moment Area Method for a Pinned Structure in Example 7.1. (a) Idealization; (b) Loads and reactions; (c) M/EI diagram; (d) Angle change and displacements; (e) Displaced shape.

which is equal to the product of the rotation θ and the horizontal distance between the point of rotation and the point of interest. The rotation at point D is equal to θ the rotation due to the applied moment at point B. The displacements of any number of discrete rotations due to the moment distribution are additive if they occur between a point of interest and a reference location. These displacements apply for determinate structures with fixed constraints. A corrective rigid body rotation must be applied at the point of constraint for determinate structures with hinged or pinned constraints. This corrective rotation must also be accounted for when determining actual displacements at points of interest throughout the structure.

Application of the moment area method to the structure of Figure 7.5(a) which assumes fixity at A will result in a vertical displacement at D shown in Figure 7.5(d). Since vertical motion at point D is constrained, a clockwise rigid body rotation at point A with the magnitude

$$\theta_A = \frac{\Delta y}{L} \qquad (7.2)$$

superposed on the system of Figure 7.5(d) will give correct displacements (translations and rotations) throughout the structure.

EXAMPLE 7.1.   Determine the displacements of point D for the structure given in Figure 7.5(a).

The loading and reactions for the determinate system illustrated in Figure 7.5(b) are used to construct the M/EI diagram in Figure 7.5(c). Discrete rotations θ_1 and θ_2 located at the centroid of area 1 and 2 respectively are summed to obtain θ_D, the rotation at point D, Figure 7.5(d). Correct rotations are obtained even though the displaced shape does not represent the true configuration. Discrete rotations are relative to the fixed reference at point A in Figure 7.5(d). Relationships in Table 7.1 facilitate the determinations of discrete rotations (areas) and their centroidal location.

The displacements of point D due to discrete rotations θ_1 and θ_2 become

Rotation		x	$\Delta y = x\theta_i$		y	$\Delta x = y\theta_i$
θ_1	$Ph^2/(2EI)$ ↻	L	$Ph^2L/(2EI)$ ↑		$h/3$	$Ph^3/(6EI)$ ←
θ_2	$PhL/(4EI)$ ↻	$2L/3$	$PhL^2/(6EI)$ ↑		0	0
$(\theta)_D = \dfrac{Ph}{4EI}(2h + L)$ ↻			$(\Delta y)_D = \dfrac{PhL}{6EI}(3h + L)$ ↑		$(\Delta x)_D = Ph^3/(6EI)$ ←	

If the rotation of point C should be required, it would be determined by discrete rotations due to the M/EI diagram along segment AC. The horizontal translation of segment CD is also determined only by rotations along segment AC.

Structural compatibility is established by a rigid body rotation about point A since this point has been initially fixed. A clockwise rotation of the entire structure about point A is required to restore the vertical movement of point D to its constrained position. Following (7.2), the rotation required is

$$\theta_A = Ph(3h + L)/(6EI) ↺$$

The total rotation and translation at D, which include the contribution from the M/EI diagram and the rigid body rotation, become

$$\theta_D = (\theta)_D - \theta_A = PhL/(12EI) ↻$$

$$\Delta x_D = \theta_A h - (\Delta x)_D = Ph^2(2h + L)/(6EI) →$$

These results are illustrated using an exaggerated displaced shape of the structure in Figure 7.5(e).   Q.E.D.

The direction of Δx and Δy translations at a point of interest associated with a discrete rotation at another location may be visualized as shown in Figure 7.6. A radius from the origin of rotation to the point of interest rotates about the origin in accordance with the curvature developed by the moment area. The x and y components of the repositioned point establish the directions of corre-

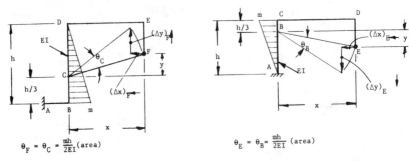

$$\theta_F = \theta_C = \frac{mh}{2EI}(area) \qquad\qquad \theta_E = \theta_B = \frac{mh}{2EI}(area)$$

Figure 7.6. Translations at a Point of Interest (Moment Area Method).

sponding translations. Displacements of indeterminate models can also be determined using this moment area method. Parameters representing unknown reaction loads, e.g. R_{FH} and R_{FV} of Figure 7.3(c), are retained in the development of θ, Δx and Δy at the point of the applied reaction load. The tabular arrangement for displacements at a desired location is generated as in Example 7.1. Relationships containing parameters of the unknown reactions are included. Parameter quantities are determined using geometric conditions of attachment at locations corresponding to these unknown reactions.

Geometric conditions corresponding to reaction loads, R_{FH} and R_{FV} in Figure 7.3(c) for example, are defined by Figure 7.3(a) as $(\Delta x)_F = (\Delta y)_F = 0$ respectively. Corresponding tabular sums are equated to these conditions, giving the equations necessary for determining reaction quantities.

EXAMPLE 7.2.   A structure idealized as shown in Figure 7.2(a) supports a distributed static load $wL = 4$ lb (17.8 N) and a concentrated load $P = 2$ lb (8.9 N). Determine the horizontal translation of point C and the maximum moment using

$L = 5$ in (127 mm) $\qquad\qquad\qquad h_1 = 1.5$ in (38.1 mm)

$h = 3$ in (76.2 mm) $\qquad\qquad\qquad h_2 = 2.0$ in (50.8 mm)

$EI = 500$ lb in^2 (1.435 $\times$ 10^6 (N·mm^2)

Structural displacements and moments are determined by eight discrete rotations defined by the moment areas shown in Figure 7.2(b) and (c). A tabular listing of translations and rotations of point F is developed in order to evaluate the unknown reaction R_{FH}. Once this quantity is known, the moment distribution of the structure can be determined directly using Figures 7.2(b) and (c).

Initially, the configuration in Figures 7.2(b) and (c) is made determinate by fixing point A. Compatibility of the structure will be restored by a rigid body rotation about point A once the vertical displacement of point F has been determined. The displacements of point F due to discrete rotations become as shown in Table 7.2.

The vertical movement of point F, $(\Delta y)_F$, is restored to its constrained position,

Table 7.2. Displacements at Point F (Figure 7.2) Due to Discrete Rotations with Point A Constrained

	Rotation, rad		x		$\Delta y = x\theta_i$		y		$\Delta x = y\theta_i$	
	English	(metric)	in	(mm)	in	(mm)	in	(mm)	in	(mm)
θ_1	0.0045 ↘	(0.0045 ↘)	5	(127)	0.0225↓	(0.572↓)	1.5	(38.1)	0.0067←	(0.171←)
θ_2	0.018 ↘	(0.018 ↘)	5	(127)	0.090↑	(2.286↑)	1	(25.4)	0.018→	(0.457→)
θ_3	0.015 ↘	(0.015 ↘)	3.33	(84.7)	0.050↑	(1.271↑)	2	(50.8)	0.030→	(0.762→)
θ_4	0.0167 ↘	(0.01667 ↘)	2.5	(63.5)	0.0418↑	(1.060↑)	2	(50.8)	0.033→	(0.848→)
θ_5	0.009R_{FH} ↙	(0.002F_{FH} ↙)	5	(127)	0.045R_{FH}↓	(0.254R_{FH}↓)	1	(25.4)	0.009R_{FH}←	(0.051R_{FH}←)
θ_6	0.005R_{FH} ↙	(0.0011R_{FH} ↙)	3.33	(84.7)	0.0167R_{FH}↓	(0.093R_{FH}↓)	2	(50.8)	0.010R_{FH}←	(0.056R_{FH}←)
θ_7	0.02R_{FH} ↙	(0.0045R_{FH} ↙)	2.5	(63.5)	0.050R_{FH}↓	(0.286R_{FH}↓)	2	(50.8)	0.040R_{FH}←	(0.229R_{FH}←)
θ_8	0.004R_{FH} ↙	(0.0009R_{FH} ↙)	0	0	0	0	1.33	(33.8)	0.0053R_{FH}←	(0.030R_{FH}←)
	$\overline{(\theta)_F = \Sigma\theta_i}$				$\overline{(\Delta y)_F = \Sigma\,\Delta y_i}$				$\overline{(\Delta x)_F = \Sigma\,\Delta x_i}$	

$(\Delta y)_F = 0$, by a rigid body rotation about point A which can be determined from (7.2) as

$$\theta_A = -\left(\frac{0.1593\uparrow + 0.1117R_{FH}\downarrow}{5}\right)$$

$$= 0.0319\downarrow + 0.0223R_{FH}\rangle \text{ rad (English)}$$

$$\theta_A = -\left(\frac{4.045\uparrow + 0.633R_{FH}\downarrow}{127}\right)$$

$$= 0.0319\downarrow + 0.0049R_{FH}\rangle \text{ rad (metric)}$$

The minus sign is used to develop a rotation opposite to the sense of computed y displacements. If required, the total rotation at F can be determined from $\theta_F = (\theta)_F + \theta_A$. The total horizontal translation at F is determined similarly as

$$\Delta x_F = y\theta_A + (\Delta x)_F$$

$$= 1.0(0.0319\downarrow + 0.0223R_{FH}\rangle) + (0.0743\rightarrow + 0.0643R_{FH}\leftarrow)$$

$$\Delta x_F = 0.1062\rightarrow + 0.0866R_{FH}\leftarrow \text{ in } \text{(English)}$$

$$\Delta x_F = 25.4(0.0319\downarrow + 0.0049R_{FH}\rangle) + (1.896\rightarrow + 0.336R_{FH}\leftarrow)$$

$$\Delta x_F = 2.706\rightarrow + 0.49R_{FH}\leftarrow \text{ mm } \text{(metric)}$$

Since $\Delta x_F = 0$ as defined by Figure 7.2(a), the reaction load becomes

$$R_{FH} = 0.1062/0.0866 = 1.226 \text{ lb } \text{(English)}$$

$$R_{FH} = 2.70/0.49 = 5.51 \text{ N } \text{(metric)}$$

The maximum moment may be determined by substituting quantities into the expressions in Figure 7.2(a). The cumulative contribution of member moments shown in Figures 7.2(b) and (c) is determined in Table 7.3.

The horizontal translation of point C is determined using a tabular approach which includes the contributions of discrete rotations along segment AC and the rigid body rotation at A. Displacements at point C become as shown in Table 7.4, where

$$\Delta x_C = 0.0799\rightarrow + 0.0579R_{FH}\leftarrow \text{ in } \text{(English)}$$

$$\Delta x_C = 2.0399\rightarrow + 0.3222R_{FH}\leftarrow \text{ mm } \text{(metric)}$$

Substituting the quantity for the horizontal reaction at point F gives the direction and magnitude of the horizontal displacement at C as

$$\Delta x_C = 0.0089 \rightarrow \text{in } \text{(English)}$$

$$\Delta x_C = 0.255 \rightarrow \text{mm } \text{(metric)} \qquad \text{Q.E.D.}$$

Table 7.3. Moments at the Nodal Points of Figure 7.2(a)

Point		Moment	
		lb·in	N·mm
B	$(P - R_{FH})(h - h_1)$	$(2 - 1.226)(3 - 1.5) = 1.16$	$(8.9 - 5.51)(76.2 - 38.1) = 129.2$
C	$R_{FH}h - P(h - h_1)$	$1.226(3) - 2(1.5) = 0.68$	$5.51(76.2) - 8.9(38.1) = 80.8$
D	$\dfrac{WL^2}{8} - \dfrac{R_{FH}}{2}(h + h_2)$ $+ \dfrac{P}{2}(h - h_1)$	$2.5 - 3.065 + 1.5 = 0.935$	$282.6 - 349.9 + 169.5 = 102.2$
E*	$R_{FH}h_2$	$1.226(2) = 2.45$	$5.51(50.8) = 279.9$

*Location of maximum moment.

The moment area method using discrete rotations may be applied successfully to most structural configurations encountered in electronic packaging. Any of the end conditions in Table 6.1 in addition to internal hinges may be included in the idealization. Use of symmetry and antisymmetry, as described in Section 6.2 when applicable, can decrease the effort of obtaining displacements and moments. The discrete rotation method also simplifies the development of parametric expressions which can provide a means of selecting optimum design quantities.

Relationship Between Board Flexure and Component Lead Stresses.
Structural configurations developed by components attached to circuit boards by their electrical leads, Figure 6.2(b), can be examined using this method. Board and component deflections are produced as a result of applied acceleration loads that may be encountered during handling, transport or in the operational environment. The attendant stresses that develop can cause fatigue failures in vibrating environments and catastrophic failure when the equipment experiences shock loads.

Table 7.4. Displacements at Point C (Figure 7.2) Due to Discrete Rotations

	Rotation, rad		y		$\Delta x = y\theta_i$	
	English	(metric)	in	(mm)	in	(mm)
θ_1	$0.0045\,\rangle$	$(0.0045\,\rangle)$	0.5	(12.7)	$0.00225\rightarrow$	$(0.0571\rightarrow)$
θ_2	$0.018\,\rangle$	$(0.018\,\rangle)$	1.0	(25.4)	$0.018\leftarrow$	$(0.4572\leftarrow)$
θ_5	$0.009R_{FH}\,\rangle$	$(0.002R_{FH}\,\rangle)$	1.0	(25.4)	$0.009R_{FH}\rightarrow$	$(0.0508R_{FH}\rightarrow)$
	$0.0319\,\rangle$	$(0.0319\,\rangle)$	3.0	(76.2)	$0.0957\rightarrow$	$(2.431\rightarrow)$
θ_A	$0.0223R_{FH}\,\rangle$	$(0.0049R_{FH}\,\rangle)$	3.0	(76.2)	$0.0669R_{FH}\leftarrow$	$(0.373R_{FH}\leftarrow)$
	$\theta_C = \Sigma\theta_i$				$\Delta x_C = \Sigma\,\Delta x_i$	

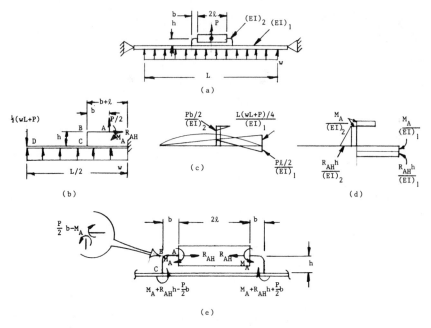

Figure 7.7. Maximum Component Lead Moments of a Card Guide Supported Circuit Board. (a) Circuit board and component configuration; (b) Idealization based on symmetry; (c) M/EI due to loads; (d) M/EI due to reactions; (e) Component lead moment resultants.

Highest stresses occur on components mounted at the center of printed circuit boards where the change in slope is the greatest. Because of lead configurations and body dimensions, some components may experience higher stresses than others at this location. Stresses in the attachments of components with body dimensions more than 1 inch long that may be mounted near the board center should be examined to assure that material allowables are not exceeded. Highly stressed conditions may be alleviated by altering the lead configuration or by fastening the component body to the board. The acceptability of these modifications is based on a subsequent evaluation of an appropriate idealization incorporating the revised design features.

Circuit boards commonly used in electronic equipment may be represented by the configuration shown in Figure 7.7. Hinge supports are developed by card guides which may have the configuration shown in Table 3.2 (A) through (C). The acceleration load g acts on individual components Pg through their CG's developing the distributed loading $W_{assy}g$ on the board. The component lead with elastic modulus $(EI)_2$, experiences movements which accommodate the developed curvature of the circuit board, $(EI)_1$, and the strain developed by the individual component load. Worst-case lead stresses are developed on the centralized component shown in Figure 7.7(a).

The idealization in Figure 7.7(b), which is based on the symmetry of the configuration in Figure 7.7(a), represents the load displacement behavior of the entire assembly. Assuming a rigid component body, two unknown reactions M_A and R_{AH} result at the lead to body interface at point A. Moments due to the applied loads and unknown reactions are illustrated in Figures 7.7(c) and (d) respectively.

The unknown reactions are determined from the geometrical conditions at point A. The only possible displacement along the axis of symmetry is Δy_A which gives $\theta_A = \Delta x_A = 0$. Through the method of discrete rotations in a tabular format, two equations for the unknown reactions result. These may be written in matrix form and inverted by the methods described in Appendix A, giving

$$
\begin{Bmatrix} M_A \\ R_{AH} \end{Bmatrix} = \frac{1}{|D|} \begin{bmatrix} h\left(\dfrac{h}{3} + \xi B\right) & -h\left(\dfrac{h}{2} + \xi B\right) \\ -\left(\dfrac{h}{2} + \xi B\right) & (b + h + \xi B) \end{bmatrix} \begin{Bmatrix} \alpha + P\left(\dfrac{b}{2}\right)^2\left(1 + \dfrac{h}{b}\right) \\ \alpha \end{Bmatrix}
$$

(7.3)

where

$$
|D| = h\left[\left(\frac{h}{3} + \xi B\right)(b + h + \xi B) - \left(\frac{h}{2} + \xi B\right)^2\right];
$$

$$
\alpha = P\left[\left(\frac{b}{2}\right)^2\left(\xi + \frac{h}{b}\right) - \xi\left(\frac{\ell}{2}\right)^2\right] + \xi\left(\frac{B}{2}\right)^2\left[(wL + P)\left(\frac{L}{B} - 1\right)\right.
$$

$$
\left. + \frac{w}{6}(3L - 4B)\right];
$$

$$
\xi = (EI)_2/(EI)_1;
$$

$$
B = b + \ell.
$$

Compact expressions for M_A and R_{AH} can be developed from (7.3) using a first order approximation for $\xi < 0.001$. This condition is indicative of a broad range of component lead and circuit board configurations. The moment M_A and axial lead force R_{AH} at the component body, point A, in terms of the applied acceleration g, are derived from (7.3) as

$$
M_A = \frac{g}{2(4b + h)}\left[Pb(h + 2b) - W_{\text{assy}}(b + \ell)L\xi\right]
$$

(7.4a)

$$R_{AH} = \frac{3g}{2h(4b + h)} [Pb^2 + W_{assy}L\xi(b + \ell)(1 + 2b/h)] \quad (7.4b)$$

where g equals the load due to acceleration acting on an element divided by its weight. Positive moments acting on the component leads, Figure 7.7(e), are determined by summing quantities shown in Figue 7.7(c) and (d). The effect of the component body length, ℓ, on lead stresses at the lead-to-component body interface, for example, decreases as the component body length is increased. Forces on the lead to the circuit board interface, however, increase for longer component body lengths. Fatigue of the solder joint, lead or plated through holes may be the result of excessive lead-to-board stresses developed by long components mounted near the center of the board.

EXAMPLE 7.3.   A circuit revision is required in order to enhance the capabilities of an electronic controller. A new layout of a 5 × 6 in (127 × 152 mm) printed circuit board assembly is required to incorporate the component changes. Most of the components with case dimensions greater than 1 inch are positioned adjacent to and oriented parallel with the card guides along the 5 in (127 mm) edge. Vibration tests performed on the original printed circuit board configuration have exposed a 160 Hz fundamental resonance with an amplification of 13. Both a resistor and capacitor in the same size cases, Figure 7.8(a), but with copper and nickel leads respectively, are being considered for mounting near the center of the assembly. Fatigue considerations limit the copper and nickel flexural stresses to a maximum of 19000 psi (131 N/mm^2) and 46000 psi (317 N/mm^2) for a 10^6 cycle design point. Determine the acceptability of placing these 0.0126 lb (0.056 N) components near the center if the equipment is exposed to a 5 G sinusoidal acceleration.

An expression for $W_{assy}L\xi$, used in (7.4), in terms of the resonant frequency for a simply supported uniformly weighted assembly may be obtained by combining δ_{max} and f_n for case V of Table 6.5, giving

$$W_{assy}L\xi = \frac{384}{5}\left(\frac{3.52}{f_nL}\right)^2 (EI)_2 \text{ lb·in}   \text{(English)}$$

$$W_{assy}L\xi = \frac{384}{5}\left(\frac{17.7}{f_nL}\right)^2 (EI)_2 \text{ N·mm}   \text{(metric)}$$

The value of $W_{assy}L\xi$ for the copper and nickel leads based on quantities from Figure 7.8(a) becomes

$$\begin{aligned} W_{assy}L\xi &= 3.57 \times 10^{-4} \text{ lb·in} \\ &= 4.03 \times 10^{-2} \text{ N·mm} \end{aligned} \Bigg\} \text{ copper}$$

$$\begin{aligned} W_{assy}L\xi &= 6.23 \times 10^{-4} \text{ lb·in} \\ &= 7.04 \times 10^{-2} \text{ N·mm} \end{aligned} \Bigg\} \text{ nickel}$$

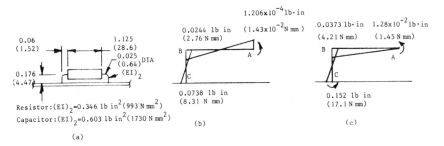

Figure 7.8. Case Configuration and Component Lead Moments for Example 7.3. (a) Component configuration; (b) Copper lead; (c) Nickel lead.

The load factor due to the amplification of the input vibration becomes

$$g = QG = 13(5G) = 65$$

The moment and axial lead force at point A, Figure 7.7(b), are given by (7.4) for the copper lead in English units as

$$M_A = 65 \left[\frac{0.0126(0.06)(0.296) - 3.57(10^{-4})(0.6225)}{2(0.416)} \right] = 1.206 \times 10^{-4} \text{ lb·in}$$

$$R_{AH} = \frac{3(65)}{2(0.176)(0.416)} \left[0.0126(0.06^2) + 3.57(10^{-4})(0.6225) \left(1 + \frac{0.12}{0.176} \right) \right]$$

$$R_{AH} = 0.558 \text{ lb}$$

In metric units, these quantities become

$$M_A = 65 \left[\frac{0.056(1.524)(7.52) - 4.03(10^{-2})(15.81)}{2(10.56)} \right] = 1.43 \times 10^{-2} \text{ N·mm}$$

$$R_{AH} = \frac{3(65)}{2(4.47)(10.56)} \left[0.056(1.524^2) + 4.03(10^{-2})(15.81) \left(1 + \frac{3.04}{4.47} \right) \right]$$

$$R_{AH} = 2.479 \text{ N}$$

Force quantities for the nickel lead are obtained similarly as $M_A = -1.282 \times 10^{-2}$ lb·in (-1.45 N·mm) and $R_{AH} = 0.9289$ lb (4.13 N). The moment distributions, Figures 7.8(b) and (c), were determined by evaluating the moment at points A, B and C using Figure 7.7(e). The moments at points B and C require the evaluation of $Pb/2$, which for the case of acceleration loading becomes

$$Pbg/2 = 0.0126(0.06)65/2 = 0.0245 \text{ lb·in} \quad \text{(English)}$$

$$= 0.056(1.524)65/2 = 2.774 \text{ N} \quad \text{(metric)}$$

Maximum lead stresses occur at the lead-to-board interface at point C. Since the lead stresses at this location are due to both flexural and axial forces, (6.17b) becomes

$$\sigma = \frac{M_c}{I} + \frac{F}{A} \tag{7.5}$$

where A = lead cross sectional area;
   $F = Pg/2$ (half the load due to acceleration is acting on each lead).

The maximum stress in the copper lead from (7.5) is

$$\sigma_{max} = \frac{0.0738(0.0126)}{2.011(10^{-8})} + \frac{0.0126(65)}{(2)(5.027)(10^{-4})} = 47054 \text{ lb/in}^2 \quad \text{(English)}$$

$$\sigma_{max} = \frac{(8.31)(0.32)}{8.37(10^{-3})} + \frac{0.056(65)}{2(0.324)} = 323.3 \text{ N/mm}^2 \quad \text{(metric)}$$

Maximum stresses in the nickel lead capacitor which occur at the same location are determined as 96051 lb/in^2 (662 N/mm^2). Since the lead stresses of the resistor and capacitor exceed design allowables, the mounting of these components at the board center is not permitted. It should be noted that these stresses also exceed the ultimate strength of the lead material.   Q.E.D.

Closed Structural Configurations.   Structures in electronic equipment provide a means of attaching both fixed and removable assemblies into a functionally organized system. The structure facilitates servicing while assuring that both handling and environmental loads do not damage or reduce the equipment's ability to perform its intended function. Removable attachments are generally idealized as pinned or hinged joints if rotation can occur at the interface. Area contact devices such as wedge lock fasteners develop a near-fixed condition provided the reaction interface supporting the clamping load is not compliant.

Structures formed by the assembly of substructures develop framework configurations similar to those shown in Figures 7.2(a) and 7.3(a) in addition to closed systems as illustrated by Figure 7.9(a). The closed systems considered may have any shape with any number of sides provided only a single cell is formed. Any number or type of loading may be imposed. Only pinned attachments to surrounding structural systems apply since fixity converts the system to a framework.

Closed systems are conveniently evaluated by the discrete rotation method which simplifies evaluation. The procedure is the same as that used for frameworks except that a tabular development is not required. A cut at any convenient location creates a determinate system. Internal forces at the cut are self-equilibrating, developing three unknown reactions. The coefficients of the three

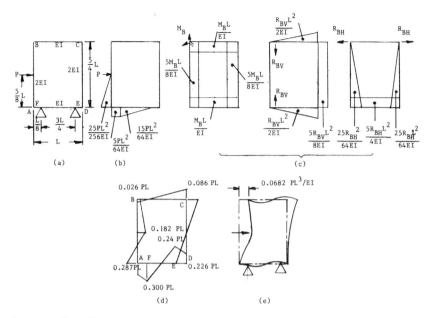

Figure 7.9. Closed System Evaluation of Example 7.4. (a) Idealized assembly; (b) M/EI area due to P; (c) M/EI area due to reactions M_B, R_{BV} and R_{BH}; (d) Moment distribution; (e) Displaced shape.

equations for these reaction quantities are determined by equating the sum of the M/EI areas and their first moments about any two orthogonal axes to zero. The M/EI areas are the result of the imposed loads and unknown self-equilibrating reactions. The composite moment distribution for the closed system is developed as the sum of these moment quantities.

EXAMPLE 7.4. An assembly composed of stiff side walls which is subjected to a side load may be idealized as shown in Figure 7.9(a). Determine the location of maximum stress and the magnitude of side sway developed.

The M/EI areas due to the load, Figure 7.9(a), and self-equilibrating reactions, Figure 7.9(c), are the result of a cut at point B. Summing the M/EI areas for M_B, R_{BV}, R_{BH} and load P and equating to zero give

$$\frac{13M_B L}{4EI} - \frac{13R_{BV}L^2}{8EI} + \frac{65R_{BH}L^2}{32EI} - \frac{105PL^2}{256EI} = 0$$

The first moments of the M/EI areas can be determined about any axis. Some effort may be saved if the selected axis corresponds to a side that experiences moments due to each of the unknown reactions and external loads. First moments of M/EI areas about the selected side are zero. The first moment due to R_{BH} about axis $AFED$ may be determined from Figure 7.9(c) as

$$\sum (Ay)_{R_{BH}} = 2\left(\frac{25R_{BH}L^2}{64EI}\right)\left(\frac{5L}{4}\right)\frac{1}{3} = \frac{125R_{BH}L^3}{384EI}$$

The first moment due to R_{BV} about axis CD which is orthogonal to $AFED$ becomes

$$\sum (Ax)_{R_{BV}} = -2\left(\frac{R_{BV}L^2}{2EI}\right)\left(\frac{L}{3}\right) = -\frac{R_{BV}L^3}{3EI}$$

The negative sign signifies an M/EI area drawn on the outside of the closed system. For a closed system $\Sigma(Ay) = \Sigma(Ax) = 0$. Grouping the coefficients of the summation of first moments of M/EI areas about the $AFED$ and CD axis and the sums of the M/EI areas due to M_B, R_{BV}, R_{BH} and P into a matrix gives

$$
\begin{bmatrix}
\dfrac{13L}{4} & -\dfrac{13L^2}{8} & \dfrac{65L^2}{32} \\[2mm]
\dfrac{13L^2}{8} & -\dfrac{L^2}{3} & \dfrac{65L^3}{64} \\[2mm]
\dfrac{65L^2}{32} & -\dfrac{65L^3}{64} & \dfrac{125L^3}{384}
\end{bmatrix}
\begin{Bmatrix} M_B \\[2mm] R_{BV} \\[2mm] R_{BH} \end{Bmatrix}
= P
\begin{Bmatrix} \dfrac{105L^2}{256} \\[2mm] \dfrac{265L^3}{1024} \\[2mm] \dfrac{125L^3}{6144} \end{Bmatrix}
$$

This matrix equation may be solved by the methods of Appendix A to give

$$
\begin{Bmatrix} M_B \\[2mm] R_{BV} \\[2mm] R_{BH} \end{Bmatrix}
= P
\begin{Bmatrix} 0.0259L \\[2mm] 0.1121 \\[2mm] 0.25 \end{Bmatrix}
$$

These results are used to compute moment quantities for the distributions in Figure 7.9(c). The composite distribution in Figure 7.9(d) depicts the sum of moment contributions due to P, M_B, R_{BV} and R_{BH}. The maximum moment (stress) occurs at the mounting attachment, point F in Figure 7.9(a).

Displacements at any location may be determined using the method of discrete rotations in conjunction with the load and reaction moment diagrams in Figure 7.9(b) and (c). The rotations at points of attachment which establish a reference may be determined with a tabular approach. Assuming that point F is initially fixed, the M/EI areas between points F and E, shown in Figures 7.9(b) and (c), are used to determine the rotation and displacement of point E. A rigid body rotation at point F restores point E to its constrained position. Table 7.5 summarizes the results of this procedure. The rigid body rotation at F from (7.2) is

Table 7.5. Displacements at Point E (Figure 7.9)
Due to Discrete Rotations with Point F
Constrained

M/EI Diagram	Rotation $(\theta)_E$	x	$(\Delta y)_E$
M_B	$0.0195\ PL^2/EI$	$3L/8$	$0.0073\ PL^3/EI\downarrow$
R_{BV}	$0.0105\ PL^2/EI$	$3L/8$	$0.0039\ PL^3/EI\uparrow$
R_{BV}	$0.0315\ PL^2/EI$	$L/4$	$0.0079\ PL^3/EI\uparrow$
R_{BH}	$0.2344\ PL^2/EI$	$3L/8$	$0.0879\ PL^3/EI\downarrow$
P	$0.2344\ PL^2/EI$	$L/2$	$0.1172\ PL^3/EI\uparrow$
	$0.0225\ PL^2/EI$		$0.0338\ PL^3/EI\uparrow$

$$\theta_F = \frac{0.0338 PL^3/EI}{0.75L} = 0.045 PL^2/EI$$

The rotation at E becomes

$$\theta_E = \theta_F - (\theta)_E = 0.0225 PL^2/EI$$

The side sway is defined as the horizontal displacement of point B or C. The displacement and rotation of point C are determined by the M/EI areas from Figure 7.9(b) and (c) between point E and point C in addition to the rotation at point E. The results are shown in Table 7.6. Figure 7.9(e) illustrates the displaced shape and side sway of the assembly.   Q.E.D.

Table 7.6. Rotation and Side Sway at Point C
(Figure 7.9) Due to Discrete Rotations

M/EI Diagram	Rotation $(\theta)_C$	y	$(\Delta x)_C$
Point E	$0.0225\ PL^2/EI$	$5L/4$	$0.0281\ PL^3/EI\rightarrow$
M_B	$0.0032\ PL^2/EI$	$5L/4$	$0.0040\ PL^3/EI\rightarrow$
M_B	$0.0162\ PL^2/EI$	$5L/8$	$0.0101\ PL^3/EI\rightarrow$
R_{BV}	$0.0123\ PL^2/EI$	$5L/4$	$0.0153\ PL^3/EI\leftarrow$
R_{BV}	$0.0009\ PL^2/EI$	$5L/4$	$0.0011\ PL^3/EI\leftarrow$
R_{BV}	$0.0700\ PL^2/EI$	$5L/8$	$0.0438\ PL^3/EI\leftarrow$
R_{BH}	$0.0039\ PL^2/EI$	$5L/4$	$0.0049\ PL^3/EI\rightarrow$
R_{BH}	$0.0976\ PL^2/EI$	$5L/6$	$0.0813\ PL^3/EI\rightarrow$
	$0.0602\ PL^2/EI$		$0.0682\ PL^3/EI\rightarrow$ (side sway)

7.2 FLEXIBILITY METHODS

Flexibility methods make it possible to evaluate the behavior of idealized structures systematically. An analytical procedure is used to develop coefficients
which describe the displacement response of a structure to applied loads at
node points. Flexibility coefficients for structures with many degrees of freedom are conveniently arranged in matrix $[\mathbf{F}]$ which transforms applied loads
$\{\mathbf{P}\}$ to coordinate displacements $\{\delta\}$

$$\{\delta\} = [\mathbf{F}] \{\mathbf{P}\} \tag{7.6}$$

The displacement response to applied loads evaluated by the method of discrete
rotations in Section 7.1 is also expressed in this form. Each of these methods
requires the development of moment diagrams that project the influence of the
applied loads onto the elements of a determinate structure. Indeterminate
structures are made determinate by defining constraint conditions using
unknown reactions or by introducing cuts having self-equilibrating internal
forces which are also unknown. These force and reaction components are
actually considered as unknown applied loads for the purpose of developing
flexibility coefficients. The magnitude and direction of the unknown reactions
and forces are determined from compatibility conditions of the structure. At
cuts, for example, compatibility requires that relative displacements be non-
existent. Approaches for developing a determinate configuration and constructing moment distributions are the same for both discrete rotation and flexibility
methods.

Element Flexibility Matrices. The difference lies in the ability to characterize
the behavior of discrete elements which can be assembled together to represent
the configuration and response characteristics of the entire structure. Elements
are characterized by flexibility coefficients that account for internal forces and

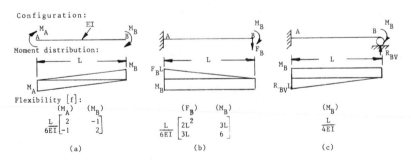

Figure 7.10. Element Flexibility Matrices. (a) Simple support; (b) Cantilever; (c) Propped.

applied loadings. An element library of commonly used configurations can be maintained to expedite idealization and evaluation of any structure. Element flexibility coefficients [f] may be determined using discrete rotations.

Consider the statically determinate beam element in Figure 7.10(a) subject to end moments. Based on the corresponding moment distribution in Figure 7.10(a), the method of discrete rotations with end A initially fixed gives the displacements (rotation and translation) at point B as

Rotation		x	$(\Delta y)_B = x\theta_i$
θ_1	$M_A L/(2EI)$ ⟩	$2L/3$	$M_A L^2/(3EI)$↑
θ_2	$M_B L/(2EI)$ ⟩	$L/3$	$M_B L^2/(6EI)$↓
$(\theta)_B = \dfrac{L}{2EI}(M_B - M_A)$ ⟩		$(\Delta y)_B = \dfrac{L^2}{6EI}(2M_A - M_B)$↑	

The rigid body rotation at A to restore compatibility at point B, $(\Delta y)_B = 0$, is

$$\theta_A = -(\Delta y)_B/L = \frac{L}{6EI}(2M_A - M_B)\,\rangle$$

The rotation at B becomes

$$\theta_B = \theta_A + (\theta)_B = \frac{L}{6EI}(2M_B - M_A)\,\rangle$$

In matrix form, the rotation in terms of the applied moments become

$$\begin{Bmatrix} \theta_A \\ \theta_B \end{Bmatrix} = \frac{L}{6EI}\begin{bmatrix} 2 & -1 \\ -1 & 2 \end{bmatrix}\begin{Bmatrix} M_A \\ M_B \end{Bmatrix} \tag{7.7}$$

which may be expressed compactly in terms of element flexibility [f] as

$$\{\theta\} = [\mathbf{f}]\,\{\mathbf{M}\}$$

Element flexibility matrices for three element configurations are given in Figure 7.10. Any structural configuration may be represented by these elements, which may be assembled end-to-end. Internal forces acting at the joints in accordance with Figure 7.10 are determined from statics and the moment distribution representing the applied loads on the structure. The loads {P} acting on the structure are transformed to forces {Q} on each element in this manner. This transformation may be conveniently represented by a load-force transformation matrix $[b]$.

Load-Force Relationships.   Any number of elements may be used to idealize the actual structure. A change in section properties or orientation requires use of different elements. Specific knowledge of the behavior of the structure at a particular location also requires the use of additional elements since joints (nodes) are the only means of obtaining these characteristics. The size of the load-force transformation matrix $[b]$ is a consequence of the number of elements used and the quantity of applied loads and unknown reactions. The relationship between these quantities becomes

$$
\begin{Bmatrix} Q_1 \\ Q_2 \\ Q_3 \\ \vdots \\ Q_i \end{Bmatrix} = [b_0] \begin{Bmatrix} R_1 \\ R_2 \\ R_3 \\ \vdots \\ R_j \end{Bmatrix} + [b_1] \begin{Bmatrix} P_1 \\ P_2 \\ P_3 \\ \vdots \\ P_k \end{Bmatrix} \tag{7.8a}
$$

which may be expressed in compact notation as

$$
\{Q\} = \left[\, b_0 \,\vdots\, b_1 \,\right] \begin{Bmatrix} R \\ \cdots \\ P \end{Bmatrix} \tag{7.8b}
$$

where

$$
[b] = \left[\, b_0 \,\vdots\, b_1 \,\right] \tag{7.8c}
$$

The subscripts i, j and k in (7.8a) represent the number of internal forces, the number of unknown reactions and the number of externally applied structural loads in matrices $\{Q\}$, $\{R\}$ and $\{P\}$ respectively.

Coefficients of the rectangular matrices $[b_0]$ and $[b_1]$ in (7.8a) may be obtained from conditions of equilibrium. Coefficients in $[b]$ are determined column by column by first setting R_1 equal to a unit load to obtain corresponding internal forces Q_i with all other R_j and P_k held to zero. Coefficients in the remaining columns of $[b]$ are determined successively in a similar manner.

A node must be placed at point B on the propped cantilever of Figure 7.1(a) if the displacements at the load are required. Also, nodes are required at points A and C to accommodate reactions. The use of three nodes develops a two element configuration for characterizing the behavior of the structure. The load-force transformation matrix $[b]$ for the configuration of Figure 7.1(a) may be developed using the moment distribution of either Figure 7.1(b) or (c). Assuming that only translation at the load is required, Equation (7.8a) using flexibility elements of Figure 7.10(a) and the moment distribution of Figure 7.1(b) becomes

$$
\begin{bmatrix} M_1 \\ M_2 \\ M_3 \\ M_4 \end{bmatrix} = \begin{bmatrix} L \\ -b \\ b \\ 0 \end{bmatrix} \{R_C\} + \begin{bmatrix} -a \\ 0 \\ 0 \\ 0 \end{bmatrix} \{P\}
$$

which in the compact notation of (7.8c) is

$$
[b] = \begin{bmatrix} L & -a \\ -b & 0 \\ b & 0 \\ 0 & 0 \end{bmatrix} \tag{7.9}
$$

Figure 7.11(b) illustrates the application of the flexibility element of Figure 7.10(a) to the configuration of Figure 7.1(a). The application of the flexibility element of Figure 7.10(b) used separately or combined with the element of Figure 7.10(a) is also illustrated in Figure 7.11. Identical results are obtained with any of the force-load transformation matrices $[b]$ of Figure 7.11 when they are used with corresponding element flexibility matrices $[f]$ of Figure 7.10.

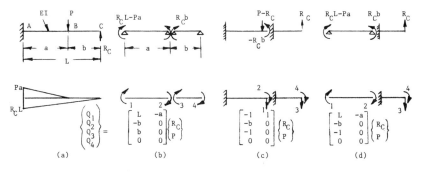

Figure 7.11. Load Force Transformation Representations. (a) Idealized assembly; (b) Simple support elements; (c) Cantilever elements; (d) Combined elements.

Matrix Development of An Idealized Assembly. The structure flexibility matrix [**F**] is developed by analytically combining the flexibilities [**f**] of the individual elements comprising the idealization with the force load transformation matrix [*b*]. This operation may be expressed as

$$[\mathbf{F}] = [b]^T[f_i][b] \tag{7.10}$$

Once [**F**] has been developed, reactions {**R**} may be determined from the constraint conditions and the displacements may be computed from (7.6). The matrix transpose $[b]^T$ is defined in Appendix A.

Matrix $[f_i]$ is a diagonal matrix composed of element matrices [*f*] set in the order used when developing [*b*]. Since element *AB* has been used to generate the first two rows of (7.9) and element *BC* the last two rows, matrix $[f_i]$ for the configuration of Figure 7.11(b) becomes

$$
[f_i] =
\begin{bmatrix}
\dfrac{a}{6EI}\begin{bmatrix} 2 & -1 \\ -1 & 2 \end{bmatrix} & 0 \\[2em]
0 & \dfrac{b}{6EI}\begin{bmatrix} 2 & -1 \\ -1 & 2 \end{bmatrix}
\end{bmatrix}
$$

$$
= \frac{1}{6EI}
\begin{bmatrix}
2a & -a & \vdots & & \\
-a & 2a & \vdots & & 0 \\
\cdots & \cdots & \cdots & \cdots & \cdots \\
& & \vdots & 2b & -b \\
& 0 & \vdots & -b & 2b
\end{bmatrix}
\tag{7.11}
$$

The flexibility matrix [**F**] for the structure of Figure 7.1(a) is determined from (7.10) using (7.9) and (7.11) as

$$[\mathbf{F}] = \frac{1}{6EI}\begin{bmatrix} 2L^3 & -a^2(2L+b) \\ -a^2(2L+b) & 2a^3 \end{bmatrix} \tag{7.12a}$$

System displacements are given by (7.6) as

$$\begin{Bmatrix} y_C \\ y_B \end{Bmatrix} = \frac{1}{6EI}\begin{bmatrix} 2L^3 & -a^2(2L+b) \\ -a^2(2L+b) & 2a^3 \end{bmatrix}\begin{Bmatrix} R_C \\ P \end{Bmatrix} \tag{7.12b}$$

Constraint imposed at point *C* of Figure 7.1(a) requires $y_C = 0$. The vertical load at point *C*, therefore, is given by the first equation of (7.12b) as

$$R_C = \frac{a^2(2L + b)P}{2L^3}$$

The vertical translation at the load is then given by the second equation of (7.12b) as

$$y_B = \frac{a^3 P}{12EI}\left[4 - \frac{a(2L + b)^2}{L^3}\right]$$

Equation (7.6) determines system displacements that correspond to the direction and location of applied structural loads which are included in the development of $[b]$. This happens even though rotations and translations also occur throughout the structure. Rotational displacements θ_B and θ_C at points B and C which also result from the applied load P in Figure 7.11(a) are not determined by (7.12b) because moments have not been imposed at these locations in the development of (7.9). Fictitious loads can be used when actual loads do not occur at a location for which displacement information is desired. A linear load develops a translation whereas a torque develops rotation at the point of application. Fictitious loads develop explicit relationships in (7.10) which account for the balance between work and strain that exists at the applied location. Application of fictitious loads is illustrated in Example 7.5.

Structure and element flexibility matrices are always symmetrical about the principal diagonal. This useful property provides a check on the development of matrices for complicated structures. The systematic evaluation of a structure by the flexibility method may be summarized as follows:

1. Evaluate determinancy of the configuration and define reaction quantities to establish determinancy as required.
2. Place nodes at locations where displacements (translations and rotations) are desired and at points where section properties or the direction of the structural configuration changes.
3. Define internal force systems for each element of the structure for which an element matrix $[f]$ has been determined.
4. Establish the force-load transformation matrix $[b]$ relating internal forces (moment and shear) to the applied loads and unknown reactions.
5. Generate the system flexibility matrix $[F]$ from element flexibility matrices $[f]$ and transformation matrix $[b]$ using (7.10).
6. Impose compatibility (geometrical conditions of constraint) to determine reaction quantities $\{R\}$ for indeterminate configurations.
7. Determine system displacements $\{\delta\}$ using (7.6).
8. Determine internal forces $\{Q\}$ using (7.8b).

9. Determine internal displacements, if desired, from

$$\{\Delta\} = [f_i]\{Q\} \tag{7.13}$$

The flexibility method characterizes the behavior of a structural configuration on the basis of the composite behavior of each of its elements.

EXAMPLE 7.5. Use flexibility to determine the displacements of point D for the structure of Figure 7.5(a) used in Example 7.1.

Since the structure is determinate, there are no unknown reaction quantities, $\{R\} = 0$. Three nodes located at points A, C and D are sufficient to represent the geometry and behavior of the structure. An internal force system corresponding to the element flexibility in Figure 7.10(a) is shown in Figure 7.12(c).

The force-load transformation matrix representing the configuration in Figure 7.12(c) based on the moment distribution of Figure 7.12(b) is

$$[b] = \begin{bmatrix} 0 \\ -h \\ h \\ 0 \end{bmatrix}$$

Element flexibility matrices $[f]$ are assembled using the same row order for the internal moments used in generating rows of $[b]$.

$$[f_i] = \begin{bmatrix} \dfrac{h}{6EI}\begin{bmatrix} 2 & -1 \\ -1 & 2 \end{bmatrix} & 0 \\ 0 & \dfrac{L}{12EI}\begin{bmatrix} 2 & -1 \\ -1 & 2 \end{bmatrix} \end{bmatrix}$$

$$= \frac{1}{6EI}\begin{bmatrix} 2h & -h & \vdots & \\ -h & 2h & \vdots & 0 \\ \cdots & \cdots & \vdots & \cdots \\ & & \vdots & L & -L/2 \\ & 0 & \vdots & -L/2 & L \end{bmatrix}$$

The system flexibility matrix using (7.10) becomes

$$[F] = \frac{h^2}{6EI}(2h + L)$$

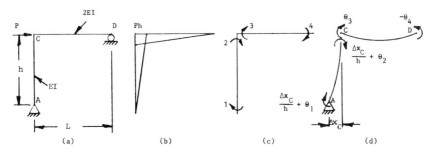

Figure 7.12. Flexibility Method for a Determinate Structure. (a) Structural system; (b) Moment distribution; (c) Internal moments; (d) Translations and rotations.

Following (7.6), the displacement at point D is

$$\Delta x_D = \frac{Ph^2}{6EI}(2h + L)$$

Internal rotations at A, C and D are determined by using (7.13), which may be expressed as

$$\{\theta\} = [f_i][b]\{P\}$$

giving

$$\begin{Bmatrix} \theta_1 \\ \theta_2 \\ \theta_3 \\ \theta_4 \end{Bmatrix} = \frac{Ph}{6EI} \begin{Bmatrix} h \\ -2h \\ L \\ -L/2 \end{Bmatrix}$$

The system rotation at D is the same as the internal rotation θ_4 since element CD does not experience a rigid body rotation. The rotation at D becomes

$$\theta_D = \frac{PhL}{12EI}$$

The sign of θ_4 indicates rotation in a direction opposite to that assumed in Figure 7.10(a). The rotation at A is developed by joint rotation θ_1 and the rigid body rotation of member AC due to Δx_c which may be determined as

$$\theta_A = \theta_1 + \Delta x_c/h = Ph(3h + L)/(6EI)$$

The rotation at joint C can be determined similarly using the internal rotation θ_2. Fictitious torsional loads at joints A, C and D provide an alternative approach. The force-load transformation matrix assuming fictitious clockwise torques at A and D becomes

$$[b] = \begin{matrix} (P) & (M_A) & (M_D) \\ \begin{bmatrix} 0 & 1 & 0 \\ -h & -1 & 0 \\ h & 1 & 0 \\ 0 & 0 & 1 \end{bmatrix} \end{matrix}$$

Equation (7.10) gives the system flexibility matrix as

$$[F] = \frac{1}{6EI} \begin{matrix} (P) & (M_A) & (M_D) \\ \begin{bmatrix} h^2(2h + L) & h(3h + L) & -hL/2 \\ h(3h + L) & 6h + L & -L/2 \\ -hL/2 & -L/2 & L \end{bmatrix} \end{matrix}$$

System displacements are determined from (7.6) using only the column developed by load P since fictitious loads do not have magnitude.

$$\begin{Bmatrix} \Delta X_D \\ \theta_A \\ \theta_D \end{Bmatrix} = \frac{Ph}{6EI} \begin{Bmatrix} h(2h + L) \\ 3h + L \\ -L/2 \end{Bmatrix}$$

This approach provides accurate results systematically.   Q.E.D.

Load Relationships Between An Assembly and Its Idealization.   The flexibility method is also suited to structures with distributed loads and/or concentrated loads between joints (nodes). These configurations result when the overall behavior of the structure is required and specific information at the point of loading is not needed. The solution is obtained by transforming loads applied between nodes to equivalent loads acting on the member joints. Equivalent loads are determined by the fixed end moments for the type of loading considered. These loads develop the same end displacements as the original loading.

The transformation of loads applied along a member to equivalent loads acting at its ends is illustrated by the free body diagram in Figure 7.13. Translation and rotation of points A and B in Figure 7.13 are the result of equivalent concentrated shear and torque loads representing each applied loading acting collectively at the ends of segment AB. The equivalent end loads for any

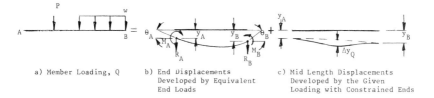

a) Member Loading, Q

b) End Displacements Developed by Equivalent End Loads

c) Mid Length Displacements Developed by the Given Loading with Constrained Ends

Figure 7.13. Dissection of Element Displacements.

applied load configuration can be developed as the sum of specific loadings listed in Table 7.7.

The equivalent end loads for the distributed loadings of cases X and XI in Table 7.7 are determined by summing the contributions for Cases IV and V. Alternatively, Case IX can be used to determine the equivalent end loads for these configurations.

EXAMPLE 7.6.   Use flexibility methods to determine the horizontal translation of point C and the maximum moment for the structural configuration studied in Example 7.2.

The equivalent end loads for the concentrated and distributed loadings shown in Figure 7.14(a) are depicted in Figure 7.14(b). The internal moments for each element are shown in Figure 7.14(c). Moment distributions representing the applied moments at joints A, C and E and the concentrated horizontal shear load at C due to equivalent end loads may be constructed to aid in the development of the load-force transformation matrix $[b]$. Considering the loads at each joint separately including the unknown reaction R_{FH}, Figure 7.2(c), enables the development of matrix $[b]$ in (7.8b) as

$$\begin{Bmatrix} M_1 \\ M_2 \\ M_3 \\ M_4 \\ M_5 \\ M_6 \end{Bmatrix} = \begin{bmatrix} 0 & 1 & 0 & 0 & 0 \\ h & -1 & 0 & 0 & -h \\ -h & 1 & 1 & 0 & h \\ h_2 & 0 & 0 & -1 & 0 \\ -h_2 & 0 & 0 & 0 & 0 \\ 0 & 0 & 0 & 0 & 0 \end{bmatrix} \begin{Bmatrix} R_{FH} \\ M_A \\ M_C \\ M_E \\ P_{CH} \end{Bmatrix} \quad (7.14)$$

where $M_A = \frac{3}{4}$ lb·in (84.8 N·mm)$\rangle$, $M_E = \frac{5}{3}$ lb·in (188.4 N·mm)$\rangle$, $M_0 = M_E - M_A = \frac{11}{12}$ lb·in (103.6 N·mm)$\rangle$ and $P_{CH} = 1.0$ lb (4.448 N)$\rightarrow$. System displacements at A, C and E will result from (7.6) since loads are applied at these joints. A fictitious torque at F must be appended to $[b]$ if the angular rotation of that joint is required.

Element flexibilities $[\mathbf{f}]$ are assembled using the same order for the internal moments used in generating rows of $[b]$ which for constant EI becomes

Table 7.7. Equivalent End Loads

$$M_A \overset{\curvearrowright}{\underset{A}{\rule{2cm}{0.4pt}}}\overset{M_B}{\underset{B}{}}$$
$$R_A \quad R_B$$
$$|\leftarrow L \rightarrow|$$

Case	Load Configuration	M_A	M_B	R_A	R_B
I		$\dfrac{Pab^2}{L^2}$	$\dfrac{Pa^2b}{L^2}$	$\dfrac{Pb^2(L+2a)}{L^3}$	$\dfrac{Pa^2(L+2b)}{L^3}$
II		$\dfrac{wL}{12}\left(\dfrac{b}{L}\right)^3(L+3a)$	$\dfrac{wb^2}{12}\left[1+2\left(\dfrac{a}{L}\right)+3\left(\dfrac{a}{L}\right)^2\right]$	$\dfrac{w}{2}\left(\dfrac{b}{L}\right)^3(L+a)$	$wb-R_A$
III		$\dfrac{wa^2}{12}\left[1+2\left(\dfrac{b}{L}\right)+3\left(\dfrac{b}{L}\right)^2\right]$	$\dfrac{wL}{12}\left(\dfrac{a}{L}\right)^3(L+3b)$	$wa-R_B$	$\dfrac{w}{2}\left(\dfrac{a}{L}\right)^3(L+b)$
IV		$\dfrac{wa^2}{60}\left[3+4\left(\dfrac{b}{L}\right)+3\left(\dfrac{b}{L}\right)^2\right]$	$\dfrac{wL}{60}\left(\dfrac{a}{L}\right)^3(2L+3b)$	$\dfrac{wa}{2}-R_B$	$\dfrac{w}{20}\left(\dfrac{a}{L}\right)^3(2b+3L)$
V		$\dfrac{wa^2}{30}\left[1+3\left(\dfrac{b}{L}\right)+6\left(\dfrac{b}{L}\right)^2\right]$	$\dfrac{wL}{20}\left(\dfrac{a}{L}\right)^3(L+4b)$	$\dfrac{wa}{2}-R_B$	$\dfrac{w}{20}\left(\dfrac{a}{L}\right)^3(8b+7L)$
VI		$\dfrac{Mb(2a-b)}{L^2}$	$\dfrac{Ma(2a-a)}{L^2}$	$-\dfrac{6abM}{L^3}$	$\dfrac{6abM}{L^3}$
VII		$\dfrac{6EI\,\Delta}{L^2}$	$-\dfrac{6EI\,\Delta}{L^2}$	$\dfrac{12EI\,\Delta}{L^3}$	$-\dfrac{12EI\,\Delta}{L^3}$
VIII	Parabolic	$\dfrac{wL^2}{15}$	$\dfrac{wL^2}{15}$	$\dfrac{wL}{3}$	$\dfrac{wL}{3}$
IX		$\dfrac{1}{L^2}\displaystyle\int_0^L w_x x b_x^2\,dx$	$\dfrac{1}{L^2}\displaystyle\int_0^L w_x x^2 b_x\,dx$	$\dfrac{1}{L^3}\displaystyle\int_0^L w_x b_x^2(L+2x)\,dx$	$\dfrac{1}{L^3}\displaystyle\int_0^L w_x x^2(3L-2x)\,dx$
X		$\dfrac{L^2}{60}(3w_1+2w_2)$	$\dfrac{L^2}{60}(2w_1+3w_2)$	$\dfrac{L}{2}(w_1+w_2)-R_B$	$\dfrac{L}{20}(3w_1+7w_2)$
XI		$\dfrac{5wL^2}{96}$	$\dfrac{5wL^2}{96}$	$\dfrac{wL}{4}$	$\dfrac{wL}{4}$

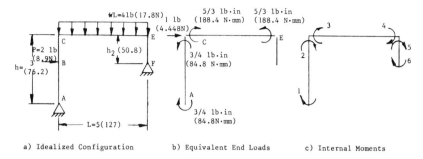

Figure 7.14. Loads and Forces for the Flexibility Solution of Example 7.6.

$$
[f_i] = \frac{1}{6EI}
\begin{bmatrix}
2h & -h & & & & \\
-h & 2h & & & & \\
& & 2L & -L & & \\
& & -L & 2L & & \\
& & & & 2h_2 & -h_2 \\
& & & & -h_2 & 2h_2
\end{bmatrix}
$$

System displacements (7.6) are determined using the system flexibility matrix (7.10), giving

$$
\begin{Bmatrix} \Delta x_F \\ \theta_A \\ \theta_B \\ \theta_C \\ \Delta X_C \end{Bmatrix} = \frac{1}{6EI}
\begin{bmatrix}
2(h^3 + h_2^3) + 2L(h^2 + hh_2 + h_2^2) & & & & \text{Symmetrical} \\
-3h^2 - L(2h + h_2) & 6h + 2L & & & \\
-L(2h + h_2) & 2L & 2L & & \\
-L(h + 2h_2) & L & L & 2L & \\
-2h^3 - Lh(2h + h_2) & h(3h + 2L) & 2hL & Lh & 2h^2(h + L)
\end{bmatrix}
\begin{Bmatrix} R_{FH} \\ M_A \\ M_C \\ M_E \\ P_{CH} \end{Bmatrix}
$$

$$(7.15)$$

The flexibility method permits algebraic solutions which are suited to parametric evaluation of structural displacements in terms of the structure geometry for a given loading.

Since $\Delta X_F = 0$ for the configuration of Figure 7.14(a), the reaction load is defined by the first equation in (7.15) as

$$
R_{FH} = \frac{[3h^2 + L(2h + h_2)]M_A + M_c L(2h + h_2) + M_E L(h + 2h_2) + [2h^3 + Lh(2h + h_2)]P_{CH}}{2(h^3 + h_2^3) + 2L(h^2 + hh_2 + h_2^2)}
$$

Substituting quantities gives

$$R_{FH} = \frac{67(3/4) + 40(11/12) + 35(5/3) + 174(1)}{260} = 1.228 \text{ lb} \text{ (English)}$$

$$R_{FH} = \frac{43226(84.8) + 25806(103.6) + 22581(188.4) + 2.855 \times 10^6(4.448)}{4.261 \times 10^6}$$

$$= 5.47 \text{ N} \text{ (metric)}$$

The reaction R_{FH} acts along the assumed direction in Figure 7.2(c) as indicated by a positive sign. The horizontal translation of C is defined by the last equation in (7.15) as

$$\Delta X_C = \frac{-[2h^3 + Lh(2h + h_2)]R_{FH} + M_Ah(3h + 2L) + 2hLM_C + M_ELh + 2h^2(h + L)P_{CH}}{6EI}$$

$$\Delta X_C = \frac{-174(1.228) + 57(3/4) + 30(11/12) + 5(5/3) + 144(1)}{6(500)}$$

$$= 0.00853 \text{ in} \text{ (English)}$$

$$\Delta X_C = \frac{-2.855 \times 10^6(5.47) + 36774(84.8) + 19355(103.6) + 9677(188.4) + 2.359 \times 10^6(4.448)}{6(1.435 \times 10^6)}$$

$$\Delta X_C = 0.212 \text{ mm} \text{ (metric)}$$

Internal moments are obtained from (7.14) after removing the equivalent end moment contributions shown in Figure 7.14(b). Equivalent end moments have been added to the structure in (7.14) to obtain the displacements due to the applied loads. The contribution of these moments must be removed in order to obtain the internal forces resisting the applied loads. This operation is performed by reversing the sense of the applied moments in Figure 7.14(b) and adding the result to the internal moments whose positive directions are shown in Figure 7.14(c). The internal moments using (7.14) become

$$\begin{Bmatrix} M_1 \\ M_2 \\ M_3 \\ M_4 \\ M_5 \\ M_6 \end{Bmatrix} = \begin{bmatrix} 0 & 1 & 0 & 0 & 0 \\ 3 & -1 & 0 & 0 & -3 \\ -3 & 1 & 1 & 0 & 3 \\ 2 & 0 & 0 & -1 & 0 \\ -2 & 0 & 0 & 0 & 0 \\ 0 & 0 & 0 & 0 & 0 \end{bmatrix} \begin{Bmatrix} 1.228 \\ 3/4 \\ 11/12 \\ 5/3 \\ 1.0 \end{Bmatrix} + \begin{Bmatrix} -3/4 \\ 3/4 \\ -5/3 \\ 5/3 \\ 0 \\ 0 \end{Bmatrix} = \begin{Bmatrix} 0 \\ 0.684 \\ -0.684 \\ 2.456 \\ -2.456 \\ 0 \end{Bmatrix} \begin{matrix} \text{lb·in} \\ \\ \\ \text{(English)} \end{matrix}$$

$$
\begin{Bmatrix} M_1 \\ M_2 \\ M_3 \\ M_4 \\ M_5 \\ M_6 \end{Bmatrix}
=
\begin{bmatrix}
0 & 1 & 0 & 0 & 0 \\
76.2 & -1 & 0 & 0 & -76.2 \\
-76.2 & 1 & 1 & 0 & 76.2 \\
50.8 & 0 & 0 & -1 & 0 \\
-50.8 & 0 & 0 & 0 & 0 \\
0 & 0 & 0 & 0 & 0
\end{bmatrix}
\begin{Bmatrix} 5.47 \\ 84.8 \\ 103.6 \\ 188.4 \\ 4.448 \end{Bmatrix}
+
\begin{Bmatrix} -84.8 \\ 84.8 \\ -188.4 \\ 188.4 \\ 0 \\ 0 \end{Bmatrix}
=
\begin{Bmatrix} 0 \\ 77.9 \\ -77.9 \\ 277.9 \\ -277.9 \\ 0 \end{Bmatrix}
\begin{array}{l} \text{N}\cdot\text{mm} \\ \\ \text{(metric)} \end{array}
$$

The maximum moment occurs at joint E as indicated by the magnitude of the internal moments M_4 and M_5.

The reduction method of Appendix A.6 may be used to obtain structural displacements from (7.15) without explicitly determining R_{FH}. The reduction process constrains the displacement of the referenced row of a flexibility matrix. When applied to the first row of (7.15), the displacement ΔX_F is set to zero. The reduced coefficients of the matrix with the first row and column removed reflects a correction that accounts for the added constraint. Structural displacements are determined directly using the reduced matrix.   Q.E.D

Simplifications in Model Development.   As discussed in Chapter 6, members of a structure develop internal forces that consist of axial, shear, torsion and flexural moments. These forces act at the ends of the member when idealized as elements. It is common practice to neglect force quantities that are small relative to others in the structural system. In this respect, axial deformation in Examples 7.5 and 7.6 is ignored since these quantities are much smaller than transverse displacements.

Neglecting relatively small force quantities reduces the size of the problem by removing insignificant degrees of freedom and unnecessary complexity. The attendant simplification produces results that are in agreement with classical solutions. Relatively small forces often occur along the length of members subjected to flexural loads. Some loadings are resisted by axial tension, compression or torsion along the member's length. Element flexibilities from Figures 7.10(d) and (e) are included in the development of the system matrix when the behavior of the structure is based on the effect of these axial forces.

EXAMPLE 7.7.   Transverse loads parallel to the plane of circuit boards develop component loading and deformation shown in Figure 7.15. Side sway of this nature may result in impacts between closely spaced components. This condition can affect performance and cause fatigue failures in a vibrating environment. Determine the displacement of the component body and the stress at the lead to body interface.

Flexural characteristics of the component body may be ignored if $(EI)_{\text{body}} \gg$

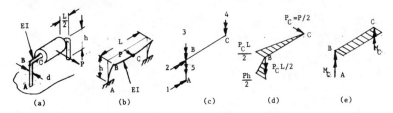

Figure 7.15. Component Idealization and Loading for Example 7.7. (a) Configuration; (b) Idealization; (c) Internal forces; (d) Applied loading; (e) Reaction forces.

$(EI)_{lead}$, which is the case for many devices. The idealization in Figure 7.15(b) represents the behavior of the normally loaded, symmetrical configuration. The assumed direction and location of the internal forces for each element are shown in Figure 7.15(c). Moment and torque distribution representing the applied loading and unknown reaction moment at point C are shown in Figures 7.15(d) and (e) respectively. The load force transformation (7.8b) is developed by considering the loading in Figures 7.15(d) and (e) separately as

$$
\begin{Bmatrix} M_1 \\ M_2 \\ M_3 \\ M_4 \\ T_{q5} \end{Bmatrix} = \begin{bmatrix} 0 & -h \\ 0 & 0 \\ 1 & -L/2 \\ -1 & 0 \\ -1 & L/2 \end{bmatrix} \begin{Bmatrix} M_C \\ P_C \end{Bmatrix} \tag{7.16}
$$

Only the horizontal translation at point C can be determined from (7.6) using (7.16) since fictitious loads at defined joints have not been specified. Element flexibilities $[f]$ are assembled using the arrangement shown in Figure 7.10, and the order established when constructing (7.16), which for constant EI becomes

$$
[f_i] = \frac{1}{6EI} \begin{bmatrix} 2h & -h & \vdots & & 0 \\ -h & 2h & \vdots & & \\ \cdots & \cdots & \cdots & \cdots & \\ & & \vdots & L & -L/2 & \vdots \\ & 0 & \vdots & -L/2 & L & \vdots \\ & & & \cdots & \cdots & \cdots \\ & & & & & \vdots 6hEI/JG \end{bmatrix}
$$

System displacements are determined from (7.6) and (7.10) as

$$\begin{Bmatrix} \theta_C \\ \Delta x_C \end{Bmatrix} = \frac{1}{6EI} \begin{bmatrix} 3L + 6hEI/JG & -3L^2/4 - 3hLEI/JG \\ -3L^2/4 - 3hLEI/JG & h^3 + L^3/8 + 3hL^2EI/4JG \end{bmatrix} \begin{Bmatrix} M_C \\ P_C \end{Bmatrix}$$

$$(7.17)$$

Since $\theta_C = 0$ for the configuration of Figure 7.15(b), the reaction load M_C is defined by the first equation in (7.17) as

$$M_C = \frac{PL}{4} \left[\frac{L/2 + 2hEI/JG}{L + 2hEI/JG} \right] \qquad (7.18)$$

where $P_C = P/2$.

The horizontal translation at point C is defined by the second equation in (7.17) as

$$\Delta x_C = \frac{1}{6EI} \left[-3LM_C \left(\frac{L}{4} + \frac{hEI}{JG} \right) + \frac{P}{2} \left(h^3 + \frac{L^3}{8} + \frac{3hL^2EI}{4JG} \right) \right]$$

The stress at the lead to component body interface is given by (6.17b), using M_3 from (7.16), as

$$\sigma = \frac{M_3 c}{I} = \frac{PL^2 d}{16I(L + 2hEI/JG)}$$

for lead diameter d.   Q.E.D.

Symmetry permits the removal of the rigid body segment in Example 7.7 without affecting the deformed shape. When the configuration or loading is unsymmetrical, however, the effect of rigid elements must be included. Consider the component of Figure 7.16 mounted asymmetrically between

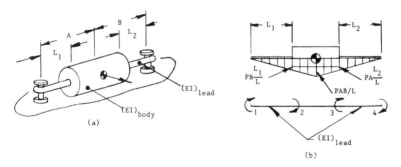

Figure 7.16. Flexibility of Elements with Infinitely Stiff Sections. (a) Configuration; (b) Internal moments and applied moment distribution.

standoffs. The element representing the component body may be assumed infinitely stiff when $(EI)_{body} \gg (EI)_{leads}$.

The moment distribution for an acceleration load imposed at the center of gravity of the component is unaffected by member stiffness. Only the flexible members are used, however, when developing the load force transformation matrix $[b]$. Based on the internal moments of Figure 7.16(b) matrices $[b]$ and $[f_i]$ become

$$[b] = \begin{bmatrix} 0 \\ -L_1 B/L \\ L_2 A/L \\ 0 \end{bmatrix} \quad ; \quad [f_i] = \frac{1}{6EI} \begin{bmatrix} 2L_1 & -L_1 & \vdots & & \\ -L_1 & 2L_1 & \vdots & & 0 \\ \cdots & \cdots & \cdots & \cdots & \cdots \\ & & \vdots & 2L_2 & -L_2 \\ 0 & & \vdots & -L_2 & 2L_2 \end{bmatrix}$$

The translation of the center of gravity is determined from (7.6) and (7.10) as

$$\Delta x_C = \frac{P}{3EI} \left[L_1^3 \left(\frac{B}{L} \right)^2 + L_2^3 \left(\frac{A}{L} \right)^2 \right] \tag{7.19}$$

When symmetry is imposed, $A = B$ and $L_1 = L_2 = L/2$, which brings (7.19) into a form representing a centrally located concentrated force on a simply supported member, giving

$$\Delta x_C = \frac{PL^3}{48EI}$$

Flexibility matrices of commonly used configurations with infinitely stiff sections can be catalogued for future application if desired. Flexibility methods may also be used to study the behavior of a structure subjected to variations in temperature.

Thermal Elastic Considerations. Member elongations occur without developing internal axial forces when a statically determinate structure experiences a change in temperature. In a statically indeterminate structure, however, internal axial member forces can occur because of constraints imposed by the structural configuration or attachments. Member motion may be inhibited or prevented which may cause large compressive or tensile stresses. The axial displacement of a member of length L subjected to load p and temperature change Δt is

$$\delta = \frac{pL}{AE} + \alpha L \, \Delta T \qquad (7.20a)$$

where α is the coefficient of expansion of the member material. If this displacement is prevented entirely by rigid attachments, the force exerted on the attachments becomes

$$P = \frac{AE}{L} \delta = p + AE\alpha \, \Delta T \qquad (7.20b)$$

The member stress may be determined from

$$\sigma = P/A = p/A + E\alpha \, \Delta T \qquad (7.20c)$$

Consider the assembly of Figure 7.17(a) which experiences a temperature change, ΔT. The axial displacement of the screws and the component are the same since each element constrains the other. The displacement of the component will be inhibited by the screws, which experience a tensile force for $\alpha_S < \alpha_c$ so that the net translation is the same. Moreover, the compressive force on the component and the tensile force on the screws are equal. With the idealization of the screws as one element and the component as another, the load force transformation $[b]$ and the element flexibility matrices for the model of Figure 7.17(b) using n screws to retain the component become

$$[b] = \begin{bmatrix} -1 \\ 1 \end{bmatrix} ; \quad [f_i] = \begin{bmatrix} \left(\dfrac{L}{AE}\right)_c & 0 \\ 0 & \left(\dfrac{L}{nAE}\right)_s \end{bmatrix}$$

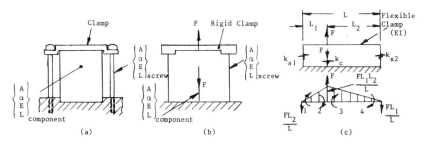

Figure 7.17. Thermoelastic Model of an Assembly. (a) Configuration; (b) Idealization; (c) Asymmetric configuration with an elastic clamp.

The displacement due to the unknown reaction F is determined using (7.6) and (7.10) as

$$\delta_F = F\left[\left(\frac{L}{AE}\right)_c + \left(\frac{L}{nAE}\right)_s\right]$$

This displacement as a result of a change in the assembly temperature is given by

$$\delta_{\Delta T} = [b]^T \begin{Bmatrix} (\alpha L)_c \\ (\alpha L)_s \end{Bmatrix} \Delta T = [(\alpha L)_s - (\alpha L)_c] \, \Delta T$$

Since the clamp and the component must always be in contact to provide proper support, $\delta_F + \delta_{\Delta T} = 0$ which gives

$$F = \frac{[(\alpha L)_c - (\alpha L)_s] \, \Delta T}{\left(\dfrac{L}{AE}\right)_c + \left(\dfrac{L}{nAE}\right)_s} \tag{7.21}$$

The reaction force defined in (7.21) may also be expressed as (5.5). The compressive stress in the component and the tensile stress in the screws may be determined using (5.6) as $\sigma_c = F/A_c$ and $\sigma_s = F/A_s$ respectively. These results are comparable to using (5.8) with $W = 0$.

Asymmetric configurations and assemblies with flexural behavior are readily accommodated using flexibility methods. The model in Figure 7.17(c) may be used to evaluate the behavior of the assembly for an asymmetric support configuration with an elastic clamp exhibiting flexural displacements. The load force transformation and element flexibility matrices for a configuration with two screws is

$$[b] = \begin{bmatrix} 0 \\ L_1L_2/L \\ -L_1L_2/L \\ 0 \\ L_2/L \\ -1 \\ L_1/L \end{bmatrix} \; ; \; [f_i] = \frac{1}{6EI} \begin{bmatrix} 2L_1 & -L_1 & & & & & \\ -L_1 & 2L_1 & & & & 0 & \\ & & 2L_2 & -L_2 & & & \\ & & -L_2 & 2L_2 & & & \\ & & & & 6EI/k_{s1} & & \\ & 0 & & & & 6EI/k_c & \\ & & & & & & 6EI/k_{s2} \end{bmatrix}$$

where $k_i = (AE/h)_i$.

Based on (7.6) and (7.10), the displacement due to the unknown reaction F becomes

$$\delta_F = F\left[\left(\frac{L_1 + L_2}{3EI}\right)\left(\frac{L_1 L_2}{L}\right)^2 + \left(\frac{L_2}{L}\right)^2 \Big/ k_{s1} + \left(\frac{L_1}{L}\right)^2 \Big/ k_{s2} + 1/k_c\right]$$

The displacement as a result of a change in the assembly temperature is given by

$$\delta_{\Delta T} = [b]^T \left\{\begin{array}{c} 0 \\ 0 \\ 0 \\ 0 \\ (\alpha h)_{s1} \\ (\alpha h)_c \\ (\alpha h)_{s2} \end{array}\right\} \quad \Delta T = \left[\frac{L_2}{L}(\alpha h)_{s1} + \frac{L_1}{L}(\alpha h)_{s2} - (\alpha h)_c\right]\Delta T$$

Since $\delta_F + \delta_{\Delta T} = 0$ to provide proper component support, the unknown reaction becomes

$$F = -\frac{\Delta T\left[\dfrac{L_2}{L_1}(\alpha h)_{s1} + \dfrac{L_1}{L}(\alpha h)_{s2} - (\alpha h)_c\right]}{\left(\dfrac{L_1 + L_2}{3EI}\right)\left(\dfrac{L_1 L_2}{L}\right)^2 + \left(\dfrac{L_2}{L}\right)^2 \Big/ k_{s1} + \left(\dfrac{L_1}{L}\right)^2 \Big/ k_{s2} + 1/k_c}$$

$$(7.22)$$

Maximum moment in the clamp and maximum tensile force in the screws are determined using $[b]$ as $M_{max} = FL_1 L_2/L$ and FL_1/L for $L_1 > L_2$ respectively. Equation (7.22) reduces to (7.21) for conditions of symmetry and a rigid clamp configuration.

Curved Structure Idealization. Curved structural configurations may be used in the design of spring fittings, contact interfaces, connector pins and brackets. The flexibility method may be used to evaluate the behavior of structures through the use of curved element matrices. Curved elements may be used individually or in combination with the element flexibilities shown in Figure 7.10. This provides the design engineer the ability to adapt flexibility elements to virtually any structural configuration encountered in equipment design.

Table 7.8 Curved Element Flexibility Matrices

Case	Configuration	Flexibility	Force Transformation
I		$$\begin{Bmatrix} \delta_T \\ \delta_N \\ \theta \end{Bmatrix}_A = \frac{r}{EI}\begin{bmatrix} \left(\frac{3\alpha}{2}-2S+\frac{SC}{2}\right)r^2 & & \text{Symm} \\ (1-C-S^2)\,r^2 & (\alpha-SC)\dfrac{r^2}{2} & \\ (S-\alpha)r & (C-1)r & \alpha \end{bmatrix}\begin{Bmatrix} R_T \\ R_N \\ M \end{Bmatrix}_A$$	$$\begin{Bmatrix} R_T \\ R_N \\ M \end{Bmatrix}_B = \begin{bmatrix} -C & S & 0 \\ -S & -C & 0 \\ r(1-C) & rS & -1 \end{bmatrix}\begin{Bmatrix} R_T \\ R_N \\ M \end{Bmatrix}_A$$
II		$$\begin{Bmatrix} \delta_T \\ \delta_N \\ \theta \end{Bmatrix}_B = \frac{r}{EI}\begin{bmatrix} \left(\frac{3\alpha}{2}-2S+\frac{SC}{2}\right)r^2 & & \text{Symm} \\ \left(\frac{S^2}{2}+C-1\right)r^2 & (\alpha-SC)\dfrac{r^2}{2} & \\ (s-\alpha)r & (1-C)r & \alpha \end{bmatrix}\begin{Bmatrix} R_T \\ R_N \\ M \end{Bmatrix}_B$$	$$\begin{Bmatrix} R_T \\ R_N \\ M \end{Bmatrix}_A = \begin{bmatrix} -C & -S & 0 \\ S & -C & 0 \\ r(1-C) & -rS & -1 \end{bmatrix}\begin{Bmatrix} R_T \\ R_N \\ M \end{Bmatrix}_B$$
III		$$\begin{Bmatrix} \delta \\ \theta_T \\ \theta_N \end{Bmatrix}_A = r\begin{bmatrix} \dfrac{r^2a}{EI}+\dfrac{r^2}{JG}(a+b-2S) & & \text{Symm} \\ \dfrac{ra}{EI}+\dfrac{r}{JG}(b-S) & \dfrac{a}{EI}+\dfrac{b}{JG} & \\ \dfrac{rS^2}{2EI}+\dfrac{r}{JG}\left(1-C-\dfrac{S^2}{2}\right) & \dfrac{S^2}{2}\left(\dfrac{1}{EI}-\dfrac{1}{JG}\right) & \dfrac{b}{EI}+\dfrac{a}{JG} \end{bmatrix}\begin{Bmatrix} R \\ M_T \\ M_N \end{Bmatrix}_A$$	$$\begin{Bmatrix} R \\ M_T \\ M_N \end{Bmatrix}_B = \begin{bmatrix} -1 & 0 & 0 \\ r(1-C) & -C & S \\ -rS & -S & -C \end{bmatrix}\begin{Bmatrix} R \\ M_T \\ M_N \end{Bmatrix}_A$$
IV		$$\begin{Bmatrix} \delta \\ \theta_T \\ \theta_N \end{Bmatrix}_B = r\begin{bmatrix} \dfrac{r^2a}{EI}+\dfrac{r^2}{JG}(a+b-2S) & & \text{Symm} \\ \dfrac{ra}{EI}+\dfrac{r}{JG}(b-S) & \dfrac{a}{EI}+\dfrac{b}{JG} & \\ -\dfrac{rS^2}{2EI}-\dfrac{r}{JG}\left(1-C-\dfrac{S^2}{2}\right) & \dfrac{S^2}{2}\left(\dfrac{1}{JG}-\dfrac{1}{EI}\right) & \dfrac{b}{EI}+\dfrac{a}{JG} \end{bmatrix}\begin{Bmatrix} R \\ M_T \\ M_N \end{Bmatrix}_B$$	$$\begin{Bmatrix} R \\ M_T \\ M_N \end{Bmatrix}_A = \begin{bmatrix} -1 & 0 & 0 \\ r(1-C) & -C & -S \\ rS & S & -C \end{bmatrix}\begin{Bmatrix} R \\ M_T \\ M_N \end{Bmatrix}_B$$

$a = \tfrac{1}{2}(\alpha - SC)$
$b = \tfrac{1}{2}(\alpha + SC)$
$S = \sin \alpha$
$C = \cos \alpha$

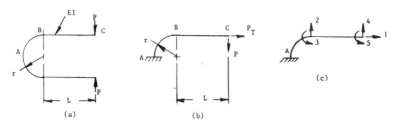

Figure 7.18. Flexibility Model of a Retaining Spring. (a) Configuration; (b) Idealization; (c) Internal forces.

Curved uniform section elements for in-plane and out-of-plane loadings are given in Table 7.8. Force transformation matrices that relate forces at the free end of the element to forces at the fixed end are also included.

EXAMPLE 7.8. Determine the maximum flexural stress and spreading displacement for the retaining spring of Figure 7.18.

The idealized configuration based on symmetry is shown in Figure 7.18(b). Coordinate references for the internal forces Figure 7.18(c) are based on using element configurations of Table 7.3 case II and Figure 7.10(b). The load force transformation matrix $[b]$ and the element flexibility matrix $[f_i]$ become

$$
\begin{Bmatrix} R_T \\ R_N \\ M_B \\ R_C \\ M_C \end{Bmatrix} = \begin{matrix} 1 \\ 2 \\ 3 \\ 4 \\ 5 \end{matrix} \begin{bmatrix} 0 & 1 \\ -1 & 0 \\ -L & 0 \\ -1 & 0 \\ 0 & 0 \end{bmatrix} \begin{Bmatrix} P \\ P_T \end{Bmatrix} ; [f_i] = \frac{1}{EI} \begin{bmatrix} (3\pi/4 - 2)r^3 & -r^3/2 & (1 - \pi/2)r^2 & \vdots & \\ -r^3/2 & \pi r^3/4 & r^2 & \vdots & 0 \\ (1 - \pi/2)r^2 & r^2 & \pi r/2 & \vdots & \\ \cdots & \cdots & \cdots & \cdots & \cdots \\ & 0 & & \vdots & L^3/3 & L^2/2 \\ & & & \vdots & L^2/2 & L \end{bmatrix}
$$

where $\alpha = 90°$. The fictitious load P_T is introduced to obtain the longitudinal displacement of the spring due to load P. The lateral spreading displacement Δy and longitudinal displacement Δx at point C are determined from (7.6) and (7.10) as

$$
\begin{Bmatrix} \Delta y \\ \Delta x \end{Bmatrix}_C = \frac{Pr^3}{EI} \begin{Bmatrix} \pi/4 + 2(L/r) + \pi(L/r)^2/2 + (L/r)^3/3 \\ 1/2 + (\pi/2 - 1)(L/r) \end{Bmatrix}
$$

Forces at point A are determined from the force transformation matrix for case II of Table 7.8. With the use of the last row of the force transformation matrix and the appropriate rows from $[b]$, the maximum moment becomes

$$M_A = \{r \quad -r \quad -1\} \begin{bmatrix} 0 \\ -1 \\ -L \end{bmatrix} \{P\} = (r + L)P$$

The maximum flexural stress is given by (6.17b).   Q.E.D.

7.3   STIFFNESS METHODS

Stiffness methods provide a systematic alternative to the flexibility method for the evaluation of loads which correspond to the displaced shape of a structure. Stated succinctly,

$$\{P\} = [K] \{\delta\} \tag{7.23}$$

Comparing (7.23) with (7.6), it is apparent that stiffness [K] is the reciprocal of flexibility [F] or

$$[K] = [F]^{-1} \tag{7.24}$$

Since [F] is symmetrical, so is its inverse [K]. The product of these matrices results in a unit matrix which may be depicted as

$$[K]^T[F] = [K][F] = I \tag{7.25}$$

Analogous to the development of the system flexibility matrix (7.10), the structure stiffness matrix may be developed from the product of a displacement transformation matrix [C] and the element stiffness matrix $[k_i]$ expressed as

$$[K] = [C]^T[k_i][C] \tag{7.26}$$

Joint displacements $\{\delta\}$ acting on the structure are transformed to element end displacements $\{\Delta\}$ by the displacement transformation matrix [C]. The quantity of joint displacements $\{\delta\}$ corresponds to the degrees of freedom represented by the idealized configuration.

The degrees of freedom may be assessed by summing the number of independent displacements possible due to joint motion. Each unrestrained joint in an idealized planar structure experiences up to three independent displacements including axial contributions. Usually, axial deformation is much smaller than transverse displacements and can be neglected. When axial deformation is neglected, one degree of freedom is removed for each member in the structure. This condition alters the independence of the translational displacements between the joints of the structure. The number of independent translations occurring in a planar structure may be determined from

$$N_t = 2n_j - (2n_f + 2n_p + n_g + n_m) \qquad (7.27a)$$

given n_j = total number of joints;
n_f = number of fixed joints;
n_p = number of pinned joints;
n_g = number of pin-guided joints;
n_m = number of members.

Similarly, the number of independent translations of a spatial structure after the neglect of axial effects becomes

$$N_t = 3n_j - (3n_f + 3n_p + n_m) \qquad (7.27b)$$

Joint rotations are not affected when axial deformation is ignored. The construction of [C] is straightforward for planar systems and unwieldy for spatial systems. Matrix [C] is generated one column at a time by applying a unit displacement to each independent degree of freedom and determining the magnitude of the internal displacements at the ends of each member.

The element stiffness matrix for rotations at the member ends may be determined by applying (7.24) to the flexibility matrix of Figure 7.10(a), giving

$$[k] = \frac{2EI}{L} \begin{bmatrix} 2 & 1 \\ 1 & 2 \end{bmatrix} \qquad (7.28)$$

Matrix $[k_i]$ of (7.26) is assembled using element matrices $[k]$ representing the internal displacements of each member of the idealization in a manner analogous to the development of $[f_i]$ in (7.11). Matrix $[k_i]$ for the idealization in Figure 7.19 becomes

$$[k_i] = \begin{bmatrix} \dfrac{2EI}{a} \begin{bmatrix} 2 & 1 \\ 1 & 2 \end{bmatrix} & \vdots & 0 \\ \cdots\cdots\cdots\cdots\cdots\cdots\cdots \\ 0 & \vdots & \dfrac{2EI}{b} \begin{bmatrix} 2 & 1 \\ 1 & 2 \end{bmatrix} \end{bmatrix} = 2EI \begin{bmatrix} 2/a & 1/a & \vdots & 0 \\ 1/a & 2/a & \vdots & \\ \cdots\cdots\cdots\cdots\cdots \\ & \vdots & 2/b & 1/b \\ 0 & \vdots & 1/b & 2/b \end{bmatrix}$$

$$(7.29)$$

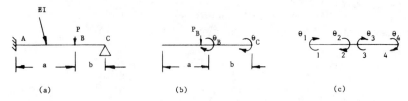

Figure 7.19. Displacement Characteristics for Stiffness. (a) Idealization; (b) System displacement (degrees of freedom); (c) Internal rotations.

The transformation of system displacements θ_B, θ_C and y_B to member displacements θ_1, θ_2, θ_3 and θ_4 for the configuration in Figure 7.19 becomes

$$\begin{Bmatrix} \theta_1 \\ \theta_2 \\ \theta_3 \\ \theta_4 \end{Bmatrix} = \begin{bmatrix} 0 & 0 & -1/a \\ 1 & 0 & -1/a \\ 1 & 0 & 1/b \\ 0 & 1 & 1/b \end{bmatrix} \begin{Bmatrix} \theta_B \\ \theta_C \\ y_B \end{Bmatrix} = [C] \begin{Bmatrix} \theta_B \\ \theta_C \\ y_B \end{Bmatrix} \qquad (7.30)$$

The stiffness matrix $[k]$ using (7.26) becomes

$$[k] = 2EI \begin{bmatrix} 2\left(\dfrac{1}{a} + \dfrac{1}{b}\right) & & \text{Symmetrical} \\[2ex] \dfrac{1}{b} & \dfrac{2}{b} & \\[2ex] -3\left(\dfrac{1}{a^2} - \dfrac{1}{b^2}\right) & \dfrac{3}{b^2} & 6\left(\dfrac{1}{a^3} + \dfrac{1}{b^3}\right) \end{bmatrix} \qquad (7.31a)$$

System loads are determined from (7.23) using (7.31a) as

$$\begin{Bmatrix} M_B \\ M_C \\ P \end{Bmatrix} = 2EI \begin{bmatrix} 2\left(\dfrac{1}{a} + \dfrac{1}{b}\right) & \dfrac{1}{b} & -3\left(\dfrac{1}{a^2} - \dfrac{1}{b^2}\right) \\[2ex] \dfrac{1}{b} & \dfrac{2}{b} & \dfrac{3}{b^2} \\[2ex] -3\left(\dfrac{1}{a^2} - \dfrac{1}{b^2}\right) & \dfrac{3}{b^2} & 6\left(\dfrac{1}{a^3} + \dfrac{1}{b^3}\right) \end{bmatrix} \begin{Bmatrix} \theta_B \\ \theta_C \\ y_B \end{Bmatrix} \qquad (7.31b)$$

Unknown displacements may be determined in terms of the applied system loading using (7.24) and (7.6) or by the reduction method described in Appen-

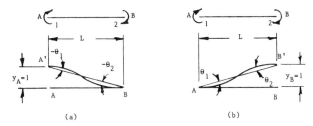

Figure 7.20. Internal Rotations Corresponding to Positive Unit Translations. (a) Negative rotations; (b) Positive rotations.

dix A.6. Displacements corresponding to the loading in Figure 7.19(a) require the determination of θ_B, θ_C and y_B in terms of P from (7.31b). Since M_B and M_C are zero, the first two equations of (7.31b) provide the solution for θ_B and θ_C. These quantities are substituted into the third equation to obtain the translation y_B in terms of P.

The stiffness method requires the determination of two unknown displacements versus the determination of one unknown using the flexibility method (7.12b) for the configuration in Figure 7.19. This provides the basis of selecting either the stiffness or flexibility method. The ease and speed of determining the desired displacements depends on the number of unknowns. It is usually desirable to select a method that results in fewer unknowns. These unknowns correspond to the unknown reactions in the flexibility method and the structural degrees of freedom in the stiffness method. The flexibility method is the preferred approach for the structure in Figure 7.14(a) since there is one unknown reaction compared to seven degrees of freedom. The degrees of freedom are described by five rotations and two independent translations of the structures joints.

The transformation of a unit translation to internal rotations at the ends of a member in the development of [C] is based on applying one displacement at a time and the assumption of small displacements. The development of these quantities is illustrated in Figure 7.20. The determination of internal rotations using Figure 7.20 is cumbersome if a member is inclined relative to the reference axes of the structure. For these members, an instant center can be established that provides the relationships between the applied unit translation and internal member rotations. Inclined members do not affect joint rotations.

EXAMPLE 7.9.   Develop the stiffness matrix relating applied loads to system displacements for the spring configuration in Figure 7.21(a).

The idealized configuration in Figure 7.21(a) consists of two members, two free

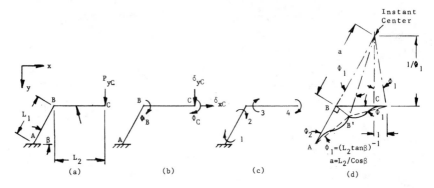

Figure 7.21. Displacement Characteristics of a Cantilevered Spring. (a) Idealization; (b) System displacements; (c) Internal displacements; (d) Displaced shape due to $\delta_{xC} = 1$.

joints and one fixed joint. The number of independent translations is obtained from (7.27a) as

$$N_T = 2(3) - [2(1) + 2] = 2$$

Since only two joints can experience rotations, the behavior of the structure is described by the four degrees of freedom shown in Figure 7.21(b). The effects of a unit translation x_C on the internal rotations shown in Figure 7.21(c) are obtained using an instant center for member BC. These geometric relationships are developed in Figure 7.21(d). The instant center is located along a projection of member AB at the intersection of a normal to the unit translation at C. On the basis of small deflections, member AB rotates about A and member BC rotates about the instant center. Translation BB' normal to AB equals L_1 times its rotation about A or the length of the projection of L_1 times its rotation about the instant center. This is expressed as

$$\phi_2 L_1 = \phi_1 a = \overline{BB'}$$

giving ϕ_2 as

$$\phi_2 = \frac{a}{L_1} \phi_1$$

The dependent translation y_B is a result of the rotation of AB about the instant center. The dependent displacements at B may be expressed in terms of x_C using the displaced geometry of Figure 7.21(d) as

$$\left\{ \begin{array}{c} y \\ x \end{array} \right\}_B = \left[\begin{array}{c} L_2\phi_1 \\ 1 \end{array} \right] \{x\}_C$$

Following Figure 7.21(d), the displacement transformation and element stiffness matrices for the internal rotations of Figure 7.21(c) are developed:

$$
[C] = \begin{array}{c} \\ 1 \\ 2 \\ 3 \\ 4 \end{array}
\begin{array}{cccc} (\theta_B) & (\theta_C) & (x_C) & (y_C) \end{array}
\left[\begin{array}{cccc}
0 & 0 & -\phi_2 & 0 \\
1 & 0 & -\phi_2 & 0 \\
1 & 0 & \phi_1 & -1/L_2 \\
0 & 1 & \phi_1 & -1/L_2
\end{array} \right]
\quad ; \quad
[k_i] = 2EI \left[\begin{array}{cccc}
2/L_1 & 1/L_1 & \vdots & \\
1/L_1 & 2/L_1 & \vdots & 0 \\
\cdots & \cdots & \cdots & \cdots \\
& & \vdots & 2/L_2 & 1/L_2 \\
0 & & \vdots & 1/L_2 & 2/L_2
\end{array} \right]
$$

These results are used to construct the stiffness matrix using (7.26). The applied load in terms of system displacements which is expressed compactly as (7.23) becomes

$$
\left\{ \begin{array}{c} M_B \\ M_C \\ P_{xC} \\ P_{yC} \end{array} \right\} = 2EI
\left[\begin{array}{cccc}
2\left(\dfrac{1}{L_1}+\dfrac{1}{L_2}\right) & \dfrac{1}{L_2} & -\dfrac{3\phi_2}{L_1}+\dfrac{3\phi_1}{L_2} & -\dfrac{3}{L_2^2} \\[2mm]
\dfrac{1}{L_2} & \dfrac{2}{L_2} & \dfrac{3\phi_1}{L_2} & -\dfrac{3}{L_2^2} \\[2mm]
-\dfrac{3\phi_2}{L_1}+\dfrac{3\phi_1}{L_2} & \dfrac{3\phi_1}{L_2} & \dfrac{6\phi_2^2}{L_1}+\dfrac{6\phi_1^2}{L_2} & -\dfrac{6\phi_1}{L_2^2} \\[2mm]
-\dfrac{3}{L_2^2} & -\dfrac{3}{L_2^2} & -\dfrac{6\phi_1}{L_2^2} & \dfrac{6}{L_2^3}
\end{array} \right]
\left\{ \begin{array}{c} \theta_B \\ \theta_C \\ x_C \\ y_C \end{array} \right\}
\tag{7.33}
$$

Q.E.D.

Matrix Development Based on Strain Energy.   An alternative approach using strain energy develops [K] directly without resorting to (7.26). Since [C] and the matrix products in (7.26) are not required, considerable time and effort may be saved.

Linear elastic behavior implies that plane cross-sections remain plane, based on the assumption that longitudinal strains in an element are linearly distributed at any cross section. The stress at a cross-section (7.5) can be generalized as the sum of stress components due to moments acting in both principal axes in addition to an axial tensile load, giving

$$
\sigma = -\frac{M_y x}{I_y} + \frac{M_x y}{I_x} + \frac{F}{A}
\tag{7.34}
$$

The internal energy developed by the displaced shape due to the application of these loads may be expressed as *strain energy*. The strain energy of an element

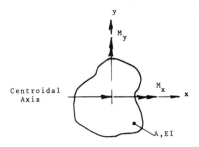

Figure 7.22. Moments Applied to Principal Axes of a Member.

of length L and cross-sectional area A may be developed using Hooke's law, $\sigma = E(\Delta L/L)$, as

$$U = \frac{1}{2E} \int_0^L ds \int \sigma^2 \, dA \tag{7.35}$$

where ds is an increment of the element length. Since x and y in Figure 7.22 are principal axes of inertia $\int x \, dA$, $\int y \, dA$ and $\int xy \, dx$ all equal zero. In this case, (7.35) and (7.34) for varying internal forces and and cross-sectional properties become

$$U = \frac{1}{2E} \int_0^L \left(\frac{M_y^2}{I_y} + \frac{M_x^2}{I_x} + \frac{F}{A} \right) ds \tag{7.36}$$

Equation (7.36) shows that the effects of flexure in both principal axes and the tensile load contribute additively to the element strain energy. Considering only flexure about one principal axes, (7.36) reduces to

$$U = \frac{1}{2E} \int_0^L \frac{M^2}{I} \, ds \tag{7.37}$$

Equation (6.71) may be used to develop a similar expression for torsion, giving

$$U = \frac{JG\theta^2}{2L} = \frac{LT^2}{2JG} \tag{7.38}$$

where $U = \int T \, d\theta$. Equation (7.38) and a similar expression $U = EA\delta^2/2L$ representing transverse shear may be added to the strain energy components in (7.36) to describe complicated loadings.

The bending moment using the curvature $1/r$ of the deformed centroidal axis of a straight element from elementary beam theory becomes

$$M = EI/r \qquad (7.39)$$

For constant section properties, (7.37) and (7.39) give

$$U = \frac{EI}{2} \int_0^L \frac{ds}{r^2} \qquad (7.40)$$

This equation may be expressed in terms of the derivatives of the arc length s as

$$U = \frac{EI}{2} \int_0^L [(x'')^2 + (y'')^2] \, ds \qquad (7.41)$$

This expression is valid for displacements that are much greater than the cross-sectional thickness of the member provided there is no yielding. For small displacements, s is approximately equal to x and x'' is approximately zero. If the x axis represents the undeformed centroidal axis of the member, the strain energy from (7.41) in terms of the transverse displacement y reduces to

$$U = \frac{EI}{2} \int_0^L (y'')^2 \, dx \qquad (7.42)$$

The displacements of a member with constant section properties subjected to moment and shear loads at each end, Figure 7.23(a), may be determined

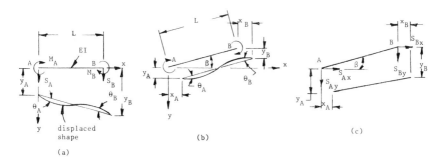

Figure 7.23. Displaced Shapes of a Member Subjected to End Loads Relative to System Axes. (a) Parallel or orthogonal; (b) Inclined; (c) Axial loads.

from (6.7) as

$$EI\theta = M_A x + S_A x^2/2 + EI\theta_A \tag{7.43a}$$

$$EIy = M_A x^2/2 + S_A x^3/6 + EI\theta_A x + EIy_A \tag{7.43b}$$

Expressions for M_A and S_A in terms of the displacements θ_A, y_A and θ_B, y_B may be determined from (7.43) as

$$M_A = \frac{2EI}{L} (2\theta_A + \theta_B - 3\phi)$$

$$\tag{7.44}$$

$$S_A = \frac{6EI}{L^2} (\theta_A + \theta_B - 2\phi)$$

where $\phi = (y_B - y_A)/L$.

The transverse displacement at any point along the member in terms of end displacements may be determined from (7.43b) and (7.44) as

$$y(x) = y_A + \theta_A x - (2\theta_A + \theta_B - 3\phi)\frac{x^2}{L} + (\theta_A + \theta_B - 2\phi)\frac{x^3}{L^2} \tag{7.45}$$

The strain energy of the deformed shape in Figure 7.23(a) is determined from (7.42) and (7.45) as

$$U = \frac{2EI}{L} [\theta_A^2 + \theta_A\theta_B + \theta_B^2 - 3\phi(\theta_A + \theta_B) + 3\phi^2] \tag{7.46}$$

A general expression that is applicable to inclined members as well as those subjected to torsion defined by (7.38) becomes

$$U = \frac{2EI}{L} [\theta_A^2 + \theta_A\theta_B + \theta_B^2 - 3(\phi_y \cos \beta + \phi_x \sin \beta)(\theta_A + \theta_B)$$

$$+ 3(\phi_y \cos \beta + \phi_x \sin \beta)^2] + \frac{JG}{2L} (\alpha_B - \alpha_A)^2 \tag{7.47}$$

given $\phi_x = (x_B - x_A)/L;$
$\phi_y = (y_B - y_A)/L;$
$\alpha_i =$ angle of twist at end i.

Equations (7.46) and (7.47) provide the basis of developing stiffness coefficients for an idealized structural system. These coefficients may be obtained

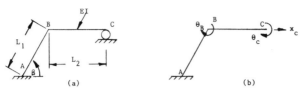

Figure 7.24. Displacements of a Pin Guided Spring. (a) Idealization; (b) System displacements.

from the partial derivative with respect to a particular displacement or degree of freedom. Once the mechanics of this process are understood, the stiffness coefficients may be determined from (7.46) or (7.47) by inspection.

The partial derivative of (7.46) with respect to θ_A gives

$$\frac{\partial U}{\partial \theta_A} = \frac{2EI}{L} (2\theta_A + \theta_B - 3\phi) = M_A \qquad (7.48a)$$

which in the form of (7.23) becomes

$$\{M_A\} = \frac{2EI}{L} \begin{bmatrix} 2 & 1 & -\dfrac{3}{L} & \dfrac{3}{L} \end{bmatrix} \begin{Bmatrix} \theta_A \\ \theta_B \\ y_A \\ y_B \end{Bmatrix} \qquad (7.48b)$$

EXAMPLE 7.10.   Use strain energy to develop the stiffness matrix of the spring configuration in Figure 7.21(a) with point C constrained vertically by a guided support shown in Figure 7.24.

The strain energy of member AB is determined from (7.47) as

$$U_{AB} = \frac{2EI}{L_1} \left[\theta_B^2 - \frac{3}{L_1} (y_B \cos \beta + x_B \sin \beta) \theta_B \right.$$

$$\left. + \frac{3}{L_1^2} (y_B \cos \beta + x_B \sin \beta)^2 \right] \qquad (7.49a)$$

Since y_B is dependent upon x_C as defined by (7.32), this expression may be expressed using the geometry in Figure 7.21(c) as

$$U_{AB} = \frac{2EI}{L_1} [\theta_B^2 - 3\phi_2 x_C \theta_B + 3(x_C \phi_2)^2] \qquad (7.49b)$$

The strain energy of member BC in Figure 7.24 is determined from (7.46) as

$$U_{BC} = \frac{2EI}{L_2} \left[\theta_B^2 + \theta_B\theta_C + \theta_C^2 + 3\frac{y_B}{L_2}(\theta_B + \theta_C) + 3\left(\frac{y_B}{L_2}\right)^2 \right] \quad (7.50a)$$

which may be expressed in terms of the independent translation x_C using (7.32) giving

$$U_{BC} = \frac{2EI}{L_2} [\theta_B^2 + \theta_B\theta_C + \theta_C^2 + 3\phi_1 x_C(\theta_B + \theta_C) + 3(\phi_1 x_C)^2] \quad (7.50b)$$

The stiffness matrix may be developed from (7.49b) and (7.50b) by taking partial derivatives with respect to each displacement and summing the results or by inspection. The stiffness matrix for the configuration of Figure 7.24 is

$$[k] = 2EI \begin{bmatrix} 2\left(\dfrac{1}{L_1} + \dfrac{1}{L_2}\right) & \dfrac{1}{L_2} & -\dfrac{3\phi_2}{L_1} + \dfrac{3\phi_1}{L_2} \\[3mm] \dfrac{1}{L_2} & \dfrac{2}{L_2} & \dfrac{3\phi_1}{L_2} \\[3mm] -\dfrac{3\phi_2}{L_1} + \dfrac{3\phi_1}{L_2} & \dfrac{3\phi_1}{L_2} & \dfrac{6\phi_2^2}{L_1} + \dfrac{6\phi_1^2}{L_2} \end{bmatrix} \quad (7.51)$$

with column headings (θ_B), (θ_C), (x_C).

This result is also given by (7.33) for $y_C = 0$.   Q.E.D.

A strain energy relationship for members subjected to axial deformation, Figure 7.23(c), may be developed as

$$U_{AB} = \frac{AE}{2L} [(y_A - y_B)\sin\beta - (x_A - x_B)\cos\beta]^2 \quad (7.52)$$

Stable structural systems composed of pin connected members may be studied using (7.52). Members that experience axial deformations in addition to flexural deformations are accommodated by a stiffness matrix with coefficients developed as the sum of the axial contributions from (7.52) and the flexural contributions from (7.47).

EXAMPLE 7.11. Define the characteristics of a cross-section that assures flexural behavior of the spring in Figure 7.25.

The strain energy due to flexural deformation from (7.47) becomes

$$U_{AB} = \frac{2EI}{L} \left[\theta_B^2 - \frac{3y_B\theta_B \cos\beta}{L} + \frac{3y_B^2 \cos^2\beta}{L^2} \right] \quad (7.53)$$

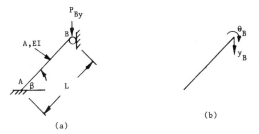

(a)

(b)

Figure 7.25. Stiffness Development of the Spring in Example 7.11. (a) Spring configuration; (b) System displacement.

The stiffness matrix describing flexural behavior is developed from (7.53) as

$$[K]_{\text{flex}} = \frac{2EI}{L} \begin{bmatrix} 2 & -\dfrac{3\cos\beta}{L} \\ \\ -\dfrac{3\cos\beta}{L} & \dfrac{6\cos^2\beta}{L^2} \end{bmatrix} \quad \begin{matrix} (\theta_B) & (y_B) \end{matrix}$$

(7.54)

Equation (7.52) describes the strain energy for axial deformation as

$$U_{AB} = \frac{AE}{2L} y_B^2 \sin^2\beta$$

(7.55)

The stiffness matrix describing axial behavior is developed from (7.55) as

$$[K]_{\text{axial}} = \frac{AE}{L} \sin^2\beta$$

(7.56)

Expressing the result of adding stiffness coefficients due to flexural (7.54) and axial (7.56) displacements in the form of (7.23) gives

$$\left\{ \begin{matrix} M \\ \\ P_y \end{matrix} \right\}_B = \frac{EI}{L} \begin{bmatrix} 4 & -\dfrac{6\cos\beta}{L} \\ \\ -\dfrac{6\cos\beta}{L} & \dfrac{12\cos^2\beta}{L^2} + \dfrac{A}{I}\sin^2\beta \end{bmatrix} \left\{ \begin{matrix} \theta \\ \\ y \end{matrix} \right\}_B$$

Since $M_B = 0$, reduction by the methods of Appendix A.6 gives

$$P_{By} = \frac{3EI}{L^3} \left[\cos^2 \beta + \frac{L^2}{3} \left(\frac{A}{I} \right) \sin^2 \beta \right] y_B \qquad (7.57)$$

Flexural behavior is assured if $A \ll I$ so that the contribution of the second term in (7.57) is negligible.   Q.E.D.

Loads and Forces.   Internal forces may be obtained using (7.46) or (7.47) once member end rotations and translations have been determined. Moment and shear forces correspond to the nature and location of the displacements in the strain energy expression. These forces are determined by the partial derivative with respect to a particular displacement. The internal moment at A of the member in Figure 7.23(a) is determined by substituting known displacements into (7.48). Internal forces are determined from the strain energy expressions whether or not the stiffness matrix has been developed from (7.26) or (7.46).

The stiffness method is also suited to structures with distributed and/or concentrated loads between joints. A solution for the displaced shape and internal forces is based on transforming loads applied between joints to equivalent loads acting on the member joints. The procedure is the same as was used in the flexibility method.

EXAMPLE 7.12.   Determine the displacement and flexural moment at the center of the structure shown in Figure 7.26(a).

The displaced shape of the structure may be represented by one independent translation from (7.27a) and three rotations when idealized using two members of length L.

The strain energy of the deformed shape is developed from (7.46) and Figure 7.26(b) as

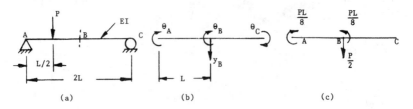

(a)   (b)   (c)

Figure 7.26. Structural Configuration of Example 7.12. (a) Idealized structure; (b) System displacements; (c) Equivalent loads.

$$U_{AB} = \frac{2EI}{L}\left[\theta_A^2 + \theta_A\theta_B + \theta_B^2 - 3\frac{y_B}{L}(\theta_A + \theta_B) + 3\left(\frac{y_B}{L}\right)^2\right]$$

$$U_{BC} = \frac{2EI}{L}\left[\theta_B^2 + \theta_B\theta_C + \theta_C^2 + 3\frac{y_B}{L}(\theta_B + \theta_C) + 3\left(\frac{y_B}{L}\right)^2\right]$$

(7.58)

Structural loads expressed in terms of system displacements by (7.23) may be determined by inspection of (7.58) giving

$$\left\{\begin{array}{c} M_A \\ M_B \\ M_C \\ P_B \end{array}\right\} = \frac{2EI}{L}\left[\begin{array}{cccc} 2 & 1 & 0 & -3/L \\ 1 & 4 & 1 & 0 \\ 0 & 1 & 2 & 3/L \\ -3/L & 0 & 3/L & 12/L^2 \end{array}\right]\left\{\begin{array}{c} \theta_A \\ \theta_B \\ \theta_C \\ y_B \end{array}\right\} = \left\{\begin{array}{c} PL/8 \\ -PL/8 \\ 0 \\ P/2 \end{array}\right\}$$

(7.59)

where the equivalent loads Figure 7.26(c), are developed using Table 7.7, case I.

System displacements in (7.59) may be determined by inverting the stiffness matrix (7.24) or by the reduction method in Appendix A.6. Using reduction, the results become

$$\left[\begin{array}{cccc} 1 & \frac{1}{2} & 0 & -3/2L \\ 0 & \frac{1}{2} & 1 & 3/2L \\ 0 & 1 & 2 & 3/L \\ 0 & 3/2L & 3/L & 15/2L^2 \end{array}\right]\left\{\begin{array}{c} \theta_A \\ \theta_B \\ \theta_C \\ y_B \end{array}\right\} = \frac{L}{2EI}\left\{\begin{array}{c} PL/16 \\ -3PL/16 \\ 0 \\ 11P/16 \end{array}\right\}$$

(7.60a)

$$\left[\begin{array}{cccc} 1 & \frac{1}{2} & 0 & -3/2L \\ 0 & 1 & \frac{3}{7} & 3/7L \\ 0 & 0 & \frac{12}{7} & 18/7L \\ 0 & 0 & 18/7L & 48/7L^2 \end{array}\right]\left\{\begin{array}{c} \theta_A \\ \theta_B \\ \theta_C \\ y_B \end{array}\right\} = \frac{L}{2EI}\left\{\begin{array}{c} PL/16 \\ -3PL/56 \\ 3PL/56 \\ 43P/56 \end{array}\right\}$$

(7.60b)

The last two equations of (7.60b) may be easily inverted, giving

$$\left\{\begin{array}{c} \theta_C \\ y_B \end{array}\right\} = \frac{PL^3}{96EI}\left\{\begin{array}{c} -15/L \\ 11 \end{array}\right\}$$

(7.60c)

The second equation in (7.60b) may be recast as

$$\{\theta_B\} = -\left[\begin{array}{cc} \frac{2}{7} & \frac{3}{7L} \end{array}\right]\left\{\begin{array}{c} \theta_C \\ y_B \end{array}\right\} - \frac{3PL^2}{112EI}$$

(7.61)

Substituting (7.60c) into (7.61) gives

$$\theta_B = -\frac{PL^2}{32EI}$$

The internal moment at B may be obtained from (7.48a) and applied to member BC, giving

$$M_B = \frac{2EI}{L}\left(2\theta_B + \theta_C + \frac{3y_B}{L}\right)$$

$$M_B = 2PL\left[2\left(-\frac{1}{32}\right) + \left(-\frac{15}{96}\right) + 3\left(\frac{11}{96}\right)\right]$$

$$M_B = PL/4$$

This example illustrates the systematic nature of the stiffness method, which can be applied to complicated structures.   Q.E.D.

The stiffness method may be used to evaluate structures that consist of mixed EIs by assigning a different member for each EI. If the EIs of some members are considerably greater than others in the structure, they may be considered as rigid to expedite the determination of displacements and internal forces. The stiffness solution for these configurations, (7.23), may be developed using a displacement transformation as illustrated for two components in Example 6.14.

Stiffness matrices for commonly used components or structural configurations may be catalogued and used repeatedly. The coefficients of various matrices may be summed for common displacement quantities to represent structures with complicated geometries.

7.4 FINITE DIFFERENCE METHODS

The finite difference method provides a numerical approach to the solution of the differential equations (6.2) and (6.7) governing the behavior of many structures. In this method, these equations are replaced with equivalent difference expressions which are applied at selected points distributed along a structure. The solution for displacements and internal forces (stresses) reduces to the simultaneous solution of a set of algebraic equations. The number of equations is a result of the quantity of points used to describe the behavior of the structure. Accuracy of the approximation is assured provided displacements are small in comparison with other dimensions so that the curvature is approximated by the second derivative $d^2y/dx^2 = M/EI$.

Approximations for the displaced shape in terms of the derivatives at equally

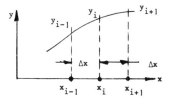

Figure 7.27. Point Spacing for a One Dimensional Structure.

spaced points may be obtained from Taylor series expansions. At points $i - 1$ and $i + 1$ of Figure 7.27, the expansions are

$$y_{i-1} = y_i - \Delta x y_i' + \frac{\Delta x^2}{2} y_i'' - \frac{\Delta x^3}{6} y_i''' + \cdots \qquad (7.62a)$$

$$y_{i+1} = y_i + \Delta x y_i' + \frac{\Delta x^2}{2} y_i'' + \frac{\Delta x^3}{6} y_i''' + \cdots \qquad (7.62b)$$

An expression for $(d^2y/dx^2)_i$ is determined by adding (7.62a) and (7.62b), giving

$$y_i'' = \frac{1}{\Delta x^2}(y_{i+1} - 2y_i + y_{i-1}) - \frac{\Delta x^2}{12} y_i^{IV} + \cdots \qquad (7.63)$$

The error in representing d^2y/dx^2 by the first term in (7.63) is given by the excluded higher order terms which are of the order $\Delta x^2 y_i^{IV}/12$.

By means of five successive equidistant points, an expression for $(d^4y/dx^4)_i$ may be developed as

$$y_i^{IV} = \frac{1}{\Delta x^4}(y_{i-2} - 4y_{i-1} + 6y_i - 4y_{i+1} + y_{i+2}) \qquad (7.64)$$

which has an error of order $\Delta x^2 y_i^{V}/6$. From (6.2), (6.7b) and (7.64)

$$(y_{i-2} - 4y_{i-1} + 6y_i - 4y_{i+2} + y_{i+2}) = \frac{w_i \Delta x^4}{EI} \qquad (7.65)$$

which defines the distributed load w_i at point i in terms of the surrounding displacements. Equation (7.65) applies at any internal point on a one dimensional structure subjected to a distributed loading. Modifications to (7.65) that account for the effects of various attachments are listed in Table 7.9. These equations are used at pivotal points at or near the constraint.

Table 7.9. Difference Equations Near Attachments

Attachment and Conditions	Displacement Coefficients

I   Pinned End

$y_k = 0;\ M_k = 0$

$n \quad m \quad i \quad j \quad k$ (Pinned)

$$1 \quad -4 \quad \boxed{6} \quad -4 \quad 0 = \frac{w_i\,\Delta x^4}{EI} \qquad (7.66a)$$

$$1 \quad -4 \quad \boxed{5} \quad 0 = \frac{w_j\,\Delta x^4}{EI} \qquad (7.66b)$$

II   Fixed End

$y_K = 0;\ (dy/dx)_k = 0$

(fixed)

$$1 \quad -4 \quad \boxed{6} \quad -4 \quad 0 = \frac{w_i\,\Delta x^4}{EI} \qquad (7.67a)$$

$$1\tfrac{1}{3} \quad -6 \quad \boxed{12} \quad 0 = \frac{w_j\,\Delta x^4}{EI} \qquad (7.67b)$$

III   Guided

$(dy/dx)_k = 0;\ (dM/dx)_k = 0$

(guided)

$$1 \quad -4 \quad \boxed{6} \quad -4 \quad 1 = \frac{w_i\,\Delta x^4}{EI} \qquad (7.68a)$$

$$1\tfrac{1}{3} \quad -6 \quad \boxed{12} \quad -\tfrac{22}{3} = \frac{w_j\,\Delta x^4}{EI} \qquad (7.68b)$$

$$2 \quad -8 \quad \boxed{6} = \frac{w_k\,\Delta x^4}{EI} \qquad (7.68c)$$

IV   Free End

$M_k = 0;\ (dM/dx)_k = 0$

(free)

$$1 \quad -4 \quad \boxed{6} \quad -4 \quad 1 = \frac{w_i\,\Delta x^4}{EI} \qquad (7.69a)$$

$$1 \quad -4 \quad \boxed{5} \quad -2 = \frac{w_j\,\Delta x^4}{EI} \qquad (7.69b)$$

$$2 \quad -4 \quad \boxed{2} = \frac{w_k\,\Delta x^4}{EI} \qquad (7.69c)$$

Accuracy increases with greater fineness of the subdivisions used. The number of subdivisions also determines the quantity of algebraic equations. Usually, only a small number of subdivisions are necessary since the order of the error for (7.64) is small. Symmetry should be used whenever possible to minimize the number of subdivisions.

One Dimensional Systems.   Equation (7.65) and those in Table 7.9 have been developed on the basis of an applied distributed load, which is often encountered in equipment design. The difference method also applies to nonuniform load distributions provided the variation of the load intensity from point to point is not excessive. Stepped or concentrated loadings require a finer point spacing to minimize the error than is necessary for uniform loadings. The equivalent load intensity for a concentrated load P may be determined using

$$w_i = P_i/\Delta x \tag{7.70}$$

Members with variable cross-sections can also be accommodated by accounting for EI in the development of (7.65). The difference equation for an internal point becomes

$$(EI)_{i-1}y_{i-2} - 2[(EI)_{i-1} + (EI)_i]y_{i-1} + [(EI)_{i-1} + 4(EI)_i$$
$$+ (EI)_{i+1}]y_i - 2[(EI)_i + (EI)_{i+1}]y_{i+1} + (EI)_{i+1}y_{i+2} = w_i \, \Delta x^4 \tag{7.71}$$

The equation for the moment at point i is given by (7.63) as

$$M = \frac{(EI)_i}{\Delta x^2}(-y_{i-1} + 2y_i - y_{i+1}) \tag{7.72a}$$

The angular rotation and the shear force developed at i may be determined from (7.62) as

$$\theta_i = \frac{\Delta x}{2}(-y_{i-1} + y_{i+1}) \tag{7.72b}$$

$$S_i = \frac{1}{2\Delta x^3}\{-(EI)_{i-1}y_{i-2} + 2(EI)_{i-1}y_{i-1}$$
$$+ [(EI)_{i+1} - (EI)_{i-1}]y_i - 2(EI)_{i+1}y_{i+1} + (EI)_{i+1}y_{i+2}\} \tag{7.72c}$$

Equations (7.72) may be added to or subtracted from (7.71) to obtain expressions near attachments similar to those for constant EI in Table 7.9. The moments at or near attachments are determined using (7.72a) modified by adding or subtracting (7.72b) if necessary to obtain the conditions in Table 7.9.

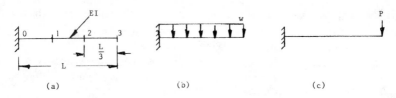

Figure 7.28. Difference Method for the Beam in Example 7.13. (a) Subdivided structure; (b) Uniform load; (c) Concentrated end load.

EXAMPLE 7.13.   Determine the fixed end moment and the deflection at each point of the subdivided cantilevered assembly in Figure 7.28 for (a) a distributed load w, and (b) a concentrated end load P.

(a) *Distributed Loading.* The load in terms of displacements is developed at each point using (7.65) and Table 7.9. At point 1, (7.67b) becomes

$$[12 \; -6 \; 1\tfrac{1}{3}] \begin{Bmatrix} y_1 \\ y_2 \\ y_3 \end{Bmatrix} = \frac{w_1}{EI} \left(\frac{L}{3}\right)^4 \tag{7.73a}$$

At point 2, (7.69b) applies with $y_0 = 0$ giving

$$[-4 \; 5 \; -2] \begin{Bmatrix} y_1 \\ y_2 \\ y_3 \end{Bmatrix} = \frac{w_2}{EI} \left(\frac{L}{3}\right)^4 \tag{7.73b}$$

Organizing (7.73) and a similar expression for point 3 based on (7.69c) into a system matrix gives

$$\begin{bmatrix} 12 & -6 & 1\tfrac{1}{3} \\ -4 & 5 & -2 \\ 2 & -4 & 2 \end{bmatrix} \begin{Bmatrix} y_1 \\ y_2 \\ y_3 \end{Bmatrix} = \frac{w}{EI} \left(\frac{L}{3}\right)^4 \begin{Bmatrix} 1 \\ 1 \\ 1 \end{Bmatrix} \tag{7.74}$$

where $w_1 = w_2 = w_3 = w$. The displacements may be determined by reduction or inversion (see Appendix A) as

$$\begin{Bmatrix} y_1 \\ y_2 \\ y_3 \end{Bmatrix} = \frac{wL^4}{EI} \begin{Bmatrix} 0.0230 \\ 0.0727 \\ 0.1204 \end{Bmatrix} \tag{7.75}$$

The fixed end moment at point 0 is determined from (7.72a) with $y_0 = (dy/dk)_0 = 0$ as

$$M_0 = -2y_1 EI \left(\frac{3}{L}\right)^2 \tag{7.76}$$

$$M_0 = -0.414wL^2$$

A comparison with the analytical solution indicates errors of 3.7 percent and 17.2 percent for the end displacement and fixed end moment respectively. By increasing the number of subdivisions, accuracy can be improved.

(b) *Concentrated Loading.* Since the configuration of the structure is independent of the applied loading, the matrix in (7.74) remains the same. The load intensity for a concentrated load at point 3 is determined from (7.70) as

$$w_3 = \frac{6P}{L}$$

where $\Delta x = \frac{1}{2}(L/3)$ at the end point. The displacements for the concentrated end load configuration may be determined from

$$\begin{bmatrix} 12 & -6 & 1\frac{1}{3} \\ -4 & 5 & -2 \\ 2 & -4 & 2 \end{bmatrix} \begin{Bmatrix} y_1 \\ y_2 \\ y_3 \end{Bmatrix} = \frac{2PL^3}{27EI} \begin{Bmatrix} 0 \\ 0 \\ 1 \end{Bmatrix}$$

as

$$\begin{Bmatrix} y_1 \\ y_2 \\ y_3 \end{Bmatrix} = \frac{PL^3}{EI} \begin{Bmatrix} 0.0494 \\ 0.1728 \\ 0.3333 \end{Bmatrix}$$

The fixed end moment at point 0 using (7.76) becomes

$$M_0 = -0.889PL$$

A comparison with the analytical solution indicates an error of 11.1 percent for the fixed end moment. The exact solution is obtained for the displaced shape.   Q.E.D.

The difference method enables direct determination of approximate translations and internal forces without requiring the evaluation of rotations or other quantities associated with determinate or indeterminate structures. Configurations with multiple supports, hinges and various end conditions can be evaluated quickly and with less effort using this method.

The use of uniformly spaced points introduces a limitation. This requires the selection of a spacing that best fits the location of support interfaces and attachments represented by the actual design. Displacements at intermediate locations can be determined by interpolation. The difference method is also limited to structures that can be idealized as beams or plates.

Two Dimensional Systems.   Circuit boards and panels of various shapes, mounting arrangements and load configurations can be evaluated with the difference method. Approximate numerical techniques are used to expedite a solution or in those cases where analytical solutions of complicated configurations are not possible. As in the treatment of beams, the governing differential equation of the plate including boundaries is replaced by equivalent difference equations. The solution for the displaced shape reduces to the solution of simultaneous algebraic equations written for each point on the configuration.

The differential equation describing the displaced shape of thin plates is given by

$$\frac{\partial^4 w}{\partial y^4} + 2\frac{\partial^4 w}{\partial x^2 \partial y^2} + \frac{\partial^4 w}{\partial x^4} = \frac{q}{D} \tag{7.77}$$

where q is the applied uniform load intensity, lb/in^3 (N/mm^2) and D is the flexural rigidity defined by

$$D = \frac{Et^3}{12(1 - v^2)} \tag{7.78}$$

where v is the Poisson ratio for the material used. The flexural and twisting moments in terms of the displaced shape becomes

$$M_x = -D\left(\frac{\partial^2 w}{\partial x^2} + v\frac{\partial^2 w}{\partial y^2}\right) \tag{7.79a}$$

$$M_y = -D\left(\frac{\partial^2 w}{\partial y^2} + v\frac{\partial^2 w}{\partial x^2}\right) \tag{7.79b}$$

$$M_{xy} = -D(1 - v)\frac{\partial^2 w}{\partial x \partial y} \tag{7.79c}$$

The difference equation equivalent of (7.77) and (7.79) is obtained from the Taylor series expansions (7.62) for the rectangular point mesh of Figure 7.29(a). Derivatives $(\partial^4 w/\partial x^4)_0$ and $(\partial^4 w/\partial y^4)_0$ in (7.77) are defined by (7.64) for point 0 in Figure 7.29(a) with respect to the x and y axes respectively. The

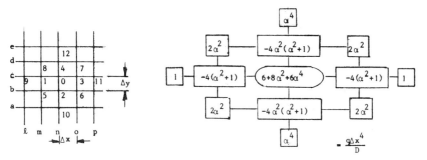

a) Rectangular Point Distribution   b) Difference Equivalent of (7.77) for $\Delta x = \alpha \Delta y$

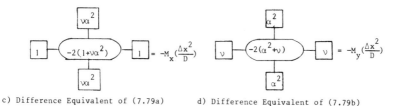

c) Difference Equivalent of (7.79a)   d) Difference Equivalent of (7.79b)

Figure 7.29. Difference Coefficients for Planar Surface.

mixed derivative $[\partial^4 w/(\partial x^2 \partial y^2)]_0$ is derived from $[\partial^2(\partial^2 w/\partial x^2)/\partial y^2]_0$ using (7.63) as

$$\left(\frac{\partial^4 w}{\partial x^2 \partial y^2}\right)_0 = \frac{1}{\Delta y^2}(w_2'' - 2w_0'' + w_4'')$$

where, for example,

$$w_2'' = \frac{1}{\Delta x^2}(w_5 - 2w_2 + w_6)$$

The difference coefficient pattern for an interior pivotal point in a rectangular grid of points (point 0 in Figure 7.29(a)) is shown in Figure 7.29 for $\Delta x = \alpha \Delta y$. The distributed load in terms of the displaced shape, for example, may be determined from Figure 7.29(b). The coefficients of the difference equation for interior points must be modified to account for the effects of attachments to be applicable to points at or near the constraint. The modifications are performed by adding or subtracting the coefficients representing the constraint conditions at the attachments. The difference coefficient patterns for various edge constraints developed by the method described are listed in Table 7.10.

Table 7.10. Coefficient Patterns for Various Edge Conditions

$$\left(\text{pivotal} \;\boxed{\text{point}}\; \text{loading} = \frac{q}{D}\,\Delta x^4 \right)$$

Pinned Edge

	α^4			α^4				α^4	
$2\alpha^2$	$\alpha^2 b$		$2\alpha^2$	$\alpha^2 b$	$2\alpha^2$		$2\alpha^2$	$\alpha^2 b$	
1 b	$\boxed{a}$		1 b	$\boxed{a+1-\alpha^4}$ b	1		1 b	$a-\alpha^4$	
$2\alpha^2$	$\alpha^2 b$								
	α^4								

(a) $w_0 = (M_x)_0 = 0$ (b) $w_b = (M_y)_b = 0$ (c) $w_0 = w_b = (M_x)_0 = (M_y)_b = 0$

Fixed Edge

	α^4			$\tfrac{2}{3}\alpha^4$			$\tfrac{2}{3}\alpha^4$	
$2\alpha^2$	$\alpha^2 b$		$2\alpha^2$	$2\alpha^2 d^*$ $2\alpha^2$			$2\alpha^2$	$2\alpha^2 d^*$
$\tfrac{2}{3}$ $2d$	$\boxed{a+7}$		1 b	$\boxed{a+1+6\alpha^4}$ b 1			$\tfrac{2}{3}$ $2d$	$\boxed{a+6\alpha^4+7}$
$2\alpha^2$	$\alpha^2 b$							
	α^4							

(d) $w_0 = (\partial w/\partial x)_0 = 0$ (e) $w_b = (\partial w/\partial y)_b = 0$ (f) $w_0 = w_b = \partial w/\partial x)_0 = (\partial w/\partial y)_b = 0$

Free Edge

		$\alpha^4 f$ $\uparrow$			$2\alpha^4$				α^4 $\uparrow$	
	$2e$	$\alpha^2 h^*$		$2e$	$-4(e+\alpha^4)$	$2e$		$2\alpha^2$	$\alpha^2 b$	e
2	$-4(e+1)$	$\boxed{g}$		$\leftarrow f$ h	$\boxed{g^*}$	h $f\rightarrow$		1 b	$\boxed{a}$	$-2(e+1)$
	$2e$	$\alpha^2 h^*$						$2\alpha^2$	$\alpha^2 b$	e
		$\alpha^4 f$ $\downarrow$						α^4	$\downarrow$	

(g) $(M_x)_n = (V_x)_n = 0$ (h) $(M_y)_c = (V_y)_c = 0$ (i) $(M_x)_0 = 0$

	α^4				α^4	$\uparrow$				$\alpha^4 f$ $\uparrow$	
$2\alpha^2$	$\alpha^2 b$	$2\alpha^2$		$2\alpha^2$	$\alpha^2 b$	e			$2e$	$\alpha^2 h$	
1 b	$\boxed{a+1-\alpha^4}$ b 1			1 b	$\boxed{a-\alpha^4}$	$-2(e+1)$		2	$-4(e+1)$	$\boxed{g-\alpha^4 f}$	
$\leftarrow e$	$-2(e+\alpha^4)$ e $\rightarrow$			$\leftarrow e$	$-2(e+\alpha^4)$	$-\alpha^2 i$		$\leftarrow$	$2e$	$\alpha^2(h/2+i)$	

(j) $(M_y)_b = 0$ (k) $(M_x)_0 = (M_y)_b = 0$ (l) $(M_x)_n = (M_y)_b = (V_x)_n = 0$

364

Table 7.10. Coefficient Patterns for Various Edge Conditions (cont.)

$$\left(\text{pivotal} \boxed{\text{point}} \text{ loading} = \frac{q}{D}\Delta x^4\right)$$

$\uparrow$

	$2\alpha^4$		$\uparrow$				$2\alpha^4 f$	
	$2e$	$-4(e+\alpha^4)$	$2e$			$4\alpha^2 i$	$\alpha^2 h^* + 2\alpha^2 i$	
$\leftarrow f$	h	$\boxed{g^* - f}$	$h/2 + \alpha^2 i$		$\leftarrow 2f$	$h + 2\alpha^2 i$	$\boxed{g^* - 4f - 2\alpha^4 v^2}$	

(m) $(M_x)_0 = (M_y)_c = (V_y)_c = 0$ (n) $(M_x)_n = (M_y)_c = (V_x)_n = (V_y)_c = 0$

$\uparrow$ $\uparrow$

	$\leftarrow e$	$-2(e+\alpha^4)$	$e\rightarrow$			$\leftarrow -\alpha^2 i$	$-2(e+\alpha^4)$	$-\alpha^2 i\rightarrow$
1	b	$\boxed{a + 1 - 2\alpha^4}$	b	1		$-2(e+1)$	$\boxed{a - 1 - 2\alpha^4}$	$-2(e+1)$
	$\leftarrow e$	$-2(e+\alpha^4)$	$e\rightarrow$			$\leftarrow -\alpha^2 i$	$-2(e+\alpha^4)$	$-\alpha^2 i\rightarrow$

$\downarrow$ $\downarrow$

(o) $(M_y)_d = (M_y)_b = 0$ (p) $(M_y)_d = (M_y)_b = (M_x)_0 = (M_x)_m = 0$

Corner Bounded by Free and Clamped Edges

		α^4					$2\alpha^4$	
	$2\alpha^2$	$\alpha^2 b$				$2e$	$-4(e+\alpha^4)$	
$\tfrac{2}{3}$	$-6 - 4\alpha^2$	$\boxed{a + 7 - \alpha^4}$			$\leftarrow \tfrac{2}{3} - v^2$	$h - 2$	$\boxed{a + 7 - v^2 - j}$	
	$\leftarrow e$	$-2(e+\alpha^4)$						

(q) $w_0 = (\partial w \partial x)_0 = (M_y)_b = 0$ (r) $w_0 = (\partial w/\partial x)_0 = (M_y)_c = (V_y)_c = 0$

$\uparrow$

	$\tfrac{1}{3}\alpha^4$		$\uparrow$			$\alpha^4(\tfrac{2}{3} - v^2)$	
	$2\alpha^2$	$-6\alpha^4 - 4\alpha^2$	e			$2e$	$\alpha^2(h^* - 2\alpha^2)$
1	b	$\boxed{a + 6\alpha^4}$	$-2(e+1)$		2	$-4(e+1)$	$\boxed{a + 1 + \alpha^4(6 - v^2) -- j^*}$

(s) $w_b = (\partial w/\partial y)_b = (M_x)_0 = 0$ (t) $w_b = (\partial w/\partial y)_b = (M_x)_n = (V_x)_n = 0$

Parameters

$\alpha = \Delta x/\Delta y$ $f = 1 - v^2$

$a = 5 + 8\alpha^2 + 6\alpha^4$ $g = 6\alpha^4(1 - v^2) + 8\alpha^2(1 - v) + 2$

$b = -4(1 + \alpha^2)$ $g^* = 6(1 - v^2) + 8\alpha^2(1 - v) + 2\alpha^4$

$c = -(\tfrac{22}{3} + 4\alpha^2)$ $h = 4[(v^2 - 1) + \alpha^2(v - 1)]$

$c^* = -(22\alpha^2/3 + 4)$ $h^* = 4[\alpha^2(v^2 - 1) + (v - 1)]$

$d = -(3 + 2\alpha^2)$ $i = 2(v - 1)$

$d^* = -(3\alpha^2 + 2)$ $j = 4(\alpha^2 + v)^2$

$e = \alpha^2(2 - v)$ $j^* = 4(1 + v\alpha^2)$

Once the surface has been divided by a grid of lines that best fits the geometry, attachments and load distribution, the load deflection relationship at each mesh point is determined by an appropriate pattern. The pattern selected is based on the location of the point relative to a constraint and the type of constraint. Coefficients in the applicable pattern are then transcribed to corresponding mesh points. Each point is suitably numbered, accounting for symmetry. The coefficients at each point may be grouped into a matrix that represents the stiffness characteristics of the surface under study. The larger the number of mesh points, the more accurate the approximation will be. A fine mesh will generally require the use of automated methods to invert or reduce the matrix.

The difference patterns in Figure 7.29(b) and Table 7.10 have been developed on the basis of an applied distributed load. This case is often encountered in equipment design. Stepped or concentrated loadings may also be applied provided an equivalent load intensity at mesh point i is determined from

$$q_i = P_i/\Delta x^2 \tag{7.80}$$

Variations in the flexural modulus D can be accommodated provided these effects are included in the derivation of difference patterns at or near constraints. Variations of D at internal points may be accounted by setting $D = D_i$ in the pattern of Figure 7.29(b).

The coefficient patterns in Table 7.10 may be applied to most rectangular planar configurations encountered in practice. Surfaces with curved boundaries can also be accommodated depending on the fineness of the mesh. Triangular and polar coefficient patterns may be developed to model surfaces conveniently with irregular or circular boundaries.

EXAMPLE 7.14.   Determine the maximum displacement of the glass epoxy ($\nu = 0.12$) uniformly loaded circuit board described in Figure 7.30(a).

The surface is divided by a uniform rectangular grid with an aspect ratio determined as

$$\alpha = \frac{\Delta x}{\Delta y} = \frac{L/4}{5L/18} = 0.9$$

The point numbering in Figure 7.30(b) accounts for symmetry. Coefficient patterns for each point are developed using Table 7.10. At point 1, Table 7.10(h) gives

$$
\begin{array}{ccccc}
0.986 & -6.794 & \boxed{12.928} & -6.794 & 0.986 \\
 & 3.046 & -8.716 & 3.046 & \\
 & & 1.312 & &
\end{array}
= \frac{q_1}{D}\Delta x^4 \tag{7.81}
$$

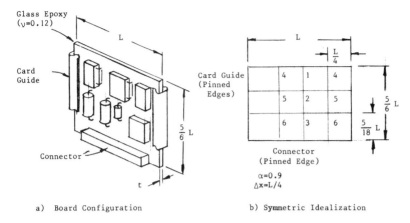

a) Board Configuration b) Symmetric Idealization

Figure 7.30. Difference Model for Circuit Board Displacements.

The coefficient patterns for points 2 and 3 are obtained from Tables 7.10(j) and (b) respectively as

$$
\begin{array}{ccc}
& 1.523 & -4.358 & 1.523 \\
1 & -7.24 & \boxed{15.761} & -7.24 & 1 = \dfrac{q_2}{D}\,\Delta x^4; \\
& 1.62 & -5.864 & 1.62 \\
& & 0.656
\end{array}
\qquad
\begin{array}{ccc}
& & 0.656 \\
& 1.62 & -5.864 & 1.62 \\
1 & -7.24 & \boxed{15.761} & -7.24 & 1
\end{array}
= \dfrac{q_3}{D}\,\Delta x^4
$$

The coefficient patterns for the remaining points may be similarly obtained. These coefficients may be assembled into a matrix which expresses the load in terms of the point displacements. The first row of this matrix is obtained from (7.81) and Figure 7.30(b) as

$$
[12.928 \quad -8.716 \quad 1.312 \quad -13.588 \quad 6.092 \quad 0]
\left\{
\begin{array}{c}
w_1 \\ w_2 \\ w_3 \\ w_4 \\ w_5 \\ w_6
\end{array}
\right\}
= \frac{\Delta x^4}{D}\{q_1\}
$$

Completing the matrix by taking each point successively gives

$$\begin{bmatrix} 12.928 & -8.716 & 1.312 & -13.588 & 6.092 & 0 \\ -4.358 & 15.761 & -5.864 & 3.046 & -14.480 & 3.240 \\ 0.656 & -5.864 & 15.761 & 0 & 3.240 & -14.480 \\ -6.794 & 3.046 & 0 & 12.928 & -8.716 & 1.312 \\ 1.523 & -7.240 & 1.62 & -4.358 & 15.761 & -5.864 \\ 0 & 1.620 & -7.24 & 0.656 & -5.864 & 15.761 \end{bmatrix} \begin{Bmatrix} w_1 \\ w_2 \\ w_3 \\ w_4 \\ w_5 \\ w_6 \end{Bmatrix} = \frac{qL^4}{256D} \begin{Bmatrix} 1 \\ 1 \\ 1 \\ 1 \\ 1 \\ 1 \end{Bmatrix}$$

where $q_1 = q_2 = \cdots q_6 = q$ and $\Delta x = (L/4)^4$. The displacements may be obtained using the reduction method described in Appendix A.6 as

$$\begin{Bmatrix} w_1 \\ w_2 \\ w_3 \\ w_4 \\ w_5 \\ w_6 \end{Bmatrix} = \frac{qL^4}{256D} \begin{Bmatrix} 2.6062 \\ 2.0577 \\ 1.2362 \\ 1.8683 \\ 1.4782 \\ 0.8920 \end{Bmatrix}$$

The maximum displacement occurs along the free edge of the circuit board at point 1.   Q.E.D.

General solutions may also be developed for configurations that are used repetitively. Consider the component board in Figure 7.31(a) which is retained at the four corners. This configuration can be idealized by the grid shown in Figure 7.31(b) for uniform loading. The matrix solution for maximum displacements may be obtained from the coefficient patterns in Table 7.10 as

$$\begin{bmatrix} 4(1 + \alpha^2)^2 & 4\alpha^2(\nu - 2 - \alpha^2) & 4\alpha^2(\nu - 2) - 4 \\ 4\alpha^2(\nu - 2 - \alpha^2) & g^* + 2\alpha^4 - 2(1 - \nu^2) & 4\alpha^2(2 - \nu) \\ 4\alpha^2(\nu - 2) - 4 & 4\alpha^2(2 - \nu) & 4\alpha^4 f - 2b - 4(1 + 2\alpha^2\nu) \end{bmatrix} \begin{Bmatrix} w_1 \\ w_2 \\ w_3 \end{Bmatrix} = \frac{qL^4}{16D} \begin{Bmatrix} 1 \\ 1 \\ 1 \end{Bmatrix}$$

$$(7.82)$$

where g^* and f are defined in Table 7.10.

Flexural stress at any point on the surface may be determined from

$$\sigma_i = \frac{6M_i}{t^2} \tag{7.83}$$

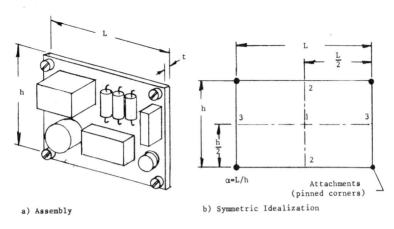

a) Assembly

b) Symmetric Idealization

Figure 7.31. Difference Model of a Point Supported Assembly.

where M_i is given by (7.72a) for points along a free edge and by the coefficient pattern of Figure 7.29(c) and (d) for points elsewhere. At point 1 in Figure 7.30(b), for example, the moment determined from (7.72a) becomes

$$M_1 = \frac{2D}{\Delta x^2}(w_1 - w_4)$$

where D replaces EI for these surfaces. Along a vertical free edge, Δx^2 would be replaced by Δy^2. The flexural stress at point 1 in Figure 7.30(b) is determined from (7.83) as

$$\sigma_1 = \frac{12D}{(t\,\Delta x)^2}(w_1 - w_4)$$

8

Dynamic Characteristics of Electronic Equipment

The relationships which determine the static displacements and stresses in an assembly also form the basis of predicting its dynamic behavior. The motions an assembly can ascribe are also dependent upon damping and inertia forces within the configuration and the nature of the disturbance. Expressions relating these quantities are used to describe the response. The response may be developed in either the frequency or time domains. These tools provide the means of minimizing dynamic displacements and stresses in order to prevent collisions between adjacent assemblies and premature fatigue failures.

8.1 FUNDAMENTALS OF MOTION

The equation of motion representing an assembly describe the balance of forces acting on the system. The equation for a spring mass system has been developed in Appendix B.1. Many simple structures can be idealized by an equivalent spring mass system. These systems have one degree of freedom since only one coordinate, x, is required to define the position of the mass at a given time, t. The reduction of a particular configuration to an equivalent dynamic system involves the principle of *dynamic similarity*. This requires that the work, strain energy and kinetic energy of the original and idealized configuration be identical. Equivalent mass and concentrated loads for single degree of freedom representations are listed in Table B.1-1. Flexural stresses may be determined by using Table 6.5.

The general equation of forced motion (B.1-2a) can be expressed as

$$M\ddot{x} + c\dot{x} + kx = P(t) \tag{8.1}$$

where M, k and c represent the mass, stiffness and damping respectively.

If the disturbing force, $P(t)$, is zero, an equation for damped free vibration is obtained. Free vibration in an assembly occurs after an initial abrupt disturbance. This disturbance may be due to a rapid change in displacement or an initial velocity as a result of an impact that can occur during handling or transport of the equipment.

Since (8.1) is linear, the time response to different disturbances can be summed to represent a combination of disturbances. The response motion will be the result of damped free vibration plus forced motion. Resonance occurs when the disturbance is similar to the free vibration characteristics of the equipment. These factors also apply to equipment configurations with many degrees of freedom.

The motions defined by (8.1) have the same character as the disturbance. Both stiffness and damping quantities also influence response motions, as illustrated in Figures B.1-2 and 3. This is shown by considering an equation developed from (8.1) representing a sinusoidal acceleration disturbance at the attachment of a spring mass system

$$M\ddot{x} + c\dot{x} + kx = M\ddot{x}_f \cos \omega_f t \qquad (8.2)$$

where $x = x_m - x_f$.

Methods of Appendix B are used to develop the general solution as

$$x = C_1 e^{-\alpha t} \sin \omega_D t + C_2 e^{-\alpha t} \cos \omega_D t$$

$$+ \frac{\ddot{x}/\omega^2}{(\Omega/Q)^2 + (1 - \Omega^2)^2} \left[(1 - \Omega^2) \cos \omega_f t + \frac{\Omega}{Q} \sin \omega_f t \right] \qquad (8.3)$$

where $\Omega = \omega_f/\omega$, $\omega = \sqrt{k/M}$, the damped time constant $1/\alpha = 2M/c = 2Q/\omega$ and the damped fundamental frequency $\omega_D = \omega\sqrt{4Q^2 - 1}/2Q$. The response (8.3) is the sum of the damped resonant frequency of the spring mass system and the sinusoidal frequency of the disturbance. The system will tend to vibrate at its natural frequency, ω_D, as well as follow the frequency of the disturbance, ω_f. The free vibration portion will slowly damp out developing a response that vibrates at the frequency of the disturbance. This condition represents the steady-state vibration of the system.

At low frequencies, $\Omega^2 \ll 1$; the motion simply follows along in phase with the disturbance at an amplitude which is nearly independent of the disturbance frequency, $x \simeq \ddot{x}/\omega^2$. At very high frequencies, $\Omega^2 \gg 1$, the motion is 180° out of phase with the disturbance regardless of the disturbance frequency. At high frequencies, the motion of the mass does not follow those of the disturbance.

When the disturbing frequency is nearly equal to the fundamental frequency of the system, $\Omega^2 = 1 - 0.5/Q^2$, the amplitude and the kinetic energy of the mass is maximum. This condition which develops a large output in the presence of a small input

$$\frac{x}{\ddot{x}/\omega^2} = \frac{Q^3\sqrt{1 - 0.5/Q^2}}{Q^2 - 0.25} \simeq Q \tag{8.4}$$

is known as resonance. At this frequency, the energy furnished by the disturbance is mostly stored in the mass and spring with little dissipated by damping. The magnitude of the response to the disturbance for any frequency ratio Ω is obtained from (8.3) as

$$\frac{x}{\ddot{x}/\omega^2} = \left[\frac{1}{(\Omega/Q)^2 + (1 - \Omega^2)^2}\right]^{1/2} \tag{8.5}$$

This relationship, shown graphically in Figure B.6-1, expresses the ratio of the force in the spring to the inertia force that would be imposed on the mass if it were rigidly attached to the moving support structure.

Where the disturbance acceleration in (8.2) is expressed in terms of the displacement of the supporting structure $(x_f\omega_f^2 \cos \omega_f t)$, the magnitude of the steady state vibration becomes

$$\frac{x}{x_f} = \left[\frac{1 + (\Omega/Q)^2}{(\Omega/Q)^2 + (1 - \Omega^2)^2}\right]^{1/2} \tag{8.6}$$

This relationship, shown graphically in Figure B.1-4, expresses the ratio of the absolute displacement (or acceleration) of the mass to the displacement (or acceleration) of the disturbance. Various transfer functions relating response behavior to the disturbance are listed in Table B.1-2.

The undamped fundamental frequency of the spring mass system in Figure B.1-1 may be determined from (8.1) as

$$f_n = \frac{1}{2\pi}\sqrt{\frac{k}{M}} \tag{8.7}$$

Expressing the spring stiffness k in terms of the static deflection and substituting the result $k = Mg/\delta_{st}$ into (8.7) gives

$$f_n = \frac{1}{2\pi}\left(\frac{g}{\delta_{st}}\right)^{1/2} \tag{8.8}$$

A similar expression for the fundamental frequency in terms of the dynamic displacement Δ for undamped free vibration may be obtained from (8.1) with $c = P(t) = 0$ giving

$$M\ddot{x} + kx = 0$$

or
$$\ddot{x} + \omega^2 x = 0 \tag{8.9}$$

The equation of motion defined by (8.9) is satisfied by

$$x = \Delta \sin \omega t \tag{8.10}$$

The acceleration experienced by the mass is obtained by substituting (8.10) into (8.9) as

$$\ddot{x} = -\omega^2 \Delta \sin \omega t \tag{8.11}$$

Maximum acceleration expressed in nondimensional gravitational units becomes

$$\ddot{x} = \omega^2 \Delta / g$$

or
$$G = (2\pi f_n)^2 \Delta / g \tag{8.12}$$

Recasting (8.12) in the form of (8.8) gives

$$f_n = \frac{1}{2\pi} \left(\frac{gG}{\Delta} \right)^{1/2} \tag{8.13}$$

From (8.8) and (8.13) the input dynamic displacement and response displacement for a spring mass system respectively become

$$\Delta = G_{in}\delta_{ST} \tag{8.14a}$$

$$\Delta_{res} = QG_{in}\delta_{ST} \tag{8.14b}$$

Relationships that define δ_{st} in terms of f_n for various idealized equipment configurations are listed in Table 6.5. The dynamic displacements of the represented equipment may be determined from (8.14) once the input acceleration loading G_{in} is defined and amplification Q has been determined.

Damping. The amplification, Q, is dependent on the damping characteristics of the configuration. Damping describes a dissipative process that may be a result of friction, viscous forces and material hysteresis. The energy dissipated by damping steadily reduces the amplitude of free vibratory motion.

Friction is generally due to relative motion at structural joints that are designed for disassembly. Joints that can rub or slide dissipate energy during vibration. Friction damping is relatively constant, being unaffected by the amplitudes or velocities developed by vibration. The effectiveness of this form of damping is more apparent at small displacements. Friction damping may introduce stiction that can inhibit static repositioning when restoring forces are insufficient to return the systems to its original position.

A vibrating system also induces motion in the surrounding medium and in the supporting structure. The vibratory energy dissipated by this means depends on the frequencies involved and the nature of the surrounding medium and design characteristics of the mounting structure.

Hysteresis losses may be due to internal strains that develop during the flexural response to rapid, repeated loading. The hysteresis loop for a particular material is related to the rate and frequency of the applied loading. The energy dissipated per cycle is a function of the enclosed area when drawn in the stress-strain plane of the response.

Dissipative processes in electronic equipment may be approximated by viscous damping or linear drag, $c\dot{x}$ in (8.1). This approach simplifies the development and solution of the equations of motion representing equipment configurations giving a good practical estimate of dynamic behavior. Damping properties of an assembly are usually expressed in terms of the response amplification, Q, which occurs at resonances.

In general, Q is proportional to the square root of the fundamental frequency of an assembly. Particular multiplying factors that increase or decrease this quantity are a function of geometry, attachment methods, weight distribution and construction. Clamped aluminum core printed circuit boards, for example, can experience an amplification $Q \simeq 2\sqrt{f_n}$ whereas the amplification of a conventional glass epoxy circuit board can be represented by $Q \simeq \sqrt{f_n}$. Amplification is also dependent on the magnitude of the disturbance. A higher Q occurs at lower acceleration loadings and as the magnitude of the applied loading increases, the amplification decreases; this characteristic has been discussed in Chapter 1.

The damping and stiffness properties of resilient materials are also influenced by temperature. The amplification, Q, and the stiffness of these materials increases at low temperatures and decrease at high temperatures.

Heat generated as a consequence of the energy loss which gradually reduces the amplitudes of free vibratory motion may increase component temperatures and adversely affect performance and the vibration lifetime of the equipment. The heat generated by damping may be determined from

$$q = 28QMf_n(\Delta f)^2 \text{ watts} \qquad (8.15a)$$

which in metric units becomes

$$q = QM\omega_n(\Delta\omega)^2 \text{ watts} \qquad (8.15b)$$

		(English)	(metric)
given f_n, ω_n	= assembly resonant frequency	Hz	rad/s
f, ω	= disturbance frequency	Hz	rad/s
M	= assembly mass	lb sec^2/in	kg
Δ	= input dynamic displacement	in	m
Q	= amplification (Δ_{res}/Δ)	dimensionless	

Amplification Q is usually obtained empirically since many factors which contribute to its composition are difficult to evaluate. Data obtained from tests performed on representative physical models may be used to update the idealizations and improve the predictions. These tests can be performed early in the design cycle.

EXAMPLE 8.1.   The resonant frequency of a 0.3 lb (0.136 kg) glass epoxy circuit board assembly 0.06 in (0.76 mm) thick 5 in (127 mm) long mounted in card guides has been determined as 225 Hz (1413.7 rad/s). Determine the heat generated due to damping, the response displacement and the flexural stress in the copper artwork trace when the assembly is (a) vibrated at resonance with an input acceleration of 2G and (b) mounted to a 350 Hz (2199 rad/s) structural support which is subjected to a vibratory acceleration of 2G.
(a) *Resonant Excitation at 225 Hz.* The amplification of the card guide supported 225 Hz assembly may be estimated as

$$Q \simeq \sqrt{f_n} = 15$$

The idealized configuration of the assembly represented by case V of Table 6.5 and (8.14a) is used to obtained an expression for the input dynamic displacement

$$\Delta = 12.39 G_{in}/f_n^2 \tag{8.16}$$

$$\Delta = 12.39(2)/225^2 = 4.895 \times 10^{-4} \text{ in } (1.243 \times 10^{-5} \text{ m})$$

The response displacement and the flexural stress in the copper artwork trace are determined from (8.14b) and case V of Table 6.5 respectively as

$$\Delta_{res} = 15\Delta = 7.34 \times 10^{-3} \text{ in } (1.865 \times 10^{-4} \text{ m})$$

$$\sigma = 9.6(17.6 \times 10^6)(0.03)7.34 \times 10^{-3}/5^2 = 1488 \text{ lb/in}^2 \quad \text{(English)}$$

$$\sigma = 9.6(1.21 \times 10^5)(0.76)0.1865 \text{ mm}/127^2 = 10.21 \text{ N/mm}^2 \quad \text{(metric)}$$

The heat generated in the circuit card by energy losses due to damping is determined from (8.15) as

$$q = 28(15)(0.3/386)(225) \, [4.895 \times 10^{-4}(225)]^2 = 0.891 \text{W} \quad \text{(English)}$$

$$q = 15(0.136)(1413.7) \, [1.243 \times 10^{-5}(1413.7)]^2 = 0.891 \text{W} \quad \text{(metric)}$$

(b) *Support Excitation at 350 Hz.* The displacement (or acceleration) transmitted to the 225 Hz circuit board due to a sinusoidal disturbance at 350 Hz is determined from Figure B.1-4. The transmissibility at a frequency ratio $f/f_n = 350/225 = 1.56$ for a circuit board amplification of 15 at resonance, following Figure B.1-4, is

$$\Delta_{res}/\Delta = 0.8 = Q$$

The input dynamic displacement for a 2G acceleration loading at 350 Hz using (8.16) is

$$\Delta = 12.39(2)/350^2 = 2.023 \times 15^4 \text{ in } (5.138 \times 10^{-6} \text{ m})$$

The response displacement and the copper artwork flexural stress are determined from (8.14b) and case V of Table 6.5 respectively as

$$\Delta_{res} = 0.8\Delta = 1.618 \times 10^{-4} \text{ in } (4.11 \times 10^{-6} \text{ m})$$

$$\sigma = 9.6(17.6 \times 10^6)(0.03)1.618 \times 10^{-4}/5^2 = 32.8 \text{ lb/in}^2 \quad \text{(English)}$$

$$\sigma = 9.6(1.21 \times 10^5)(0.76)5.138 \times 10^{-3} \text{ mm}/127^2 = 0.281 \text{ N/mm}^2 \quad \text{(metric)}$$

The heat generated in the circuit card by energy lost due to damping is determined from (8.15) as

$$q = 28(0.8)(0.3/386)(225) \ [2.023 \times 10^{-4}(350)]^2 = 0.0196\text{W} \quad \text{(English)}$$

$$q = 0.8(0.136)(1413.7) \ [5.138 \times 10^{-6}(2199)]^2 = 0.0196\text{W} \quad \text{(metric)}$$

Input vibrations that closely resemble the resonant frequencies of the equipment should be avoided in order to minimize response displacements, stresses, vibratory heating and fatigue failures.   Q.E.D.

Fundamental Resonances of Equipment and Components.   Estimates of the fundamental frequencies of many equipment configurations can be determined from Table B.1-1 with $f_n = (k_{eq}/M_{eq})^{1/2}/2\pi$. The fundamental frequencies for other idealized configurations may be estimated from the displaced shape due to static loading using Rayleigh's method. Following this method, the natural frequency can be determined by equating the maximum kinetic energy to the maximum potential energy of a freely vibrating system. The accuracy depends on how closely the displaced shape due to a static loading approximates the dynamic displacement. If the dynamic displacement x of a harmonically vibrating configuration can be represented by

$$x = y \cos \omega t \qquad (8.17)$$

the kinetic energy of an element dl with distributed mass per unit length m becomes

$$d(KE) = \tfrac{1}{2}\dot{x}^2 \, dm \qquad (8.18a)$$

or using (8.17)

$$d(KE) = \frac{(\omega y)^2}{2} \, mdl \qquad (8.18b)$$

Considering all elements of the system, the maximum kinetic energy becomes

$$KE = \tfrac{1}{2}m\omega^2 \int y^2 dl \qquad (8.19a)$$

If the idealization also includes concentrated masses M_1, M_2, $\ldots$, M_n that develop dynamic displacement y_1, y_2, $\ldots$, y_n at corresponding locations, the total maximum kinetic energy is then expressed as

$$KE = \tfrac{1}{2}m\omega^2 \int y^2 dl + \sum_{i=1}^{n} M_i\omega^2 y_i^2 \qquad (8.19b)$$

The total maximum potential energy is of the same form as (7.42) with the variation of curvature y'' accounting for the effects of the concentrated masses. The general expression for the fundamental frequency ω by equating KE and U becomes

$$\omega^2 = \frac{\tfrac{1}{2}\int wy\,dl + \sum\limits_{i=1}^{n} W_i y_i}{\tfrac{1}{2}m\int y^2 dl + \sum\limits_{i=1}^{n} m_i y_i^2} \qquad (8.20)$$

where the flexural strain energy is equal to the external work of the statically applied distributed weight.

EXAMPLE 8.2. Determine the approximate fundamental frequency of the cantilever beam with the distributed load shown in Table B.1-1 ($M_{\text{conc}} = 0$).
    The static displacement at a distance x from the free end of the beam is given by the shape function case IV of Table 6.6. The shape function normalized to the end displacement y_m becomes

$$y = \frac{y_m}{3} (3 - 4\xi + \xi^4)$$

where $y_m = wL^4/8EI$ and $\xi = x/L$. Substituting this expression into (8.20) in the absence of concentrated masses gives

$$\omega^2 = \frac{g \int_0^1 yd\xi}{\int_0^1 y^2 d\xi} = \frac{2gy_m/5}{104y_m^2/405}$$

or
$$\omega^2 = \frac{162EIg}{13wL^4}$$

The fundamental frequency computed by the Rayleigh method is 3.3 percent greater than frequency developed by the parameters given in Table B.1-1, case II.   Q.E.D.

Rayleigh's method generally results in somewhat higher frequencies than the exact quantities due to the constraints imposed in defining the dynamic displacement. This method provides a convenient basis for estimating the fundamental frequencies of rectangular configurations with various edge conditions.

When Rayleigh's method is applied to planar surfaces such as printed circuit boards, dynamic deflection functions which satisfy edge conditions are selected. Either trigonometric or polynomial functions may be used. The results of applying the Rayleigh method to various configurations are summarized in Table 8.1.

EXAMPLE 8.3.   Determine the fundamental frequency of the glass epoxy circuit board shown in Figure 7.30(a) when $L = 6$ in (152.4 mm) and $t = 0.09$ in (2.29 mm).

Edge support provided by the card guides and connector may be considered as simply supported. This edge support configuration is described in Table 8.1, case 12, which gives an expression for the fundamental frequency as

$$f_n = \frac{\pi}{2}\left[\frac{Dab}{M}\left(\frac{0.927}{a^4} + \frac{0.65}{a^2b^2} + \frac{0.31}{b^4}\right)\right]^{1/2} \text{ Hz}$$

Based on the elastic modulus of glass epoxy (Appendix C), $E = 2.36 \times 10^6$ lb/in^2 (1.627×10^4 N/mm^2), the fundamental frequency in terms of the circuit card mass for $a = L$ and $b = 5L/6$ becomes

$$f_n = 2.27 \sqrt{\frac{D}{L^2M}}$$

$$f_n = 2.27 \sqrt{\frac{157.5}{6^2M}} = \frac{4.75}{\sqrt{M}} \text{ Hz   (English)}$$

$$f_n = 2.27 \sqrt{\frac{17.89 \times 10^6}{152.4^2M}} = \frac{63}{\sqrt{M}} \text{ Hz   (metric)}$$

where $D = 157.5$ lb·in (17.89×10^6 kg·mm^2/s^2).   Q.E.D.

Table 8.1. Fundamental Frequencies of Various Uniformly Loaded Rectangular Surfaces

Case	A_1	A_2	B_1	B_2	C_o	P_o
		Edge Configuration				
1	F	F	F	F	$\pi/2$	$[2.08(ab)^{-2}]^{1/2}$
2	F	F	F	C	0.55	a^{-2}
3	F	F	C	C	3.55	a^{-2}
4	F	F	C	S	2.45	a^{-2}
5	F	F	S	S	$\pi/2$	a^{-2}
6	F	C	C	F	$\pi/5.42$	$[a^{-4} + 3.2(ab)^{-2} + b^{-4}]^{1/2}$
7	F	S	S	F	$\pi/11$	$(a^{-2} + b^{-2})$
8	F	S	C	F	$\pi/2$	$[0.138a^{-4} + 0.251(ab)^{-2}]^{1/2}$
9	F	S	C	C	$\pi/2$	$(0.25b^{-2} + 2.25a^{-2})$
10	F	C	S	S	$\pi/2$	$[a^{-4} + 0.608(ab)^{-2} + 0.1266b^{-4}]^{1/2}$
11	F	C	C	C	$\pi/3$	$[12a^{-4} + 2.25(ab)^{-2} + 0.31b^{-4}]^{1/2}$
12	F	S	S	S	$\pi/2$	$[0.927a^{-4} + 0.65(ab)^{-2} + 0.31b^{-4}]^{1/2}$
13	F	C	C	S	$\pi/2$	$[2.44a^{-4} + 0.707(ab)^{-2} + 0.127b^{-4}]^{1/2}$
14	F	S	C	S	$\pi/2$	$[2.56a^{-4} + 0.57(ab)^{-2}]^{1/2}$
15	S	S	S	S	$\pi/2$	$(a^{-2} + b^{-2})$
16	C	C	C	C	$\pi/1.5$	$[2.91a^{-4} + 1.663(ab)^{-2} + 2.91b^{-4}]^{1/2}$
17	S	S	C	C	$\pi/3.53$	$[16a^{-4} + 7.7(ab)^{-2} + 3b^{-4}]^{1/2}$
18	C	C	C	S	$\pi/2$	$[2.45a^{-4} + 2.9(ab)^{-2} + 5.13b^{-4}]^{1/2}$
19	S	C	C	S	$\pi/2$	$[2.45a^{-4} + 2.68(ab)^{-2} + 2.45b^{-4}]^{1/2}$
20	S	S	C	S	$\pi/2$	$[2.43a^{-4} + 2.33(ab)^{-2} + 0.985b^{-4}]^{1/2}$

$$F = \text{Free edge}$$

$$S = \text{Simple support}$$

$$C = \text{Clamped support}$$

$$f_n = C_o P_o \sqrt{\frac{Dab}{M}} \text{ Hz}$$

$$D = \frac{Et^3}{12(1 - \nu^2)}$$

$$M = W/g \text{ (mass)}$$

Coupled Motion. Motion can become coupled if an assembly can move in more than one direction. The assembly in Figure 8.1 can experience translation along the x, y and z axes and torsion about the x and z axis. Coupling can occur between each of these motions. The behavior at a particular frequency could be composed of components of each motion. Coupling can also occur between components mounted on a common structure or between assemblies in a chassis. To minimize coupling, the fundamental frequencies of assemblies and subassemblies within a common enclosure should be separated by a factor ≥ 2 during the development of each design.

Coupling becomes apparent when the effect of a load or torque acting on the center of gravity of an assembly is examined. If the assembly translates without rotating as a result of applying a load through the center of gravity, then the

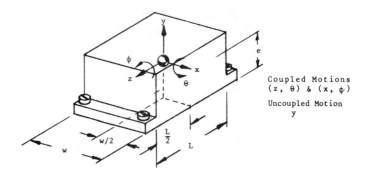

Figure 8.1. Motions of an Assembly.

translational motion in the direction of the applied load is uncoupled. If a torque is applied and rotation occurs without translation, then the rotational motion is uncoupled about the axis of applied loading.

Coupling may be a consequence of the stiffness, damping or inertial properties of the configuration. Coupling is indicated by the off-diagonal terms in the dynamic matrix development of the idealized configuration. Equation (6.35) is an example of an uncoupled system exhibiting three different motions. Pure inertial coupling is apparent in (6.36). Both stiffness and inertial coupling are indicated in (6.37). Design characteristics that influence off-diagonal quantities may be selected to minimize coupling if desired. Suitably selected element stiffnesses uncouple the y motion in (6.37), giving the result shown in (6.39).

Coupled motion can be visualized by plotting the frequency response of a particular coordinate. Appendix B.6 describes various methods of constructing the frequency response which can be readily applied to an idealized configuration. The response shown in Figure B.6-7 illustrates different degrees of coupling for a system with two degrees of freedom. Figure B.6-7(b) represents a higher degree of coupling than does the system in Figure B.6-7(a). This difference is due to a off-diagonal quantity of greater magnitude in the matrix representation for the system in Figure B.6-7(b) as compared to the off-diagonal quantity for the system in Figure B.6-7(a).

Each resonance in Figure B.6-7(a) represents a vibration mode of a system with two degrees of freedom. Each mode is associated with a particular fundamental frequency of the system and each mode represents a degree of freedom. A system idealized using four degrees of freedom has four modes of vibration and four resonant frequencies. The lowest resonant frequency is referred to as the fundamental frequency of the system. At this frequency, the displacements and attendant stresses experienced by the system are usually the greatest.

In a multimode system, peak displacements and stresses at modes above the lowest resonant frequency decrease about 3 db/octave below the maximum amplitude at the fundamental frequency. The effect of increased coupling between modes therefore, reduces the response displacements at modes above the fundamental. Displacements and stresses can be minimized by selecting design characteristics that assure separation between modes by a factor equal to or greater than two. Additionally, resonant frequencies should be selected during equipment design that avoid known peak displacements or accelerations in the excitation spectra.

Fatigue.   One of the objectives of a design engineer is to predict likely sources of failure in an electronic assembly and develop alternative designs for their avoidance. Fatigue is a particular kind of failure that accompanies cyclic stresses developed in assemblies exposed to vibratory disturbances. Every element of an assembly that is subjected to alternating loading accumulates stress cycles. Some materials can withstand a greater accumulation than others. When sufficient stress cycles have been accumulated, failure by fatigue is imminent. Stress cycles accumulate rapidly when excitation frequencies coincide with equipment resonances. Two minutes of vibration at a resonance of 150 Hz can produce 18,000 cycles. Component leads, solder connections, connector pins, wiring and assembly attachment interfaces are more prone to fatigue than other equipment elements.

The vibration lifetime of an assembly can be increased by examining only elements along major load paths since it is unlikely that every facet of a configuration can be studied due to schedule and resource constraints. High stresses in component leads and attachments can be reduced by relocating the component closer to and parallel with supported edges of the mounting surface. Bonding or adding damping between the component body and its mounting surface minimizes relative displacement that may alleviate stresses at the attachments. Relative displacements in an assembly are considerably smaller at locations adjacent to subassembly or assembly attachments. Use of additional support or attachments and reduction of unsupported spans will reduce the displacements and stresses in an assembly. Alternatives include the use of gussets, stiffening ribs, increased structural thickness or a reduction in the supported mass.

Tensile or compressive tests on controlled specimens provide a basis for predicting the yield and ultimate stresses at which the materials used in an assembly will likely fail under static load conditions. Allowable stresses considerably below material ultimate quantities are used during equipment design to avoid conditions that are known to develop static failures. Presumably, this approach enables the equipment to perform a long time without obvious damage while exposed to vibratory disturbances. Sudden loss of performance may occur,

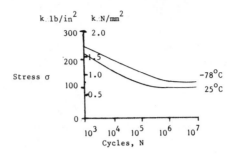

Figure 8.2. Fatigue Strength of 18-8 Stainless Steel.

however, as a result of fractured elements due to thousands of cyclic load reversals. This kind of failure occurs when the cumulative fatigue limit of a material is reached.

Alternating stresses may be induced by cyclic tension, compression or by twist about the axis of a member. For many materials there exists a stress below which fatigue is not likely to occur. This limiting stress, known as the *endurance limit,* is usually determined by means of rotating beam tests using controlled specimens. In equipment, material surface conditions, composition, temperature and geometric characteristics can affect the endurance limit. Knicks, scratches, holes, abrupt transitions and surface finish can contribute to premature fatigue failure. Very small surface disparities in the form of stretch marks or cracks at stressed locations are indications of impending fatigue.

The stress-cycle (S-N) diagram, Figure 8.2, indicates how long a material will resist different levels of cyclic stress. An approximation for predicting the vibration lifetime of an element that experiences stresses of different magnitudes may be determined from

$$\sum \frac{n_i}{N_i} = A \qquad (8.21)$$

where A varies between 0.75 and 1.0 for electronics equipment depending on the quantity of cycles at each stress level, n_i is the number of cycles the element experiences at stress σ_i and N_i is the number of cycles required for failure at σ_i. If the hypotheses represented by (8.21) holds, the element uses up separate fractions of its life at different stress levels. If at a certain stress, 10^6 cycles are required to cause failure, then approximately 25 percent of the element's life would be consumed after 250K cycles at that stress level.

The S-N diagram for different materials which represents the mean of a statistical distribution of failure data at particular stress levels can be approximately determined using

Table 8.2. Fatigue Parameters of Various Materials

		B	
Material	**b**	**(lb/in²-Cycles)**	**(N/mm²-Cycles)**
Aluminum 2024 T6	9.10	2.307×10^{47}	4.938×10^{27}
6061 T4	14.06	7.755×10^{66}	3.156×10^{36}
6061 T6	14.34	4.722×10^{68}	4.769×10^{37}
BeCu Annealed	7.35	3.438×10^{41}	4.461×10^{25}
Hard	7.65	3.572×10^{43}	1.041×10^{27}
Copper	9.57	9.314×10^{46}	1.922×10^{26}
Kovar	15.80	1.893×10^{80}	1.335×10^{46}
Nickel	16.61	2.756×10^{83}	3.452×10^{47}
Silver	6.13	9.116×10^{30}	5.128×10^{17}
Solder 80/20	5.91	9.708×10^{25}	1.632×10^{13}
50/50	10.68	2.242×10^{43}	1.845×10^{20}
37/63	9.85	1.968×10^{41}	1.008×10^{20}
Stainless Stl 301 1/4 Hard	8.74	2.506×10^{46}	3.218×10^{27}
301 3/4 Hard	9.55	1.713×10^{52}	3.904×10^{31}
302 Annealed	10.05	1.778×10^{52}	3.366×10^{30}
302 3/4 Hard	11.48	1.549×10^{62}	2.378×10^{37}
Steel 1020	13.61	1.073×10^{67}	4.101×10^{37}
G10 Glass Epoxy	11.36	7.253×10^{55}	2.023×10^{31}

$$N\sigma^b = B \qquad (8.22)$$

where constants b and B are properties of the material given in Table 8.2.

EXAMPLE 8.4.   The circuit board assembly in Example 8.3 is to be subjected to an endurance test using a 10G vibratory dwell excitation at its fundamental frequency. Determine the maximum allowable weight of the assembly to assure a 60 minute useful vibration lifetime of the copper trace interconnect.

The maximum stress in the copper trace on the surface of the circuit board for an amplification $Q = \sqrt{f_n}$ may be determined using case V of Table 6.5 and the results of Example 8.3, giving

$$\sigma = 11.51 EcGM^{0.75}/L^2 \text{ lb/in}^2 \quad \text{(English)}$$

$$\sigma = 11.51(17.2 \times 10^6)(0.045)(10)M^{0.75}/6^2 = 2.475 \times 10^6 M^{0.75}$$

$$\sigma = 6.06 EcGM^{0.75}/L^2 \text{ N/mm}^2 \quad \text{(metric)}$$

$$\sigma = 6.06(1.186 \times 10^5)(1.14)(10)M^{0.75}/152.4^2 = 352.8 M^{0.75}$$

where from Appendix C, $E_{cu} = 17.6 \times 10^6$ lb/in² $(1.186 \times 10^5$ N/mm²). To assure a useful life of 60 minutes (3600 s), (8.21) gives

$$\frac{n}{N} = \frac{tf_n}{N} = 0.75$$

Recasting this expression using (8.22) provides a relationship for the vibration lifetime of the copper circuit trace as

$$t = \frac{0.75N}{f_n} = \frac{0.75B}{f_n \sigma^b} = 3600$$

Substituting f_n from Example 8.3 and the material fatigue properties from Table 8.2 gives

$$\frac{0.75(9.314 \times 10^{46})M^{0.5}}{4.75(2.475 \times 10^6 M^{0.75})^{9.75}} = 3600$$

or

$$M = (1.878 \times 10^{-20})^{0.1468} = 0.0013 \text{ lb} \cdot \text{s}^2/\text{in}$$

The maximum assembly weight becomes

$$W = Mg = 0.0013(386)$$

$$W = 0.5 \text{ lb (max)}$$

In metric units

$$\frac{0.75(1.922 \times 10^{26})M^{0.5}}{63(352.8 M^{0.75})^{9.75}} = 3600$$

or

$$M = (9.221 \times 10^{-5})^{0.1468} = 0.25 \text{ kg (max)}$$

To assure a useful life of 60 minutes for the assembly, critical electronic component structures and attachment interfaces must be similarly examined.   Q.E.D.

8.2 EQUIPMENT MOTIONS

Aesthetics or constraints imposed by the user, application, heat transfer and function usually dictates the configuration of the electronic equipment and its attachments. These factors also influence equipment static and dynamic behavior and response to loadings developed during transport or operational use.

High aspect ratio configurations with narrow cross-sections and attachments on a small face, Figure 8.1, exhibit torsional motion coupled with bending. The remote location of the center of gravity relative to a plane through the attachments can contribute to the development of large displacements, stresses and

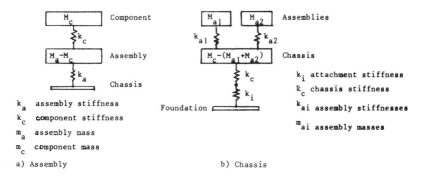

Figure 8.3. Spring Mass Idealizations.

a reduced fundamental frequency. Attachments that are located on a plane through the center of gravity uncouples motions which reduce displacements and stresses.

The method of attachment can decrease the apparent fundamental frequency of an assembly. Attachments and adjacent structure act as a spring support which is in series with the spring characteristics of the assembly. Inoperability, distortion and fracture at equipment attachments may often be traced to insufficient stiffness.

An initial assessment of behavior of an assembly can be obtained using a spring mass representation, Figure 6.26. The idealization is based on the direction of excitation. Each spring mass represents the probable load path to the idealized element. Paths of greater stiffness carry a larger portion of the load. The idealization can also include the attachment mechanization. The development of the spring mass idealization is shown in Figure 8.3.

The displaced shape at a particular frequency is referred to as the *mode shape*. Mode shapes are obtained from the dynamic equation of the system. The mode shapes of the model in Figure 6.28(b) are obtained by writing (6.37) as

$$
\begin{bmatrix}
Ms^2 + k_3 & 0 & Mas^2 - k_3 h/2 \\
0 & Ms^2 + k_1 + k_2 & Mbs^2 + (k_1 - k_2)L/2 \\
Mas^2 - k_3 h/2 & Mbs^2 + (k_1 - k_2)L/2 & [I_{cg} + M(a^2 + b^2)]s^2 + (k_1 + k_2)(L/2)^2 + k_3(h/2)^2
\end{bmatrix}
\begin{Bmatrix}
x_s \\
y_s \\
\theta_s
\end{Bmatrix} = 0
$$

$$
(8.22)
$$

By selecting any one of the degrees of freedom as a reference, one can write all but one of the equations with the selected degree of freedom on the right hand side. Since any equation can be omitted, (8.22) becomes

$$\begin{bmatrix} 0 & Mas^2 - k_3h/2 \\ Ms^2 + k_1 + k_2 & Mbs^2 + (k_1 + k_2)L/2 \end{bmatrix} \begin{Bmatrix} y_s \\ \theta_s \end{Bmatrix} = \begin{bmatrix} Ms^2 + k_3 \\ 0 \end{bmatrix} \{x_s\}$$

$$(8.23)$$

The mode shapes are determined from the transfer functions of (8.23) as

$$\frac{y}{x} = \frac{(Ms^2 + k_3)[Mbs^2 + (k_1 + k_2)L/2]}{(Mas^2 - k_3h/2)(Ms^2 + k_1 + k_2)}$$

$$\frac{\theta}{x} = \frac{(Ms^2 + k_3)}{(Mas^2 - k_3h/2)}$$

where $s^2 = -\omega_i^2$, one of the roots of (8.22). Mode shapes may also be determined by an alternative procedure described in Appendix B.3.

Usually, the behavior of only a few coordinates of an idealized system is desired. The dynamic equations of a system can be transformed to represent the desired degrees of freedom using the matrix reduction method of Appendix A.6. If the system stiffness matrix is partitioned so that the desired coordinates δ_D to be retained are suitably grouped, then (7.23) becomes

$$\begin{Bmatrix} P_D \\ P_R \end{Bmatrix} = \begin{bmatrix} k_{11} & k_{12} \\ k_{21} & k_{22} \end{bmatrix} \begin{Bmatrix} \delta_D \\ \delta_R \end{Bmatrix} \qquad (8.24)$$

Reduction, using the method of Appendix A.6, gives an expression for the applied loads in terms of the displacements at the retained degrees of freedom as

$$P_D = (k_{11} - k_{12}k_{22}^{-1}k_{21})\delta_D \qquad (8.25)$$

None of the structural properties or behavioral characteristics of the original system is lost since the contribution of all elements are retained. Reduction amounts to a transformation

$$\begin{Bmatrix} \delta_D \\ \delta_R \end{Bmatrix} \begin{bmatrix} I \\ -k_{22}^{-1}k_{21} \end{bmatrix} \{\delta_D\}$$

or

$$\{\delta\} = [\mathbf{T}] \{\delta_D\} \tag{8.26}$$

If the kinetic and potential energies may be written as

$$KE = \tfrac{1}{2}\dot{\delta}^T M \dot{\delta} \tag{8.27a}$$

$$U = \tfrac{1}{2}\delta^T K \delta$$

then the application of (8.26) gives

$$KE = \tfrac{1}{2}\dot{\delta}_D^T T^T M T \dot{\delta}_D \tag{8.27b}$$

$$U = \tfrac{1}{2}\delta_D^T T^T K T \delta_D$$

Expressing (8.25) using the results obtained in (8.27b) or $\overline{\mathbf{K}} = \mathbf{T}^T \mathbf{K} \mathbf{T}$ gives

$$P_D = (\mathbf{T}^T \mathbf{K} \mathbf{T}) \delta_D \tag{8.28}$$

If the partitioned mass matrix corresponding to (8.24) is expressed as

$$[\mathbf{M}] = \begin{bmatrix} m_{11} & m_{12} \\ m_{21} & m_{22} \end{bmatrix}$$

then the reduced mass matrix using $\overline{\mathbf{M}} = \mathbf{T}^T \mathbf{M} \mathbf{T}$ from (8.27b) becomes

$$\overline{\mathbf{M}} = m_{11} - k_{12} k_{22}^{-1} m_{21} - m_{12} k_{22}^{-1} k_{21} + k_{12} k_{22}^{-1} m_{22} k_{22}^{-1} k_{21} \tag{8.29}$$

It is seen that combinations of mass and stiffness coefficients appear. The system frequencies and modal displacements are determined using the reduced mass $\overline{\mathbf{M}}$ and stiffness $\overline{\mathbf{K}}$ matrices from

$$[-\omega^2 \overline{\mathbf{M}} + \overline{\mathbf{K}}] \{\delta_D\} = 0 \tag{8.30}$$

The modal displacements of the reduced or discarded coordinates may be recovered from

$$\{\delta_R\} = - (-\omega^2 m_{22} + k_{22})^{-1} (-\omega^2 m_{21} + k_{21}) \{\delta_D\}$$

or

$$\{\delta_R\} = - (k_{22}^{-1} + \omega^2 k_{22}^{-1} m_{22} k_{22}^{-1}) (-\omega^2 m_{21} + k_{21}) \{\delta_D\} \tag{8.31}$$

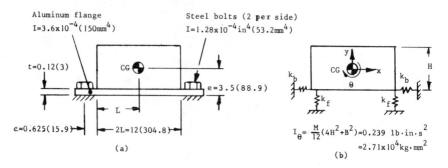

Figure 8.4. Influence of Attachments. (a) Assembly; (b) Model.

EXAMPLE 8.5.   Determine the in-plane resonant frequencies and mode shapes of the flange-mounted 15 lb (6.81 kg) assembly in Figure 8.4(a).

The influence of the attachments may be determined using the rigid body model shown in Figure 8.4(b). Motion along the y axis is uncoupled due to symmetry. The vertical fundamental frequency is given by

$$f_y = \frac{1}{2\pi} \sqrt{\frac{2k_f}{M}} \tag{8.32}$$

where k_f is the stiffness of the flange support on one end of the assembly. Using configuration parameters from Figure 8.4(a), the flange and bolt stiffness is determined as

$$k_f = \frac{3EI}{c^3} \quad \text{(flange)}$$

$$k_f = \frac{3(10^7)3.6 \times 10^{-4}}{0.625^3} = 4.424 \times 10^4 \text{ lb/in} \quad \text{(English)}$$

$$k_f = \frac{3(6.89 \times 10^4)150}{15.9^3} = 7.713 \times 10^3 \text{ N/mm} \quad \text{(metric)}$$

$$k_b = \frac{3EI}{t^3} \text{(bolt)}$$

$$k_b = \frac{3(30 \times 10^6)1.28 \times 10^{-4}}{0.12^3} = 6.66 \times 10^6 \text{ lb/in}$$

$$k_b = \frac{3(2.07 \times 10^5)53.2}{3^3} = 1.22 \times 10^6 \text{ N/mm}$$

where $E_{al} = 10^7$ lb/in^2 (6.89 $\times$ 10^4 N/mm^2) and $E_{stl} = 30 \times 10^6$ lb/in^2 (2.07 $\times$ 10^5 N/mm^2). Substituting quantities in (8.32) gives the vertical frequency as

$$f_y = \frac{1}{2\pi} \left[\frac{2(4.42 \times 10^4)}{15/386} \right]^{1/2} = 240 \text{ Hz} \quad \text{(English)}$$

$$f_y = \frac{1}{2\pi} \left[\frac{2(7.713 \times 10^3)1000 \text{ mm/m}}{6.81} \right]^{1/2} = 240 \text{ Hz} \quad \text{(metric)}$$

Even though the y motion is uncoupled from the x and θ motions, the freedom to vibrate in all modes simultaneously is not restricted. Following methods of Section 6.3, the dynamic equation for x and θ motions due to a foundation acceleration excitation $\ddot{U}$ becomes

$$\begin{bmatrix} M & 0 \\ 0 & I_\theta \end{bmatrix} \begin{Bmatrix} \ddot{x} \\ \ddot{\theta} \end{Bmatrix} + \begin{bmatrix} 2k_b & 2k_b e \\ 2k_b e & 2(k_b e^2 + k_f c^2) \end{bmatrix} \begin{Bmatrix} x \\ \theta \end{Bmatrix} = \begin{Bmatrix} -\ddot{U}M \\ 0 \end{Bmatrix} \qquad (8.33)$$

Fundamental frequencies are the roots of the characteristic equation which may be developed from the s form of (8.33):

$$\begin{bmatrix} Ms^2 + 2k_b & 2k_b e \\ 2k_b e & I_\theta s^2 + 2(k_b e^2 + k_f c^2) \end{bmatrix} \begin{Bmatrix} x_s \\ \theta_s \end{Bmatrix} = 0 \qquad (8.34)$$

Following Appendix B.3, the characteristic equation of (8.34) becomes

$$MI_\theta s^4 + [2k_b I_\theta + 2(k_b e^2 + k_f c^2)M] s^2 + 4k_b k_f c^2 = 0 \qquad (8.35)$$

Substituting quantities in (8.35) and solving as a quadratic in terms of s^2 gives

$$s_1^2 = -4.41 \times 10^6$$
$$s_2^2 = -1.03 \times 10^9$$

Since $s^2 = -\omega^2$ and $\omega = 2\pi f$, the resonant frequencies become

$$f_1 = 335 \text{ Hz}$$
$$f_2 = 5117 \text{ Hz}$$

The mode shapes are determined from the first equation in (8.34) as

$$\frac{\theta}{x} = \frac{M\omega^2 - 2k_b}{2k_b e} \qquad (8.36)$$

At 335 Hz (8.36) becomes

$$\frac{\theta}{x} = \frac{15(4.41 \times 10^6)/386 - 2(6.66 \times 10^6)}{2(6.66 \times 10^6)3.5} = -0.282 \text{ rad/in} \quad \text{(English)}$$

$$\frac{\theta}{x} = \frac{6.81(4.41 \times 10^6) - 2(1.22 \times 10^6)1000 \text{ mm/m}}{2(1.22 \times 10^6)(88.9)1000 \text{ mm/m}} = -0.011 \text{ rad/mm} \quad \text{(metric)}$$

If the assembly experiences a maximum displacement $x = 0.01$ in (0.25 mm), the rotation about the CG would be $\theta = 0.0028$ rad(max). If only the displacement along the x axis is of interest, the reduced stiffness, from (8.25) or Appendix A.6 and the reduced mass from (8.29) using (8.33) become

$$\overline{K} = 2k_b - \frac{2(k_b e)^2}{k_b e^2 + k_f c^2}$$

$$\overline{K} = 2(6.66 \times 10^6) - \frac{2[(6.66 \times 10^6)3.5]^2}{(6.66 \times 10^6)3.5^2 + (4.424 \times 10^4)6^2}$$
$$= 2.55 \times 10^5 \text{ lb/in} \quad \text{(English)}$$

$$\overline{K} = 2(1.22 \times 10^6) - \frac{2[(1.22 \times 10^6)88.9]^2}{(1.22 \times 10^6)88.9^2 + (7.713 \times 10^3)152.4^2}$$
$$= 4.45 \times 10^4 \text{ N/mm} \quad \text{(metric)}$$

$$\overline{M} = M + \frac{(k_b e)^2 I_\theta}{(k_b e^2 + k_f c^2)^2}$$

$$\overline{M} = \frac{15}{386} + \frac{[(6.66 \times 10^6)3.5]^2 0.2396}{[(6.66 \times 10^6)3.5^2 + (4.424 \times 10^4)6^2]^2} = 0.0577 \text{ lb} \cdot \text{s}^2/\text{in} \quad \text{(English)}$$

$$\overline{M} = 6.81 + \frac{[(1.22 \times 10^6)88.9]^2 2.71 \times 10^4}{[(1.22 \times 10^6)88.9^2 + (7.713 \times 10^3)152.4^2]^2} = 10.1 \text{ kg} \quad \text{(metric)}$$

The fundamental frequency along the x axis is

$$f_x = \frac{1}{2\pi} \sqrt{\frac{\overline{K}}{\overline{M}}}$$

$$f_x = \frac{1}{2\pi} \sqrt{\frac{2.55 \times 10^5}{0.0577}} = 335 \text{ Hz} \quad \text{(English)}$$

$$f_x = \frac{1}{2\pi} \sqrt{\frac{(4.45 \times 10^4)1000 \text{ mm/m}}{10.1}} = 334 \text{ Hz} \quad \text{(metric)} \qquad \text{Q.E.D.}$$

Factors that Influence Coordinate Selection. The coordinates used to represent the geometry and mass properties of the idealized equipment describe

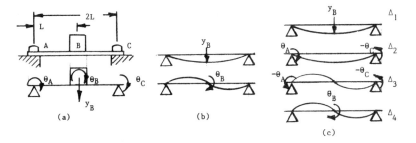

Figure 8.5. Symmetrical and Asymmetrical Motions. (a) Assembly and idealization; (b) Mode shapes due to y_B and θ_B; (c) Mode shapes in terms of coordinate displacements.

the displacements due to applied static loads in addition to the mode shapes corresponding to the resonant frequencies. The number of mode shapes that can be obtained from an idealization corresponds to the quantity of coordinates with inertia. As an example, consider a six degree of freedom idealization with inertial characteristics defined at only two of the coordinates. In this case, only two resonant frequencies and their corresponding mode shapes can be determined.

Inertia coordinates may be used to visualize the mode shapes that are developed by an idealization. A mode shape approximates the displaced shape at resonance. A knowledge of the displaced shape provides a basis for the design engineer to alter the observed behavior. The mode shape indicates where additional stiffness or compliance would be effective in achieving a desired displacement or eliminating an undesirable behavioral characteristic. The mode shapes shown in Figure 8.5(b) are the result of inertia coordinates y_B and θ_B in the four degree of freedom model in Figure 8.5(a). Excitations at frequencies different from the resonances that correspond to the shapes in Figure 8.5(b) result in a response which is composed of a combination of both motions.

The dynamic equation representing the motions developed by an idealization may be obtained directly by the flexibility method, Section 7.2, or by the stiffness method, Section 7.3. The dynamic equation for the model in Figure 8.5(a) based on the stiffness matrix (7.59) is

$$
\begin{bmatrix} I_A & & & 0 \\ & I_B & & \\ & & I_C & \\ 0 & & & M_B \end{bmatrix}
\begin{Bmatrix} \ddot{\theta}_A \\ \ddot{\theta}_B \\ \ddot{\theta}_C \\ \ddot{y}_B \end{Bmatrix}
+ \frac{2EI}{L}
\begin{bmatrix} 2 & 1 & 0 & -3/L \\ 1 & 4 & 1 & 0 \\ 0 & 1 & 2 & 3/L \\ -3/L & 0 & 3/L & 12/L^2 \end{bmatrix}
\begin{Bmatrix} \theta_A \\ \theta_B \\ \theta_C \\ y_B \end{Bmatrix}
= 0
$$

$$(8.37)$$

Since $I_A = I_C = 0$, the reduction method of Appendix A.6, or alternatively (8.30), gives

$$\begin{bmatrix} I_B & 0 \\ 0 & M_B \end{bmatrix} \begin{Bmatrix} \ddot{\theta}_B \\ \ddot{y}_B \end{Bmatrix} + \frac{EI}{L} \begin{bmatrix} 6 & 0 \\ 0 & 6/L^2 \end{bmatrix} \begin{Bmatrix} \theta_B \\ y_B \end{Bmatrix} = 0 \qquad (8.38)$$

Equation (8.38) describes the uncoupled translational and rotational motions shown in Figure 8.5(b) with the corresponding frequencies expressed as

$$f_{yB} = \frac{1}{2\pi} \sqrt{\frac{6EI}{M_B L^3}} \; ; \quad f_{\theta B} = \frac{1}{2\pi} \sqrt{\frac{6EI}{I_B L}}$$

Configurations with geometric and inertial symmetry develop both symmetrical and asymmetrical motions. These motions are illustrated in Figure 8.5(c) for the coordinate displacements of Figure 8.5(a) acting individually or in pairs. A transformation matrix [C] may be developed that relates the symmetric or asymmetric shape to the participating system coordinates. Based on the coordinate sequence from (8.37), the transformation to the mode shapes in Figure 8.5(c) becomes

$$\begin{Bmatrix} \theta_A \\ \theta_B \\ \theta_C \\ y_B \end{Bmatrix} = \begin{bmatrix} 0 & 1 & -1 & 0 \\ 0 & 0 & 0 & 1 \\ 0 & -1 & -1 & 0 \\ 1 & 0 & 0 & 0 \end{bmatrix} \begin{Bmatrix} \Delta_1 \\ \Delta_2 \\ \Delta_3 \\ \Delta_4 \end{Bmatrix}$$

or

$$\{\delta\} = [C]\{\Delta\} \qquad (8.39)$$

The inertia and stiffness matrices in terms of the model shapes are obtained from (8.39) as

$$[\overline{K}] = [C]^T [K] [C] \qquad (8.40)$$
$$[\overline{M}] = [C]^T [M] [C]$$

Applying (8.40) to (8.37) gives

$$\begin{bmatrix} M_B & 0 & \vdots & 0 & 0 \\ 0 & I_A + I_C & \vdots & I_C - I_A & 0 \\ \cdots & \cdots & \cdots & \cdots & \cdots \\ 0 & I_C - I_A & \vdots & I_A + I_C & 0 \\ 0 & 0 & \vdots & 0 & I_B \end{bmatrix} \begin{Bmatrix} \ddot{\Delta}_1 \\ \ddot{\Delta}_2 \\ \ddot{\Delta}_3 \\ \ddot{\Delta}_4 \end{Bmatrix}$$

$$+ \frac{2EI}{L} \begin{bmatrix} 12/L^2 & -6L & \vdots & \\ -6/L & 4 & \vdots & 0 \\ \cdots & \cdots & \cdots & \cdots \\ & & \vdots & 4 & -2 \\ & 0 & \vdots & -2 & 4 \end{bmatrix} \begin{Bmatrix} \Delta_1 \\ \Delta_2 \\ \Delta_3 \\ \Delta_4 \end{Bmatrix} = 0$$

which reduces to (8.38) for $I_A = I_c = 0$. The transformation (8.39) uncouples the dynamic equations for conditions of symmetry, reducing the effort required to obtain the system frequencies.

Evaluating Equipment Behavior.   The dynamic equation is composed of the stiffness matrix, developed in the same way as for static analysis, Section 7.3, in addition to a mass and damping matrix. The mass matrix is developed using the inertial characteristics of the actual system apportioned to the coordinates in the idealization. Coordinates without inertia are reduced out of the mass and stiffness matrices using the transformation (8.29) and (8.28) or methods of Appendix A.6. The damping matrix for the retained coordinates can be derived in a manner analogous to a spring mass system

$$[c] = \frac{1}{Q} [M^{1/2}KM^{1/2}]^{1/2} \tag{8.41}$$

in which $\sqrt{M}$ is a term by term operation.

EXAMPLE 8.6.   Component placement in an assembly of mass m can produce any of the symmetric mass distributions shown in Figure 8.6(a). Neglecting damping, determine the three translational frequencies and mode shapes of the configuration representing a uniform distribution. Illustrate the frequency response of this configuration in terms of the transmissibility (transfer function) $y_1/\ddot{U}$ for support excitation. Compare the resonances of each distribution.

The flexibility matrix in terms of the three translational coordinates can be obtained directly using the methods of Section 7.2:

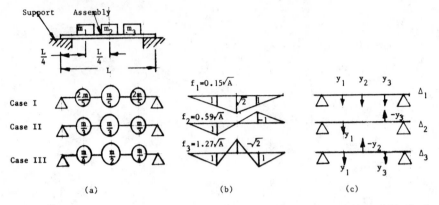

Figure 8.6. Dynamic Motions for the Assembly in Example 8.6. (a) Assembly and idealized distributions; (b) Mode shapes for Case II; (c) Transformed mode shape coordinates.

$$[\mathbf{F}] = \frac{L^3}{768EI} \begin{matrix} & (P_1) & (P_2) & (P_3) \\ & \begin{bmatrix} 9 & 11 & 7 \\ 11 & 16 & 11 \\ 7 & 11 & 9 \end{bmatrix} \end{matrix} \qquad (8.42)$$

Reduction methods of Appendix A.6 must be applied to the stiffness matrix determined by using the methods of Section 7.3, to obtain

$$[\mathbf{K}] = \frac{192EI}{7L^3} \begin{matrix} & (y_1) & (y_2) & (y_3) \\ & \begin{bmatrix} 23 & -22 & 9 \\ -22 & 32 & -22 \\ 9 & -22 & 23 \end{bmatrix} \end{matrix} \qquad (8.43)$$

since the initial development contains five rotations in addition to the three translations. As an alternative, (8.43) could have been obtained from (7.24) using (8.42) and the inversion methods of Appendix A.7.

The dynamic equation for the idealized uniform distribution, case II of Figure 8.6(a) becomes

$$\begin{bmatrix} 1 & 0 & 0 \\ 0 & 1 & 0 \\ 0 & 0 & 1 \end{bmatrix} \begin{Bmatrix} \ddot{y}_1 \\ \ddot{y}_2 \\ \ddot{y}_3 \end{Bmatrix} + A \begin{bmatrix} 23 & -22 & 9 \\ -22 & 32 & -22 \\ 9 & -22 & 23 \end{bmatrix} \begin{Bmatrix} y_1 \\ y_2 \\ y_3 \end{Bmatrix} = 0 \qquad (8.44a)$$

where $A = 3(192EI)/7mL^3$. Equation (8.44a) may be written using the s operator as

$$\begin{bmatrix} s^2 + 23A & -22A & 9A \\ -22A & s^2 + 32A & -22A \\ 9A & -22A & s^2 + 23A \end{bmatrix} \begin{Bmatrix} y_{1s} \\ y_{2s} \\ y_{3s} \end{Bmatrix} = 0 \quad (8.44b)$$

The characteristic equation of (8.44b) based on the methods of Appendix B.3, becomes

$$s^6 + 78As^4 + 952A^2s^2 + 784A^3 = 0 \quad (8.45)$$

The roots of (8.45) give the frequencies as

$$\begin{aligned} s_1^2 &= -0.8873A \\ s_2^2 &= -14.0A \\ s_3^2 &= -63.1127A \end{aligned} \quad (8.46a)$$

Sine $s^2 = -\omega^2$ and $\omega = 2\pi f$ (8.46a) may be written as

$$\begin{aligned} f_1 &= 0.150\sqrt{A} \\ f_2 &= 0.596\sqrt{A} \quad \text{Hz} \\ f_3 &= 1.265\sqrt{A} \end{aligned} \quad (8.46b)$$

The mode shapes are developed from an equation analogous to (8.23) developed from (8.44b) as

$$\begin{bmatrix} s^2 + 32A & -22A \\ -22A & s^2 + 23A \end{bmatrix} \begin{Bmatrix} y_{2s} \\ y_{3s} \end{Bmatrix} = \begin{Bmatrix} 22A \\ -9A \end{Bmatrix} y_{1s} \quad (8.47)$$

Solution of (8.47) at values of s_i^2 from (8.46a) gives

$$\begin{array}{ccc} (s_1^2) & (s_2^2) & (s_3^2) \end{array}$$

$$\frac{y_{2s_i}}{y_{1s_i}} = \frac{22As_i^2 + 308A^2}{s_i^4 + 55As_i^2 + 252A^2} = \quad \sqrt{2} \quad \ 0 \quad -\sqrt{2}$$

$$\frac{y_{3s_i}}{y_{1s_i}} = \frac{-9As_i^2 + 196A^2}{s_i^4 + 55As_i^2 + 252A^2} = \quad 1 \quad -1 \quad \ 1$$

The mode shapes in Figure 8.6(b) are the result of setting $y_{1s} = 1$.

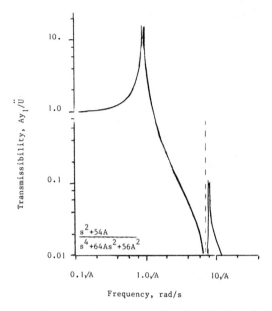

Figure 8.7. Frequency Response of y_1 Coordinate in Example 8.6.

The transfer function in terms of support acceleration $\ddot{U}$ may be derived from (8.44) using methods of Appendix B as

$$\frac{y_1}{\ddot{U}} = \frac{s^4 + 68As^2 + 756A^2}{s^6 + 78As^4 + 952A^2s^2 + 784A^3} \tag{8.48a}$$

The frequency response of (8.48a) developed using the unfactored graphical technique of Appendix B.6 is illustrated in Figure 8.7. Since the response depicted by Figure 8.7 indicates that support motion along the y axis does not excite the second mode, only f_1 and f_3 of (8.46b) will contribute to the dynamic behavior of the assembly. A sub-assembly with a resonance near f_2 would experience an input of approximately $0.1\ddot{U}$. This insight is a favored attribute of the frequency response and transfer function development.

The second mode in (8.48a) would be excited by support rotations however, or by translations along the plane of the assembly. Configuration symmetry is the predominant factor in uncoupling the second mode for support motion along the y axis. Since the transfer function provides a basis for evaluating the response of the assembly to random vibration and shock, the removal of uncoupled modes results in a simplification of (8.48a) to

$$\frac{y_1}{\ddot{U}} = \frac{s^2 + 54A}{s^4 + 64As^2 + 56A^2} \tag{8.48b}$$

The symmetry of the configuration of Figure 8.6(a) suggests an alternative to the determination of frequencies (8.46) from (8.44).

Equation (8.39), which transforms modal coordinates Δ_i to the idealized coordinates y_i may be obtained using Figure 8.6(c) as

$$
\begin{Bmatrix} y_1 \\ y_2 \\ y_3 \end{Bmatrix} = \begin{bmatrix} 1 & 1 & 1 \\ 1 & 0 & -1 \\ 1 & -1 & 1 \end{bmatrix} \begin{Bmatrix} \Delta_1 \\ \Delta_2 \\ \Delta_3 \end{Bmatrix}
$$

or

$$\{y\} = [C]\{\Delta\} \tag{8.49}$$

The transformed dynamic equation using $[\overline{K}]$ and $[\overline{M}]$ from (8.40) becomes

$$
\begin{bmatrix} 3 & 0 & 1 \\ 0 & 2 & 0 \\ 1 & 0 & 3 \end{bmatrix} \begin{Bmatrix} \ddot{\Delta}_1 \\ \ddot{\Delta}_2 \\ \ddot{\Delta}_3 \end{Bmatrix} + A \begin{bmatrix} 8 & 0 & 32 \\ 0 & 28 & 0 \\ 32 & 0 & 184 \end{bmatrix} \begin{Bmatrix} \Delta_1 \\ \Delta_2 \\ \Delta_3 \end{Bmatrix} = 0 \tag{8.50}
$$

The frequency of the uncoupled second mode Δ_2, Figure 8.6(c) is determined from (8.50) as

$$f_2 = \frac{1}{2\pi} \sqrt{\frac{28A}{2}} = 0.596\sqrt{A}$$

The frequencies of the first and third modes are determined from

$$
\begin{bmatrix} 3 & 1 \\ 1 & 3 \end{bmatrix} \begin{Bmatrix} \ddot{\Delta}_1 \\ \ddot{\Delta}_3 \end{Bmatrix} + A \begin{bmatrix} 8 & 32 \\ 32 & 184 \end{bmatrix} \begin{Bmatrix} \Delta_1 \\ \Delta_3 \end{Bmatrix} = 0
$$

The characteristic equation of this relationship compares to that of (8.48b)

$$s^4 + 64As^2 + 56A^2 = 0$$

giving results that are also identical to those of (8.46b).

Using modal coordinates, the resonant frequencies of each mass distribution in Figure 8.6(a) may be summarized as

	Case I	Case II	Case III
f_1	$0.158\sqrt{A}$	$0.150\sqrt{A}$	$0.141\sqrt{A}$
f_2	$0.544\sqrt{A}$	$0.596\sqrt{A}$	$0.688\sqrt{A}$
f_3	$1.416\sqrt{A}$	$1.265\sqrt{A}$	$1.266\sqrt{A}$

Lower modal frequencies result when a greater portion of the assembly mass is lumped at the coordinates corresponding to a mode shape. Q.E.D.

The techniques discussed may be applied to idealized configurations with any number of degrees of freedom.

8.3 RANDOM VIBRATION

Vibrations in electronic equipment may be the result of motion in the supporting structure or it may be due to the impingement of pressure or acoustic waves on the enclosure. If this motion varies in strength and direction in an unpredictable manner, it may be defined as *random vibration*. The evaluation of the response to random disturbances deals with the relationships between the statistical characteristics of motion and the dynamic properties of configuration.

Periodic and harmonic disturbances are characteristic of deterministic vibration since an instantaneous value for displacement, velocity or acceleration can be predicted at any time. Nonperiodic motions are a basic characteristic of random vibration. This type of motion may exhibit either narrow band or broadband characteristics. A narrow band disturbance transmits loads that occupy a narrow range of frequencies, whereas a broadband disturbance occupies a wide range of frequencies usually encompassing the major structural resonances of equipment. The loadings developed by a random disturbance are usually quantified in terms of a mean square value y^2 which characterizes the excitation providing a basis for statistical evaluation.

The Excitation Spectrum. To determine the mean square value of a random disturbance y^2, the area under the squared $y(t)$ curve divided by the record length T is computed from $t = 0$ to $t = T$. Accuracy increases with increasing T. The mean square value expressed in the limit as $T \to \infty$, even though this condition cannot be achieved empirically, becomes

$$y^2 = \lim_{T \to \infty} \frac{1}{T} \int_0^T y^2(t) \, dt \tag{8.51}$$

If the disturbance is processed by a spectrum analyzer and the fraction of the mean square value Δy^2 is recorded for frequencies in a narrow range Δf, the power spectral density for $f_c \pm \Delta f/2$ becomes

$$W_y(f) = \Delta y^2 / \Delta f \tag{8.52}$$

As Δf approaches zero, the spectrum becomes continuous and (8.52) may be written as

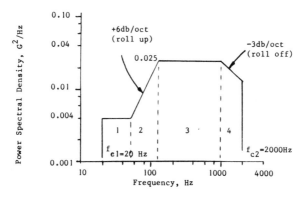

Figure 8.8. Random Vibration Spectrum.

$$W_y(f) = \frac{dy^2}{df}$$

or

$$dy^2 = W_y(f)df \qquad (8.53)$$

The power spectral density of a disturbance may be determined empirically or artificially by enveloping discrete spectra with straight line segments as shown in Figure 8.8.

At f_{c1} and f_{c2}, the cutoff frequencies in Figure 8.8, $W_y(f)$ is zero for $f < f_{c1}$ and $f > f_{c2}$. $W_y(f)$ along rollup or rolloff slopes which are drawn for m db/oct may be determined in terms of a known quantity along the slope using

$$\text{Rollup:} \quad W_{y2} = W_{y1} \left(\frac{f_2}{f_1}\right)^{m/3} \qquad (8.54a)$$

$$\text{Rolloff:} \quad W_{y2} = W_{y1} \left(\frac{f_1}{f_2}\right)^{m/3} \qquad (8.54b)$$

The overall mean square value y^2 may be determined by integrating (8.53) over the limits of the spectrum

$$y^2 = \int_{f_{c1}}^{f_{c2}} W_y(f) \, df \qquad (8.55)$$

which evaluates the area under the spectrum between the prescribed limits. The area under straight line segments drawn on log-log ordinates for various

Table 8.3. Mean Square Quantities
(for linear segments on log-log PSD graph)

Mean Square Quantity	Power Spectral Density, W	Transition Slope β, db/oct	Mean Square Relationship $m \neq \beta$	Mean Square Relationship $m = \beta$
G^2	G^2/Hz	-3	A_6B_2	$\frac{1}{2}B_7 \ln (f_2/f_1)$
	v^2/Hz	-9	$A_{10}B_1$	$\frac{1}{2}A_2B_6 \ln (f_2/f_1)$
	d^2/Hz	-15	$A_{12}B_0$	$\frac{1}{2}A_4B_5 \ln (f_2/f_1)$
v^2	G^2/Hz	3	A_7B_3	$\frac{1}{2}A_0B_8 \ln (f_2/f_1)$
	v^2/Hz	-3	A_6/B_2	$\frac{1}{2}B_7 \ln (f_2/f_1)$
	d^2/Hz	-9	$A_{11}B_1$	$\frac{1}{2}A_5B_6 \ln (f_2/f_1)$
d^2	G^2/Hz	9	A_8B_4	$\frac{1}{2}A_1B_9 \ln (f_2/f_1)$
	v^2/Hz	3	A_9B_3	$\frac{1}{2}A_3B_8 \ln (f_2/f_1)$
	d^2/Hz	-3	A_6B_2	$\frac{1}{2}B_7 \ln (f_2/f_1)$

Parameters and units

$A_0 = (g/2\pi)^2$

$A_1 = (g/(2\pi)^2)^2$

$A_2 = (2\pi/g)^2$

$A_3 = (2\pi)^{-2}$

$A_4 = ((2\pi)^2/g)^2$

$A_5 = (2\pi)^2$

$A_6 = 3/(m+3)$

$A_7 = 3A_0/(m-3)$

$A_8 = 3A_1/(m-9)$

$A_9 = 3A_3/(m-3)$

$A_{10} = 3A_2/(m+9)$

$A_{11} = 3A_5/(m+9)$

$A_{12} = 3A_4/(m+15)$

$B_0 = W_2 f^5 - W_1 f_1^5$

$B_1 = W_2 f_2^3 - W_1 f_1^3$

$B_2 = W_2 f_2 - W_1 f_1$

$B_3 = W_2/f_2 - W_1/f_1$

$B_4 = W_2/f_2^3 - W_1/f_1^3$

$B_5 = W_2 f_2^5 + W_1 f_1^5$

$B_6 = W_2 f_2^3 + W_1 f_1^3$

$B_7 = W_2 f_2 + W_1 f_1$

$B_8 = W_2/f_2 + W_1/f_1$

$B_9 = W_2/f_2^3 + W_1/f_1^3$

$g = 386.4$ in/s^2, $(9.81$m/s$^2)$

$v = $ in/s, (m/s)

$d = $ in, (m)

$f = $ Hz

power spectral density (PSD) quantities developed from (8.55) are listed in Table 8.3.

The mean square acceleration and displacement for a flat PSD ($m = 0$; $W_1 = W_2 = W$) expressed in G^2/Hz, for example, are obtained from Table 8.3 as

$$G^2 = A_6B_2 = W(f_2 - f_1) \tag{8.56a}$$

$$d^2 = A_8B_4 = \frac{W}{3}\left[\frac{g}{(2\pi)^2}\right]^2 (1/f_2^3 - 1/f_1^3) \tag{8.56b}$$

The mean square velocity for a sloped PSD rolling up at $+3$ db/oct ($m = 3$) expressed in G^2/Hz may be obtained similarly from Table 8.3 as

$$v^2 = \frac{1}{2} A_0 B_8 \ln (f_2/f_1) = \frac{1}{2} \left(\frac{g}{2\pi}\right)^2 (W_2/f_2 + W_1/f_1) \ln (f_2/f_1)$$

EXAMPLE 8.7.   Determine the overall rms acceleration; G_{rms}, for the random vibration spectrum shown in Figure 8.8.

The area under the spectrum may be developed as the sum of the four areas shown in Figure 8.8. Intercepts for the sloped regions may be determined using (8.54). For some spectrums intercepts may be determined directly from the graphical representation by considering an octave as a factor of two change in frequency and ± 3 db as a factor of two change in magnitude.

The intercept of the -3 db/oct rolloff at 2000 Hz, a change of one octave from $0.025G^2/Hz$ at 1000 Hz is therefore $\frac{1}{2} \times 0.025$ or $0.0125G^2/Hz$. Slopes which are multiples of 3 db/oct are treated accordingly. The frequency at the intercept of $0.025G^2/Hz$ and $+6$ db/oct rollup from (8.54a) is

$$f_2 = f_1 \left(\frac{W_2}{W_1}\right)^{3/m} = 50 \left(\frac{0.025}{0.004}\right)^{3/6}$$

$$f_2 = 125 \text{ Hz}$$

The mean square value of each area in Figure 8.8 is determined using the appropriate relationship from Table 8.3. The results are given in tabular form as follows:

Area	G^2 Relationship	Freq. Range, Hz f_1	f_2	PSD or Slope	G^2 Quantity
1	$A_6 B_2 = W(f_2 - f_1)$	20	50	$0.004G^2/Hz$	0.12
2	$A_6 B_2 = (W_2 f_2 - W_1 f_1)/3$	50	125	$+6$ db/oct	0.98
3	$A_6 B_2 = W(f_2 - f_1)$	125	1000	$0.025G^2/Hz$	21.88
4	$\frac{1}{2} B_7 \ln (f_2/f_1)$ $= \frac{1}{2}(W_2 f_2 + W_1 f_1) \ln (f_2/f_1)$	1000	2000	-3 db/oct	17.33
				Total $G^2 =$	40.31

The overall rms acceleration becomes

$$G = \sqrt{40.31} = 6.35 G_{rms}$$

The rms displacement D_{rms}, can be obtained using an analogous approach as $D = 0.00414$ in rms $(1.05 \times 10^{-4}$m rms$)$.   Q.E.D.

Characteristics of Equipment Response.   The random vibration spectrum of Figure 8.8 and the acoustic spectrum of Figure 8.16 characterize the nature of nonperiodic excitation. This type of excitation may be used as part of a stress screening program which can increase the reliability of electronic equipment. The response of a linear system can be thought of as a linear function of the excitation. A random disturbance simultaneously excites all the equipment resonances for which there is input. This provides an effective means of detecting defective components and workmanship-related weaknesses such as loose parts or hardware, intermittent connections and loose particles in components. This examination can be performed without affecting the equipment service life provided the PSD and duration of excitation are suitably selected.

The relationship between PSD levels and duration discussed in Chapter 1 may be used to modify the characteristics of a disturbance to obtain a more acceptable format. Those relationships were developed on the basis of comparable equipment stress levels and accumulated fatigue damage. Since different assemblies and subassemblies within an enclosure are susceptible to different acceleration levels, the stress screen PSD could be tailored to develop suitable stress levels throughout the equipment.

The transmission characteristics for any location in the equipment is defined by its transfer function H. Assuming that the dynamic characteristics of the equipment do not change with time, the response PSD for a particular location becomes

$$W_x = |H(f)|^2 W_y(f) \qquad (8.57)$$

The mean square response may be obtained by integrating over the limits of the frequency spectrum as

$$x^2 = \int_{f_{c1}}^{f_{c2}} |H(f)|^2 W_y(f) \, df \qquad (8.58)$$

The response PSD (8.57) exhibits pronounced amplifications at frequencies corresponding to the resonances of the assembly. Except in the neighborhood of the resonances, the contribution to the response at distant frequencies is quite small. Using this characteristic, the mean square response (8.58) for the coordinate represented by H becomes

$$x^2 = \int_0^\infty H(\omega)^2 W_y(\omega) \, d\omega \qquad (8.59)$$

Transfer functions relating coordinate response to input excitation may be developed for equipment idealizations using the methods of Appendix B. Equation (8.59) may be evaluated without performing integration by using a system of equations developed from the coefficients of the numerator and denominator of the transfer function expressed in the s plane. The squared magnitude of the transfer function

$$H(s) = \frac{b(s)}{a(s)} = \frac{\sum_{k=0}^{n} b_k s^k}{\sum_{k=0}^{\ell} a_k s^k} \quad n < \ell \qquad (8.60)$$

becomes

$$|H(s)|^2 = \frac{b(s)b(-s)}{a(s)a(-s)} \qquad (8.61a)$$

which is composed of only even powers of s. Equation (8.61a) may be expanded as partial fractions giving

$$|H(s)|^2 = \frac{c(s)}{a(s)} + \frac{c(-s)}{a(-s)} = 2\frac{c(s)}{a(s)} \qquad (8.61b)$$

The mean square response for a constant excitation PSD, W_0, using (8.59), (8.61b) and the properties of the Laplace transform as $s \to \infty$ becomes

$$x^2 = \frac{c_{\ell-1}W_0}{4a_\ell} \qquad (8.62)$$

where $c_{\ell-1}$ is obtained from

$$\begin{bmatrix} a_0 & 0 & 0 & 0 & 0 & \cdots \\ a_2 & -a_1 & a_0 & 0 & 0 & \cdots \\ a_4 & -a_3 & a_2 & -a_1 & 0 & \cdots \\ \cdot \\ \cdot \\ \cdot \end{bmatrix} \begin{Bmatrix} c_0 \\ c_1 \\ c_2 \\ \cdot \\ \cdot \\ \cdot \\ c_{\ell-1} \end{Bmatrix} = \begin{bmatrix} b_0 & 0 & 0 & 0 & 0 & 0 & \cdots \\ b_2 & -b_1 & b_0 & 0 & 0 & 0 & \cdots \\ b_4 & -b_3 & b_2 & -b_1 & b_0 & 0 & \cdots \\ \cdot \\ \cdot \\ \cdot \end{bmatrix} \begin{Bmatrix} b_0 \\ b_1 \\ b_2 \\ \cdot \\ \cdot \\ \cdot \\ b_n \end{Bmatrix}$$

$$(8.63)$$

or

$$[a]_{\ell \times \ell}\{c\}_{\ell} = [b]_{\ell \times (n+1)}\{b\}_{(n+1)}$$

As an example, the mean square response of the system represented by the transfer function

$$H(s) = \frac{b_1 s + b_0}{a_4 s^4 + a_3 s^3 + a_2 s^2 + a_2 s^2 + a_1 s + a_0}$$

for W_0 = constant becomes

$$x^2 = \frac{c_3 W_0}{4 a_4}$$

where c_3 is obtained from

$$
\begin{bmatrix}
a_0 & 0 & 0 & 0 \\
a_2 & -a_1 & a_0 & 0 \\
a_4 & -a_3 & a_2 & -a_1 \\
0 & 0 & a_4 & -a_3
\end{bmatrix}
\begin{Bmatrix}
c_0 \\
c_1 \\
c_2 \\
c_3
\end{Bmatrix}
=
\begin{bmatrix}
b_0 & 0 \\
0 & -b_1 \\
0 & 0 \\
0 & 0
\end{bmatrix}
\begin{Bmatrix}
b_0 \\
b_1
\end{Bmatrix}
$$

EXAMPLE 8.8.   For a spring mass idealization, Figure 8.9, subjected to support random acceleration PSD determine (a) the relative mean square displacement response and (b) the absolute mean square acceleration response.

(a) *Relative Mean Square Displacement Response.* In relative coordinates ($x_r = x - u$), the equation of motion becomes

$$m\ddot{x}_r + c\dot{x}_r + kx_r = -m\ddot{u}$$

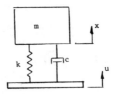

Figure 8.9. Idealized configuration for Example 8.8.

Expressed as a transfer function in the s plane,

$$\frac{x_r(s)}{\ddot{u}(s)} = \frac{-1}{s^2 + \dfrac{\omega_n}{Q} s + \omega_n^2}$$

Equation (8.63) with $\ell = 2$ and $n = 0$ becomes

$$\begin{bmatrix} \omega_n^2 & 0 \\ 1 & -\dfrac{\omega_n}{Q} \end{bmatrix} \begin{Bmatrix} c_0 \\ c_1 \end{Bmatrix} = \begin{bmatrix} -1 \\ 0 \end{bmatrix} \{-1\}$$

Using methods of Appendix A,

$$c_1 = Q/\omega_n^3$$

The mean square displacement response from (8.62) is

$$x_r^2 = \frac{Q W_0}{4\omega_n^3} \tag{8.64}$$

where the excitation PSD is given in $\dfrac{in^2/s^4}{rad/s} \left(\dfrac{m^2/s^4}{rad/s} \right)$.

(b) *Absolute Mean Square Acceleration Response.* The equation of motion in absolute coordinates becomes

$$m\ddot{x} + c\dot{x} + kx = c\dot{u} + ku$$

Taking successive derivatives, the transfer function becomes

$$\frac{\ddot{x}u(s)}{\ddot{u}(s)} = \frac{\dfrac{\omega_n}{Q} s + \omega_n^2}{s^2 + \dfrac{\omega_n}{Q} s + \omega_n^2}$$

Equation (8.63) with $\ell = 2$ and $n = 1$ becomes

$$\begin{bmatrix} \omega_n^2 & 0 \\ 1 & -\dfrac{\omega_n}{Q} \end{bmatrix} \begin{Bmatrix} c_0 \\ c_1 \end{Bmatrix} = \begin{bmatrix} \omega_n^2 & 0 \\ 0 & -\dfrac{\omega_n}{Q} \end{bmatrix} \begin{Bmatrix} \omega_n^2 \\ \dfrac{\omega_n}{Q} \end{Bmatrix}$$

Solution for c_1 gives

$$c_1 = Q\omega_n \left(1 + \frac{1}{Q^2}\right)$$

Substituting c_1 and $a_2 = 1$ from the transfer function into (8.62) gives

$$\ddot{x}^2 = \frac{Q\omega_n W_0}{4}\left(1 + \frac{1}{Q^2}\right) \qquad (8.65a)$$

If the excitation PSD is given in G^2/Hz (8.65a) may be recasted as

$$\ddot{x}^2 = \frac{\pi f_n Q W_0}{2}\left(1 + \frac{1}{Q^2}\right) \qquad (8.65b)$$

where $\omega_n = 2\pi f_n$. Equations (8.64) and (8.65) provide an excellent approximation for the response quantities even when approximately constant narrow band random excitations in the neighborhood of system resonance ω_n are considered.   Q.E.D.

The apparent resonant frequency of a system subjected to random excitation is dependent upon the number of zero crossings that occur per unit of time. The system is expected to respond with a frequency in the neighborhood of the fundamental frequency. Since two zero crossings define a cycle, the apparent frequency is given by

$$f = \frac{1}{2\pi}\left(\frac{\dot{x}}{x}\right) \qquad (8.66)$$

EXAMPLE 8.9.   Determine the response of the uniformly loaded assembly represented by case II of Figure 8.6 using (8.50) with $A = 6.4 \times 10^5 s^{-2}$ and a total mass $m = 0.005$ lb·s^2/in when exposed to the PSD of Figure 8.8.

The damping matrix (8.41) is developed by selecting an amplification $Q = 10$ based on the fundamental frequency f_1 stated in (8.46). The damping matrix for the coupled modes in (8.50) is determined by applying (8.41) to (8.44a), giving

$$[c] = \left(\frac{M}{3}\right)\frac{\sqrt{A}}{Q}[K]^{1/2} = \begin{bmatrix} 4.7958 & -4.6904 & 3.0 \\ -4.6904 & 5.6569 & -4.6904 \\ 3.0 & -4.6904 & 4.7958 \end{bmatrix}$$

which when transformed to modal coordinates by (8.49) becomes

$$[c] = \left(\frac{M}{3}\right)\frac{\sqrt{A}}{Q}\begin{bmatrix} 2.4869 & 0 & 9.9347 \\ 0 & 3.5916 & 0 \\ 9.9347 & 0 & 40.01 \end{bmatrix}$$

The response of coupled coordinates Δ_1 and Δ_3 to support acceleration $\ddot{U}$ may be determined from

$$\begin{bmatrix} 3 & 1 \\ 1 & 3 \end{bmatrix} \begin{Bmatrix} \ddot{\Delta}_1 \\ \ddot{\Delta}_3 \end{Bmatrix} + \frac{\sqrt{A}}{Q} \begin{bmatrix} 2.4869 & 9.9347 \\ 9.9347 & 40.01 \end{bmatrix} \begin{Bmatrix} \dot{\Delta}_1 \\ \dot{\Delta}_3 \end{Bmatrix}$$

$$+ A \begin{bmatrix} 8 & 32 \\ 32 & 184 \end{bmatrix} \begin{Bmatrix} \Delta_1 \\ \Delta_3 \end{Bmatrix} = \begin{bmatrix} 3 & 1 \\ 1 & 3 \end{bmatrix} \begin{Bmatrix} \ddot{U} \\ \ddot{U} \end{Bmatrix}$$

which may be written using the Laplace transform as

$$\begin{bmatrix} s^2 - 24.74s - 6.4(10^5) & -102.059s - 7.04(10^6) \\ 273.172s + 7.04(10^6) & s^2 + 1100.95s + 4.16(10^7) \end{bmatrix} \begin{Bmatrix} \Delta_{1s} \\ \Delta_{3s} \end{Bmatrix} = \begin{Bmatrix} \ddot{U}_s \\ \ddot{U}_s \end{Bmatrix}$$

The transfer function for Δ_1 becomes

$$\frac{\Delta_1}{\ddot{U}} = \frac{s^2 + 1203s + 4.864(10^7)}{s^4 + 1074.2s^3 + 4.096(10^7)s^2 + 9.0783(10^8)s + 2.2938(10^{13})} \tag{8.67}$$

Coefficients of (8.61) are determined from (8.63)

$$\begin{bmatrix} 2.2938(10^{13}) & 0 & 0 & 0 \\ {}^-4.096(10^7) & -9.0783(10^8) & 2.2938(10^{13}) & 0 \\ 1 & -1076.2 & 4.096(10^7) & -9.078(10^8) \\ 0 & 0 & 1 & -1076.2 \end{bmatrix} \begin{Bmatrix} c_0 \\ c_1 \\ c_2 \\ c_3 \end{Bmatrix} = \begin{Bmatrix} 2.3658(10^{15}) \\ 9.5833(10^7) \\ 1 \\ 0 \end{Bmatrix}$$

$$c_0 = 103.14, \quad c_1 = 13.917, \quad c_2 = 3.708(10^{-4}), \quad c_3 = 3.4455(10^{-7})$$

giving the squared magnitude as

$$\frac{1}{2} \left| \frac{\Delta_1}{\ddot{U}} \right|^2 = \frac{3.4455(10^{-7})s^3 + 3.708(10^{-4})s^2 + 13.917s + 103.14}{s^4 + 1076.2s^3 + 4.096(10^7)s^2 + 9.0783(10)^8 s + 2.2939(10^{13})} \tag{8.68a}$$

The squared magnitude of the transfer function $\dot{\Delta}_1/\ddot{U}$ developed by multiplying the numerator of (8.67) by s is similarly determined as

$$\frac{1}{2} \left| \frac{\dot{\Delta}_1}{\ddot{U}} \right|^2 = \frac{0.1957s^3 + 209.65s^2 + 7.902(10^6)s}{s^4 + 1076.2s^3 + 4.096(10^7)s^2 + 9.0783(10^8)s + 2.2938(10^{13})} \tag{8.68b}$$

The response to broadband random excitation from (8.68) using (8.62) requires only coefficient c_3

$$\Delta_1 = \tfrac{1}{2}[3.4455(10^{-7})(386^2)\,W_0]^{1/2} = 0.1133\sqrt{W_0} \text{ in rms}$$ (8.69)

$$\dot{\Delta}_1 = \tfrac{1}{2}[0.1957(386^2)\,W_0]^{1/2} = 85.38\sqrt{W_0} \text{ in/s rms}$$

given W_0 in G^2/Hz. The apparent resonant frequency is determined using (8.66) as

$$f = \frac{1}{2\pi}\frac{85.38\sqrt{W_0}}{0.1133\sqrt{W_0}} = 120 \text{ Hz}$$

This result compares with f_1 of (8.46b), indicating that the fundamental mode is dominant in the assembly response.

An approach that develops the response in terms of each of the system frequencies would be of value if the PSD random excitation is not broadband. Based on the method of Appendix B.4 to obtain the approximate factors of the characteristic equation, the denominator of (8.68), a partial fraction expansion would provide response characteristics at each mode.

Ignoring odd powers of s, the denominator of (8.68) becomes

$$s^4 + 4.096(10^7)s^2 + 2.2938(10^{13})$$

with the roots obtained as

$$\beta_1 = 5.6788(10^5)$$

$$\beta_2 = 4.0392(10^7)$$

and

$$d_1 = \beta_1 - \beta_2 = -3.9824(10^7)$$

Equation (B.4-43) becomes

$$\begin{Bmatrix} \alpha_1 \\ \alpha_2 \end{Bmatrix} = \begin{bmatrix} -2.511(10^{-8}) & 0 \\ 0 & 2.511(10^{-8}) \end{bmatrix} \begin{bmatrix} 5.6788(10^5) & -1 \\ 4.0392(10^7) & -1 \end{bmatrix} \begin{Bmatrix} 1076.2 \\ 9.0783(10^8) \end{Bmatrix}$$

giving

$$\alpha_1 = 7.449$$

$$\alpha_2 = 1068.7$$

so that the factors of the denominator are

$$[s^2 + 7.449s + 5.6788(10^5)]\,[s^2 + 1068.7s + 4.0392(10^7)]$$

The partial fraction expansion of (8.68a) becomes

$$\tfrac{1}{2}\left|\frac{\Delta_1}{U}\right|^2 = \frac{3.445(10^{-7})s + 2.552(10^{-6})}{s^2 + 7.449s + 5.6788(10^5)} + \frac{7.01(10^{-11})s + 1.017(10^{-7})}{s^2 + 1068.7s + 4.0392(10^7)}$$

The response to narrow band random excitation at the resonant frequency represented by each fraction requires only coefficient C_1 of each using (8.62)

$$\Delta_1 = \tfrac{1}{2}[3.445(10^{-7})(386^2)W_1 + 7.01(10^{-11})(386^2)W_2]^{1/2} \qquad (8.70a)$$

$$\Delta_1 = 0.1133\,[W_1 + 2.035(10^{-4})W_2]^{1/2}\ \text{in}\quad \text{rms}$$

where W_1 and W_2 in G^2/Hz are taken at 120 Hz and 1008 Hz respectively. A similar relationship for determining $\dot\Delta_1$ can be developed from the partial fraction expansion of (8.68b) as

$$\dot\Delta_1 = 85.38\,[W_1 + 2.194(10^{-4})W_2]^{1/2}\ \text{in/s}\quad \text{rms} \qquad (8.70b)$$

The PSD excitation at 120 Hz and 1008 Hz may be determined from (8.54a) and Figure 8.8 as

$$W_1 = 0.004\left(\frac{120}{50}\right)^2 = 0.025\left(\frac{120}{125}\right)^2$$

$$W_1 = 0.023G^2/Hz$$

$$W_2 = 0.025\left(\frac{1000}{1008}\right) = 0.0248G^2/Hz$$

Substituting these quantities in (8.70) gives

$$\Delta_1 = 0.0172\ \text{in rms}$$

$$\dot\Delta_1 = 12.95\ \text{in/s rms}$$

The apparent frequency is determined from (8.66) as

$$f = \frac{1}{2\pi}\left(\frac{12.95}{0.0172}\right) = 120\ \text{Hz}$$

The rms acceleration at this frequency may be determined using either displacement or velocity as

$$\ddot\Delta_1 = \Delta_1\omega_{a1}^2$$

$$\ddot\Delta_1 = 0.0172\left(\frac{12.95}{0.0172}\right)^2 = 25.3G\quad \text{rms} \qquad \text{Q.E.D.}$$

Probability of Equipment Failure. Displacement and acceleration response quantities provide a basis for defining the clearance required between assemblies and determining the stresses developed on components and supporting structure. This information can also be used to estimate the vibration lifetime of the assembly when exposed to various levels of excitation.

The distribution of peaks in random vibration may be characterized as Gaussian, which means that the motion can be defined mathematically in terms of a mean, variance and moments. Since random vibration that is stationary and ergodic has a zero mean, the variance alone sufficiently characterizes the motion. The variance, σ, a measure of the spread of the distribution is defined by the computed rms quantity. In a normal distribution, the rms quantity will occur 68.27% of the time. Peak magnitudes greater than the rms quantity, σ, occur less often and may be stated succinctly as

Magnitude (multiple of rms quantity)	Occurence (% of time)
$\pm \sigma$	68.27
$\pm \sigma$ to $\pm 2\sigma$	27.18
$\pm 2\sigma$ to $\pm 3\sigma$	4.28
$\pm 3\sigma$ to $\pm 4\sigma$	0.26

A 3σ quantity should be used to assure a design that reliably meets its random vibration requirement. Fatigue failure or performance loss due to chattering electrical contacts, impacts between parts and high stresses as a result of excessive displacements could be avoided. As an example, a peak displacement of 0.15 in (3.81 mm) would be the result of applying 3σ to an rms displacement of 0.05 in (1.27 mm) or

$$0.15 \text{ in peak} = 3(0.05 \text{ in rms})$$

$$3.81 \text{ mm peak} = 3(1.27 \text{ mm rms})$$

It is possible to estimate the probability of failure of a material since the normal distribution also applies to material properties. The probability of failure, represented by the shaded area in Figure 8.10, is a result of an overlap of the random vibration response stress distribution and the yield strength distribution of the material. If σ_y represents the rms variation of the material yield strength S_y, and σ represents the rms stress due to random vibration, a conservative estimate of the probability of failure using a uniform distribution becomes

$$P_f = \frac{S_0^2 - S_1^2 - S_2^2}{72\sigma S_y} \tag{8.71a}$$

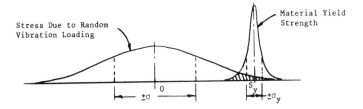

Figure 8.10. Distribution of Material Strength Versus Stresses Due to Random Vibration Loading.

given

$$S_0 = 3(\sigma + \sigma_y) - S_y$$

$$S_1 = 3(\sigma_y - \sigma) - S_y$$

$$S_2 = 3(\sigma - \sigma_y) - S_y$$

S_0, S_1 or S_2 are omitted from (8.71a) if a negative quantity results. For most metals $\sigma_y = 0.08S_y$. The probability of failure (8.71a) for equipment subjected to vibration may be simplified to

$$P_f = \frac{(3\sigma - S_y)}{6\sigma} \tag{8.71b}$$

where a negative quantity indicates an infinite equipment lifetime.

EXAMPLE 8.10.   The aluminum structure of an assembly experiences a flexural stress of 7121 lb/in^2 rms when exposed to random vibration. Determine the probability of failure if the material yield strength $S_y = 18000$ lb/in^2.
    The probability of failure is given by (8.71b) as

$$P_f = \frac{3(7121) - 18000}{6(7121)} = 0.08 \text{ or } 8\%$$

Q.E.D.

Nonlinear Considerations for Thin Circuit Boards and Panels.   When the response to a vibratory excitation causes the displacement of a circuit board or planar panel to exceed one third of its thickness, inplane stretching contributes significantly to the stress. This is a nonlinear effect. The membrane forces, which are a result of stretching, carry an increasing portion of the load with increasing displacement. When the displacement is equivalent to the surface

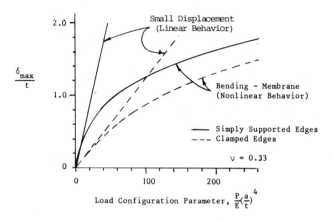

Figure 8.11. Deflection of Square Plates with a Distributed Load.

thickness, the membrane action is comparable to that developed by bending. At greater displacements, membrane actions dominate. Small displacement techniques become inappropriate for displacements greater than $t/3$ owing to increasing nonlinearity.

Large surfaces can develop substantial displacements when subjected to acoustic noise, random vibration or accelerations. The stresses that are the result of bending and stretching are combined to assess the vibration lifetime of the assembly. For loadings that develop displacements greater than $t/3$, consideration of inplane stretching results in a realistic assessment of deflections and stresses. These quantities are considerably less than those predicted by small displacement techniques for the same load. Each of these techniques are compared in Figure 8.11.

Consider the simply supported planar assembly in Figure 8.12. The maxi-

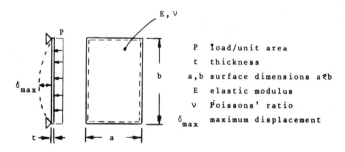

Figure 8.12. Surface Characteristics.

mum deflection at the center is given by the linear small displacement approach as

$$\left(\frac{\delta}{t}\right)_{max} = \frac{0.195(1 - \nu^2)P_b}{\left[\left(\frac{a}{b}\right)^2 + 1\right]^2 E}\left(\frac{a}{t}\right)^4 \qquad (8.72)$$

Equation (8.72) is illustrated as straight lines in Figure 8.11. The deflection representing membrane behavior is

$$\left(\frac{\delta}{t}\right)_{max} = \frac{0.195(1 - \nu^2)\frac{P_m}{E}\left(\frac{a}{t}\right)^4}{(3 - \nu^2)\left[\left(\frac{a}{b}\right)^4 + 1\right] + 4\nu\left(\frac{a}{b}\right)^2} \qquad (8.73)$$

The total load acting on the surface is the sum of the bending load P_b and the membrane load P_m acting together:

$$P = P_b + P_m \qquad (8.74)$$

Solving for P_b and P_m from (8.72) and (8.73) respectively and substituting in (8.74) gives a coupled relationship for the load parameter:

$$\frac{P}{E}\left(\frac{a}{t}\right)^4 = \frac{5.128}{(1 - \nu^2)}\left\{\left[\left(\frac{a}{b}\right)^2 + 1\right]^2\left(\frac{\delta}{t}\right)\right.$$
$$\left. + \left\{(3 - \nu^2)\left[\left(\frac{a}{b}\right)^4 + 1\right] + 4\nu\left(\frac{a}{b}\right)^2\right\}\left(\frac{\delta}{t}\right)^3\right\}$$

$$(8.75a)$$

Equation (8.75a) may be recast for metals ($\nu = 0.33$) as

$$\frac{P}{E}\left(\frac{a}{t}\right)^4 = 5.755\left[\left(\frac{a}{b}\right)^2 + 1\right]^2\left(\frac{\delta}{t}\right)$$
$$+ \left\{16.637\left[\left(\frac{a}{b}\right)^4 + 1\right] + 7.596\left(\frac{a}{b}\right)^2\right\}\left(\frac{\delta}{t}\right)^3$$

$$(8.75b)$$

Figure 8.11 illustrates the nonlinear characteristics of (8.75b). For the configuration in Figure 8.12, the maximum stress occurs at the center. This stress is based on the sum of the bending stress and the tensile stress due to stretching:

$$\sigma = \sigma_b + \sigma_m \tag{8.76}$$

given

$$\sigma_b = \frac{\pi^2 E}{2(1 - \nu^2)} \left(\frac{t}{a}\right)^2 \left[1 + \nu \left(\frac{a}{b}\right)^2\right] \left(\frac{\delta}{t}\right)_{max} \tag{8.77a}$$

$$\sigma_m = \frac{\pi^2 E}{8(1 - \nu^2)} \left(\frac{t}{a}\right)^2 \left[2 - \nu^2 + \nu \left(\frac{a}{b}\right)^2\right] \left(\frac{\delta}{t}\right)_{max}^2 \tag{8.77b}$$

Equations (8.77) may be recasted for metals ($\nu = 0.33$) as

$$\sigma_b = 5.538E \left(\frac{t}{a}\right)^2 \left[1 + 0.33 \left(\frac{a}{b}\right)^2\right] \left(\frac{\delta}{t}\right)_{max} \tag{8.78a}$$

$$\sigma_m = 1.384E \left(\frac{t}{a}\right)^2 \left[1.891 + 0.33 \left(\frac{a}{b}\right)^2\right] \left(\frac{\delta}{t}\right)_{max}^2 \tag{8.78b}$$

EXAMPLE 8.11.   A power supply is being designed as a planar assembly supported on a 6 by 10 inch simply supported 6061-T6 aluminum plate 0.08 inches thick. Determine the displacement and surface stress for an assembly weight of 2, 4 and 6 pounds. A response acceleration of 40G can be expected due to chassis and assembly coupling.

The displacement to thickness ratio (δ/t) as a function of applied load is given by (8.75b) for an aspect ratio $a/b = 0.6$ with $a \leq b$ as

$$\frac{P}{10^7} \left(\frac{6}{0.08}\right)^4 = 5.755(0.6^2 + 1)^2 \left(\frac{\delta}{t}\right) + [16.637(0.6^4 + 1) + 7.596(0.6)^2] \left(\frac{\delta}{t}\right)^3$$

or

$$3.164P = 10.64 \left(\frac{\delta}{t}\right) + 21.53 \left(\frac{\delta}{t}\right)^3 \tag{8.79}$$

where P represents the total load as follows:

$$P = \frac{\text{load}}{\text{area}} = \frac{WG}{A} = \frac{2}{3} W$$

The solution for (δ/t) in (8.79) in terms of $W = 2$, 4 and 6 pounds using Appendix B.4 is summarized:

Weight, lb	(δ/t)	Displacement, in $(t = 0.08 \text{ in})$	Stress, lb/in^2 σ_b	σ_m	σ
2	0.326	0.026	3591	525	4116
4	0.516	0.041	5684	1316	7000
6	0.645	0.052	7105	2057	9162

Inplane stretching occurs in the weight range considered since the predicted displacements exceed $t/3 = 0.026$ inches. The surface stress σ and the contribution from membrane actions σ_m determined from (8.76) and (8.78) are included in the tabular summary.   Q.E.D.

When the displacement is greater than $t/3$, the membrane effect actually results in a higher surface stiffness than would be obtained using the linear small displacement approach. A higher stiffness results in smaller displacements and stresses which extends the predicted vibration lifetime of the assembly. Small displacement techniques result in overly conservative predictions when displacements exceed $t/3$.

Heat transfer, weight, attachment stresses on supported components and peripheral attachments to the supporting structure also affect the selection of a suitable panel assembly thickness. When clamped edge conditions are used in a design, for example, maximum stress occurs along the shortest edge. These stresses are always greater than those of a simply supported surface for the same load parameter $\dfrac{P}{E}\left(\dfrac{a}{t}\right)^4$ even though displacements are considerably less. A stress concentration factor of 2 should be applied when peripheral screws or rivets are used for attachment. In this case, the design stress $= 2\sigma$ where σ is determined using (8.76).

EXAMPLE 8.12.   The side panel of the assembly shown in Figure 8.13 is subjected to the acoustic environment of Figure 8.14. Determine the probability of failure of the panel.

The fundamental frequency of the simply supported panel is determined using case 15 of Table 8.1 as

$$f_n = \frac{\pi}{2}\left(\frac{1}{7^2} + \frac{1}{15^2}\right)\sqrt{\frac{202(7)(15)}{1.632(10^{-3})}}$$

$$f_n = 141 \text{ Hz}$$

where $D = 10^7(0.06)^3/12(1 - 0.33^2) = 202 \text{ lb·in};$
$M = \rho a b t/g = 1.632(10^{-3}) \text{ lb·s}^2/\text{in};$
$\rho = \text{material density, lb/in}^3.$

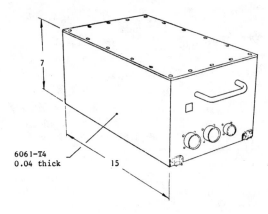

Figure 8.13. Assembly Subjected to Acoustic Load.

The sound pressure level (SPL) is determined from Figure 8.14 at the panel fundamental frequency. The acoustic environment is commonly given in a tabular or graphical format for ⅓ octave bands. The relationship between the upper and lower center frequencies of a ⅓ octave band is obtained from

$$f_n = 1.2589 f_\ell \tag{8.80}$$

Relationships for the pressure spectral density, S_p, are based on treating the panel fundamental frequency as a center frquency of a ⅓ octave band. In this case, the center frequency becomes

$$f_c = (f_\ell f_h)^{1/2} \tag{8.81}$$

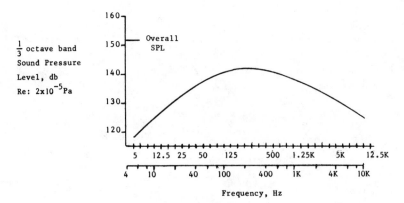

Figure 8.14. Acoustic Environment.

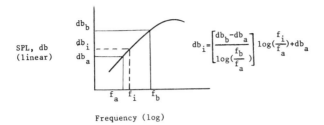

Figure 8.15. Characteristics of the Acoustic Environment.

and the bandwidth from (8.80) and (8.81) is

$$w_b = f_h - f_\ell = 0.231 f_c \qquad (8.82)$$

If only a tabular listing of the acoustic environment is available, the SPL can be obtained by interpolation using Figure 8.15.

The SPL for the panel with a fundamental frequency of 141 Hz is obtained from Figure 8.14 as $db_i = 141$. The bandwidth using (8.82) becomes

$$W_b = 0.231(141) = 32.57 \text{ Hz}$$

A graphical development of the pressure spectral density that can be used to determine the overall rms pressure due to the acoustic environment is obtained from

$$S_p = \frac{(2.9 \times 10^{-9})^2}{0.235 f_i} 10^{db_i/10} \, (\text{lb/in}^2)^2/\text{Hz} \qquad (8.83)$$

Figure 8.16 is a result of applying (8.83) at sufficient center frequencies to define the envelope. The spectral excitation at the panel frequency from (8.83) is

$$S_p = \frac{(2.9 \times 10^{-9})^2}{0.235(141)} 10^{14.1} = 3.195(10^{-5})(\text{lb/in}^2)^2/\text{Hz}$$

Assuming the panel can be represented by a spring mass idealization, a variation of (8.65b) gives the rms response due to the acoustic environment as

$$\overline{P} = \left[\frac{\pi f_n Q S_p}{2} \left(1 + \frac{1}{Q^2} \right) \right]^{1/2} = \left[\frac{\pi (141)^{3/2} (3.195 \times 10^{-5})}{2} \left(1 + \frac{1}{141} \right) \right]^{1/2}$$

$$\overline{P} = 0.291 \text{ lb/in}^2 \text{ rms}$$

where $Q = \sqrt{f_n}$.

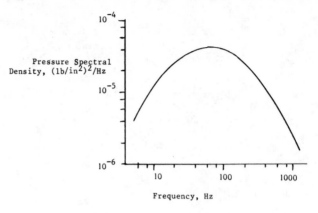

Figure 8.16. Acoustic PSD.

For an aspect ratio $a/b = 7/15$, the rms displacement is obtained from (8.75b):

$$5.39 = 8.53 \left(\frac{\delta}{t}\right) + 19.08 \left(\frac{\delta}{t}\right)^3$$

where

$$\frac{\overline{P}}{E}\left(\frac{a}{t}\right)^4 = \frac{0.291}{10^7}\left(\frac{7}{0.06}\right)^4 = 5.39 \quad \text{(dimensionless)}$$

A solution for (δ/t) using methods of Appendix B.4 gives

$$\left(\frac{\delta}{t}\right)_{\text{rms}} = 0.441$$

from which the rms displacement becomes

$$\delta = 0.441(0.06) = 0.027 \text{ in } \text{rms}$$

Since the displacement is greater than $t/3 = 0.02$ inches, panel behavior is composed of both bending and stretching. Peak displacements are obtained by substituting $P_{\text{peak}} = 3\overline{P}$ into (8.75b) giving

$$\delta_{\text{peak}} = 0.791(0.06) = 0.048 \text{ in peak}$$

Stresses at the center of the panel are determined from (8.76) using (8.78) as follows:

$$\sigma_{\text{rms}} = \sigma_b + \sigma_m = 1923 + 388$$

$$\sigma_{rms} = 2311 \text{ lb/in}^2 \text{ rms}$$

$$\sigma_{pk} = \sigma_b + \sigma_m = 3450 + 1249$$

$$\sigma_{pk} = 4699 \text{ lb/in}^2 \text{ peak}$$

A stress concentration factor would be applied to these quantities if cutouts or holes were present in a central region of the panel. The determination of the probability of failure from (8.71) uses the substitution $3\sigma = \sigma_{pk}$ since nonlinear behavior has skewed the distribution. An infinite lifetime in the acoustic environment of Figure 8.14 can be expected since negative quantities for S_0, S_1 and S_2 in (8.71a) give a zero probability of failure.

If desired, the rms response acceleration of the panel may be obtained from

$$\ddot{\delta} = \frac{\overline{P}}{t\rho} = \frac{0.291}{0.06(0.1)} = 48.5G \text{ rms}$$

Q.E.D.

Equipment Random Vibration Lifetime.   Random excitation excites all modes of an assembly simultaneously. The significant response, however, occurs at the apparent frequency (8.66) which is strongly influenced by the fundamental frequency and includes contributions from the other modes. In the narrow bandwidth of the Gaussian response at the apparent frequency, the distribution of the peaks is Rayleigh, based on the assumption that the response resembles a sine wave of varying amplitude and phase. The magnitude of a majority of the peaks in the Rayleigh distribution is represented by the computed rms quantity.

The vibration lifetime of an assembly may be considered as the period T, required for the gradual propagation of cracks at locations of high stress to cause failure. The number of cycles in a Rayleigh distribution of peak stresses of a Gaussian narrow band response is

$$n = \int_0^\infty fT \frac{s}{\sigma^2} \text{EXP} \left[-\frac{1}{2} \left(\frac{s}{\sigma} \right)^2 \right] ds \qquad (8.84)$$

since the distribution is continuous. If N from (8.22) represents the number of cycles of stress σ which cause failure, then the average damage resulting from cycles of all stress levels occurring during period T is obtained from (8.21) as

$$T = \frac{AB}{F \int_0^\infty \sigma^{b-2} \text{EXP} \left[-\frac{1}{2} \left(\frac{s}{\sigma} \right)^2 \right] ds} \qquad (8.85a)$$

Performing the required integration

$$T = \frac{0.75 N_\sigma}{f\left(\frac{b}{2}\right)! \, 2^{b/2}}$$ (8.85b)

where N_σ is evaluated at the rms stress, A = 0.75 and

$$\left(\frac{b}{2}\right)! \simeq (\sqrt{2\pi}) e^{-b/2} \left(\frac{b}{2}\right)^{(b+1)/2} \qquad \frac{b}{2} \gg 1.0$$

Use of a Gaussian distribution for the peak stresses results in an overly optimistic estimate of the vibration lifetime. A useful approximation for (8.85b) based on a skewed Gaussian distribution becomes

$$T = \frac{N_{3\sigma}}{4f\left[0.2718\left(\frac{2}{3}\right)^b + 0.0428\right]}$$ (8.86)

where $N_{3\sigma}$ is obtained from (8.22) evaluated at 3σ.

EXAMPLE 8.13.   The apparent frequency of an assembly subjected to a random vibration environment is determined to be 100 Hz. As a result of the loading, the 6061-T4 aluminum structure experiences an rms stress of 8500 lb/in². Determine the probability of failure for a yield strength S_y = 18,000 lb/in² and an estimate of the vibration lifetime of the structure.

The probability of failure is obtained from (8.71b) as

$$P_f = \frac{[(3)(8500) - (18000)]}{6(8500)}$$

$$P_f = 0.147, \text{ or } 14.7\%$$

The number of cycles at stress 3σ which cause failure is obtained from (8.22) using fatigue properties for the material from Table 8.2:

$$N_{3\sigma} = \frac{7.755(10^{66})}{[(3)(8500)]^{14.06}} = 85827 \text{ cycles}$$

Substituting this quantity into (8.86) at an apparent frequency of 100 Hz gives

$$T = \frac{85827}{(4)(100)\left[0.2718\left(\dfrac{2}{3}\right)^{14.06} + 0.0428\right]} = 4909\text{s}$$

for the expected vibration lifetime of the structure as a result of the applied loading. Other components and attachments of the assembly should also be examined so that the limiting element of the design can be identified and corrected if necessary to extend the lifetime of the equipment.   Q.E.D.

8.4. SHOCK

Any sudden change that affects the position, velocity, acceleration or forces applied to the system may be considered as shock. Shock can occur during assembly, handling or transport of the equipment. Shock can result when two adjacent vibrating assemblies contact each other due to excessive displacements. Near-miss explosions and loads due to wind gusts are also examples of shock-producing environments.

Shock can develop significant internal forces that diminish rapidly. After the application of shock loading, motion consists of damped harmonic vibration. Characteristics that differentiate shock from static loading involve the time required for the response to reach maximum values of displacement, velocity and acceleration. The intensity of the response to the applied load depends on how close the fundamental frequencies of the assembly are to the frequency of the loading.

All modes of an assembly respond simultaneously experiencing transient vibration. Damping and fatigue do not influence peak response quantities or equipment lifetime since only a small number of oscillations occur. The work developed by the applied shock load is dissipated by damped transient motion at the frequencies of the assembly. Loads on components can be reduced if the components can be mounted to avoid the direction of the highest-intensity response. The intensity of the inplane response may be considerably less than the normal-to-plane response, for example. Brackets and gussets or other structural entities may be used to change the characteristics of the response near sensitive components.

The internal forces and stresses developed by shock are evaluated in a manner similar to that illustrated in Chapter 7 for static loads with the added complexity of a time dependent loading. The time dependent load displaces the assembly while developing internal forces in its structural members that are themselves time dependent. As in the evaluation of static loads, the vector sum of external loads and reaction loads at any instant are zero.

The evaluation of response quantities may be expedited if the assembly can be idealized as a one dimensional system along the axis of excitation. Table

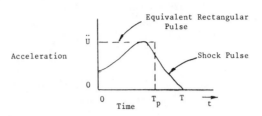

Figure 8.17. Shock Loading Time History.

B.1-1 illustrates several configurations that may be used to reduce the overall degrees of freedom of an assembly. The representation of an assembly by an equivalent idealization with a reduced number of degrees of freedom must maintain dynamic similarity with the actual equipment. This imposes the requirement that the work, strain energy and kinetic energy of the idealization be identical respectively with those of the actual design. The accuracy of the results obtained from the equivalent system depend on the validity of the assumptions employed in developing the idealization.

In general, all loads acting on an assembly are treated as if they act simultaneously. Superposition is not valid if individual shock loads develop maximum displacements that do not occur at the same time. The total displacement is not necessarily equal to the sum of the maximum displacements produced by different time dependent loads. In this case, the phase difference between individual loads must be considered.

Impact and Acceleration. The limiting response of a single degree of freedom idealization with fundamental frequency f_n to a given shock pulse time history, Figure 8.17, can be determined from the product of $f_n T_p$ as

(a) Impact ($\Delta \dot{U}$ velocity step): $f_n T_p < \frac{1}{6}$
(b) Static load ($\ddot{U}$ acceleration): $f_n T_p > 3$

where

$$T_p = \frac{\Delta \dot{U}}{\ddot{U}} \tag{8.87}$$

$$\Delta \dot{U} = \int_0^T \ddot{U}(t) \, dt \tag{8.88}$$

A force pulse can be transformed to an acceleration pulse by dividing the force amplitude by the mass of the idealized system, giving

$$\ddot{U} = F/M \qquad (8.89)$$

The velocity step $\Delta \dot{U}$ represents the area under the acceleration pulse. The general response to pulse or step impacts may be determined from Figure 8.18. A different approach is required for $f_n T_p > 1/3$ shocks since peak impact response characteristics would result in an overly conservative design.

Peak displacements and velocities may be determined from the peak acceleration response $\ddot{x}_{max}$ using the fundamental frequency of the idealized system

Peak acceleration $= \ddot{x}_{max}$

Peak velocity $\quad = \ddot{x}_{max}/\omega \qquad (8.90)$

Peak displacement $= \ddot{x}_{max}/\omega^2$

Only approximate peak velocities are obtained from (8.90) since the maxima occurs at a different time than for acceleration.

A static acceleration that is equivalent to a steadily applied acceleration

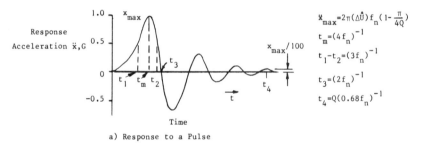

a) Response to a Pulse

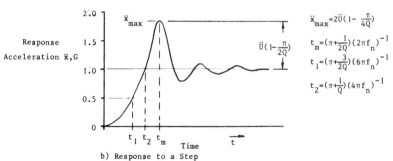

b) Response to a Step

Figure 8.18. Generalized Impact Response Characteristics.

results for $f_nT_p > 3$. If the static displacement due to gravational forces is determined as

$$\delta_{st} = \frac{mg}{k} = \frac{g}{\omega_n^2} \tag{8.91}$$

the maximum displacement resulting from the action of shock becomes

$$\delta_{max} = \ddot{U}\delta_{st} \tag{8.92}$$

EXAMPLE 8.14.   Determine the maximum response acceleration and displacement of an idealized assembly that has a fundamental frequency of 40 Hz subjected to a 50G half-sine shock with a period of (a) 0.011s and (b) 0.11s.

(a) *Response to a 0.011s Half-sine Shock Pulse.* The velocity step due to the shock pulse from (8.88) is

$$\Delta \dot{U} = \ddot{U} \int_0^T \sin\left(\frac{\pi t}{T}\right) dt = \frac{2}{\pi} \ddot{U}T \tag{8.93}$$

The shock parameter f_nT_p may be determined from (8.93) and (8.87) as

$$f_nT_p = \frac{2}{\pi} Tf_n = \frac{2}{\pi} (0.011)40 = 0.28 \tag{8.94}$$

The shock can be considered an impact since $f_nT_p < \frac{1}{4}$. Maximum response acceleration and displacement is given by Figure 8.18(a), (8.93) and (8.90) for $Q = \sqrt{f_n}$ as

$$\ddot{x}_{max} = 2\pi(40) \left[\frac{2}{\pi}(50)(0.011)\right]\left[1 - \frac{\pi}{4\sqrt{40}}\right]$$

$$\ddot{x}_{max} = 77G$$

$$x_{max} = \frac{77\left(386\ \frac{\text{in/s}^2}{g}\right)}{[2\pi(40)]^2} = 0.47\ \text{in}$$

(b) *Response to a 0.11s Half-sine Shock Pulse.* The shock parameter f_nT_p from (8.94) is

$$f_nT_p = \frac{2}{\pi}(0.11)(40) = 2.8$$

Acceleration will be considered since this value is sufficiently close to the limiting quantity. The maximum displacement from (8.92) using (8.91) becomes

$$x_{max} = \frac{50(386)}{[2\pi(40)]^2} = 0.31 \text{ in}$$

where the maximum acceleration is 50G.   Q.E.D.

When the shock parameter $\frac{1}{3} < f_n T_p < 3$, a solution of the differential equations for the applied loading is required. Direct application of differential equations or their Laplace transforms may be readily used if the applied loading is a simple mathematical function. Usually, however, shock loadings are sufficiently complex to require a numerical approach that uses step-by-step integration.

The Response Time History.   Direct application of the governing differential equations gives an exact closed form solution for the response of a single degree of freedom system if the applied loading can be developed using step, ramp and parabolic functions. The response for an applied support acceleration or force at the center of gravity of the mass for a single degree of freedom idealization becomes

$$\left\{ \begin{matrix} x \\ \dot{x} \end{matrix} \right\}_{i+1} = \left\{ \begin{matrix} B_0 \\ B_1 \end{matrix} \right\} + \Delta\theta_{\Delta t} \left\{ \begin{matrix} x_i + B_2 \\ x_i + B_3 \end{matrix} \right\}$$

given

$$\Delta\theta_{\Delta t} = \frac{\omega}{\omega_D} e^{-\omega\Delta t/2Q} \left[\begin{matrix} \cos(\omega_D\Delta t - \sigma) & \dfrac{\sin \omega_D\Delta t}{\omega} \\ \\ -\omega \sin \omega_D\Delta t & \cos(\omega_D\Delta t + \sigma) \end{matrix} \right];$$

$$\omega_D = \omega \sqrt{1 - \frac{1}{4Q^2}};$$

$$\sigma = \tan^{-1}\left(\frac{\omega}{2Q\omega_D}\right);$$

$V_j = -\ddot{U}_j/\omega^2$ (support acceleration); $V_j = F_j/k$ (forced mass);

$a_1 = (Q\omega\Delta t)^{-1};$

$$a_2 = \frac{1 - 1/Q^2}{(\omega\Delta t)^2}.$$

For step segments:

$$B_0 = V_n$$
$$B_1 = 0$$
$$B_2 = -V_n$$
$$B_3 = 0$$

(8.96a)

For ramp segments:

$$B_0 = [V_{n+1} - a_1(V_{n+1} - V_n)]$$

$$B_1 = \frac{1}{\Delta t}(V_{n+1} - V_n)$$

$$B_2 = [-V_n + a_1(V_{n+1} - V_n)]$$

$$B_3 = \frac{1}{\Delta t}(V_n - V_{n+1})$$

(8.96b)

For parabolic segments:
Forward difference, Figure 8.19(a):

$$B_0 = \left[V_{n+1} - \frac{a_1}{2}(V_{n+2} - V_n) - a_2(V_n - 2V_{n+1} + V_{n+2}) \right]$$

$$B_1 = \frac{1}{\Delta t}\left[\frac{1}{2}(V_{n+2} - V_n) - a_1(V_n - 2V_{n+1} + V_{n+2}) \right]$$

$$B_2 = \left[-V_n + \frac{a_1}{2}(4V_{n+1} - 3V_n - V_{n+2}) + a_2(V_n - 2V_{n+1} + V_{n+2}) \right]$$

$$B_3 = \frac{1}{\Delta t}\left[\frac{1}{2}(3V_n - 4V_{n+1} + V_{n+2}) + a_1(V_n - 2V_{n+1} + V_{n+2}) \right]$$

(8.96c)

Central difference, Figure 8.19(b):

$$B_0 = \left[V_{n+1} - \frac{a_1}{2}(V_{n-1} - 4V_n + 3V_{n+1}) - a_2(V_{n-1} - 2V_n + V_{n+1}) \right]$$

$$B_1 = \frac{1}{\Delta t}\left[\frac{1}{2}(V_{n-1} - 4V_n + 3V_{n+1}) - a_1(V_{n-1} - 2V_n + V_{n+1}) \right]$$

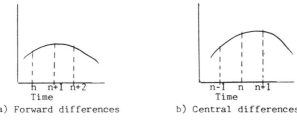

a) Forward differences   b) Central differences

Figure 8.19. Parabolic Differences Used in (8.96c).

$$B_2 = \left[-V_n + \frac{a_1}{2}(V_{n+1} - V_{n-1}) + a_2(V_{n-1} - 2V_n + V_{n+1}) \right]$$

$$B_3 = \frac{1}{\Delta t}\left[\frac{1}{2}(V_{n-1} + V_{n+1}) + a_1(V_{n-1} - 2V_n + V_{n+1}) \right]$$

$$(8.96d)$$

After the applied loading ceases, the damped harmonic response can be determined at successive Δt's using

$$\begin{Bmatrix} x \\ \dot{x} \end{Bmatrix}_{i+1} = \Delta\theta_{\Delta t} \begin{Bmatrix} x \\ \dot{x} \end{Bmatrix}_i \qquad (8.97)$$

The response for an applied support velocity becomes

$$\begin{Bmatrix} x \\ \dot{x} \end{Bmatrix}_{i+1} = -\begin{Bmatrix} B_1 \\ B_4 \end{Bmatrix} + \Delta\theta_{\Delta t} \begin{Bmatrix} x_i - B_3 \\ x_i + B_4 \end{Bmatrix} \qquad (8.98)$$

where B_1 and B_3 are obtained from (8.96) using $V_j = \dot{U}_j/\omega^2\Delta t$, and B_4 is given by

$$B_4 = \frac{1}{\Delta t}(V_n - 2V_{n+1} + V_{n+2}) \quad \text{(forward difference)}$$

or

$$B_4 = \frac{1}{\Delta t}(V_{n-1} - 2V_n + V_{n+1}) \quad \text{(central difference)}$$

The absolute acceleration response may be determined from

$$\ddot{x} = \left[-\omega^2 \quad -\frac{\omega}{Q} \right] \begin{Bmatrix} x \\ \dot{x} \end{Bmatrix} \qquad (8.99)$$

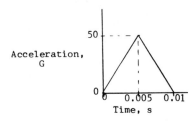

Figure 8.20. Triangular Shock Pulse.

Equations (8.95) and (8.98) may be used for damped or undamped idealizations when either an analytical or graphical definition of the shock motion is available.

EXAMPLE 8.15.   Determine the maximum acceleration response of an undamped idealized assembly that has a fundamental frequency of 50 Hz and is subjected to the triangular shock illustrated in Figure 8.20.

The shock excitation is first examined to determine whether its nature is an impact. The velocity step (8.88), which represents the area under the shock pulse is used to determine the equivalent period of a rectangular pulse giving

$$T_p = \tfrac{1}{2}T = 0.005\text{s}$$

The shock parameter $f_n T_p = 0.25$, which indicates impact. The maximum acceleration may be obtained using the relationships in Figure 8.18(a) with $Q = \infty$ as

$$\ddot{x}_{max} = 2\pi(0.25)50 = 78.5\text{G}$$

where $\Delta \dot{U} = 0.25\text{G}\cdot\text{s}$

The characteristics of the response can be obtained from (8.95) which for $\Delta t = 0.005\text{s}$ using ramp segments (8.96b) becomes

$$\begin{Bmatrix} \dot{x} \\ \ddot{x} \end{Bmatrix}_{5\text{ms}} = \begin{Bmatrix} 50 \\ 50/\Delta t \end{Bmatrix} + \begin{bmatrix} 0 & 1/\omega \\ -\omega & 0 \end{bmatrix} \begin{Bmatrix} 0 \\ -50/\Delta t \end{Bmatrix} = \begin{Bmatrix} 50\left(1 - \dfrac{1}{\omega\Delta t}\right) \\ 50/\Delta t \end{Bmatrix}$$

$$\begin{Bmatrix} \dot{x} \\ \ddot{x} \end{Bmatrix}_{5\text{ms}} = \begin{Bmatrix} 18.17G \\ 50/\Delta t G \end{Bmatrix}$$

$$\begin{Bmatrix} \dot{x} \\ \ddot{x} \end{Bmatrix}_{10\text{ms}} = \begin{Bmatrix} 0 \\ -50/\Delta t \end{Bmatrix} + \begin{bmatrix} 0 & 1/\omega \\ -\omega & 0 \end{bmatrix} \begin{Bmatrix} 18.169 - 50 \\ 50/\Delta t + 50/\Delta t \end{Bmatrix}$$

$$\begin{Bmatrix} \dot{x} \\ \ddot{x} \end{Bmatrix}_{10\text{ms}} = \begin{Bmatrix} \dfrac{100}{\omega\Delta t} \\ 0 \end{Bmatrix} = \begin{Bmatrix} 63.7G \\ 0 \end{Bmatrix} \quad (\ddot{x}_{max} \text{ since } \dot{x} = 0)$$

where $\ddot{x}_0 = \dot{x}_0 = 0$ and $V_j = \dot{U}_j$

The maximum acceleration of 63.7G determined by an exact method illustrates the conservatism of the impact evaluation for a shock parameter $f_n T_p$ approaching the limiting quantity of ⅛. Q.E.D.

EXAMPLE 8.16. Determine the acceleration response at 0.0055s and 0.011s of the idealized 40 Hz assembly evaluated by the impact method in Example 8.14(a) when $\ddot{x}(0) = \dddot{x}(0) = 0$.

Parameters used in (8.95) are determined for t = 0.0055s as

$$Q = \sqrt{40};$$

$$\omega = (2\pi)(40) = 251.33 \text{ rad/s};$$

$$\omega_D = 251.33 \left[1 - \frac{1}{4(40)} \right]^{1/2} = 250.54 \text{ rad/s};$$

$$\sigma = \tan^{-1} \frac{251.33}{2(250.54)\sqrt{40}} = 0.0793;$$

$$\Delta\theta_{0.0055} = (1.0031)(0.8965) \begin{bmatrix} 0.26878 & 0.00391 \\ -246.67 & 0.11328 \end{bmatrix} = \begin{bmatrix} 0.2417 & 0.00351 \\ -221.83 & 0.10187 \end{bmatrix};$$

$$a_1 = [251.33(0.0055)\sqrt{40}]^{-1} = 0.1144;$$

$$a_2 = \frac{1 - 1/40}{[251.33(0.0055)]^2} = 0.5103$$

Using forward differences (8.96c) where $t_n = 0$, $t_{n+1} = 0.0055$s and $t_{n+2} = 0.011$s with $V_j = \ddot{U}_j$ gives

$$B_0 = V_{n+1}(1 + 2a_2) = 101.03G$$

$$B_1 = 2a_1 V_{n+1}/\Delta t = 2080\dot{G}$$

$$B_2 = 2V_{n+1}(a_1 - a_2) = -39.59G$$

$$B_3 = -2V_{n+1}(a_1 + 1)/\Delta t = -20262\dot{G}$$

The acceleration and jerk at $t = 0.0055$s is obtained from (8.95) as

$$\begin{Bmatrix} \ddot{x} \\ \dddot{x} \end{Bmatrix}_{0.0055} = \begin{Bmatrix} 101.03 \\ 2080. \end{Bmatrix} + \begin{bmatrix} 0.2417 & 0.00351 \\ -221.83 & 0.10187 \end{bmatrix} \begin{Bmatrix} -39.59 \\ -20262 \end{Bmatrix} = \begin{Bmatrix} 20.34G \\ 8798.2\dot{G} \end{Bmatrix}$$

Central differences (8.96d) are used at $t_n = 0.0055$s to obtain the response at $t_n + \Delta t = 0.0011$s giving

$$B_0 = 2V_n(a_1 + a_2) = 62.4G$$

$$B_1 = 2V_n(a_1 - 1)/\Delta t = -16102\dot{G}$$

$$B_2 = -V_n(1 + 2a_2) = -101.03G$$

$$B_3 = -2a_1 V_n/\Delta t = -2080\dot{G}$$

The response at $t = 0.011$s is obtained from (8.95) as

$$\begin{Bmatrix} \dot{x} \\ \ddot{x} \end{Bmatrix}_{0.011} = \begin{Bmatrix} 62.4 \\ -16102 \end{Bmatrix} + \begin{bmatrix} 0.2417 & 0.00351 \\ -221.83 & 0.10187 \end{bmatrix} \begin{Bmatrix} 20.34 - 101.03 \\ 8798.2 - 2080 \end{Bmatrix} = \begin{Bmatrix} 66.47G \\ 2481.6\dot{G} \end{Bmatrix}$$

The displacement may be determined from (8.90) as

$$x_{0.011} = 66.47(386)/251.33^2 = 0.406 \text{ in}$$

Since $\ddot{x} \neq 0$, the acceleration due to $\dot{x}$ can be accounted for by (8.99) which may be expressed in terms of $\ddot{x}$ and $\dot{x}$ as

$$\ddot{x}_{\text{abs}} = -\ddot{x} - \frac{\dot{x}}{Q\omega}$$

At 0.011s, the maximum acceleration becomes

$$(\ddot{x}_{\text{abs}})_{0.011} = -68G$$

The Shock Spectrum.   Maximum response quantities can be determined by graphing (8.95) at a sufficient number of points. Maxima can also be determined from a spectra developed by a known shock pulse.

A spectrum of a specific shock pulse represents the peak acceleration response of a series of simple linear spring mass systems expressed in terms of their fundamental frequencies. *Shock spectra are dependent upon the shape and amplitude of the pulse.* Shock spectra of undamped spring-mass systems for commonly encountered pulse shapes are illustrated in Table 8.4. Relationships that define the maximum response amplitude as a function of the shock parameter $f_n T$ are also included.

Maximum amplitudes determined from Table 8.4 establish upper response limits for determining design characteristics. Shock spectra are often used to specify the equipment shock environment and the tests used to verify its behavior. Spectra of known pulse shapes may be used to envelope an unfamiliar or arbitrarily derived spectrum to determine the composition of the pulse histories required to simulate the environment.

EXAMPLE 8.17.   Use the shock spectrum to determine the maximum acceleration response; neglect damping of the 40 Hz idealized assembly that was evaluated in Example 8.14(a).

Table 8.4. Shock Spectra of Undamped Spring Mass Systems

Time History	Shock Spectrum	Spectrum Relationships

I Half-Sine

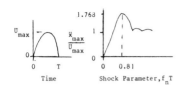

$$\frac{\ddot{x}_{max}}{\ddot{U}_{max}} = \frac{4f_n T}{1 - 4(f_n T)^2} \cos(\pi f_n T) \qquad f_n T < \tfrac{1}{2}$$

$$\frac{\ddot{x}_{max}}{\ddot{U}_{max}} = \frac{2f_n T}{2f_n T - 1} \sin\left(\frac{2\pi}{2f_n T + 1}\right) \qquad \tfrac{1}{2} < f_n T < \tfrac{3}{2}$$

$$\frac{\ddot{x}_{max}}{\ddot{U}_{max}} = \frac{2f_n T}{2f_n T - 1} \sin\left(\frac{4\pi}{2f_n T + 1}\right) \qquad \tfrac{3}{2} < f_n T < \tfrac{5}{2}$$

$$\frac{\ddot{x}_{max}}{\ddot{U}_{max}} \simeq 1.0 \text{ (equiv. accel)} \qquad f_n T > \tfrac{5}{2}$$

II Terminal Sawtooth

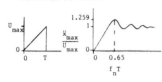

$$\frac{\ddot{x}_{max}}{\ddot{U}_{max}} = \left[1 + \frac{1 - \cos 2\pi f_n T}{2(\pi f_n T)^2} - \frac{\sin 2\pi f_n T}{\pi f_n T}\right]^{1/2}$$

$$\frac{\ddot{x}_{max}}{\ddot{U}_{max}} \simeq 1.0 \text{ (equiv. accel.)} \quad f_n T > 3$$

III Rectangular

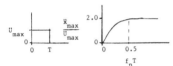

$$\frac{\ddot{x}_{max}}{\ddot{U}_{max}} = 2 \sin(\pi f_n T) \quad f_n T < \tfrac{1}{2}$$

$$\frac{\ddot{x}_{max}}{\ddot{U}_{max}} \simeq 2.0 \text{ (equiv. accel.)} \quad f_n T \geq \tfrac{1}{2}$$

IV Step Ramp

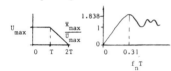

$$\frac{\ddot{x}_{max}}{\ddot{U}_{max}} = \left[\frac{\sin \pi f_n T}{\pi f_n T} - \frac{2}{\pi f_n T} \sin(\pi f_n T) \cos(3\pi f_n T) + 1\right]^{1/2}$$

V Ramp Step

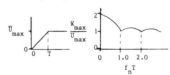

$$\frac{\ddot{x}_{max}}{\ddot{U}_{max}} = 1 + \left|\frac{\sin \pi f_n T}{\pi f_n T}\right|$$

$$\frac{\ddot{x}_{max}}{\ddot{U}_{max}} \simeq 1.0 \text{ (equiv. accel)} \quad f_n T > 3$$

VI Exponential

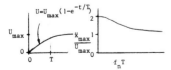

$$\frac{\ddot{x}_{max}}{\ddot{U}_{max}} = 1 + \frac{1}{[1 + (2\pi f_n T^2]^{1/2}}$$

431

Table 8.4 Shock Spectra of Undamped Spring Mass Systems (Cont.)

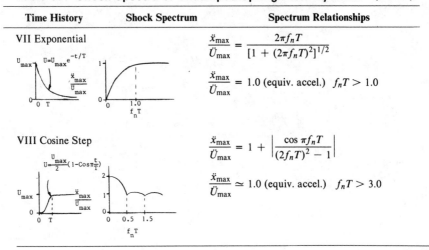

Time History	Shock Spectrum	Spectrum Relationships
VII Exponential		$\dfrac{\ddot{x}_{max}}{\ddot{U}_{max}} = \dfrac{2\pi f_n T}{[1 + (2\pi f_n T)^2]^{1/2}}$
		$\dfrac{\ddot{x}_{max}}{\ddot{U}_{max}} = 1.0$ (equiv. accel.) $f_n T > 1.0$
VIII Cosine Step		$\dfrac{\ddot{x}_{max}}{\ddot{U}_{max}} = 1 + \left\lvert \dfrac{\cos \pi f_n T}{(2f_n T)^2 - 1} \right\rvert$
		$\dfrac{\ddot{x}_{max}}{\ddot{U}_{max}} \simeq 1.0$ (equiv. accel.) $f_n T > 3.0$

The shock parameter based on the pulse period $T = 0.015s$ and the fundamental frequency of the system $f_n = 40$ Hz becomes

$$f_n T = (0.015)(40) = 0.6$$

The maximum acceleration response is determined from case I of Table 8.4 as

$$\ddot{x}_{max} = \frac{50(2)(0.6)}{2(0.6) - 1} \sin \left[\frac{2\pi}{2(0.6) + 1} \right] = 84.5G$$

This result may be taken conservatively as the limiting condition for a design. It is unlikely that the actual acceleration would exceed this value. Q.E.D.

The spectrum also provides a convenient method for evaluating the upper bound of the maximum response of idealizations with several degrees of freedom. The shock response is determined from the peak acceleration corresponding to the frequencies of the retained degrees of freedom weighted in proportion to their effect on the dynamic behavior of the system. Maximum acceleration and displacement responses are obtained by summing these contributions as

$$\ddot{x} = \sum_{i}^{n} |\phi_i \ddot{x}_i p_i| \qquad (8.100a)$$

$$\dot{x} = \sum_{i}^{n} \left\lvert \phi_i \left(\frac{\ddot{x}}{\omega^2} \right)_i p_i \right\rvert \qquad (8.100b)$$

given

$$p_i = \frac{\sum_j W_j \phi_{ij}}{\sum_j W_j \phi_{ij}^2} \quad \text{(modal participation factor);}$$

$$W_j = \text{weight at coordinate } j; \tag{8.100c}$$

$$\phi_{ij} = j\text{th modal displacement of mode } i$$

EXAMPLE 8.18. Determine the maximum response of a uniformly weighted 2 pound assembly idealized as case II of Figure 8.6(a) when subjected to a 20G terminal sawtooth shock with a period of 0.011 second using the modal frequencies of 80, 318 and 675 Hz corresponding to the mode shapes shown in Figure 8.6(b).

The mode shapes shown in Figure 8.6(b) may be assembled into a modal matrix with columns corresponding to the ascending frequencies

$$
\begin{array}{ccc}
(80) & (318) & (675)
\end{array}
$$

$$[\phi] = \begin{bmatrix} 1 & 1 & 1 \\ \sqrt{2} & 0 & -\sqrt{2} \\ 1 & -1 & 1 \end{bmatrix}$$

Modal participation factors are determined from (8.100c) using the columns of $[\phi]$ and $W_i = W/3$ as

$$p_{80} = \frac{(W/3)(2 + \sqrt{2})}{(W/3)(4)} = 0.853$$

$$p_{318} = 0$$

$$p_{675} = \frac{(W/3)(2 - \sqrt{2})}{(W/3)(4)} = 0.146$$

The acceleration amplification ratios $\ddot{x}/\ddot{u}$ are obtained from case II of Table 8.4 for $f_n T = 0.88, 3.5$ and 7.4 corresponding to the frequencies of 80, 318 and 675 respectively:

$$\left(\frac{\ddot{x}}{\ddot{U}}\right)_{80} = 1.12; \quad \left(\frac{\ddot{x}}{\ddot{U}}\right)_{318} \simeq \left(\frac{\ddot{x}}{\ddot{U}}\right)_{655} = 1.0$$

The peak response of each coordinate to the 20G shock is obtained from (8.100):

$$(\ddot{x}_1)_{\text{peak}} = |(1)(1.12 \times 20)(0.853)| + 0 + |(1)(20)(0.146)| = 22G$$

$$(\ddot{x}_2)_{\text{peak}} = |\sqrt{2}(1.12 \times 20)(0.853)| + 0 + |(-\sqrt{2})(20)(0.146)| = 31.1G$$

$$(\ddot{x}_3)_{\text{peak}} = |(1)(1.12 \times 20)(0.853)| + 0 + |(1)(20)(0.146)| = 22G$$

$$(x_1)_{\text{peak}} = \left| \frac{(1)(1.12 \times 20)(0.853)}{[2\pi(80)]^2} \right| + 0 + \left| \frac{(1)(20)(0.146)}{[2\pi(675)]^2} \right| = 7.58 \times 10^{-5} \text{ in}$$

The peak displacements of coordinates 2 and 3 are similarly obtained as

$$(x_2)_{\text{peak}} = 1.07 \times 10^{-4} \text{ in}$$

$$(x_3)_{\text{peak}} = 7.58 \times 10^{-5} \text{ in} \hspace{3cm} \text{Q.E.D.}$$

Considerations That Limit Displacements and Stresses. The upper bound of the shock response developed from the shock spectrum can be used to define configuration characteristics that will assure survivability and provide a basis for establishing clearances between adjacent assemblies. Stresses at the attachments of assemblies and components evaluated at the upper bound acceleration and displacement response assures that actual quantities will be within the allowable limits for the materials used.

Stresses may be reduced by increasing stiffness or flexibility at appropriate locations determined from the idealized configuration. Increasing stiffness is more effective than an increase in flexibility for reducing peak response stresses. Stresses can also be reduced by increasing the section modulus (I/C) of the structure, by avoiding concentration of mass through the employment of several smaller components rather than a large single device, or by adding additional support or improving the constraints provided by existing attachments. Fasteners should be positioned to reduce the moments (stresses) developed by the loading. The combined stresses on these fasteners and the effect of mass changes as a result of design changes should also be considered.

Snubbers can be used to limit deflections with the result of increased acceleration loads that incur greater stresses. Snubbing occurs when the clearance around an assembly is less than the response displacement due to the applied loads. At the instant of contact, the velocity attained by the responding motion of the assembly is reversed incurring a velocity change analogous to impact. The developed impact acceleration is dependent upon the elastic characteristics at the location of contact and the ratio of the available design clearances Δ_d to the displacement δ required for free vibratory motion. The maximum acceleration at impact can be obtained from the response illustrated in Figure 8.18(a) as

$$\ddot{x}_{\text{snubbed}} = 2\pi(1 + r)\dot{U}\left[1 - \left(\frac{\Delta_d}{\delta}\right)^2 \right]^{1/2} f_n \sqrt{\frac{k_s}{k}} \left(1 - \frac{\pi}{4Q_s} \right) \hspace{1cm} (8.101)$$

where k_s represents the snubber stiffness or the equivalent stiffness of the adjacent assembly relative to the point of impact and r represents the coefficient of restitution at the contacting interface. For most configurations where inelastic contact occurs, $r = 0.6$ to 0.8. A value of $r = 1$ represents an elastic contact with perfect rebound.

The increased acceleration loading due to snubbing may be visualized by evaluating the ratio of $\ddot{x}_{snubbed}$ to the acceleration response due to free vibration $\ddot{x} = 2\pi U f_n$

$$\frac{\ddot{x}_{snubbed}}{\ddot{x}} = (1 + r)\left[1 - \left(\frac{\Delta_d}{\delta}\right)^2\right]^{1/2}\sqrt{\frac{k_s}{k}}\left(1 - \frac{\pi}{4Q_s}\right) \qquad (8.102)$$

for a snubbed limited clearance of $\Delta_d = 0.03$ inches with the following configuration characteristics:

Assembly: $Q = 10$, $k = 5000$ and $\delta = 0.1$

Snubber: $Q_s = 30$, $k_s = 25000$ and $r = 0.8$

Substituting these quantities into (8.102) gives

$$\frac{\ddot{x}_{snubbed}}{\ddot{x}} = (1 + 0.8)\left[1 - \left(\frac{0.03}{0.1}\right)^2\right]^{1/2}\sqrt{\frac{25000}{5000}}\left[1 - \frac{\pi}{4(30)}\right]$$

$$\frac{\ddot{x}_{snubbed}}{\ddot{x}} = 3.74$$

An amplification of 3.74 due to snubbing will affect structural and attachment stresses throughout the assembly.

The velocity change that characterizes impact may be applied to develop the response history of a spring mass system to an acceleration shock pulse of any shape. The necessary velocity change simulating impact is obtained by dividing the given pulse into uniform increments Δt that are small compared to the period of the shock T and to the period of the spring mass idealization $1/f_n$. The undamped response due to increment Δt with acceleration $\ddot{U}$ becomes

$$\ddot{x}(t) = \omega \Delta t \ddot{U} \sin \omega t \qquad (8.103)$$

Because of linearity, the response (8.103) for a particular increment is independent of the motions developed by other increments. A composite summation of the motions developed by all the increments represents the total response which at any time t may be expressed as

$$\ddot{x}(t) = \omega\Delta t \sum_{n=0}^{T/\Delta t-1} \ddot{U}_{n+1} \sin \omega(t - n\Delta t) \qquad (8.104)$$

The maximum acceleration for any spring mass frequency ω and hence the shock spectra for the given pulse may be determined by evaluating the magnitude at the end of the pulse ($t = T$) using

$$\ddot{x}_{max} = \left\{ [\ddot{x}(t)]^2 + \left[\frac{\ddot{x}(T)}{\omega} \right]^2 \right\}^{1/2} \qquad (8.105)$$

where $\ddot{x}(T)$ is determined from (8.104) and

$$\frac{\ddot{x}}{\omega}(T) = \omega\Delta t \sum_{n=0}^{T/\Delta t-1} \ddot{U}_{n+1} \cos \omega(T - n\Delta t)$$

EXAMPLE 8.19.   Determine the maximum acceleration experienced by a 50 Hz undamped system to the shock pulse described in Figure 8.20.
    The shock pulse is conveniently divided by an increment $\Delta t = 0.0025$s. Acceleration amplitudes $\ddot{U}_i$ for each increment are determined at the midpoint of each increment. Substituting quantities into (8.104) gives the response at any time t as

$$\ddot{x}(t) = 0.785 [12.5 \sin 314(t) + 37.5 \sin 314(t - 0.0025)$$
$$+ 37.5 \sin 314(t - 0.005) + 12.5 \sin 314(t - 0.0075)]$$

The acceleration at $t = 0.01$s becomes

$$\ddot{x}(T) = 57.23G$$

Determining $\ddot{x}(T)/\omega$ similarly and substituting into (8.105) gives

$$\ddot{x}_{max} = [(57.23)^2 + (-23.64)^2]^{1/2} = 61.9G$$

This maximum acceleration compares closely with the result obtained in Example 8.15.   Q.E.D.

    Use of increments Δt provides the basis of performing numerical integration of the equations of motion for idealizations with any number of degrees of freedom. If increments are selected which satisfy $\Delta t f_n < 0.05$, an approximation using step by step integration becomes

$$\left[\frac{6M}{\Delta^2} + \frac{3C}{\Delta} + K\right] \{x\}_{n+1} = \left[\frac{6M}{\Delta^2} + \frac{3C}{\Delta}\right] \{x\}_n$$

$$+ \left[\frac{6M}{\Delta} + 2C\right] \{\dot{x}\}_n + \left[2M + \frac{\Delta C}{2}\right] \{\ddot{x}\}_n + F_{n+1} \quad (8.106)$$

where $\Delta = \Delta t$

$$\dot{x}_{n+1} = \frac{3}{\Delta} (x_{n+1} - x_n) - 2\dot{x}_n - \frac{\Delta}{2} \ddot{x}_n$$

$$\ddot{x}_{n+1} = \frac{6}{\Delta^2} (x_{n+1} - x_n) - \frac{6}{\Delta} \dot{x}_n - 2\ddot{x}_n$$

F_{n+1} equals imposed loading at $t + \Delta t$ (equivalent ramp time history); M, C and K representing the mass, damping or stiffness respectively in (8.106) may be scalar or matrix quantities.

EXAMPLE 8.20.   A 40 pound assembly which is vibration-isolated at 20 Hz is subjected to a ramp shock defined by $F(t) = 100t$ pounds. Determine the transient response during the first 0.01 seconds neglecting damping.
   The suspension stiffness determined from the weight and fundamental frequency becomes

$$k = \frac{40}{386} [2\pi(20)]^2 = 1635 \text{ lb/in}$$

For uniform time increments of $\Delta t = 0.0025$s ($\Delta t f_n = 0.05$), the displacement response is obtained from (8.106) as

$$[99456 + 1635] \{x\}_{n+1} = [99456] \{x\}_n + [248.6] \{\dot{x}\}_n + [0.2072] \{\ddot{x}\}_n + F(t)$$

which may be recasted as

$$\{x\}_{n+1} = [0.9838] \{x\}_n + [0.0025] \{\dot{x}\}_n$$

$$+ [2.083(10^{-6})] \{\ddot{x}\} + 1.006(10^{-5})F(t) \quad (8.107)$$

The response at $t = 0.0025$s for $F_{n+1} = 0.25$ lb becomes

$$\begin{array}{ccc} & (8.107) & (\text{exact}) \\ \left\{ \begin{array}{c} x \\ \dot{x} \\ \ddot{x} \end{array} \right\}_{0.0025s} = & \left\{ \begin{array}{ll} 2.515(10^{-6}) & \text{in} \\ 0.0030 & \text{in/s} \\ 2.4144 & \text{in/s}^2 \end{array} \right\} & \left\{ \begin{array}{ll} 2.500(10^{-6}) & \text{in} \\ 0.003 & \text{in/s} \\ 2.3727 & \text{in/s}^2 \end{array} \right\} \end{array}$$

where $x_0 = \dot{x}_0 = \ddot{x}_0 = 0$. Substituting these response characteristics into (8.107) for $F_{0.005} = 0.5$ lb gives the motion at $t = 0.005$s as

$$
\begin{array}{cc}
(8.107) & (\text{exact})
\end{array}
$$

$$
\left\{ \begin{array}{c} x \\ \dot{x} \\ \ddot{x} \end{array} \right\}_{0.005s} = \left\{ \begin{array}{ll} 2.003(10^{-5}) & \text{in} \\ 0.012 & \text{in/s} \\ 4.7889 & \text{in/s}^2 \end{array} \right\} \left\{ \begin{array}{ll} 1.9708(10^{-5}) & \text{in} \\ 0.0117 & \text{in/s} \\ 4.5134 & \text{in/s}^2 \end{array} \right\}
$$

Successive steps give

$$
\begin{array}{cc}
(8.107) & (\text{exact})
\end{array}
$$

$$
\left\{ \begin{array}{c} x \\ \dot{x} \\ \ddot{x} \end{array} \right\}_{0.0075s} = \left\{ \begin{array}{ll} 6.7226(10^{-5}) & \text{in} \\ 0.0266 & \text{in/s} \\ 6.9302 & \text{in/s}^2 \end{array} \right\} \left\{ \begin{array}{ll} 6.489(10^{-5}) & \text{in} \\ 0.02519 & \text{in/s} \\ 6.2127 & \text{in/s}^2 \end{array} \right\}
$$

$$
\left\{ \begin{array}{c} x \\ \dot{x} \\ \ddot{x} \end{array} \right\}_{0.010s} = \left\{ \begin{array}{ll} 1.5713(10^{-4}) & \text{in} \\ 0.0460 & \text{in/s} \\ 8.6099 & \text{in/s}^2 \end{array} \right\} \left\{ \begin{array}{ll} 1.4859(10^{-4}) & \text{in} \\ 0.0422 & \text{in/s} \\ 7.3045 & \text{in/s}^2 \end{array} \right\}
$$

Results from the exact solution $x(t) = 100[t - \sin(\omega t)/\omega]/k$ are included for comparison. Greater accuracy is achieved with smaller increments since each successive result is based on the previously determined response history. Q.E.D.

REFERENCES

Acton, F. S., *Numerical Methods That Work,* Harper and Row, New York (1956).

Ausman, J. S., "Amplitude Frequency Response Analysis and Synthesis of Unfactored Transfer Functions," *Trans. ASME (Basic Engineering),* Mar. 1964.

Baker, E., "Calculation of Thermally Induced Mechanical Stresses in Incapsulated Assemblies," *IEEE Trans.,* Vol. PMP-6, No. 4 (Dec. 1970), p. 121.

Baker, E., "Some Effects of Temperature on Material Properties and Device Reliability," *IEEE Trans.,* Vol. PMP-8, No. 4 (Dec. 1970), p. 121.

Becker G., "Testing and Results Related to the Mechanical Strength of Solder Joints," IPC Fall Meeting, Sept. 1979.

Bikerman, J. J., "Theory of Peeling Through a Hookean Solid," *J. App. Physics,* Vol. 28, No. 12 (Dec. 1957), p. 1484.

Burgreen, D., "The Thermal Ratchet Mechanism," *Trans. ASME (Basic Engineering),* Sept. 1968, p. 319.

Cannon, R. H., *Dynamics of Physical Systems,* McGraw-Hill Book Company, New York, 1967.

Chen, W. T., and Nelson C. W., "Thermal Stress in Bonded Joints," *IBM Jour. Res. and Dev.,* Vol. 23, No. 2 (Mar. 1979), p. 179.

Chu, T. Y., "A Hydrostatic Model of Solder Fillets," *The Western Electric Engineer,* Vol. 19, No. 2 (Apr. 1975).

Crandall, S. H., *Engineering Analysis,* McGraw-Hill Book Company, New York, 1956.

Crandall, S. H., and Mark, W. O., *Random Vibration,* Academic Press, New York, 1973.

Crede, C. E., Vibration and Shock Isolation, John Wiley & Sons, New York, 1951.

Crede, C., and Harris, M., *Shock and Vibration Handbook,* McGraw-Hill Book Company, New York, 1961.

Doetsh, G., *Guide to the Application of Laplace Transforms,* W. B. Saunders, Phildelphia, 1966.

Eshleman, A. L., Jr; Van Dyke, J. D., Jr; and Belcher, R. M., "A Procedure for Designing and Testing Aircraft Structures Loaded by Jet Noise," *ASME 59-AV-48, Aviation Conference,* 1959.

Gatewood, B. E., *Thermal Stresses,* McGraw-Hill Book Company, New York, 1957.

Gutkowski, R. M., *Structures,* Van Nostrand Reinhold, New York, 1981.

Kuhn, P., *Stresses in Aircraft and Shell Structures,* McGraw-Hill Book Company, New York, 1953.

Langhaar, H. L., *Energy Methods in Applied Mechanics,* John Wiley & Sons, New York, 1962.

Leissa, A. W., "Vibration of Plates," *NASA SP-160,* 1969.

Meyers, G. E., *Analytical Methods in Conduction Heat Transfers,* McGraw-Hill Book Company, New York, 1971.

Miller, D. R., "Thermal-Stress Ratchet Mechanism in Pressure Vessels," *Trans. ASME (Basic Engineering),* June 1959, p. 190.

Mohler, J. B., "Solder Joints Vs. Time and Temperature," *Machine Design,* Apr. 15, 1971, p. 84.

Munford, J. W., "The Influence of Several Design and Material Variables on the Propensity for Solder Joint Cracking," *IEEE Trans.,* Vol. PHP-11, No. 4 (Dec. 1975), p. 296.

Mustin, G. S., "Theory and Practice of Cushion Design," The Shock and Vibration Information Center, United States Department of Defense, 1968.

Newland, D. E., *Random Vibration and Spectral Analysis,* Longman, London, 1975.

Meek, J. L., *Matrix Structural Analysis,* McGraw-Hill Book Company, New York, 1971.

Nixon, F. E., *Handbook of Laplace Transformation,* Prentice-Hall, Englewood Cliffs, N.J., 1965.

Roberts, G. E., and Kaufman, H., *Table of Laplace Transforms,* W. B. Saunders, Philadelphia, 1966.

Rubinstein, M. F., *Matrix Computer Analysis of Structures,* Prentice-Hall, Englewood Cliffs, N.J., 1966.

Steinberg, D. S., *Cooling Techniques for Electronic Equipment,* John Wiley & Sons, New York, 1980.

Szilard, R., *Theory and Analysis of Plates,* Prentice-Hall, Englewood Cliff, N.J., 1974.

Taylor, T. C., and Yuan, F. G., "Thermal Stress and Fracture in Shear-Constrained Semiconductor Device Structures," *IRE Trans.,* Vol. ED-6 (July 1959), p. 303.

Thom, A., and Apelt, C. J., *Field Computations in Engineering and Physics,* Van Nostrand Reinhold, New York, 1961.

Thomson, W. T., *Introduction to Space Dynamics,* John Wiley & Sons, New York, 1961.

Timoshenko, T.; Young, D. H.; and Weaver, W., Jr., *Vibration Problems in Engineering,* John Wiley & Sons, New York, 1974.

Van Santen, G. W., *Introduction to a Study of Mechanical Vibration,* MacMillan, New York, 1958.

Wang, Chu-Kia, *Statically Indeterminate Structures,* McGraw-Hill Book Company, New York, 1953.

APPENDICES

A. Matrix Mechanics

B. Transfer Functions

C. Properties of Materials, Solids and Shells

D. Conversions

Appendix A
Matrix Mechanics

The matrix is one of the many tools available to the designer for organizing the parameters of a problem under study and performing the operations and reasoning to obtain a solution in an accurate rapid way. The matrix is a shorthand representation of the algebraic relationships describing the physical representation of actual equipment.

Complete algebraic descriptions and the analyses of modeled systems may consist of mathematical representations involving many variables. The operations on these variables may entail numerous transformations from one framework to another. The effort required in generating the sets of equations necessary for mathematical development, with all the possibility of error, led to the formulation of a contracted notation in which a rectangular pattern of coefficients is detached from their variables. This *matrix of coefficients* can be considered an operator acting on the variables.

Sets of simultaneous equations are the inevitable result of the analysis of complex systems. The use of matrices and determinants provides the advantage of speed, accuracy and versatility, offering a high degree of insurance against loss of critical descriptive data that otherwise might occur in other solution processes of these equations.

A.1 MATRIX DEFINITION

Consider generating the algebraic relationships that describe the thermal model of Figure A.1-1. The relationships which are based on a heat balance at each node for the unknown temperatures become:

T_1 (Node 1): $G_1(T_1 - T_{s1}) + G_2(T_1 - T_2) - q_1 = 0$

T_2 (Node 2): $G_2(T_2 - T_1) + G_3(T_2 - T_{s2}) + G_4(T_2 - T_3) = 0$

T_3 (Node 3): $G_4(T_3 - T_2) + G_5(T_3 - T_{s3}) - q_3 = 0$

$$(A.1a)$$

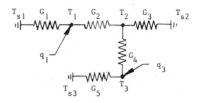

Figure A.1-1. Thermal Conductance Model.

Collecting conductances for the variables, T_1, T_2 and T_3 for each expression gives

Node 1: $(G_1 + G_2)T_1 - G_2T_2$ $\qquad = q_1 + G_1T_{s1}$

Node 2: $-G_2T_1 + (G_2 + G_3 + G_4)T_2 - G_4T_3 = G_3T_{s2}$ $\qquad$ (A.1b)

Node 3: $-G_4T_2 + (G_4 + G_5)T_3$ $\qquad = q_3 + G_5T_{s3}$

The simultaneous equations (A.1b) or (A.1a) describe the behavior of the model shown in Figure A.1-1. The solution for T_1, T_2 and T_3 can be determined by direct substitution and elimination. This method for anything but rudimentary sets of equations is impractical and is conducive to error.

The matrix is a special shorthand both to represent the sets of equations and manipulate them efficiently. The set of equations (A.1b), for example, could be written in the matrix form

$$[G]\{T\} = \{Q\} \qquad (A.2)$$

The meaning of the matrix notation used in (A.2) is almost evident from a comparison of (A.2) and (A.1b). To see the logic, consider the set (A.1b) with each variable T_1, T_2 and T_3 written at the top of the appropriate column instead of writing them in the equation

$$
\begin{array}{ccc}
T_1 & T_2 & T_3
\end{array}
$$
$$
\begin{bmatrix}
(G_1 + G_2) & -G_2 & 0 \\
-G_2 & (G_2 + G_3 + G_4) & -G_4 \\
0 & -G_4 & (G_4 + G_5)
\end{bmatrix}
=
\begin{Bmatrix}
q_1 + G_1T_{s1} \\
G_3T_{s2} \\
q_3 + G_5T_{s3}
\end{Bmatrix}
$$

$$(A.3)$$

Next, we move the row of variables, T_i around the end of the bracketed coefficients, $[G]$.

$$\begin{bmatrix} (G_1 + G_2) & -G_2 & 0 \\ -G_2 & (G_2 + G_3 + G_4) & -G_4 \\ 0 & -G_4 & (G_4 + G_5) \end{bmatrix} \begin{Bmatrix} T_1 \\ T_2 \\ T_3 \end{Bmatrix} = \begin{Bmatrix} q_1 + G_1 T_{s1} \\ G_3 T_{s2} \\ q_3 + G_5 T_{s3} \end{Bmatrix}$$

$$\text{(A.4)}$$

The equations (A.1b) are now in matrix form. The bracketed coefficients, $[G]$, is referred to as a 3×3 matrix for obvious reasons. The braced terms $\{T\}$ and $\{Q\}$ are called *column vectors*. In (A.2), $\{Q\}$ is a compact representation for the the nodal energy vector

$$\{Q\} = \begin{Bmatrix} q_1 + G_1 T_{s1} \\ G_3 T_{s2} \\ q_3 + G_5 T_{s3} \end{Bmatrix} \qquad \text{(A.5)}$$

It is important to recognize that column "vectors" are not physical vectors like velocity or force; column vectors are vectors only in the mathematical sense. A column vector is merely a set of numbers which have physical meaning. In (A.2), $\{T\}$ represents node temperatures. In a mechanical system, $\{\dot{x}\}$ represents the column vector for node velocities.

The solution for the node temperatures in (A.2) can be obtained algebraically by dividing through by $[G]$. In matrix notation the results becomes

$$\{T\} = [G]^{-1}\{Q\} \qquad \text{(A.6)}$$

In (A.6) $[G]^{-1}$ symbolizes the inverse of the $[G]$ matrix. The matrix inverse is discussed in Section A.7. Equation (A.6) represents the solution for the unknown node temperatures which can be obtained directly by matrix multiplication, discussed in Section A.3.

A matrix is an array of numerical quantities identified by the number of rows and columns. The numerical quantities may be represented by mathematical symbols. If matrix $[A]$ equals $[G]$ of (A.2), we have

$$[A]_{3\times 3} = \begin{bmatrix} a_{11} & a_{12} & a_{13} \\ a_{21} & a_{22} & a_{23} \\ a_{31} & a_{32} & a_{33} \end{bmatrix} = \begin{bmatrix} (G_1 + G_2) & -G_2 & 0 \\ -G_2 & (G_2 + G_3 + G_4) & -G_4 \\ 0 & -G_4 & (G_4 + G_5) \end{bmatrix}$$

$$\text{(A.7)}$$

Thus, $a_{11} = (G_1 + G_2)$, $a_{12} = -G_2$, $a_{13} = 0$, etc., are the coefficients of matrix $[\mathbf{A}]$.

The subscripts indicate the (row) $\times$ (column) location of each coefficient; a_{32} is located in the third row and second column of $[\mathbf{A}]$. In another matrix, b_{24} is the coefficient found in the second row and fourth column.

1. The subscript of $[\mathbf{A}]_{3\times3}$ defines the matrix size in terms of its number of rows and columns. A matrix may either be square or rectangular. $[\mathbf{B}]_{4\times5}$ is a rectangular matrix with 4 rows and 5 columns. The order of a square matrix is given by its number of rows.

2. A *diagonal matrix* is a square matrix with numerical quantities only along the diagonal. Matrix $[\mathbf{C}]$, equation (A.8), is a diagonal matrix of the fourth order.

$$[\mathbf{C}] = \begin{bmatrix} C_{11} & & & \\ & C_{22} & & \\ & & C_{33} & \\ & & & C_{44} \end{bmatrix} \tag{A.8}$$

3. A *unit* or *identity matrix* is a diagonal matrix with all coefficients equal to 1. The unit matrix is usually identified with the symbol $[\mathbf{I}]$. Therefore,

$$[\mathbf{I}] = \begin{bmatrix} 1 & 0 \\ 0 & 1 \end{bmatrix} \tag{A.9}$$

is a unit matrix of the second order.

4. A *symmetric matrix* is one in which $a_{ij} = a_{ji}$. Matrix $[\mathbf{G}]$ of equation (A.7) is an example of a symmetric matrix.

A.2 MATRIX ADDITION AND SUBTRACTION

The sum (or difference) of two matrices of the same structure (order) is found by adding (or subtracting) similarly placed elements. The result is a new matrix. Only matrices of the same structure, that is, those containing the same number of rows and columns, may be added. Defining matrices $[\mathbf{B}]$ and $[\mathbf{C}]$ of the third order, we have

$$(1) \text{ Addition: } [\mathbf{A}] = [\mathbf{B}] + [\mathbf{C}] = [\mathbf{C}] + [\mathbf{B}] \tag{A.9}$$

EXAMPLE A.2-1. Determine the sum of matrices $[\mathbf{B}]$ and $[\mathbf{C}]$ given

$$[\mathbf{B}] = \begin{bmatrix} 5 & 2 \\ 2 & 1 \end{bmatrix} \qquad [\mathbf{C}] = \begin{bmatrix} -1 & 2 \\ -1 & 2 \end{bmatrix}$$

$$[\mathbf{A}] = \begin{bmatrix} (5-1) & (2+2) \\ (2-1) & (1+2) \end{bmatrix} = \begin{bmatrix} 4 & 4 \\ 1 & 3 \end{bmatrix}$$

Q.E.D.

(2) Subtraction: $[\mathbf{A}] = [\mathbf{B}] - [\mathbf{C}] = -[\mathbf{C}] + [\mathbf{B}]$ (A.10)

EXAMPLE A.2-2. Determine the difference of matrices $[\mathbf{B}]$ and $[\mathbf{C}]$ of Example A.2-1

$$[\mathbf{A}] = \begin{bmatrix} 5-(-1) & 2-2 \\ 2-(-1) & 1-2 \end{bmatrix} = \begin{bmatrix} 6 & 0 \\ 3 & -1 \end{bmatrix}$$

Q.E.D.

A.3 MATRIX MULTIPLICATION

(1) Multiplication of a matrix by a factor is equivalent to multiplying each element of the matrix by the factor. The product of factor s and matrix $[\mathbf{C}]$ becomes

$$[\mathbf{A}]_{n \times n} = s[\mathbf{C}]_{n \times n} = \begin{bmatrix} sc_{11} & sc_{12} & \cdots & sc_{1n} \\ sc_{21} & sc_{22} & \cdots & sc_{2n} \\ \vdots & \vdots & & \vdots \\ sc_{n1} & sc_{n2} & \cdots & sc_{nn} \end{bmatrix} \qquad (A.11)$$

EXAMPLE A.3-1. Determine the product of $s = -2$ and matrix $[\mathbf{C}]$ of Example A.2-1:

$$[\mathbf{A}] = \begin{bmatrix} (-2)(-1) & (-2)(2) \\ (-2)(-1) & (-2)(2) \end{bmatrix} = \begin{bmatrix} 2 & -4 \\ 2 & -4 \end{bmatrix}$$

Q.E.D.

(2) The multiplication of two matrices is contingent on the correspondence of their row–column structures and position. The multiplication of matrix $[\mathbf{C}]$

by matrix $[B]$ implies $[B][C]$ since generally, $[B][C] \neq [C][B]$. Matrix multiplication is valid only when the number of columns in $[B]$ is equal to the number of rows in $[C]$. Thus

$$[A]_{m \times p} = [B]_{m \times n}[C]_{n \times p} \qquad (A.12)$$

Equations (A.4) and (A.6) meet the condition for multiplication since the number of columns of $[G]^{-1}$ is equal to the number of rows of matrix $\{Q\}$. This is readily seen by expressing (A.6) using the notation of (A.12) as follows:

$$\{T\}_{3 \times 1} = [G]^{-1}_{3 \times 3}\{Q\}_{3 \times 1} \qquad (A.13)$$

EXAMPLE A.3-2.   Determine the product of matrices $[D]$ and $[E]$, given

$$[D] = \begin{bmatrix} 3 & 1 & 2 \\ 1 & 2 & 3 \end{bmatrix}, [E] = \begin{bmatrix} 1 & 2 & 0 \\ 3 & 4 & 6 \\ 5 & 1 & 2 \end{bmatrix}$$

By inspection, the number of columns of $[D]$ equals the number of rows of $[E]$ satisfying the condition for multiplication. With the use of offset matrices, the first column of the matrix product becomes

(1st Col)

$$[A] = [D][E] = \begin{bmatrix} 3 & 1 & 2 \\ 1 & 2 & 3 \end{bmatrix} \quad \overset{\begin{bmatrix} 1 & \cdot & \cdot \\ 3 & \cdot & \cdot \\ 5 & \cdot & \cdot \end{bmatrix}}{} = \begin{bmatrix} (1 \times 3) + (3 \times 1) + (5 \times 2) & \cdot & \cdot \\ (1 \times 1) + (3 \times 2) + (5 \times 3) & \cdot & \cdot \end{bmatrix}$$

(1st Col)

The procedure is repeated for the second column of the product matrix

(2nd Col)

$$[A] = [D][E] = \begin{bmatrix} 3 & 1 & 2 \\ 1 & 2 & 3 \end{bmatrix} \quad \overset{\begin{bmatrix} \cdot & 2 & \cdot \\ \cdot & 4 & \cdot \\ \cdot & 1 & \cdot \end{bmatrix}}{} = \begin{bmatrix} 16 & (2 \times 3) + (4 \times 1) + (1 \times 2) & \cdot \\ 22 & (2 \times 1) + (4 \times 2) + (1 \times 3) & \cdot \end{bmatrix}$$

(2nd Col)

The third column of the product matrix is similarly developed. The solution becomes

$$[A] = \begin{bmatrix} 3 & 1 & 2 \\ 1 & 2 & 3 \end{bmatrix} \cdot \begin{bmatrix} 1 & 2 & 0 \\ 3 & 4 & 6 \\ 5 & 1 & 2 \end{bmatrix} = \begin{bmatrix} 16 & 12 & 10 \\ 22 & 13 & 18 \end{bmatrix}$$

Q.E.D.

The procedure illustrated in Example A.3-2 may be used to obtain the matrix product of square or rectangular matrices of any order provided the column-row requirement is satisfied. Once the technique is mastered, the product matrix can be written directly without portraying the offset multiplicand.

A.4 MATRIX TRANSPOSE

The *transpose* of matrix $[A]$ is a new matrix $[A]^T$ whose rows and columns are interchanged with those of matrix $[A]$. The transpose of $[A]_{m \times p}$ is $[A]^T_{p \times m}$ where the superscript indicates the transpose and each coefficient $a^T_{ji} = a_{ij}$. The transpose operation is similar to pivoting the original matrix $[A]$ about the diagonal. For a symmetric matrix, the transpose equals the original matrix: $[A]^T = [A]$.

EXAMPLE A.4-1.   Determine the transpose of matrix $[D]$ of example A.3-2.

$$[D] = \begin{bmatrix} 3 & 1 & 2 \\ 1 & 2 & 3 \end{bmatrix}, [D]^T = \begin{bmatrix} 3 & 1 \\ 1 & 2 \\ 2 & 3 \end{bmatrix}$$

Q.E.D.

A.5 CHARACTERISTICS OF A DETERMINANT

The *determinant* is a useful tool for solving a system of simultaneous equations, developing the transfer function of a modeled system (Appendix B) or developing the matrix inverse (Appendix A.7). Consider, for example, the following set of simultaneous equations similar to (A.1b):

$$a_{11}x_1 + a_{12}x_2 = b_1 \tag{A.14a}$$

$$a_{21}x_1 + a_{22}x_2 = b_2 \tag{A.14b}$$

A solution for x_1 and x_2 in terms of the coefficients a_{ij} and loads b_i is sought. If equation (A.14a) is multiplied by a_{22} and (A.14b) is multiplied by a_{12}, we have

$$a_{22}a_{11}x_1 + a_{22}a_{12}x_2 = a_{22}b_1 \qquad \text{(A.15a)}$$

$$a_{12}a_{21}x_1 + a_{12}a_{22}x_2 = a_{12}b_2 \qquad \text{(A.15b)}$$

Since $a_{22}a_{12} = a_{12}a_{22}$, subtracting (A.15b) from (A.15a) eliminates the variable x_2, giving

$$(a_{22}a_{11} - a_{12}a_{21})x_1 = a_{22}b_1 - a_{12}b_2 \qquad \text{(A.16)}$$

The variable x_1 can be found directly from equation (A.16). Following a similar approach, an equation for determining x_2 becomes

$$(a_{22}a_{11} - a_{12}a_{21})x_2 = a_{11}b_2 - a_{21}b_1 \qquad \text{(A.17)}$$

Writing (A.14) as a matrix equation similar to (A.2) or (A.4), we obtain a square 2×2 matrix of coefficients [A] as

$$\begin{bmatrix} a_{11} & a_{12} \\ a_{21} & a_{22} \end{bmatrix} \begin{Bmatrix} x_1 \\ x_2 \end{Bmatrix} = \begin{Bmatrix} b_1 \\ b_2 \end{Bmatrix} \qquad \text{(A.18)}$$

The determinant of second or third order square arrays of elements is commonly known. The determinant for the second order array of (A.18), is expressed as

$$\begin{vmatrix} a_{11} & a_{12} \\ a_{21} & a_{22} \end{vmatrix} = (a_{22}a_{11} - a_{12}a_{21}) \qquad \text{(A.19)}$$

The determinant (A.19) is identical to the coefficient of x_1 and x_2 of equations (A.16) and (A.17). The right-hand coefficient of (A.16) and (A.17) can be recovered from the determinant as well. For x_1, replace the coefficient column corresponding to the x_1 variable with the $\{b\}$ vector and evaluate the determinant. This procedure gives the determinant

$$\begin{vmatrix} b_1 & a_{12} \\ b_2 & a_{22} \end{vmatrix} = a_{22}b_1 - a_{12}b_2 \qquad \text{(A.20)}$$

which is the same as the right side of (A.16). The determinant of the array with the x_2 coefficients replaced by vector $\{b\}$ is identical to the right side of

(A.17). The reasoning used to solve the two simultaneous equations (A.18), applies equally to a system containing any number of equations. Characteristics and properties of a determinant are described in the following sections.

(1) Bookkeeping constants have been defined which facilitate the manipulations involved with larger numbers of equations. One such constant is the minor M. The minor for a 2×2 array is the determinant (A.19). Using subscripts to designate the row-column intercepts, equation (A.19) expressed as a minor becomes

$$M_{1 \cdot 2, 1 \cdot 2} = a_{22} a_{11} - a_{12} a_{21} \tag{A.21}$$

The use of subscripts for M becomes clearer for larger arrays. Consider the general array with four coefficients located at the intercepts of rows r, s and columns p, q. The second order (2×2) minor becomes

$$
M_{r \cdot s, p \cdot q} = r
\begin{array}{c}
\quad p \quad q \\
\left|
\begin{array}{ccccc}
\cdot & \cdot & \cdot & \cdot & \cdot \\
\cdot & \cdot & \cdot & \cdot & \cdot \\
\cdot & \cdot & a_{rp} & a_{rq} & \cdot \\
\cdot & \cdot & \cdot & \cdot & \cdot \\
\cdot & \cdot & a_{sp} & a_{sq} & \cdot \\
\cdot & \cdot & \cdot & \cdot & \cdot
\end{array}
\right|
\end{array}
= a_{rp} a_{sq} - a_{rq} a_{sp} \tag{A.22}
$$

Each element of array (A.22) can be considered a first order minor. Using the same notation, the minor at location rp is

$$M_{rp} = a_{rp} \tag{A.23}$$

(2) The number of minors that could be developed depends on the order of the minor, m, and the larger of the number of rows or columns, n, for which the minors are sought. The number of minors are determined from

$$N_m = \frac{n!}{m!(n - m)!} \tag{A.24}$$

The number of minors that could be determined from the first two columns of a fifth order (5×5) array using (A.24) becomes

$$N_m = \frac{5!}{2!(5 - 2)!} = 10$$

Since two columns were specified, $m = 2$ and because there are five rows, $n = 5$.

(3) In any array, the complement of a minor is itself a minor. The complement designated M^* is the determinant of an array which does not have any rows or columns common to M. Consider the fourth order array

$$
\begin{vmatrix}
a & b & c & d \\
e & f & g & h \\
i & k & l & m \\
n & o & p & q
\end{vmatrix}
\tag{A.25}
$$

The second order minor $M_{1\cdot2,1\cdot2}$ has the complementary minor $M^*_{3\cdot4,3\cdot4}$. The first order minor $M_{2\cdot3}$ has a third order complementary minor $M^*_{1\cdot3\cdot4,1\cdot2\cdot4}$. The complementary minor is easily visualized by striking out the intercepts for the minor M. Consider the minor $M_{2\cdot3}$ and line out its intercepts, row 2, column 3:

$$
\begin{vmatrix}
a & b & \cancel{c} & d \\
\cancel{e} & \cancel{f} & \cancel{g} & \cancel{h} \\
i & k & \cancel{l} & m \\
n & o & \cancel{p} & q
\end{vmatrix}
\tag{A.26}
$$

The complement is composed of the remaining elements, which determine its row-column intercepts. Therefore, according to (A.26) we have

$$
M_{2\cdot3} = |g|; \quad M^*_{1\cdot3\cdot4,1\cdot2\cdot4} =
\begin{vmatrix}
a & b & d \\
i & k & m \\
n & o & q
\end{vmatrix}
\tag{A.27}
$$

(4) The determinant of any array may be determined in terms of its minors and their complements according to

$$
|D| = \sum_{1}^{N_m} (-1)^{r+s+..+p+q+..} M_{a\cdot b..,e\cdot f..} M^*_{r\cdot s..,p\cdot q..}
$$

$$
\text{Rows:} \quad a \cdot b \ldots \neq r \cdot s \ldots
\tag{A.28}
$$

$$
\text{Columns:} \; e \cdot f \ldots \neq p \cdot q \ldots
$$

Equation (A.28) shows the determinant to be the sum of the products of the minor and its complement with the sign governed by the sum of the row-column intercept of the complement.

EXAMPLE A.5-1.   Using (A.28), evaluate the determinant

$$\begin{vmatrix} 8 & 6 & 4 \\ 5 & 4 & 2 \\ 3 & 2 & 1 \end{vmatrix}$$

The last two columns are used to determine second order minors. The number of minors from (A.24) are

$$N_m = \frac{3!}{2!(3 - 2)!} = 3$$

The minors and their complements evaluated by inspection become

(1) $M_{1\cdot2,2\cdot3} = 6(2) - 4(4) = -4$ $M^*_{3,1} = 3$

(2) $M_{1\cdot3,2\cdot3} = 6(1) - 4(2) = -2$ $M^*_{2,1} = 5$

(3) $M_{2\cdot3,2\cdot3} = 4(1) - 2(2) = 0$ $M^*_{1,1} = 8$

From (A.28) the determinant becomes

$$|D| = \sum_1^3 (-1)^{r+1} M M^* = (-1)^4 (-4)(3) + (-1)^3 (-2)(5) + (-1)^2 (0)(8)$$

$$|D| = \qquad -2 \qquad = \qquad -12 \qquad\qquad +10 \qquad\qquad + 0$$

Q.E.D.

EXAMPLE A.5-2.   Using (A.28), evaluate the determinant

$$\begin{vmatrix} 8 & 7 & 5 & 4 \\ 6 & 6 & 4 & 3 \\ 4 & 5 & 2 & 1 \\ 3 & 3 & 1 & 1 \end{vmatrix}$$

The first two rows are used to evaluate the second order minors. The number of minors from (A.24) are

$$N_m = \frac{4!}{2!(4-2)!} = 6$$

The minors and their complements determined by inspection are

(1) $M_{1\cdot2,1\cdot2} = 8(6) - 7(6) = 6$ $M^*_{3\cdot4,3\cdot4} = 2(1) - (1)(1) = 1$

(2) $M_{1\cdot2,1\cdot3} = 8(4) - 5(6) = 2$ $M^*_{3\cdot4,2\cdot4} = 5(1) - (1)(3) = 2$

(3) $M_{1\cdot2,1\cdot4} = 8(3) - 4(6) = 0$ $M^*_{3\cdot4,2\cdot3} = 5(1) - (2)(3) = -1$

(4) $M_{1\cdot2,2\cdot3} = 7(4) - 5(6) = -2$ $M^*_{3\cdot4,1\cdot4} = 4(1) - (1)(3) = 1$

(5) $M_{1\cdot2,2\cdot4} = 7(3) - 4(6) = -3$ $M^*_{3\cdot4,1\cdot3} = 4(1) - (2)(3) = -2$

(6) $M_{1\cdot2,3\cdot4} = 5(3) - 4(4) = -1$ $M^*_{3\cdot4,1\cdot2} = 4(3) - 5(3) = -3$

From (A.27) the determinant becomes

$$|D| = \sum^{6} (-1)^{7+p+q} M M^* = (-1)^{14}(6)(1) + (-1)^{13}(2)(2) + (-1)^{12}(0)(-1)$$
$$+ (-1)^{12}(-2)(1) + (-1)^{11}(-3)(-2) + (-1)^{10}(-1)(-3)$$
$$|D| = 6 - 4 + 0 - 2 - 6 + 3 = -3$$

Q.E.D.

(5) Interchanging any row or column with its neighboring row or column reverses the sign of the determinant.

EXAMPLE A.5-3. Evaluate the determinant of Example A.5-1 after interchanging the first two rows. Using (A.28) and minors from the second and third columns

$$|D| = \begin{vmatrix} 5 & 4 & 2 \\ 8 & 6 & 4 \\ 3 & 2 & 1 \end{vmatrix} = (-1)^4(4)(3) + (-1)^3(0)(8) + (-1)^2(-2)(5) = 2$$

Q.E.D.

An interchange of column three with column one in the determinant of Example A.5-2 involves two adjacent column interchanges. Since an even number of adjacent column interchanges has occurred, the sign of the determinant remains unchanged. For the determinant of Example A.5-2

$$
\begin{vmatrix} 8 & 7 & 5 & 4 \\ 6 & 6 & 4 & 3 \\ 4 & 5 & 2 & 1 \\ 3 & 3 & 1 & 1 \end{vmatrix} = \begin{vmatrix} 5 & 7 & 8 & 4 \\ 4 & 6 & 6 & 3 \\ 2 & 5 & 4 & 1 \\ 1 & 3 & 3 & 1 \end{vmatrix} \qquad \text{(A.29)}
$$

(6) Multiplication or division of the determinant by a constant is equivalent to multiplying or dividing any single row or column by the same constant. This determinant property is illustrated using the determinants of Examples A.5-1 and A.5-3.

$$
\begin{vmatrix} 8 & 6 & 4 \\ 5 & 4 & 2 \\ 3 & 2 & 1 \end{vmatrix} = - \begin{vmatrix} 5 & 4 & 2 \\ 8 & 6 & 4 \\ 3 & 2 & 1 \end{vmatrix} = \begin{vmatrix} -5 & 4 & 2 \\ -8 & 6 & 4 \\ -3 & 2 & 1 \end{vmatrix} \qquad \text{(A.30)}
$$

(7) The value of the determinant is unchanged if elements of any row or column are added to or subtracted from the elements of any other row or column.

EXAMPLE A.5-4.   Determine the value of the rightmost determinant of (A.30) after adding the second column to the first.

$$
|D| = \begin{vmatrix} -1 & 4 & 2 \\ -2 & 6 & 4 \\ -1 & 2 & 1 \end{vmatrix} = (-1)^4(4)(-1) + (-1)^3(0)(-2) + (-1)^2(-2)(-1) = -2
$$

Q.E.D.

(8) Equations developed from physical models have special characteristics that could be used advantageously to obtain solutions. The diagonal coefficients are composed of all off-diagonal as well as boundary coefficients. This characteristic was observed in the coefficient matrix of (A.4) which determines the temperature distribution for the thermal model of Figure A.1-1.

Figures A.5-1(a) and A.5-1(b) illustrate the same coefficient structure for determining heat flow in a thermal model and displacement distribution in a spring mass system.

The relationship between the diagonal and off-diagonal coefficients for each variable, i.e., each row, can be used to reduce the effort of obtaining the determinant. The expansion (A.28) contains negative terms that are cancelled by

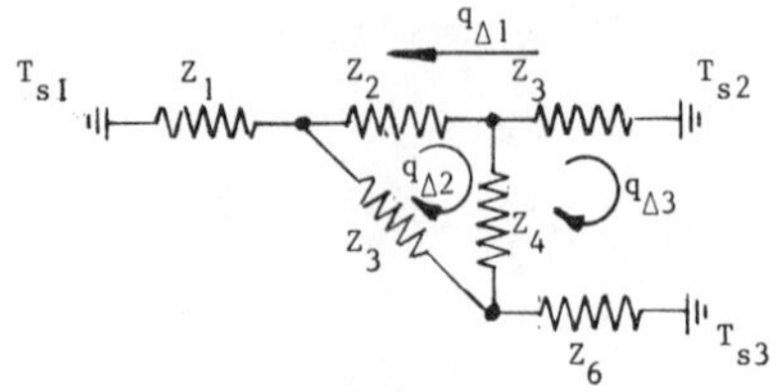

$$\begin{bmatrix} Z_1+Z_2+Z_5 & -Z_2 & -Z_5 \\ -Z_2 & Z_2+Z_3+Z_4 & -Z_4 \\ -Z_5 & -Z_4 & Z_4+Z_5+Z_6 \end{bmatrix} \begin{Bmatrix} q_{\Delta 1} \\ q_{\Delta 2} \\ q_{\Delta 3} \end{Bmatrix}$$

a) Coefficient Matrix for Heat Flow

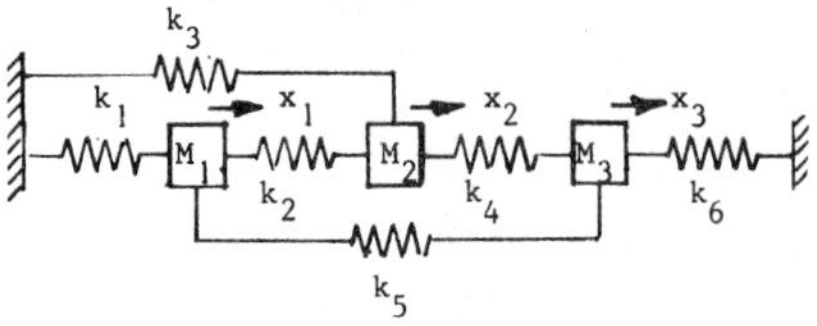

$$\begin{bmatrix} k_1+k_2+k_5 & -k_2 & -k_5 \\ -k_2 & k_2+k_3+k_4 & -k_4 \\ -k_5 & -k_4 & k_4+k_5+k_6 \end{bmatrix} \begin{Bmatrix} x_1 \\ x_2 \\ x_3 \end{Bmatrix}$$

b) Coefficient Matrix for Displacements

Figure A.5-1. Diagonal and Off-Diagonal Coefficient Relationships.

positive terms. By manipulating the coefficients in accordance with Sections A.5(5) through A.5(7), all negative terms may be precancelled. The following procedure develops the determinant either algebraically or numerically directly without redundant terms. This facilitates parametric development for optimizing the design configuration.

Step 1. To the first column add the sum of all the other columns. As shown in Section A.5(7), the value of the determinant remains unchanged. Selecting the coefficient matrix illustrated in Figure A.5-1, the result of step 1 becomes

Applying Step 1

$$\begin{vmatrix} K_1 + K_2 + K_5 & -K_2 & -K_5 \\ -K_2 & K_2 + K_3 + K_4 & -K_4 \\ -K_5 & -K_4 & K_4 + K_5 + K_6 \end{vmatrix}$$

$$= \begin{vmatrix} K_1 & -K_2 & -K_5 \\ K_3 & K_2 + K_3 + K_4 & -K_4 \\ K_6 & -K_4 & K_4 + K_5 + K_6 \end{vmatrix} \qquad \text{(A.31a)}$$

A comparison of minors M_{11} shows that a considerable simplification has already occurred.

Step 2. Develop first order minors M, using the first column and their complements M^* as described in Sections A.5(2)–A.5(3). For first order minors, $m = 1$ and since (A.31a) has three rows, $n = 3$, and (A.24) gives

$$N_m = \frac{3!}{1!(3 - 1)!} = 3$$

The minors and complements of (A.31a) can be determined by inspection to be

(1) $\quad M_{1,1} = |K_1| \qquad M^*_{2 \cdot 3, 2 \cdot 3} = \begin{vmatrix} K_2 + K_3 + K_4 & -K_4 \\ -K_4 & K_4 + K_5 + K_6 \end{vmatrix}$

$$\text{(A.31b)}$$

(2) $\quad M_{2,1} = |K_3| \qquad M^*_{1 \cdot 3, 2 \cdot 3} = \begin{vmatrix} -K_2 & -K_5 \\ -K_4 & K_4 + K_5 + K_6 \end{vmatrix} \qquad \text{(A.31c)}$

(3) $\quad M_{3,1} = |K_6| \qquad M^*_{1 \cdot 2, 2 \cdot 3} = \begin{vmatrix} -K_2 & -K_5 \\ K_2 + K_3 + K_4 & -K_4 \end{vmatrix} \qquad \text{(A.31d)}$

The determinant at this point as determined by (A.28) is

$$|D| = (-1)^{10} M_{1,1} M^*_{2 \cdot 3, 2 \cdot 3} + (-1)^9 M_{2,1} M^*_{1 \cdot 3, 2 \cdot 3} + (-1)^8 M_{3,1} M^*_{1 \cdot 2, 2 \cdot 3}$$

$$\text{(A.31e)}$$

Step 3. The summation of elements represented by the diagonal coefficients of the original matrix must be on the diagonal for each minor complement. This requires a column exchange, Section A.5(5), for the complement (A.31d) in addition to the sign incorporation, A.5(6), to preserve the positive summation of (A.31e). Since $(-1)^9$ is negative, Section A.5(6) is applied to (A.31c) to establish positive summing. Finally, step 1 is reapplied to the complement (A.31b). The minors and complements have become

(1) $\quad M_{1,1} = |K_1| \qquad M^*_{2 \cdot 3, 2 \cdot 3} = \begin{vmatrix} K_2 + K_3 & -K_4 \\ K_5 + K_6 & K_4 + K_5 + K_6 \end{vmatrix}$

$$\text{(A.31f)}$$

(2) $M_{2,1} = |K_3|$ $M^*_{\uparrow \cdot 3,2 \cdot 3} = \begin{vmatrix} K_2 & -K_5 \\ K_4 & K_4 + K_5 + K_6 \end{vmatrix}$ (A.31g)

(3) $M_{3,1} = |K_6|$ $M^*_{\uparrow \cdot 2,2 \cdot 3} = \begin{vmatrix} K_5 & -K_2 \\ K_4 & K_2 + K_3 + K_4 \end{vmatrix}$ (A.31h)

The determinant at this point is

$$|D| = K_1 M^*_{\uparrow \cdot 3,2 \cdot 3} + K_3 M^*_{\uparrow \cdot 3,2 \cdot 3} + K_6 M^*_{\uparrow \cdot 2,2 \cdot 3}$$ (A.31i)

$$|D| = K_1[(K_2 + K_3)(K_4 + K_5 + K_6) + K_4(K_5 + K_6)] + K_3[K_2(K_4$$
$$+ K_5 + K_6) + K_4 K_5] + K_6[K_5(K_2 + K_3 + K_4) + K_2 K_4]$$

(A.31j)

Step 3 is repeated as necessary for larger arrays. The procedure is easily remembered and applied. Once mastered, the determinants of large arrays can be written directly with a minimum of effort.

EXAMPLE A.5-5.   Describe the algebraic determinant of (A.7).
   Step 1.

$$|D| = \begin{vmatrix} G_1 & -G_2 & 0 \\ G_3 & (G_2 + G_3 + G_4) & -G_4 \\ G_5 & -G_4 & (G_4 + G_5) \end{vmatrix}$$

   Steps 2 and 3.

$$|D| = G_1 \begin{vmatrix} (G_2 + G_3) & -G_4 \\ G_5 & (G_4 + G_5) \end{vmatrix} + G_3 \begin{vmatrix} G_2 & 0 \\ G_4 & (G_4 + G_5) \end{vmatrix} + G_5 \begin{vmatrix} 0 & -G_2 \\ G_4 & (G_2 + G_3 + G_4) \end{vmatrix}$$

Expanding each determinant in the sum above results in a system determinant without redundant terms:

$$|D| = G_1[(G_2 + G_3)(G_4 + G_5) + G_4 G_5] + G_2 G_3(G_4 + G_5) + G_2 G_4 G_5$$

Q.E.D.

A6. MATRIX REDUCTION

Simultaneous equations are the inevitable result of defining a physical model of an actual system. The *reduction process* provides an efficient way of dealing with these equations. Moreover, the technique can facilitate model construction. This approach also provides a means of augmenting completed models when changes or additions are desired.

It is often convenient to develop models with greater detail than just the unknowns for the characteristics or behavior sought. In these cases, reduction may be used to tailor the equations by the systematic elimination of unknowns or unnecessary detail. The reduction process facilitates removal of coordinates without applied loads in the stiffness approach and the addition of displacement constraints to coordinates in the flexibility approach. When an impedance solution matrix for a thermal network is being developed, reduction facilitates branch connections including grounding. The reduced equations retain the characteristics of the original equations. The reduction process, described in the following paragraph, enables recovery of removed data which may be pertinent to design definition.

(2) *Elimination,* a commonly used method for solving linear simultaneous equations that requires fewer operations than other techniques, is the foundation of the reduction process. The operations effectively decompose the square coefficient matrix into an upper triangular structure which facilitates use of back substitution for determining the unknowns successively. Consider the general matrix equation analogous to (A.4):

$$
\begin{bmatrix} a & b & c & d \\ e & f & g & h \\ i & k & l & m \\ n & p & q & r \end{bmatrix} \begin{Bmatrix} x_1 \\ x_2 \\ x_3 \\ x_4 \end{Bmatrix} = \begin{Bmatrix} F_1 \\ F_2 \\ F_3 \\ F_4 \end{Bmatrix} \tag{A.32}
$$

Step 1. Divide the first row including the vector by the diagonal coefficient. Next, move the first column values below the diagonal to the left of the matrix, replacing each with zero. Following (A.32), this gives

$$
\begin{matrix} \\ e \\ i \\ n \end{matrix} \begin{bmatrix} \dfrac{a}{a} = 1 & b/a & c/a & d/a \\ 0 & f & g & h \\ 0 & k & l & m \\ 0 & p & q & r \end{bmatrix} \begin{Bmatrix} x_1 \\ x_2 \\ x_3 \\ x_4 \end{Bmatrix} = \begin{Bmatrix} F_1/a \\ F_2 \\ F_3 \\ F_4 \end{Bmatrix} \tag{A.33a}
$$

Step 2. The value at the intersection of each extended row (values to the left of the matrix) and the divided coefficients to the right of the diagonal is reduced by the corresponding row-column product. Beginning with the second row-column intersection of (A.33a), we have

$$\begin{array}{c} \\ \boxed{e} \\ i \\ n \end{array} \left[\begin{array}{cccc} 1 & \boxed{b/a} & c/a & d/a \\ \hdashline 0 & [f-(e)b/a] & g & h \\ 0 & k & l & m \\ 0 & p & q & r \end{array}\right] \left\{\begin{array}{c} x_1 \\ x_2 \\ x_3 \\ x_4 \end{array}\right\} = \left\{\begin{array}{c} F_1/a \\ F_2 \\ F_3 \\ F_4 \end{array}\right\} \qquad \text{(A.33b)}$$

Completing step 2 gives

$$\left[\begin{array}{cccc} 1 & b/a & c/a & d/a \\ \hdashline 0 & [f-(e)b/a] & [g-(e)c/a] & [h-(e)d/a] \\ 0 & [k-(i)b/a] & [l-(i)c/a] & [m-(i)d/a] \\ 0 & [p-(n)b/a] & [q-(n)c/a] & [r-(n)d/a] \end{array}\right] \left\{\begin{array}{c} x_1 \\ x_2 \\ x_3 \\ x_4 \end{array}\right\} = \left\{\begin{array}{c} F_1/a \\ F_2-(e)F_1/a \\ F_3-(i)F_1/a \\ F_4-(n)F_1/a \end{array}\right\}$$

$$\text{(A.33c)}$$

Reduction algebraically diminishes coefficients of the original matrix. Some coefficients may be negative in which case their numerical values may actually increase.

Step 3. A new matrix expression of one less order is obtained as a result of applying steps 1 and 2. Referring to (A.33c) the reduced model which excludes the variable x_1, becomes

$$\left[\begin{array}{ccc} [f-(e)b/a] & [g-(e)c/a] & [h-(e)d/a] \\ [k-(i)b/a] & [l-(i)c/a] & [m-(i)d/a] \\ [p-(n)b/a] & [q-(n)c/a] & [r-(n)d/a] \end{array}\right] \left\{\begin{array}{c} x_2 \\ x_3 \\ x_4 \end{array}\right\} = \left\{\begin{array}{c} F_2-(e)F_1/a \\ F_3-(i)F_1/a \\ F_4-(n)F_1/a \end{array}\right\}$$

$$\text{(A.34a)}$$

which can be written as

$$\left[\begin{array}{ccc} f_1 & g_1 & h_1 \\ k_1 & l_1 & m_1 \\ p_1 & q_1 & r_1 \end{array}\right] \left\{\begin{array}{c} x_2 \\ x_3 \\ x_4 \end{array}\right\} = \left\{\begin{array}{c} F_{21} \\ F_{31} \\ F_{41} \end{array}\right\} \qquad \text{(A.34b)}$$

Equation (A.34b) retains all the characteristics of the original equation (A.32) with the unknown x_1 removed. The process can be repeated using (A.34b) to obtain a second order matrix equation with unknown x_2 removed. Application of steps 1 through 3 is repeated until only one unknown remains, which can then be determined by a division.

Back substitutions using the sequence of coefficients obtained in step 1 returns all unknowns. This is illustrated in (A.35) where each step 1 operation on the first row of successively decreasing order matrices are collected.

$$
\begin{bmatrix}
1 & b_1 & c_1 & d_1 \\
0 & 1 & g_2 & h_2 \\
0 & 0 & 1 & m_3 \\
0 & 0 & 0 & 1
\end{bmatrix}
\begin{Bmatrix}
x_1 \\
x_2 \\
x_3 \\
x_4
\end{Bmatrix}
=
\begin{Bmatrix}
F_{11} \\
F_{22} \\
F_{33} \\
F_{44}
\end{Bmatrix}
\tag{A.35}
$$

In (A.35), for example, b_1, c_1, ... are the result of dividing the first row of the original matrix (A.33a) by the diagonal coefficient i.e., $b_1 = b/a$, $c_1 = c/a$, etc. Similarly, g_2, h_2 ... are the result of dividing the first row of the reduced matrix (A.34b) by the diagonal coefficient, i.e., $g_2 = g_1/f_1$, $h_2 = h_1/f_1$. The value of x_4 is given by the fourth row of (A.35), i.e.,

$$
x_4 = F_{44} \tag{A.36}
$$

The value of x_3 is obtained from the third row of (A.35) which can be explicitly expressed as

$$
x_3 + m_3 x_4 = F_{33} \tag{A.37a}
$$

Substituting x_4 from (A.36) into (A.37a) and simplifying returns the value for x_3

$$
x_3 = F_{33} - m_3 F_{44} \tag{A.37b}
$$

The process of obtaining x_3 using (A.35) in terms of (A.37) describes *back substitution*. Repeating the procedure using the second row of (A.35) to obtain x_2 and the first row of (A.35) to determine x_1 completes the process.

EXAMPLE A.6-1.   Determine temperatures T_1, T_2 and T_3 for the model of Figure A.6-1.

Following (A.4), the matrix equation for the bracket is expressed as

$$
\begin{bmatrix}
0.8 & -0.5 & 0 \\
-0.5 & 1.0 & -0.3 \\
0 & -0.3 & 9.6
\end{bmatrix}
\begin{Bmatrix}
T_1 \\
T_2 \\
T_3
\end{Bmatrix}
=
\begin{Bmatrix}
40 \\
20 \\
25
\end{Bmatrix}
$$

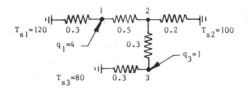

Figure A.6-1. Bracket Conductance Model.

Step 1 requires that the first row be divided by the diagonal coefficient, 0.8. Performing the division and completing step 1 gives the result:

$$
\begin{matrix}
* \\
-0.5 \\
0
\end{matrix}
\begin{bmatrix}
1 & -0.625 & 0 \\
0 & 1.0 & -0.3 \\
0 & -0.3 & 0.6
\end{bmatrix}
\begin{Bmatrix}
T_1 \\
T_2 \\
T_3
\end{Bmatrix}
=
\begin{Bmatrix}
50 \\
20 \\
25
\end{Bmatrix}
$$

With step 2 performed on the second row, the matrix equation becomes

$$
\begin{matrix}
 \\
-0.5 \\
0
\end{matrix}
\begin{bmatrix}
1 & -0.625 & 0 \\
0 & [1 - (-0.5)(-0.625)] & -0.3 \\
0 & -0.3 & 0.6
\end{bmatrix}
\begin{Bmatrix}
T_1 \\
T_2 \\
T_3
\end{Bmatrix}
=
\begin{Bmatrix}
50 \\
20 - (-0.5)(50) \\
25
\end{Bmatrix}
$$

Step 2 is completed by reducing the remaining row. Additional reductions do not occur, however, since the row intercept is zero. At this point, one application of step 1 and 2 results in a matrix with the form of (A.33c), i.e.,

$$
\begin{bmatrix}
1 & -0.625 & 0 \\
\hdashline
0 & 0.6875 & -0.3 \\
0 & -0.3 & 0.6
\end{bmatrix}
\begin{Bmatrix}
T_1 \\
T_2 \\
T_3
\end{Bmatrix}
=
\begin{Bmatrix}
50 \\
45 \\
25
\end{Bmatrix}
$$

The reduced matrix corresponding to (A.34b) is shown boxed. This matrix equation is one order less than the original matrix. Written explicitly, the equations which are free of T_1 become

$$
\begin{bmatrix}
0.6875 & -0.3 \\
-0.3 & 0.6
\end{bmatrix}
\begin{Bmatrix}
T_2 \\
T_3
\end{Bmatrix}
=
\begin{Bmatrix}
45 \\
25
\end{Bmatrix}
$$

These equations can be reinterpreted in terms of model geometry. Essentially, the coordinate (node) corresponding to T_1 has been replaced by a single conductance with the value of 0.1875. Additionally, heat is injected into node 2 such that the resulting temperature distribution is the same as the original model, Figure A.6-1. The model corresponding to the second order reduced matrix equation is shown in Figure A.6-2.

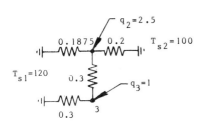

Figure A.6-2. Reduced Bracket Conductance Model.

The second order matrix equation is in the form of (A.18) which admits to solutions (A.16) and (A.17) for T_2 and T_3. Applying (A.17), T_3 becomes

$$[(0.6)(0.6875) - (-0.3)(-0.3)] T_3 = [0.6875(25) - (-0.3)(45)]$$

$$T_3 = \frac{30.6875}{0.3225} = 95.155$$

It is instructive to continue use of the reduction process, however, obtaining solutions for the node temperatures using back substitution.

An additional cycle of both steps 1 and 2 generates a reduced model similar to Figure A.6-2 with node 2 removed. Reapplying step 1 to the second order matrix gives

$$\begin{matrix} * \\ -0.3 \end{matrix} \begin{bmatrix} 1 & -0.3/0.6875 \\ 0 & 0.6 \end{bmatrix} \begin{Bmatrix} T_2 \\ T_3 \end{Bmatrix} = \begin{Bmatrix} 45/0.6875 \\ 25 \end{Bmatrix}$$

This matrix equation after applying step 2 and simplifying becomes

$$\begin{bmatrix} 1 & -0.4364 \\ \cdots\cdots\cdots\cdots\cdots\cdots \\ 0 \vdots [0.6 - (-0.3)(-0.4364)] \end{bmatrix} \begin{Bmatrix} T_2 \\ \cdots \\ T_3 \end{Bmatrix} = \begin{Bmatrix} 65.4545 \\ \cdots\cdots\cdots\cdots\cdots \\ 25 - (-0.3)(65.4545) \end{Bmatrix}$$

The simplified result is

$$\begin{bmatrix} 1 & -0.4364 \\ \cdots\cdots\cdots\cdots \\ 0 \vdots & 0.4691 \end{bmatrix} \begin{Bmatrix} T_2 \\ \cdots \\ T_3 \end{Bmatrix} = \begin{Bmatrix} 65.4545 \\ \cdots\cdots\cdots \\ 44.6364 \end{Bmatrix}$$

The boxed section of this matrix equation represents a new equation free of both T_1 and T_2. T_3 is determined by division or a final application of step 1 as

$$* \quad T_3 = \frac{44.6364}{0.4691} = 95.15$$

This is the same result obtained from (A.17).

Each application of step 1 has been labeled with an asterisk (*). Assembling the simplified results for step 1 operations gives the back substitution relationship in the form of (A.35) as

$$
\begin{bmatrix}
1 & -0.625 & 0 \\
0 & 1 & -0.4364 \\
0 & 0 & 1
\end{bmatrix}
\begin{Bmatrix}
T_1 \\
T_2 \\
T_3
\end{Bmatrix}
=
\begin{Bmatrix}
50 \\
65.4545 \\
95.15
\end{Bmatrix}
$$

From the second row, an expression similar to (A.37) determines T_2 as

$$T_2 = 65.4545 + 0.4364(95.15) = 106.98$$

Based on the first row in which the only unknown is T_1, Section A.3(2) gives

$$T_1 - 0.625T_2 = 50$$

therefore,

$$T_1 = 50 + 0.625(106.98) = 116.86$$

Q.E.D.

(2) The reduction process facilitates constraint modifications to modeled systems without requiring the equipment designer to start over from the beginning. Assume that a flexibility model for part of a chassis structure has been developed and is represented as

$$
\begin{Bmatrix}
x_1 \\
x_2 \\
x_3 \\
x_4
\end{Bmatrix}
=
\begin{bmatrix}
a & b & c & d \\
e & f & g & h \\
i & k & l & m \\
n & p & q & r
\end{bmatrix}
\begin{Bmatrix}
F_1 \\
F_2 \\
F_3 \\
F_4
\end{Bmatrix}
\tag{A.38}
$$

given x_2 as the displacement at the end of an overhanging beam. Due to excessive internal stresses, the design drawings have been reexamined, with the conclusion that support at the free end is feasible. In (A.38) this is tantamount to setting x_2 equal to zero. Reduction of the second row of (A.38) simultaneously sets the displacement to zero and accommodates the corresponding load and geometrical changes necessary. The reaction force can be recovered from (A.38) and Section A.3(2) as

$$F_2 = -\frac{(gF_3 + hF_4 + eF_1)}{f} \tag{A.39}$$

The reduced matrix equation representing the modified structure is determined from (A.38) using row-column intercepts which pass through the corresponding diagonal free-displacement coefficient. For (A.38) this becomes

$$
\begin{Bmatrix} x_1 \\ x_2 \\ x_3 \\ x_4 \end{Bmatrix} =
\begin{bmatrix} a & b & c & d \\ e & f & g & h \\ i & k & l & m \\ n & p & q & r \end{bmatrix}
\begin{Bmatrix} F_1 \\ F_2 \\ F_3 \\ F_4 \end{Bmatrix}
\tag{A.40}
$$

First, divide each coefficient of the row corresponding to the change by the diagonal quantity. Next, reduce all coefficients excluded from the affected row columns. The reduced coefficient on the first row at the fourth column has the form

$$
\begin{Bmatrix} x_1 \\ x_2/f = 0 \\ x_3 \\ x_4 \end{Bmatrix} =
\begin{bmatrix} a & b & c & [d - b(h/f)]_{\Delta} \\ e/f & 1 & g/f & h/f \\ i & k & l & m \\ n & p & q & r \end{bmatrix}
\begin{Bmatrix} F_1 \\ F_2 \\ F_3 \\ F_4 \end{Bmatrix}
\tag{A.41a}
$$

The reduced coefficients are collected into a new lower order matrix. Completing the process, the matrix equation for the unknown displacements becomes

$$
\begin{Bmatrix} x_1 \\ x_3 \\ x_4 \end{Bmatrix} =
\begin{bmatrix}
[a - b(e/f)] & [c - b(g/f)] & [d - b(h/f)] \\
[i - k(e/f)] & [l - k(g/f)] & [m - k(h/f)] \\
[n - p(e/f)] & [q - p(g/f)] & [r - p(h/f)]
\end{bmatrix}
\begin{Bmatrix} F_1 \\ F_3 \\ F_4 \end{Bmatrix}
\tag{A.41b}
$$

which can be written as

$$
\begin{Bmatrix} x_1 \\ x_3 \\ x_4 \end{Bmatrix} =
\begin{bmatrix} a_1 & c_1 & d_1 \\ i_1 & l_1 & m_1 \\ n_1 & q_1 & r_1 \end{bmatrix}
\begin{Bmatrix} F_1 \\ F_3 \\ F_4 \end{Bmatrix}
\tag{A.41c}
$$

All the characteristics of the original model modified by the imposed constraint are retained in (A.41c). The solution for the unknowns follows the process described in Section A.6(1).

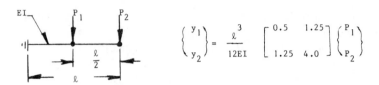

Figure A.6-3. Cantilever Bracket Model.

EXAMPLE A.6-2.   A model for the load displacements of the cantilever bracket is given in Figure A.6-3. Develop the reduced model matrix equation for a pin-ended modification.

Via reduction of the second row of the given matrix equation, the free end displacement will be converted to the pin-ended condition. Following the process of (A.40) and (A.41a), the first step for the given matrix becomes

$$\begin{Bmatrix} y_1 \\ y_2/4 \end{Bmatrix} = \frac{\ell^3}{12EI} \begin{bmatrix} 0.5 & 1.25 \\ 1.25/4 & 1 \end{bmatrix} \begin{Bmatrix} P_1 \\ P_2 \end{Bmatrix}$$

The relationship for the displacement at P_1 and the modified model configuration are shown in Figure A.6-4.

The pin-end reaction force due to P_1 is obtained in a manner similar to (A.39) as

$$P_2 = -{}^{15}\!/_{16} P_1$$

where the negative sign indicates a direction opposite to that of the applied force P_1.
Q.E.D.

Matrix equation (A.38) also represents the format for the impedance solution for temperature distribution. Application of reduction to impedance matrix equations as described above for flexibility, grounds the desired node. Conductance and stiffness matrix equations are of the form (A.31a). When reduction is applied to stiffness models, the applied force at that coordinate is equated to zero. Similarly, reduction of a conductance matrix removes and redistributes the injected heat load to other nodes while removing the coordinate.

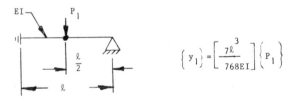

Figure A.6-4. Pin-End Supported Bracket.

(3) Additions, changes in value or deletions of impedance branches or beam (spring) elements in the flexibility matrix are easily accommodated by the reduction process.

Connections to ground or to external constraints. A circuit board mounted on standoffs has been modeled as an impedance matrix system in the form of (A.38). Excessive component temperatures at one location can be reduced with an additional standoff in the vicinity of node 3 (T_3). When row-column intercepts are drawn through the corresponding diagonal temperature-heat load coefficient for node 3, (A.38) becomes

$$
\begin{Bmatrix} T_1 \\ T_2 \\ T_3 \\ T_4 \end{Bmatrix} = \begin{bmatrix} a & b & c & d \\ e & f & g & h \\ i & k & l & m \\ n & p & q & r \end{bmatrix} \begin{Bmatrix} q_1 \\ q_2 \\ q_3 \\ q_4 \end{Bmatrix} \tag{A.42}
$$

Only the matrix of coefficients of (A.42) is involved. First, duplicate the column of coefficients along the intercept to the right of the matrix and duplicate the row intercept coefficients at the bottom of the matrix. The coefficient on the diagonal at the intercepts of (A.42) is duplicated at the junction of the newly formed row and column. This value is the node connection impedance for the model.

The new branch impedance is added to the node connection impedance on the newly formed diagonal since the selected node is to be connected to the sink. A change to an existing sink connection can be performed by adding the value of a parallel branch needed to provide the desired result to the impedance at the new row-column junction. An existing branch can be removed by subtracting its value from the node connection impedance.

Following this procedure, the new standoff impedance s is added to the node connection impedance giving the result

$$
\begin{Bmatrix} T_1 \\ T_2 \\ T_3 \\ T_4 \\ 0 \end{Bmatrix} = \begin{bmatrix} a & b & c & d & \vdots & c \\ e & f & g & h & \vdots & g \\ i & k & l & m & \vdots & l \\ h & p & q & r & \vdots & q \\ \hdotsfor{6} \\ i & k & l & m & \vdots & l+s \end{bmatrix} \begin{Bmatrix} q_1 \\ q_2 \\ q_3 \\ q_4 \\ 1 \end{Bmatrix} \tag{A.43}
$$

The new matrix obtained after reducing out the new row and column includes the model changes desired. Reduction similar to the process described for (A.41a) is performed.

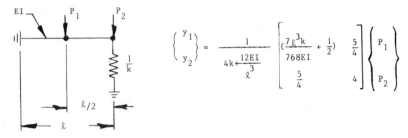

Figure A.6-5. Gusset Supported Bracket.

EXAMPLE A.6-3. The high flexural stress in the cantilevered bracket shown in Figure A.6-3 can be reduced by adding a gusset support to the free end with flexibility $1/k$. Develop the matrix equation for the resulting model.

For the support at the free end, the procedure for the y_2 coordinate of (A.42) and (A.43) develops the result

$$
\begin{Bmatrix} y_1 \\ y_2 \\ 0 \end{Bmatrix} =
\begin{bmatrix}
0.5c & 1.25c & \vdots & 1.25c \\
1.25c & 4c & \vdots & 4c \\
1.25c & 4c & \vdots & 4c + \dfrac{1}{k}
\end{bmatrix}
\begin{Bmatrix} P_1 \\ P_2 \\ 1 \end{Bmatrix}
$$

given $c = \ell^3/12EI$

The relationship for the displacements are determined by reducing out the third row and column. The resulting matrix equation and modified model configuration are shown in Figure A.6-5.

The model in Figure A.6-5 provides the basis for a parametric evaluation of the gusset stiffness k in terms of the bracket displacements y_1 and y_2.   Q.E.D.

Connections to ground or to external constraints may be summed directly into conductance or stiffness matrix equations. Reduction processes for the purposes of adding these branches to the conductance or stiffness matrix models are not applicable.

(4) *Connections between nodes or coordinates.* Requirements may necessitate the addition of a new subsystem model or additional connections between coordinates or nodes of an existing flexibility or impedance model. Since the new branch connects two existing nodes, the procedure described for (A.42) is repeated for each end point. For connections between nodes 1 and 2, for example, the matrix intercepts appear as

$$
\begin{Bmatrix} T_1 \\ T_2 \\ T_3 \\ T_4 \end{Bmatrix} = \begin{matrix} r_1 \\ r_2 \\ \\ \end{matrix} \overset{\begin{matrix} c_1 & c_2 \end{matrix}}{\left[\begin{array}{cccc} a & b & c & d \\ e & f & g & h \\ i & k & l & m \\ n & p & q & r \end{array} \right]} \begin{Bmatrix} q_1 \\ q_2 \\ q_3 \\ q_4 \end{Bmatrix} \qquad (A.44)
$$

The difference between column coefficients $(c_1 - c_2)$ form a new column to the right of the matrix. Also, the difference between row coefficients $(r_1 - r_2)$ form a new row at the bottom of the matrix. The difference between the intercept coefficients of the new row or column plus the added interconnecting branch impedance, s, become the new diagonal coefficient with the result

$$
\begin{Bmatrix} T_1 \\ T_2 \\ T_3 \\ T_4 \\ 0 \end{Bmatrix} = \begin{matrix} \\ \\ \\ \\ (r_1 - r_2) \end{matrix} \overset{(c_1 - c_2)}{\left[\begin{array}{cccc:c} a & b & c & d & (a - b) \\ e & f & g & h & (e - f) \\ i & k & l & m & (i - k) \\ n & p & q & r & (n - p) \\ \hdashline (a - e) & (b - f) & (c - g) & (d - h) & (a - b) - (e - f) + s \end{array} \right]} \begin{Bmatrix} q_1 \\ q_2 \\ q_3 \\ q_4 \\ 1 \end{Bmatrix}
$$

$$(A.45)$$

A change to an existing branch connection value can be performed by adding the value, s, of a parallel branch needed to provide the result. A negative impedance, $-s$, may be used to change or remove an existing connection.

EXAMPLE A.6-4.   The component located at node 3, Figure A.6-6, of the impedance model of a device module exceeds the design allowable. The effectiveness of a conductive strap with impedance 1.5 connected between the component node 3 and module wall node 1 is to be evaluated. Develop the resulting impedance model.

Application of (A.45) using the model of Figure A.6-6 with the added branch impedance of 1.5 gives

$$
\begin{Bmatrix} T_1 \\ T_2 \\ T_3 \\ 0 \end{Bmatrix} = \left[\begin{array}{ccc:c} 0.667 & 0.5 & 0.5 & 0.167 \\ 0.5 & 0.75 & 0.75 & -0.25 \\ 0.5 & 0.75 & 3.75 & -3.25 \\ \hdashline 0.167 & -0.25 & -3.25 & 3.417 + 1.5 \end{array} \right] \begin{Bmatrix} q_1 \\ q_2 \\ q_3 \\ 1 \end{Bmatrix}
$$

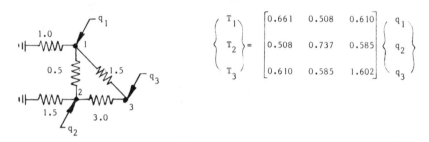

Figure A.6-6. Model Impedance Model.

$$\begin{Bmatrix} T_1 \\ T_2 \\ T_3 \end{Bmatrix} = \begin{bmatrix} 0.661 & 0.508 & 0.610 \\ 0.508 & 0.737 & 0.585 \\ 0.610 & 0.585 & 1.602 \end{bmatrix} \begin{Bmatrix} q_1 \\ q_2 \\ q_3 \end{Bmatrix}$$

Figure A.6-7. Revised Module Thermal Model.

The new impedance model and its graphic representation shown in Figure A.6-7 is the result of the fourth row and column reduction process similar to (A.41a). Q.E.D.

When either conductance or stiffness matrix models are used, branch additions, changes or deletions can be made directly. Modification of impedance or flexibility models, however, require reduction techniques for proper implementation.

A.7 MATRIX INVERSE

The matrix inverse can be used to develop the stiffness matrix from the flexibility matrix or vice versa. Similarly, since the impedance matrix is the inverse of the conductance matrix, either formulation can be developed knowing the other. Solutions to simultaneous equations may also be obtained with the matrix inverse as described in Section A.1.

The product of a matrix and its inverse produces the identity matrix Section A.1(3). This property which may be used to verify the inverse extraction process is expressed as

$$[A]^{-1}[A] = [A][A]^{-1} = [I] \tag{A.46}$$

(1) The inverse of a matrix may be obtained using complementary minors as in Section A.5(3). For large matrices, however, the number of operations becomes prohibitive. Efficient methods using a variation of the reduction process for matrices of any order are described in Section A.7(2).

To apply the method of minors, the transpose of the matrix is first obtained per Section A.4. Next, the complementary minor of each coefficient is computed and placed in a new matrix at the coefficient's row-column intercept. The sign of each complementary minor is determined by its row column intercepts. The inverse is found by dividing the new matrix by the determinant of the original matrix. Consider a transposed matrix of the third order:

$$[A]^T = \begin{bmatrix} a & b & c \\ d & e & f \\ g & h & i \end{bmatrix} \tag{A.47}$$

The signed complementary minor for coefficient a is computed as

$$M^*_{2\cdot3,2\cdot3} = (-1)^{2+3+2+3}(ei - fh) = (ei - fh) \tag{A.48}$$

where the exponent is the sum of the row-column intercepts as described in Section A.5(4). Minor (A.48) is located in a new matrix at the location of coefficient a. The completed new matrix becomes

$$\begin{bmatrix} (ei - fh) & -(di - fg) & (dh - eg) \\ -(bi - ch) & (ai - cg) & -(ah - bg) \\ (bf - ce) & -(af - cd) & (ae - bd) \end{bmatrix} \tag{A.49}$$

The determinant equals the sum of element-to-element products of any two similarly located rows or columns. Selecting the first row of (A.47) and (A.49), the determinant of $|A|$ is

$$|A| = a(ei - fh) - b(di - fg) + c(dh - eg) \tag{A.50}$$

The inverse of $[A]$ is then (A.49) divided by (A.50) or

$$[A]^{-1} = \frac{1}{|A|} \begin{bmatrix} (ei - fh) & (fg - di) & (fh - eg) \\ (ch - bi) & (ai - cg) & (bg - ah) \\ (bf - ce) & (cd - af) & (ae - bd) \end{bmatrix} \tag{A.51}$$

EXAMPLE A.7-1.   Determine the inverse of matrix $[\mathbf{B}]$ given

$$[\mathbf{B}] = \begin{bmatrix} 1 & 2 & 0 \\ -1 & 3 & 4 \\ 2 & 1 & 6 \end{bmatrix}, [\mathbf{B}]^T = \begin{bmatrix} 1 & -1 & 2 \\ 2 & 3 & 1 \\ 0 & 4 & 6 \end{bmatrix}$$

The matrix of complementary minors of $[\mathbf{B}]^T$ becomes

$$[\mathbf{C}] = \begin{bmatrix} [(\ 3)(6) - (1)(4)] & -[(2)(6) - (1)(0)] & [(2)(4) - (3)(0)] \\ -[(-1)(6) - (2)(4)] & [(1)(6) - (2)(0)] & -[(1)(4) - (-1)(0)] \\ [(-1)(1) - (2)(3)] & -[(1)(1) - (2)(2)] & [(1)(3) - (-1)(2)] \end{bmatrix} = \begin{bmatrix} 14 & -12 & 8 \\ 14 & 6 & -4 \\ -7 & 3 & 5 \end{bmatrix}$$

The determinant of $|\mathbf{B}|$ is developed from the products of each element of the first column of $[\mathbf{B}]^T$ and $[\mathbf{C}]$ as

$$|\mathbf{B}| = (1)(14) + (2)(14) + (0)(-7) = 42$$

The inverse is therefore

$$[\mathbf{B}]^{-1} = \frac{1}{42} \begin{bmatrix} 14 & -12 & 8 \\ 14 & 6 & -4 \\ -7 & 3 & 5 \end{bmatrix}$$

Q.E.D.

(2) The reduction process provides an efficient method of inverting matrices of any order. Row-column indices are drawn through the first diagonal element. Reduction of all elements not on the indices proceeds as described in (A.41a). Next, values along the column index, except for the diagonal element, have their signs reversed. Finally, this entire column is divided by the previous diagonal quantity. When this operation is performed on the first diagonal term of (A.47), the result becomes

$$\begin{bmatrix} \dfrac{1}{a} & \dfrac{b}{a} & \dfrac{c}{a} \\ -\dfrac{d}{a} & \left[e - j\left(\dfrac{b}{a}\right)\right] & \left[f - j\left(\dfrac{c}{a}\right)\right] \\ -\dfrac{g}{a} & \left[h - g\left(\dfrac{b}{a}\right)\right] & \left[i - g\left(\dfrac{c}{a}\right)\right] \end{bmatrix} = \begin{bmatrix} a_1 & b_1 & c_1 \\ d_1 & e_1 & f_1 \\ g_1 & h_1 & i_1 \end{bmatrix} \quad (A.52)$$

The same process is repeated for the second diagonal quantity and each successive diagonal element until all have been completed. For a nth order matrix, there are n diagonal elements, each requiring the reduction treatment described. When the second diagonal of (A.52) is considered, the result of the process is

$$
\begin{bmatrix}
\left[a_1 - b_1 \left(\dfrac{d_1}{e_1} \right) \right] & -\dfrac{b_1}{e_1} \left[c_1 - b_1 \left(\dfrac{f_1}{e_1} \right) \right] \\[2ex]
d_{1/e_1} & \dfrac{1}{e_1} f_{1/e_1} \\[2ex]
\left[g_1 - h_1 \left(\dfrac{d_1}{e_1} \right) \right] & -\dfrac{h_1}{e_1} \left[i_1 - h_1 \left(\dfrac{f_1}{e_1} \right) \right]
\end{bmatrix}
=
\begin{bmatrix}
a_2 & b_2 & c_2 \\[2ex]
d_2 & e_2 & f_2 \\[2ex]
g_2 & h_2 & i_2
\end{bmatrix}
\qquad (A.53)
$$

Only the diagonal with row-column indices through i_2 remain to complete the third and final process for the third order matrix. When this has been completed the result is the inverse of the original matrix.

EXAMPLE A.7-2. Determine the inverse of matrix [B] of Example A.7-1. Matrix [B] with reduction applied to the first diagonal element becomes

$$
[B] =
\begin{bmatrix}
1 & 2 & 0 \\
-1 & 3 & 4 \\
2 & 1 & 6
\end{bmatrix}
, \quad \begin{array}{c} \text{1st diagonal} \\ \text{Result:} \end{array}
\begin{bmatrix}
1 & 2 & 0 \\
1 & 5 & 4 \\
-2 & -3 & 6
\end{bmatrix}
$$

Following the same procedure for the second and third diagonal elements, gives the results

$$
\begin{bmatrix}
1 & 2 & 0 \\
1 & 5 & 4 \\
-2 & -3 & 6
\end{bmatrix}
, \quad \begin{array}{c} \text{2nd diagonal} \\ \text{Result:} \end{array}
\begin{bmatrix}
0.6 & -0.4 & -1.6 \\
0.2 & 0.2 & 0.8 \\
-1.4 & 0.6 & 8.4
\end{bmatrix}
$$

$$
\begin{bmatrix}
0.6 & -0.4 & -1.6 \\
0.2 & 0.2 & 0.8 \\
1.4 & 0.6 & 8.4
\end{bmatrix}
, \quad [B]^{-1} =
\begin{bmatrix}
0.333 & -0.2857 & 0.1905 \\
0.333 & 0.1428 & -0.0952 \\
-0.1667 & 0.714 & 0.119
\end{bmatrix}
$$

Q.E.D.

Appendix B
Transfer Functions

The *transfer function* is the ratio of the response of a system to an excitation expressed in terms of output/input. In heat transfer, the transfer function relates the thermal response to the heat loading on a system. The transfer function of a structual system describes the displacement, velocity or acceleration response in terms of the forcing excitations on a system. The transfer function describes how much of the loading or excitation on the system is transferred to a specified part of the system.

B.1 TRANSFER FUNCTION CONCEPT

The transfer function of a system is a consequence of the analytical description of a physical system. Considerable insight into the system characteristics can be obtained by the examination of the transfer function. Factors of the numerator or zeros of the transfer function indicate characteristics of the system which are insensitive to excitation or loading. Factors of the denominator or poles of the transfer function describe the sensitive characteristics of the system where forcing conditions can cause the greatest disturbance. Factors of the transfer function also indicate the time constant or damping associated with the transient response.

An equivalent spring-mass representation of many physical system configurations is often used to evaluate fundamental mode behavior and stress. Table B.1-1 lists the equivalent mass and stiffness for spring mass representations of both one and two dimensional hardware systems.

Components of the spring mass system of Figure B.1-1 are the mass M_{eq}, a spring with constant K_{eq}, system damping C, and a time dependent excitation. The excitation may be created by circuit card or electronic chassis motion, by moving vehicles, flight loads, shock, sound, etc. The system of Figure B.1-1 has one degree of freedom since only one coordinate, x_m, is required to define the position of the mass at a given time, t.

The mass M_{eq} is defined by Newton's law, which states that the product of mass and its acceleration equals the force applied to the mass. This may be

Table B.1-1. Approximate Spring Mass Equivalents for Distributed Mass Systems

Case	Configuration	Mass	Stiffness	Dynamic Load
One Dimensional Systems				
I	M_{dist}	$M_{eq} = M_{con} + \frac{1}{3}M_{dist}$	$K_{eq} = \dfrac{Gd^4}{8nD^3}$	$P_{eq} = 0.518P$
II	M_{dist}, M_{conc}	$M_{eq} = M_{con} + 0.257M_{dist}$	$K_{eq} = \dfrac{3EI}{\ell^3}$	$P_{eq} = 0.40P_{dist}$
III	M_{conc}, M_{dist}	$M_{eq} = M_{con} + 0.504M_{dist}$	$K_{eq} = \dfrac{48EI}{\ell^3}$	$P_{eq} = 0.64P_{dist}$
IV	M_{conc}, M_{dist}	$M_{eq} = M_{con} + 0.406M_{dist}$	$K_{eq} = \dfrac{192EI}{\ell^3}$	$P_{eq} = 0.533P_{dist}$
V	M_{conc}, M_{dist}	$M_{eq} = M_{con} + 0.446M_{dist}$	$K_{eq} = \dfrac{768EI}{7\ell^3}$	$P_{eq} = 0.5777P_{dist}$
Two Dimensional Systems				
VI	M_{conc}, M_{dist}, Fixed Support	$M_{eq} = M_{con} + 0.406M_{dist}$	$K_{eq} = \dfrac{16\pi Et^3}{3d^2(1-\nu^2)}$	$P_{eq} = 0.533P_{dist}$
VII	Simple Support	$M_{eq} = M_{con} + \dfrac{128(31+11\nu+\nu^2)}{315(5+\nu)^2}M_{dist}$	$K_{eq} = \dfrac{16\pi Et^3}{3d^2(1-\nu)(3+\nu)}$	$P_{eq} = \dfrac{8}{15}\left(\dfrac{6+\nu}{5+\nu}\right)P_{dist}$
VIII	a, $a\times b$, b	$M_{eq} = M_{con} + \frac{3}{4}M_{dist}$	$K_{eq} = \left[140.2 - 21.86\left(\dfrac{a}{b}\right)^2 + 60.26\left(\dfrac{a}{b}\right)^4\right]\dfrac{D}{a^2}$	$P_{eq} = \frac{1}{4}P_{dist}$
IX	$a\times b$	$M_{eq} = M_{con} + \dfrac{M_{dist}}{4}$	$K_{eq} = \left[58.12 + 5.48\left(\dfrac{a}{b}\right)^2 + 22.65\left(\dfrac{a}{b}\right)^4\right]\dfrac{D}{a^2}$	$P_{eq} = \left(\dfrac{2}{\pi}\right)^2 P_{dist}$

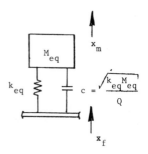

Figure B.1-1. Equivalent Spring-Mass System Representation.

expressed as

$$F = M_{eq} \frac{d^2x}{dt^2} = M_{eq}\ddot{x}_m \qquad \text{(B.1-1)}$$

The weightless spring K_{eq} produces forces only when it is deformed. Using D'Alembert's equilibrium principle to derive the equation of motion, we first displace the foundation by the amount x_f. The spring force $k_{eq}x_f$ tends to return the foundation to its original position. The inertial force $M_{eq}\ddot{x}$ acting against the dynamic force on the foundation also displaces the spring by the amount $-K_{eq}x_m$. Similar forces due to the velocity of the foundation and mass acting on the damper occur. The summation of all these forces yields

$$K_{eq}(x_f - x_m) + C(\dot{x}_f - \dot{x}_m) - M_{eq}\ddot{x}_m = 0 \qquad \text{(B.1-2)}$$

or $\qquad M_{eq}\ddot{x}_m + C\dot{x}_m + K_{eq}x_m = C\dot{x}_f + K_{eq}x_f \qquad \text{(B.1-2a)}$

which represents the general equation for forced foundation motion. Since the displacement of the mass is relative to its static equilibrium position, the effect of gravity is not included. When the disturbing force and foundation motion no longer exists, we obtain the equation of free vibration

$$M_{eq}\ddot{x}_m + C\dot{x}_m + K_{eq}x_m = 0 \qquad \text{(B.1-3)}$$

The s operator where $x(s)s^2 = \ddot{x}$ and $x(s)s = \dot{x}$ provides the most convenient way to manipulate the system equations that are derived from a model of the physical system. The s plane furnishes a convenient graphic display of the system characteristics. Substituting the s operator into (B.1-3), we obtain

$$(M_{eq}s^2 + Cs + K_{eq})x_m(s) = 0 \qquad \text{(B.1-4)}$$

Equation (B.1-4) is referred to as the characteristic equation of the single degree of freedom system of Figure B.1-1. Treating (B.1-4) as a quadratic equation, the roots are determined as

$$s = -\frac{C}{2M_{eq}} \pm \left[\left(\frac{C}{2M_{eq}}\right)^2 - \frac{K_{eq}}{M_{eq}}\right]^{1/2} \qquad \frac{K_{eq}}{M_{eq}} \leq \left(\frac{C}{2M_{eq}}\right)^2 \qquad \text{(B.1-5a)}$$

$$s = -\frac{C}{2M_{eq}} \pm j\left[\frac{K_{eq}}{M_{eq}} - \left(\frac{C}{2M_{eq}}\right)^2\right]^{1/2} \qquad \left(\frac{C}{2M_{eq}}\right)^2 \leq \frac{K_{eq}}{M_{eq}} \qquad \text{(B.1-5b)}$$

The roots are dependent on the relative magnitudes of M_{eq}, C, and K_{eq}. Equations (B.1-5) determine the character of the motion. By means of plotting the roots in the s plane as a function of a particular parameter, the effects on the dynamic behavior of the system can be seen. With the use of compact notation, (B.1-5b) becomes

$$s = -\sigma \pm j\omega_D \qquad \text{(B.1-6)}$$

where ω_D represents the damped natural frequency. The quantity $1/\sigma$ indicates the time required for the motion to damp to $(1/e)$th of its original value. This is represented by

$$\tau = 1/\sigma \text{ sec} \qquad \text{(B.1-7)}$$

which is the damping time constant of the system. Figure B.1-2 compares the response for several conditions of (B.1-6) with different values of σ. Figure B.1-2 uses the s plane and time domain to illustrate the effects of increasing damping on the system.

When $\omega_D = 0$, values of σ fall on the real axis, indicative of pure exponential behavior. This occurs when $(C/2M_{eq})^2 \geq K_{eq}/M_{eq}$ in (B.1-5b). The limiting value of system amplification for this to occur is

$$Q \leq \frac{1}{2} \qquad \text{(B.1-8)}$$

Generally, amplifications of electronic equipment and suspension systems are greater than $\frac{1}{2}$. Figure B.1-3 illustrates the effect varying stiffness K_{eq} has on response behavior. When $K_{eq} = 0$ in (B.1-5a), s has the values of 0 and $-C/M_{eq} = -2\sigma$. This condition shown on Figure B.1-3(a) is also similar to thermal response behavior. For small K_{eq}, s has two real values giving a response, Figure B.1-3(b), which is the sum of two exponentials, each with a different time constant analogous to a transient heat transfer condition. Increasing K_{eq}

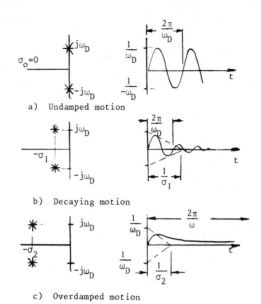

a) Undamped motion

b) Decaying motion

c) Overdamped motion

Figure B.1-2. Response Due to Varying Damping: $\sigma_0 < \sigma_1 < \sigma_2$.

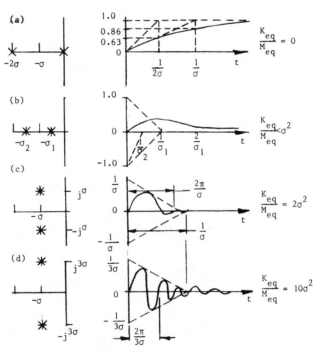

Figure B.1-3. Response Due to Varying Stiffness.

still more results in an underdamped condition, Figure B.1-3(c), with the system frequency varying in proportion to $\sqrt{K_{eq}}$, Figure B.1-3(d).

If the real part of (B.1-6) is positive, then the resulting system behavior will be unstable. This condition in any root of the characteristic equation will cause progressively increasing oscillation or diverging amplitudes. The transfer function that describes the displacement response characteristics of the system corresponding to (B.1-2a) written in terms of the s operator becomes

$$\frac{x_m}{x_f} = \frac{Cs + K_{eq}}{M_{eq}s^2 + Cs + K_{eq}} \tag{B.1-9}$$

Normalizing (B.1-9) by dividing both the numerator and denominator by M_{eq}/s^2 on the right side in addition to substituting $s = j\omega$ gives

$$\frac{x_m}{x_f} = \frac{1 + s/Q\omega_n}{1 + \left(\dfrac{s}{\omega_n}\right)^2 + \dfrac{s}{Q\omega_n}} = \frac{1 + j(\Omega/Q)}{1 - \Omega^2 + j(\Omega/Q)} \tag{B.1-10}$$

where $\Omega = \omega/\omega_n$. The equation $\omega_n = \sqrt{K_{eq}/M_{eq}}$ represents the undamped natural frequency of the system and ω corresponds to the frequency of the exciting force. In the frequency domain, (B.1-10) describes the amplification characteristics of the system. For $s = 0$ in (B.1-10), the amplification factor $x_m/x_f = 1$, indicating that the response echos the excitation. When $s \to \infty$, $x_m/x_f \to 0$ indicating attenuation at frequencies $\omega \gg \omega_n$. These response characteristics serve to guide the equipment designer in the selection of design parameters and structural section properties. Hardware designs consist of numerous interacting structures and assemblies that must be uncoupled to avoid resonant amplification. When an assembly has the same frequency as its support, $s = \omega_n$, the amplification $x_m/x_f = Q$. The assembly therefore experiences Q times the response acceleration of the support.

The property of (B.1-10), which determines the behavior of this single degree of freedom representation is given by the denominator, which is the characteristic equation previously described as (B.1-4). The frequency response represented by (B.1-10) is illustrated in Figure B.1-4. For light damping $Q > 3$, attenuation occurs at frequencies greater than $\sqrt{2}\omega_n$, rolling off from $\Omega = 1$ at -6db per octave. At ω equal to $\omega_n/\sqrt{2}$, the amplification becomes $x_m/x_f = 2$.

Equation (B.1-10) describes the system amplification in terms of the displacement ratio of the mass relative to the foundation. Similar relationships which express the output response relative to the input excitation for various combinations of behavior are listed in Table B.1-2. The absolute value of the

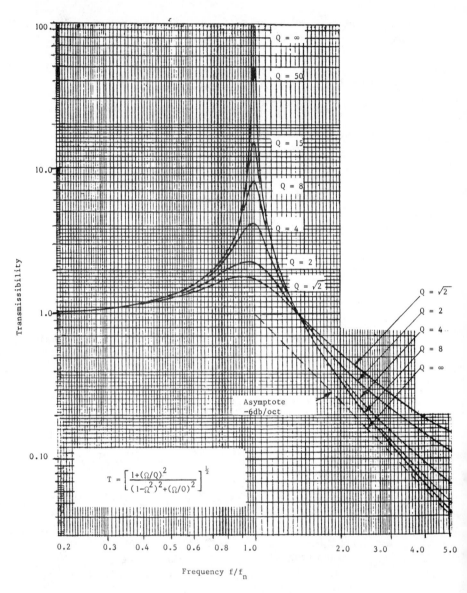

Figure B.1-4. Transmissibility of Spring Mass Systems.

Table B.1-2. Spring-Mass Transfer Functions

$$\Omega = \omega/\omega_n; \quad H_0 = \left[(1 - \Omega^2)^2 + \left(\frac{\Omega}{Q}\right)^2 \right]^{-1/2}$$

Input Excitation	Output Response	Transfer Function $H(\Omega)$	$H(\Omega)$	Equation
$\ddot{x}_f$	$(x_m - x_f)$	$-\dfrac{1/\omega_n^2}{1 - \Omega^2 + j(\Omega/Q)}$	H_0/ω_n^2	B.1-12
$\ddot{x}_f$	$\ddot{x}_m$	$\dfrac{1 + j(\Omega/Q)}{1 - \Omega^2 + j(\Omega/Q)}$	$\left[1 + \left(\dfrac{\Omega}{Q}\right)^2\right]^{1/2} H_0$	B.1-13
x_f	x	$\dfrac{1 + j(\Omega/Q)}{1 - \Omega^2 + j(\Omega/Q)}$	$\left[1 + \left(\dfrac{\Omega}{Q}\right)^2\right]^{1/2} H_0$	B.1-10
$f(t)$	x	$\dfrac{1/\omega_n^2}{1 - \Omega^2 + j(\Omega/Q)}$	H_0/ω_n^2	B.1-14
$f(t)$	$\ddot{x}$	$-\dfrac{\Omega^2}{1 - \Omega^2 + j(\Omega/Q)}$	$\Omega^2 H_0$	B.1-15

characteristic response (denominator of B.1-10) used in Table B.1-2 becomes

$$H_0 = \left[(1 - \Omega^2)^2 + \left(\frac{\Omega}{Q}\right)^2 \right]^{-1/2} \qquad \text{(B.1-11)}$$

The amplification factor, the ratio of displacements given by (B.1-10), is the same as that given for the ratio of accelerations Equation (B.1-13) of Table B.1-2. The transfer function for foundation force $f(t)$ is also included in Table B.1-2.

EXAMPLE B.1-1.   An aluminum chassis, Figure B.1-5(a), is subjected to sinusoidal vibration per Figure B.1-5(b) normal to the mounting plane. The assembly, which consists of modular components, has a distributed weight of 16 pounds (7.26 kg). The capacitor assembly supported by the chassis has a fundamental resonance of 200 Hz. Determine (a) the response characteristics of the chassis with $Q = \sqrt{f_n}$, (b) dynamic load on the flange mountings, and (c) the input acceleration to the capacitor assembly.

Solution:
   (a) The moment of inertia I about the mounting plane is developed from (6.20a) as

Element	Area, A	z	Az	Az^2	$I_{cg} = bd^3/12$
Side(s)	0.12(2)(1.2) = 0.288	0.6	0.1728	0.1037	0.0345
Bottom	0.12(6) = 0.720	0.06	0.0432	0.0026	—
	1.008		0.216	0.1063	0.0345

$$\bar{y} = \frac{\Sigma Az}{\Sigma A} = \frac{0.216}{1.008} = 0.214$$

$$I_z = \sum I_{cg} + \sum Az^2 - \frac{(\Sigma Az)^2}{\Sigma A} = 0.0345 + 0.1063 - \frac{0.216^2}{1.008} = 0.0945 \text{ in}^4$$

Equivalent stiffness and mass using case III of Table B.1-1 are determined as

$$K_{eq} = \frac{48EI}{\ell^3} = \frac{48(10^7)(0.0945)}{12^3} = 26250 \text{ lb/in}$$

$$M_{eq} = 0.504 M_{dist} = \frac{0.504(16)}{386} = \frac{0.021 \text{ lb-sec}^2}{\text{in}}$$

The natural frequency follows directly from ω_n of (B.1-10) as

$$\omega_n = \sqrt{\frac{K_{eq}}{M_{eq}}} = \sqrt{\frac{26250}{0.021}} = 1118.0 \text{ rad/sec (178.0 Hz)}$$

The chassis amplification factor becomes

$$Q\sqrt{f_n} = \sqrt{\frac{\omega_n}{2\pi}} = \sqrt{178.0} = 13.34$$

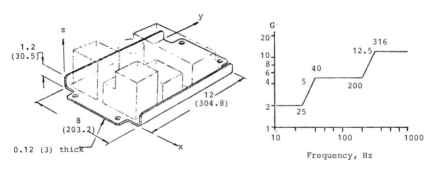

a) Chassis Configuration   b) Sinusoidal Vibration Environment

Figure B.1-5. Chassis Configuration and Environment.

The absolute value of the transfer function (B.1-13) determined as a function of the input excitation is of the form

$$|H(f)| = \left\{ \frac{1 + \left(\frac{f}{f_n Q}\right)^2}{\left[1 - \left(\frac{f}{f_n}\right)^2\right]^2 + \left(\frac{f}{f_n Q}\right)^2} \right\}^{1/2} = \left[\frac{1 + \frac{f^2}{5.672(10^6)}}{\left(1 - \frac{f^2}{3.182(10^4)}\right)^2 + \frac{f^2}{5.672(10^6)}} \right]^{1/2}$$

which describes the amplification characteristics of the chassis, Figure B.1-6(a). The chassis response, Figure B.1-6(b), is developed by multiplying the ordinates of Figure B.1-6(a) by the ordinates of Figure B.1-5(b) at corresponding frequencies.

(b) The distributed static load at chassis resonance becomes

$$P_{dist} = WG = (16 \text{ lb}) (66.86 \text{ peak}) = 1068.8 \text{ lb}$$

when vibrating at resonance only part of the mass contributes to the modal deformation. This dynamic load for the configuration of Figure B.1-5(a) is given by case III of Table B.1-1. This loading acts through the CG of the assembly and conservatively assumes that all components are in resonance and in phase at the same time. The resulting dynamic load becomes

$$P_{eq} = 0.64 P_{dist} = 0.64(1068.8) = 684 \text{ lb}$$

Each flange shares half this load or 342 lb (155.3 kg) which develops along the axis normal to the mounting surface.

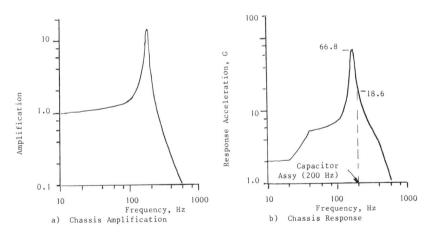

Figure B.1-6. Chassis Response Characteristics.

(c) The input acceleration of 18.6G's to the capacitor assembly is determined from the chassis response Figure B.1-6(b) at the capacitor assembly's resonant frequency. Alternatively, Figure B.1-4 could have been used by noting that $f/f_n = 200/178.4 = 1.12$, which gives an amplification of 4.8. The resultant acceleration load in the capacitor assembly of 18.6G is the product of the chassis input acceleration at 200 Hz from Figure B.1-6(b) and the amplification of 4.8.

Components within the capacitor assembly will experience an additional amplification of about $\sqrt{200}$ due to the resonance of the assembly. The component acceleration in the environment of Figure B.1-6(b) would reach $18.6\sqrt{200}$ or 263G. Q.E.D.

The transfer function concept is based on the fact that if an input or forcing function is applied to a system, the output or response of the system will be of the same form as the input although in general its magnitude and phase will be different. That is, the system will *transfer* this input to an output of the same form but modified by the system's transfer function. The phase of the response is both a function of ratio of the excitation to natural frequency ω/ω_n, and amplification of the system. At small values of ω/ω_n the response is nearly in phase with the excitation. As ω/ω_n approaches unity, the response is out of phase with the input of up to $90°$. At $\omega/\omega_n = 1$ they are out of phase by $90°$. As the frequency of excitation increases above that of the system, the condition continually changes so that at $\omega/\omega_n \gg 1$ the response motion is directly opposite to that of the excitation being in the neighborhood of $180°$ out of phase. The response, therefore, lags the input.

The concepts introduced for linear systems having one degree of freedom are readily extended to systems having any finite number of degrees of freedom. In systems of more than one degree of freedom, however, each inertial element affects the others. This interaction affects the nature of the response of each element although the characteristics of the element response are of the same form as the input.

The transfer function is of great utility in these cases, enabling a compact description of all variables which can be solved for the interaction between any two.

Thus for the system of Figure B.1-7, the matrix form of the differential equations for the absolute motion of the two coordinates is first constructed.

$$\begin{bmatrix} M_1 s^2 + (C_1 + C_2)s + (k_1 + k_2) & -(C_2 s + k_2) \\ -(C_2 s + k_2) & M_2 s^2 + C_2 s + k_2 \end{bmatrix} \begin{Bmatrix} x_1(s) \\ x_2(s) \end{Bmatrix}$$
$$= \begin{bmatrix} M_1 \\ M_2 \end{bmatrix} \begin{Bmatrix} -U(s) \end{Bmatrix} \quad \text{(B.1-16)}$$

The solution relating any two of the variables, $x_1(s)$ and $x_2(s)$, and $U(s)$ can be derived using methods of Appendix A.5(8). For example, the transfer func-

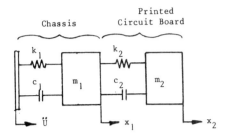

Figure B.1-7. Two Degree of Freedom System.

tion between the input $\ddot{U}(s)$ and $x_2(s)$ in (B.1-16) is obtained by

$$x_2(S) = \frac{\begin{vmatrix} -M_1\ddot{U}(s) & -(C_2s + k_2) \\ -M_2\ddot{U}(s) & M_2s^2 + C_2s + k_2 \end{vmatrix}}{\begin{vmatrix} M_1s^2 + (C_1 + C_2)s + (k_1 + k_2) & -(C_2s + k_2) \\ -(C_2s + k_2) & M_2s^2 + C_2s + k_2 \end{vmatrix}}$$

(B.1-17a)

or

$$\frac{x_2(s)}{\ddot{U}(s)} = \frac{M_2(C_2s + k_2) - M_1(M_2s^2 + C_2s + k_2)}{(M_1s^2 + C_1s + k_1)(M_2s^2 + C_2s + k_2) + M_2(C_2s + k_2)}$$

(B.1-17b)

The transfer function between $\ddot{U}(s)$ and $x_1(s)$ can be obtained in an analagous fashion.

For the system of Figure B.1-7 or a system with many more degrees of freedom, the transfer function between any variables and the input of a particular system always has the same denominator. This is the system's characteristic function which always indicates the natural motions a system can have. These motions are determined by the factors or roots of the denominator or characteristic equation (Appendix B.4). The factors of the characteristic are designated *poles* and those of the numerator are zeros of the transfer function. The poles tell what kind of natural motion the system can have regardless of how the motion is simulated.

The poles and zeros of a factored-transfer function can easily be plotted on the *s* plane similar to those shown on Figures B.1-2 and B.1-3 to determine the nature of the resulting motion and the behavior to expect. The behavior can be examined in both the frequency or time domain. Determination of the response behavior in the time domain is discussed in Appendix B.5.

Transient solutions shown on Figure B.1-2 and Figure B.1-3 are developed by assuming a solution of the form

$$x_m = Ae^{st} \tag{B.1-18}$$

where A and s are constants. Substituting (B.1-18) into (B.1-3) gives equation (B.1-4). The two roots of (B.1-4) developed as equation (B.1-5) gives the general solution of (B.1-18) as

$$x_m = A_1 e^{s_1 t} + A_2 e^{s_2 t} \tag{B.1-19}$$

where A_1 and A_2 are constants to be determined from initial conditions. Initial conditions are the values assigned to the displacement, x_m, and the velocity, $\dot{x}_m$, of the system mass at, for example, $t = 0$.

The three important cases illustrated in Figure B.1-2 depend on the relative values of $(C/2M_{eq})$ and (k_{eq}/M_{eq}) in equation (B.1-5). Although uncommon in physical systems, critical damping occurs when

$$(C/2M_{eq})^2 = k_{eq}/M_{eq} \tag{B.1-20a}$$

This condition describes motion in which the mass returns to rest from a displaced position without oscillation. An overdamped condition occurs when

$$(C/2M_{eq})^2 > (k_{eq}/M_{eq}) \tag{B.1-20b}$$

In this case motion is aperiodic, Figure B.1-3(b) where the motion is the sum of two exponentials. The third and most common motion in electronic hardware systems occurs when

$$(C/2M_{eq})^2 < (k_{eq}/M_{eq}) \tag{B.1-20c}$$

which is known as the underdamped condition having the limiting value of amplification given by (B.1-8). This form of motion is described by Figure B.1-2(b), in which case the system is lightly damped vibrating with decreasing amplitude for amplification $Q > \frac{1}{2}$.

Based on (B.1-6) and Euler's expression

$$e^{\pm j\omega_D t} = \cos \omega_D t \pm j \sin \omega_D t \tag{B.1-21}$$

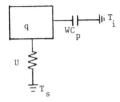

Figure B.1-8. Component Model.

Equation (B.1-19) can be transformed to

$$x_m = e^{-\sigma t}(B_1 \cos \omega_D t + B_2 \sin \omega_D t) \qquad \text{(B.1-22)}$$

where B_1 and B_2 replace the constants $(A_1 + A_2)$ and $j(A_2 - A_2)$, respectively and from (B.1-5b)

$$\omega_D = \sqrt{\frac{k_{eq}}{M_{eq}}} (1 - \zeta^2)^{1/2} = \omega_n \left(1 - \frac{1}{4Q^2} \right)^{1/2} \qquad \text{(B.1-23)}$$

The development of the transfer function and time response of a single degree of freedom thermal model follows analogous lines. Consider a component having an initial temperature t_i and heat generation rate q with conductance U to a sink with temperature t, shown in Figure B.1-8.

The incremental quantity of heat stored, q_s, per unit of time, ϕ, as a function of the component weight and specific heat product WC_p becomes

$$q_s = WC_p \frac{dt}{d\phi} \qquad \text{(B.1-24)}$$

The quantity of heat transferred from the component to the sink is a function of the temperature difference $(t - t_s)$ expressed as

$$q_u = U(t - t_s) \qquad \text{(B.1-25)}$$

A heat balance on the component accounting for heat generated, q, gives

$$q = q_s + q_u \qquad \text{(B.1-26)}$$

Substituting (B.1-24) and (B.1-25) into (B.1-26) and simplifying gives the differential equation for the component temperature:

$$WC_p \frac{dt}{d\phi} + Ut = Ut_s + q \qquad \text{(B.1-27a)}$$

or

$$\frac{dt}{d\phi} + \frac{U}{WC_p} t = \frac{Ut_s + q}{WC_p} \qquad \text{(B.1-27b)}$$

Using the s operator, the characteristic equation representing (B.1-27b) becomes

$$\left(S + \frac{U}{WC_p} \right) t(s) = 0 \qquad \text{(B.1-28)}$$

which gives the root

$$s = -\frac{U}{WC_p} \qquad \text{(B.1-29a)}$$

The real root given by (B.1-29a) is indicative of the exponential temperature response of the system and is defined as the *time constant*. The quantity WC_p/U indicates the time required for the temperature to change to $(1/e)$th of its original value. Based on the notation of (B.1-7), the thermal time constant is expressed as

$$\tau = \frac{WC_p}{U} \qquad \text{(B.1-29b)}$$

Multiple application of the time constant can be used to describe the temperature response. This is illustrated in Figure B.1-9.

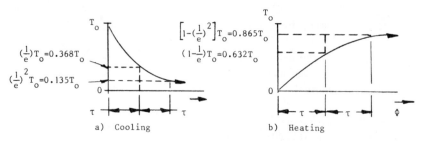

Figure B.1-9. Time Constant Development of Temperature Response.

We assume a general solution of (B.1-27b) in the form

$$t = A_1 + A_2 e^{S\phi} = A_1 + A_2 e^{-\phi/\tau} \qquad \text{(B.1-30)}$$

where A_1 and A_2 are functions of the initial temperature and the component input temperature rate due to heating given by the right side of (B.1-27b). Substituting (B.1-30) into (B.1-27b) gives

$$A_1 = \frac{Ut_s + q}{U} \qquad \text{(B.1-31a)}$$

Setting $\phi = 0$ and solving (B.1-30) with $t = t_i$ determines A_2 as follows:

$$A_2 = t_i - A_1 \qquad \text{(B.1-31b)}$$

Substituting (B.1-31) into (B.1-30), we obtain

$$t = A_1 + (t_i - A_1)e^{-\phi/\tau} \qquad \text{(B.1-32a)}$$

or

$$t = \frac{Ut_s + q}{U}(1 - e^{-U\phi/WC_p}) + t_i e^{-U\phi/WC_p} \qquad \text{(B.1-32b)}$$

for the temperature response of the component model of Figure B.1-8.

B.2 TRANSFER FUNCTION DEVELOPMENT

We have examined the development of the transfer function of equivalent spring-mass systems for foundation motion in Appendix B.1. System representations of physical hardware may use spring-mass, elastic or rigid elements for the structural entity and resistor-capacitor networks for the heat transfer idealization. Initial conditions, boundary constraints and the loads experienced by the assembled system are additional inputs which influence the design and behavior of the equipment. The transfer functions which include these inputs describe succinctly the behavioral characteristics of the system. The transfer function which expressed as the ratio of the response to its corresponding input is illustrated conceptually in Figure B.2-1.

Structural Considerations.   Consider the equivalent system of Figure B.1-1 with an initial velocity $\dot{x}_m(0)$ and displacement $x_m(0)$ subject to a foundation displacement x_f. Using reasoning analogous to the development of (B.1-2), we

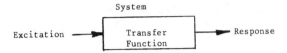

Figure B.2-1. Conceptual Representation of a Transfer Function.

again arrive at (B.1-2a) since the initial conditions do not change the force balance of the system.

$$M_{eq}\ddot{x}_m + C\dot{x}_m + k_{eq}x_m = C\dot{x}_f + k_{eq}x_f \qquad (B.1-2a)$$

The operational form of (B.1-2a) using the s operator which includes the initial conditions (B.5-39) is

$$(M_{eq}s^2 + Cs + k_{eq})x_m(s)$$

$$= (M_{eq}s + C)x_m(0) + M_{eq}\dot{x}_m(0) + (Cs + k_{eq})x_f(s) \qquad (B.2-1)$$

When solved for the operational displacement, we have

$$x_m(s) = \frac{(M_{eq}s + C)x_m(0)}{(M_{eq}s^2 + Cs + k_{eq})} + \frac{M_{eq}\dot{x}_m(0)}{(M_{eq}s^2 + Cs + k_{eq})}$$

$$+ \frac{(Cs + k_{eq})x_f(s)}{(M_{eq}s^2 + Cs + k_{eq})} \qquad (B.2-2)$$

the response displacement of (B.2-2) is a sum of three motions due to independent inputs $x_m(0)$, $\dot{x}_m(0)$ and x_f, which may be expressed as

$$x_m(s) = x_{m1}(s) + x_{m2}(s) + x_{m3}(s) \qquad (B.2-3)$$

where x_{mi} is the displacement of coordinate m due to input i. Each of these motions may be regarded as a transfer function relating the displacement to the corresponding input. The transfer function relating the displacement $x_{m2}(s)$ to the initial velocity in (B.2-2) for example becomes

$$\frac{x_{m2}(s)}{\dot{x}_m(0)} = \frac{M_{eq}}{M_{eq}s^2 + Cs + k_{eq}} \qquad (B.2-4)$$

Natural motions of the system are described by the denominator. The denominators of the corresponding transfer functions of any system are therefore always the same. The denominator is the characteristic equation.

Response acceleration can be obtained by multiplying (B.2-2) through by s^2

since $s^2 x_m(s) = \ddot{x}_m(s)$. Equation (B.2-3) then expresses the sum of three acceleration motions

$$\ddot{x}_m(s) = \ddot{x}_{m1}(s) + \ddot{x}_{m2}(s) + \ddot{x}_{m3}(s) \qquad \text{(B.2-5)}$$

The transfer function corresponding to acceleration $\ddot{x}_{m2}(s)$, is

$$\frac{\ddot{x}_{m2}(s)}{\ddot{x}_m(s)} = \frac{s^2 M_{\text{eq}}}{M_{\text{eq}} s^2 + Cs + k_{\text{eq}}} \qquad \text{(B.2-6)}$$

Multicoordinate systems are treated similarly. In this case (B.2-1) written in matrix form is

$$([\mathbf{M}]s^2 + [\mathbf{C}]s + [\mathbf{k}])\{x(s)\} = [\mathbf{B}]\{q(s)\} \qquad \text{(B.2-7)}$$

where $[\mathbf{M}]$, $[\mathbf{C}]$ and $[\mathbf{k}]_{n,n}$ represent the mass, damping and stiffness matrices respectively. $[\mathbf{B}]_{np}$ is the force transformation matrix and $\{\mathbf{q}\}_p$ is a vector of system inputs.

A solution of (B.2-7) for $x_{mi}(s)$ in terms of $q_i(x)$ expressed as the ratio of output to input determines the transfer function. This approach was used in determining (B.1-17b).

EXAMPLE B.2-1.   Derive the transfer function $x_{11}(s)/u(s)$ for the model of Figure B.2-2 subjected to a foundation displacement, u. At time $= 0$, coordinate 1 has a displacement $x_1(0)$. The system equations analogous to (B.2-7) are

$$\begin{bmatrix} M_1 s^2 + C_1 s + k_1 & -k_1 \\ -k_1 & M_2 s^2 + C_2 s + (k_1 + k_2) \end{bmatrix} \begin{bmatrix} x_1(s) \\ x_2(s) \end{bmatrix}$$

$$= \begin{bmatrix} C_1 s & M_1 s + C_1 \\ C_2 s + k_2 & 0 \end{bmatrix} \begin{bmatrix} u(s) \\ x_1(0) \end{bmatrix} \qquad \text{(B.2-8)}$$

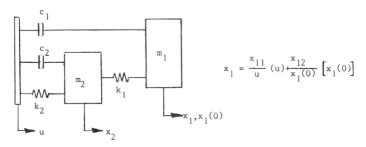

$$x_1 = \frac{x_{11}}{u}(u) + \frac{x_{12}}{x_1(0)} [x_1(0)]$$

Figure B.2-2. Two Coordinate (Degrees of Freedom) System.

Using Cramer's method and the method of Appendix A-5, the desired transfer function is

$$\frac{x_{11}(s)}{u(x)} = \frac{C_1 s (M_2 s^2 + C_2 s + k_1 + k_2) + k_1 (C_2 s + k_2)}{(M_1 s^2 + C_1 s)(M_2 s^2 + C_2 s + k_1 + k_2) + k_1 (M_1 s^2 + C_1 s)} \quad \text{(B.2-9)}$$

Q.E.D.

Multi coordinate heat transfer systems can be expressed in a form similar to (B.2-7), as

$$([C]s + [G])\{T(s)\} = [B]\{q(s)\} \quad \text{(B.2-10)}$$

where $[C]$ and $[G]_{nn}$ represent the specific capacitance and conductance matrices respectively. $[B]_{np}$ is the thermal energy transformation matrix and $\{q\}_p$ is a vector of system inputs.

As before, a solution of (B.2-10) for $T_{mi}(s)$ in terms of $q_i(s)$ expressed as a ratio of output to input determines the transfer function. The nodal temperature would then be found from

$$T_m(s) = T_{m1}(s) + T_{m2}(s) + \cdots + T_{mp}(s) = \sum_{i=1}^{p} T_{mi}(s) \quad \text{(B.2-11)}$$

where $T_{mi}(s)$ is the response of coordinate m to input i.

EXAMPLE B.2-2. Determine the operational temperature response of the component mounting strip of Figure B.2-3(a) using two nodes. Conductance G_1 and G_2 of Figure B.2-3 include the interface as well as the strip conductance. Capacitance $C_i =$

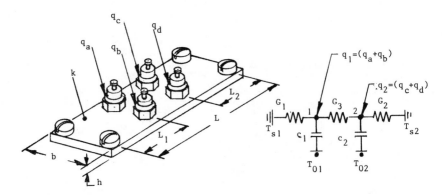

Figure B.2-3. Component Strip Temperature.

$(WC_p)_i$ includes the components in addition to part of the strip length corresponding to the nodal division.

The system equations in the form of (B.2-10) are

$$
\begin{bmatrix}
C_1 s + (G_1 + G_3) & -G_3 \\
-G_3 & C_2 s + (G_2 + G_3)
\end{bmatrix}
\begin{Bmatrix}
T_1(s) \\
T_2(s)
\end{Bmatrix}
$$

$$
=
\begin{bmatrix}
1 & 1 \\
Q_2 & T_{02} C_2 \\
Q_1 & T_{01} C_1
\end{bmatrix}
\begin{Bmatrix}
Q_1/s \\
T_{01} C_1
\end{Bmatrix}
\quad \text{(B.2-12)}
$$

where $Q_1 = T_{s1} G_1 + q_1$ and $Q_2 = T_{s2} G_2 + q_2$.

Using Cramer's method and the technique of Appendix A.5, we have

$$
T_1(s) = \frac{1}{\Delta}
\begin{bmatrix}
Q_1/s & -G_3 \\
Q_2/s & C_2 s + G_2 + G_3
\end{bmatrix}
$$

$$
+ \frac{1}{\Delta}
\begin{bmatrix}
T_{01} C_1 & -G_3 \\
T_{02} C_2 & C_2 s + G_2 + G_3
\end{bmatrix}
\quad \text{(B.2-13)}
$$

where

$$
\Delta = (C_1 s + G_1)(C_2 s + G_2 + G_3) + (C_2 s + G_2) G_3
$$

Simplifying (B.2-13) gives

$$
T_1(s) = \frac{Q_1(C_2 s + G_2 + G_3) + Q_2 G_3}{s\Delta}
$$

$$
+ \frac{T_{01} C_1(G_2 s + G_2 + G_3) + T_{02} C_2 C_3}{\Delta}
\quad \text{(B.2-14a)}
$$

By a similar approach,

$$
T_2(s) = \frac{Q_2(C_1 s + G_1 + G_3) + Q_1 G_3}{s\Delta}
$$

$$
+ \frac{T_{02} C_2(C_1 s + G_1 + G_3) + T_{01} C_1 G_3}{\Delta}
\quad \text{(B.2-14b)}
$$

Q.E.D.

Six inputs are actually imposed on the model of Figure B.2-3. These consist of the sink temperatures T_{s1} and T_{s2}, initial temperatures T_{01} and T_{02}, and injected heat quantities q_1 and q_2 representing component heat dissipation. Through structuring (B.2-12), two combined inputs result which simplifies subsequent manipulations. The transfer functions for each of the six inputs are included in the operational response of (B.2-14). As an example (B.2-14a) includes the transfer functions.

$$\frac{T_{11}(s)}{T_{s1}(s)} = \frac{G_1(C_2 s + G_2 + G_3)}{s\Delta}$$

$$\frac{T_{12}(s)}{T_{01}(s)} = \frac{C_1(C_2 s + G_2 + G_3)}{\Delta}$$

(B.2-15)

The time history of the response may be obtained by the methods of Appendix B.5. The steady state temperature distribution, however, may be obtained directly from (B.2-11) using the condition

$$T_m(\text{steady state}) = [sT_m(s)]\big|_{s=0}$$

(B.2-16)

which is equivalent to an evaluation at infinite time. Equation (B.2-16) for the steady state temperatures T_1 and T_2 of Example B.2-2, based on (B.2-14), are

$$T_1 = [sT_1(s)]\big|_{s=0} = \frac{Q_1(G_2 + G_3) + Q_2 G_3}{G_1(G_2 + G_3) + G_2 G_3}$$

$$T_2 = [sT_2(s)]\big|_{s=0} \quad \frac{Q_2(G_1 + G_3) + Q_1 G_3}{G_1(G_2 + G_3) + G_2 G_3}$$

(B.2-17)

B.3 DEVELOPMENT OF THE CHARACTERISTIC EQUATION

The characteristic equation of a system compactly describes its natural motions. For this reason, there have been many methods both algebraic and numerical designed to generate its form from the model equations. Appendix A.5(8) describes an efficient process of extracting the characteristic equation, but any approach that develops the determinant polynomial or its coefficients may be used. Among these are iterative methods, Leverrier's algorithm, Danivalski's method, and similarity transformations, to name a few.

EXAMPLE B.3-1.   Develop the characteristic equation for the two degree of freedom system of Example B.2-1 using the process of Appendix A.5(8).
    The characteristic equation is the determinant of the system equation of (B.2-8) written as

$$\begin{bmatrix} 0.07772s^2 + 0.1864s + 30 & -30 \\ -30 & 0.15544s^2 + 0.4366s + 90 \end{bmatrix} = 0 \quad \text{(B.3-1)}$$

Application of step 1 of A.5(8) results in column 1 of (B.3-1) being replaced by the sum of columns 1 and 2, giving

$$\begin{bmatrix} 0.07772s^2 + 0.1864s & -30 \\ 0.15544s^2 + 0.4366s + 60 & 0.15544s^2 + 0.4366s + 90 \end{bmatrix} = 0 \quad \text{(B.3-2)}$$

The determinant of (B.3-2) contains no redundant terms which would result if (B.3-1) were expanded. The characteristic equation of Example B.2-1 using (B.3-2) becomes

$$|D| = (0.07772s^2 + 0.1864s)(0.15544s^2 + 0.4366s + 90)$$
$$+ 30(0.15544s^2 + 0.4366s + 60) \quad \text{(B.3-3a)}$$

or

$$|D| = 0.01208s^4 + 0.06291s^3 + 11.73938s^2 + 29.874s + 1800 \quad \text{(B.3-3b)}$$

Q.E.D.

The efficiency of the process of Appendix A.5(8) is even more pronounced for larger systems. An even greater advantage is that zero mass or capacity coordinates for structural or thermal systems can be treated directly without removal by the reduction methods of Appendix A.6.

Other methods which develop the determinant polynomial or its coefficients are more restrictive requiring removal of the zero mass or capacity coordinates before proceeding. The use of zero mass or capacity coordinates permits evaluation at certain system locations without an increase in mathematical complexity or in the degrees of freedom of the solution matrix. In general, however, as the number of coordinates with mass or capacity increase, the effort required to evaluate the system increases even more rapidly, requiring automated processing.

Many electronic hardware configurations can be aptly evaluated using manual techniques based on computer methods. These originate from the system equation by transformation or matrix division. Consider the system equation for transient temperature response given by

$$[C]\{\dot{T}\} + [G]\{T\} = \{Q\} \quad \text{(B.3-4)}$$

given $[C]$ = thermal capacity matrix;
       $[G]$ = conductance matrix;
       $\{Q\}$ = coordinate energy vector.

The desired form is derived by premultiplying (B.3-4) by $[C]^{-1}$ which gives

$$\{\dot{T}\} + [A]\{T\} = [C]^{-1}\{Q\} \qquad (B.3-5)$$

where $[A] = [C]^{-1}[G]$
The characteristic form is developed from the Laplace transform of (B.3-5):

$$[Is + A]\{T(s)\} = [C]^{-1}\{Q(s)\} \qquad (B.3-6a)$$

or

$$[I]s + [A] = 0 \qquad (B.3-6b)$$

The normalized form of the matrix characteristic equation (B.3-6b) is suitable for both hand or machine development of the determinant polynomial. A damped structural system requires an additional transformation to arrive at a companion form of (B.3-6b). Consider the general equation for a n coordinate system given by

$$[M]\{\ddot{x}\} + [C]\{\dot{x}\} + [K]\{x\} = [F] \qquad (B.3-7)$$

where $[M]_{n,n}$ = mass matrix;
      $[C]_{n,n}$ = damping matrix;
      $[K]_{n,n}$ = stiffness matrix;
      $\{F\}_{n,1}$ = cordinate force vector.

Defining

$$
\begin{array}{ll}
r_1 = x_1 & r_{n+1} = \dot{x}_1 \\
r_2 = x_2 & r_{n+2} = \dot{x}_2 \\
\quad\cdot & \quad\cdot \\
\quad\cdot & \quad\cdot \\
\quad\cdot & \quad\cdot \\
r_n = x_n & r_{2n} = \dot{x}_n
\end{array}
\qquad (B.3-8)
$$

Equation (B.3-7) may be transformed using (B.3-8) to

$$[I]\{\dot{r}\} + [A]\{r\} = \begin{Bmatrix} 0 \\ I \end{Bmatrix} \{F\} \qquad (B.3\text{-}9)$$

given

$$[A] = \begin{bmatrix} 0 & -I \\ M^{-1}k & M^{-1}C \end{bmatrix}_{2n,2n} \text{ and } \{r\} = \begin{Bmatrix} x_1 \\ x_2 \\ \cdot \\ \cdot \\ \cdot \\ \dot{x}_1 \\ \dot{x}_2 \\ \cdot \\ \cdot \end{Bmatrix}$$

so that the characteristic form is given by (B.3-6b). It can be seen that a zero capacity coordinate results in an indeterminant form of (B.3-5). A zero mass coordinate affects (B.3-9) similarly.

Similarity Transformation. Recurrence relationships can be used to develop the characteristic equation if (B.3-6b) is transformed into

$$\begin{bmatrix} a_1 & b_{12} & b_{13} & \cdots & b_{1n} \\ c_2 & a_2 & b_{23} & \cdots & b_{2n} \\ 0 & c_3 & a_3 & \cdots & b_{3n} \\ 0 & 0 & c_4 & \cdots & b_{4n} \\ \cdot & \cdot & \cdot & & \cdot \\ \cdot & \cdot & \cdot & & \cdot \\ \cdot & \cdot & \cdot & & \cdot \\ 0 & 0 & 0 & \cdots c_n & a_n \end{bmatrix} \qquad (B.3\text{-}10)$$

A sequence of polynomials, the last of which is the determinant polynomial of (B.3-10), becomes

$$P_0(s) = 1$$

$$P_1(s) = a_1$$

$$P_2(s) = P_1 a_2 - b_{12} c_2$$

$$P_3(s) = P_2 a_3 - b_{23} c_3 P_1 + b_{13} c_3 c_2$$

$$P_4(s) = P_3 a_4 - b_{34} c_4 P_2 + b_{24} c_4 c_3 P_1 - b_{14} c_4 c_3 c_2 \qquad \text{(B.3-11)}$$

.

.

.

$$|D| = P_n(s) = P_{n-1} a_n - b_{n-1,n} c_n P_{n-2} + b_{n-2,n} c_n c_{n-1} P_{n-3}$$

$$- \cdots \pm [b_{1,n} c_n c_{n-1} \cdots c_2]$$

The Hessenberg form of (B.3-10) can be obtained by a series of transformations which generate the lower triangular zeros. These transformations are analogous to the elimination method of Appendix A.6(1) which maintains the value of the determinant. A transformation is used to generate the required zeros for each column by the matrix product $[T][A]$. If m columns are involved, the transformation of the $[A]$ matrix is given by

$$[T_m \cdots T_2 T_1][A] \qquad \text{(B.3-12)}$$

Consider the following system matrix

$$[Is + A] = \begin{bmatrix} a_1 & b_{12} & b_{13} & \cdots & b_{1n} \\ c_{21} & a_2 & b_{23} & \cdots & b_{2n} \\ c_{31} & c_{32} & a_3 & \cdots & b_{3n} \\ c_{41} & c_{42} & c_{43} & \cdots & b_{4n} \\ \cdot & \cdot & \cdot & & \cdot \\ \cdot & \cdot & \cdot & & \cdot \\ \cdot & \cdot & \cdot & & \cdot \\ c_{n1} & c_{n2} & c_{n3} & \cdots & a_n \end{bmatrix} \qquad \text{(B.3-13)}$$

Zeros in the first column of (B.3-13) analogous to (B.3-10) are generated using the transformation matrix

$$[\mathbf{T}_1] = \begin{bmatrix} 1 & 0 & 0 & \cdots & 0 \\ 0 & 1 & 0 & \cdots & 0 \\ 0 & -c_{31}/c_{21} & 1 & \cdots & 0 \\ 0 & -c_{41}/c_{21} & 0 & \cdots & 0 \\ \cdot & & \cdot & & \cdot \\ \cdot & & \cdot & & \cdot \\ \cdot & & \cdot & & \cdot \\ 0 & -c_{n1}/c_{21} & 0 & \cdots & 1 \end{bmatrix} \qquad (B.3\text{-}14)$$

so that (B.3-12) becomes

$$[\mathbf{T}_1][\mathbf{I}s + \mathbf{A}] = \begin{bmatrix} a_1 & b_{12} & b_{13} & \cdots & b_{1n} \\ c_{21} & a_2 & b_{23} & \cdots & b_{2n} \\ 0 & c_{32}^* & a_3^* & \cdots & b_{3n}^* \\ 0 & c_{42}^* & c_{43}^* & \cdots & b_{4n}^* \\ \cdot & \cdot & \cdot & & \cdot \\ \cdot & \cdot & \cdot & & \cdot \\ \cdot & \cdot & \cdot & & \cdot \\ 0 & c_{n2}^* & c_{n3}^* & \cdots & a_n^* \end{bmatrix} \qquad (B.3\text{-}15)$$

EXAMPLE B.3-2. Develop the characteristic equation of

$$[\mathbf{I}s + \mathbf{A}] = \begin{bmatrix} s + 4 & -3 & -1 \\ -3 & s + 6 & -2 \\ -1 & -2 & s + 3 \end{bmatrix} \qquad (B.3\text{-}16)$$

The transformation matrix developed by inspection of (B.3-16) becomes

$$[\mathbf{T}_1] = \begin{bmatrix} 1 & 0 & 0 \\ 0 & 1 & 0 \\ 0 & -\tfrac{1}{3} & 1 \end{bmatrix} \qquad (B.3\text{-}17)$$

The Hessenberg form of the system equation is then

$$[\mathbf{T}_1][\mathbf{I}s + \mathbf{A}] = \begin{bmatrix} 1 & 0 & 0 \\ 0 & 1 & 0 \\ 0 & -\frac{1}{8} & 1 \end{bmatrix} \begin{bmatrix} s+4 & -3 & -1 \\ -3 & s+6 & -2 \\ -1 & -2 & s+3 \end{bmatrix}$$

$$= \begin{bmatrix} s+4 & -3 & -1 \\ -3 & s+6 & -2 \\ 0 & -\frac{1}{8}(s+12) & s+3\frac{7}{8} \end{bmatrix} \qquad \text{(B.3-18)}$$

Using (B.3-11) with (B.3-18), the characteristic equation of (B.3-16) becomes

$$P_1(s) = s + 4$$

$$P_2(s) = (s+4)(s+6) - 9 = s^2 + 10s + 15$$

$$|D| = P_3(s) = (s^2 + 10s + 15)(s + 3\tfrac{7}{8}) - \tfrac{7}{8}(s+12)(s+4) - (s+12)$$

$$|D| = s^3 + 13s^2 + 40s + 11$$

Q.E.D.

Similarity Procedure.   A systematic approach which reduces the effort by eliminating operations and the need of noting partial results has been developed by Crandall. Operations are performed on the $[-\mathbf{A}]$ matrix with (B.3-6b) written in the form

$$[\mathbf{I}]s = -[\mathbf{A}] \qquad \text{(B.3-6c)}$$

For an undamped structural system (B.3-6c) becomes

$$[\mathbf{I}]s^2 = -[\mathbf{A}] \qquad \text{(B.3-6d)}$$

The procedure consists of augmenting the nth order $[-\mathbf{A}]$ matrix via an arbitrarily selected column $a_{i,n+1}$, $i = 1,2 \ldots n$. The augmented matrix of order $2n$ is developed as

$$\begin{bmatrix} -A_{nn} & \vdots & A_{np} \\ \cdots & \cdots & \cdots \\ A_{rn} & \vdots & A_{rp} \end{bmatrix} \qquad \text{(B.3-20)}$$

where $[A_{np}]$ = a lower triangular matrix of coefficients;
$[A_{rp}]$ = an upper triangular matrix of coefficients;
$[A_{rn}]$ = a triangular matrix of coefficients of the polynomials (B.3-11) to the $(n-1)$th order.

The operations are best described by expanding (B.3-20) in its coefficient form:

$$
\begin{array}{c}
\\
\\
\\
\\
\\
\\
\\
\\
\\
1
\end{array}
\begin{bmatrix}
a_{11} & a_{12} & \cdots & a_{1n} & \vdots & a_{1,n+1} & 0 & \cdots & 0 \\
a_{21} & a_{22} & \cdots & a_{2n} & \vdots & a_{2,n+1} & a_{2,n+2} & \cdots & 0 \\
\vdots & \vdots & & & \vdots & \vdots & & & \vdots \\
a_{n1} & a_{n2} & \cdots & a_{nn} & \vdots & a_{n,n+1} & a_{n,n+2} & \cdots & a_{n,2n} \\
\cdots & \cdots & \cdots & \cdots & \cdots & \cdots & \cdots & \cdots & \cdots \\
0 & \cdots & 0 & 1 & \vdots & a_{n+1,n+1} & a_{n+1,n+2} & \cdots & a_{n+1,2n} \\
0 & \cdots & 1 & a_{n+2,n} & \vdots & 0 & a_{n+2,n+2} & \cdots & a_{n+2,2n} \\
\vdots & & \vdots & \vdots & \vdots & \vdots & 0 & & \vdots \\
1 & \cdots & a_{2n,n-1} & a_{2n,n} & \vdots & 0 & 0 & \cdots & a_{2n,2n} \\
a_{2n+1,1} & \cdots & a_{2n+1,n-1} & a_{2n+1,n} & & & &
\end{bmatrix}
$$

with column labels a, b, $\ldots$, $2n$ over the right-hand block.

(B.3-21)

Each column a, b, $\ldots$ of (B.3-21) is developed before proceeding to the next. The first column of $[A_{np}]$ is completely arbitrary with any combination of ones or zeros that do not later lead to zero diagonal coefficients of $[A_{np}]$. Column a of (B.3-21) is easily completed by computing the product $-[\text{row } 1 \text{ of } -A_{nn}] \times [\text{col. } 1 \text{ of } A_{np}]$, i.e.,

$$
\{a_{n+1,n+1}\} = -\frac{1}{a_{1,n+1}} \{a_{11} \quad a_{12} \quad \cdots \quad a_{1n}\}
\begin{Bmatrix}
a_{1,n+1} \\
a_{2,n+1} \\
\vdots \\
a_{n,n+1}
\end{Bmatrix}
\qquad (\text{B.3-22a})
$$

Next, terms of the second column of $[A_{np}]$ in (B.3-20) are a consequence of the product $[\text{rows } 2, 3 \ldots n] \times \text{col. } [a]$, i.e.,

$$
\begin{Bmatrix}
a_{2,n+2} \\
\vdots \\
a_{n,n+2}
\end{Bmatrix}
=
\begin{bmatrix}
a_{21} & a_{22} & \cdots & a_{2n} & a_{2,n+1} \\
\vdots & \vdots & & \vdots & \vdots \\
a_{n1} & a_{n2} & \cdots & a_{nn} & a_{n,n+1}
\end{bmatrix}
\begin{Bmatrix}
a_{1,n+1} \\
a_{2,n+1} \\
\vdots \\
a_{n+1,n+1}
\end{Bmatrix}
\qquad (\text{B.3-22b})
$$

Coefficients of $[A_{rp}]$ which complete column b are determined analogously to (B.3-22a) as

$$\{a_{n+1,n+2}\} = -\frac{1}{a_{1,n+1}}\{a_{11} \quad a_{12} \quad \ldots \quad a_{1n}\} \left\{\begin{array}{c} 0 \\ a_{2,n+2} \\ \vdots \\ a_{n,n+2} \end{array}\right\}$$

$$\{a_{n+2,n+2}\} = -\frac{1}{a_{2,n+2}}\{a_{21} \quad a_{22} \quad \ldots \quad a_{2n} \quad a_{2,n+1}\} \left\{\begin{array}{c} 0 \\ a_{2,n+2} \\ \vdots \\ a_{n+1,n+2} \end{array}\right\}$$

$$(B.3-22c)$$

Notice that (B.3-22c) must be determined in order. The third column of $[A_{np}]$ is computed using $[\text{rows } 3, 4 \ldots n] \times [\text{col. } b]$, i.e.,

$$\left\{\begin{array}{c} a_{3,n+3} \\ \vdots \\ a_{n,n+3} \end{array}\right\} = \left[\begin{array}{cccccc} a_{31} & a_{32} & \ldots & a_{3n} & a_{3,n+1} & a_{3,n+2} \\ \vdots & \vdots & & \vdots & & \\ a_{n1} & a_{n2} & & a_{nn} & a_{n,n+1} & a_{n,\,n+2} \end{array}\right] \left\{\begin{array}{c} 0 \\ a_{2,n+2} \\ \vdots \\ a_{n+2,n+2} \end{array}\right\}$$

$$(B.3-22d)$$

The remaining coefficients of column c are developed sequentially similar to (B.3-22c), i.e.,

$$\{a_{n+1,n+3}\} = -\frac{1}{a_{1,n+1}}\{a_{11} \quad a_{12} \quad \ldots \quad a_{1n}\} \left\{\begin{array}{c} 0 \\ 0 \\ a_{3,n+3} \\ \vdots \\ a_{n,n+3} \end{array}\right\}$$

$$\vdots \qquad\qquad\qquad \vdots \qquad\qquad\qquad\qquad (B.3-22e)$$

$$\{a_{n+3,n+3}\} = -\frac{1}{a_{3,n+3}} \{a_{31} \quad a_{32} \quad \cdots \quad a_{3n} \quad a_{3,n+1} \quad a_{3,n+2}\} \begin{Bmatrix} 0 \\ 0 \\ a_{3,n+3} \\ \vdots \\ a_{n+2,n+3} \end{Bmatrix}$$

Remaining columns of $\begin{bmatrix} A_{np} \\ \cdots \\ A_{rp} \end{bmatrix}$ are completed in a manner analogous to (B.3-22d) and (B.3-22e).

The coefficients of $[A_{rn}]$ are developed from $[A_{rp}]$. The first column becomes

$$\{a_{n+2,n}\} = a_{n+1,n+1}$$

$$\{a_{n+3,n}\} = \begin{Bmatrix} 1 \\ a_{n+2,n} \end{Bmatrix}^T \begin{Bmatrix} a_{n+1,n+2} \\ a_{n+2,n+2} \end{Bmatrix} = \{1 \quad a_{n+2,n}\} \begin{Bmatrix} a_{n+1,n+2} \\ a_{n+2,n+2} \end{Bmatrix}$$

$$\{a_{n+4,n}\} = \begin{Bmatrix} 1 \\ a_{n+2,n} \\ a_{n+3,n} \end{Bmatrix}^T \begin{Bmatrix} a_{n+1,n+3} \\ a_{n+2,n+3} \\ a_{n+3,n+3} \end{Bmatrix} \qquad \text{(B.3-22f)}$$

$$\vdots \qquad\qquad \vdots$$

$$\{a_{2n+1,n}\} = \begin{Bmatrix} 1 \\ \vdots \\ a_{2n,n} \end{Bmatrix}^T \begin{Bmatrix} a_{n+1,2n} \\ \vdots \\ a_{2n,2n} \end{Bmatrix}$$

The coefficients of the second column of $[A_{rn}]$ are determined from the first column and the elements of $[A_{rp}]$:

$$\{a_{n+3,n-1}\} = a_{n+2,n+2} + a_{n+2,n}$$

$$\{a_{n+4,n-1}\} = \begin{Bmatrix} 1 \\ a_{n+3,n-1} \end{Bmatrix}^T \begin{Bmatrix} a_{n+2,n+3} \\ a_{n+3,n+3} \end{Bmatrix} + a_{n+3,n}$$

$$\text{(B.3-22g)}$$

$$\{a_{2n+1,n-1}\} = \left\{\begin{array}{c} 1 \\ \vdots \\ a_{2n,n-1} \end{array}\right\}^{T} \left\{\begin{array}{c} a_{n+2,2n} \\ \vdots \\ a_{2n,2n} \end{array}\right\} + a_{2n,n}$$

The coefficients of the third column of $[A_{rn}]$ are computed in an analogous fashion using the second column and elements of $[A_{rp}]$:

$$\{a_{n+4,n-2}\} = a_{n+3,n+3} + a_{n+3,n-1}$$

$$\{a_{n+5,n-2}\} = \left\{\begin{array}{c} 1 \\ a_{n+4,n-2} \end{array}\right\}^{T} \left\{\begin{array}{c} a_{n+3,n+4} \\ a_{n+4,n+4} \end{array}\right\} + a_{n+4,n-1}$$

$$\vdots \qquad\qquad \vdots \qquad\qquad\qquad \text{(B.3-22h)}$$

$$\{a_{2n+1,n-2}\} = \left\{\begin{array}{c} 1 \\ \vdots \\ a_{2n,n-2} \end{array}\right\}^{T} \left\{\begin{array}{c} a_{n+3,2n} \\ \vdots \\ a_{2n,2n} \end{array}\right\} + a_{2n,n-1}$$

The characteristic equation of (B.3-6c) becomes

$$s^{n} + a_{2n+1,1}s^{n-1} + \cdots + a_{2n+1,n} \qquad \text{(B.3-22i)}$$

and likewise for (B.3-6d)

$$s^{2n} + a_{2n+1,1}s^{2n-1} + \cdots + a_{2n+1,n} \qquad \text{(B.3-22j)}$$

Equation (B.3-21) features a convenient method of checking the results of developing $[A_{np}]$ and $[A_{np}]$. The sums of each column of $[-A_{nn} \vdots A_{np}]$ expressed as a row vector multiplied by the last augmented column $2n$, should equal zero. For column b, for example

$$\left[\sum_{1}^{n} a_{i,1} \quad \sum_{1}^{n} a_{i,2} \quad \cdots \quad \sum_{1}^{n} a_{i,2n}\right] \left\{\begin{array}{c} 0 \\ \vdots \\ a_{n,2n} \\ a_{n+1,2n} \\ a_{n+2,2n} \\ \vdots \end{array}\right\} = 0 \qquad \text{(B.3-23)}$$

Except for column $2n$, each column is augmented with -1 in place of the first zero. When (B.3-23) does not equal zero due to round-off errors, a residual which is a small percentage of the working products is acceptable.

EXAMPLE B.3-3. Resolve Example B.3-2 for the characteristic equation using the similarity procedure. Validate the procedure using (B.3-23).

From (B.3-16)

$$[-A] = \begin{bmatrix} -4 & 3 & 1 \\ 3 & -6 & 2 \\ 1 & 2 & -3 \end{bmatrix} \tag{B.3-24}$$

The first column of $[A_{np}]$ is chosen as $\{1 \quad 0 \quad 0\}^T$ so that the initial form of (B.3-21) with $a_{44} = -(-4)$ from (B.3-22a) becomes

$$\begin{array}{ccc} & a & b & c \end{array}$$

$$\begin{bmatrix} -4 & 3 & 1 & \vdots & 1 & 0 & 0 \\ 3 & -6 & 2 & \vdots & 0 & a_{25} & 0 \\ 1 & 2 & -3 & \vdots & 0 & a_{35} & a_{36} \\ \cdots & \cdots & \cdots & & \cdots & \cdots & \cdots \\ 0 & 0 & 1 & \vdots & 4 & a_{45} & a_{46} \\ 0 & 1 & a_{53} & \vdots & 0 & a_{55} & a_{56} \\ 1 & a_{62} & a_{63} & \vdots & 0 & 0 & a_{66} \end{bmatrix} \tag{B.3-25a}$$

The coefficients a_{ij} of (B.3-25a) are determined using (B.3-22). Coefficients of second column of $[A_{np}]$ are determined next followed by those of $[A_{rp}]$ before proceding to column c. Based on (B.3-22b), we obtain

$$\begin{Bmatrix} a_{25} \\ a_{35} \end{Bmatrix} = \begin{bmatrix} 3 & -6 & 2 & 0 \\ 1 & 2 & -3 & 0 \end{bmatrix} \begin{Bmatrix} 1 \\ 0 \\ 0 \\ 4 \end{Bmatrix} = \begin{Bmatrix} 3 \\ 1 \end{Bmatrix} \tag{B.3-25b}$$

Then (B.3-22c) gives

$$\{a_{45}\} = -1\{-4 \quad 3 \quad 1\} \begin{Bmatrix} 0 \\ 3 \\ 1 \end{Bmatrix} = \{-10\} \tag{B.3-25c}$$

$$\{a_{55}\} = -\tfrac{1}{8}\{3 \;\; -6 \;\; 2 \;\; 0\} \begin{Bmatrix} 0 \\ 3 \\ 1 \\ -10 \end{Bmatrix} = \{^{16}\!/\!_8\} \qquad \text{(B.3-25d)}$$

Substituting (B.3-25b), (B.3-25c) and (B.3-25d) into (B.3-21) then completing the procedure gives (B.3-25a):

$$
\begin{array}{ccccccc}
& & & a & b & c & \\
\begin{bmatrix}
-4 & 3 & 1 & \vdots & 1 & 0 & 0 \\
3 & -6 & 2 & \vdots & 0 & 3 & 0 \\
1 & 2 & -3 & \vdots & 0 & 1 & {}^{25}\!/\!_3 \\
\hdashline
0 & 0 & 1 & \vdots & 4 & -10 & -{}^{25}\!/\!_3 \\
0 & 1 & a_{53} & \vdots & 0 & {}^{16}\!/\!_3 & -{}^{50}\!/\!_9 \\
1 & a_{62} & a_{63} & \vdots & 0 & 0 & {}^{11}\!/\!_3
\end{bmatrix}
\end{array}
\qquad \text{(B.3-25e)}
$$

Since the verification of (B.3-25e) does not involve $\{A_{rn}\}$ of (B.3-20) it is worthwhile to check these results before proceeding. Based on (B.3-23), the product of a vector formed from the sums of columns of $[-A_{nn} \vdots A_{np}]$ and column $\{c\}$ of (B.3-25e) is

$$\{0 \;\; -1 \;\; 0 \;\; 1 \;\; 4 \;\; {}^{25}\!/\!_3\} \begin{Bmatrix} 0 \\ 0 \\ {}^{25}\!/\!_3 \\ -{}^{25}\!/\!_3 \\ -{}^{50}\!/\!_9 \\ {}^{11}\!/\!_3 \end{Bmatrix} = 0 \qquad \text{(B.3-26)}$$

If an error was detected using column $\{c\}$, then (B.3-23e) could be used with either columns $\{a\}$ or $\{b\}$ to locate the problem. Equation (B.3-25e) is then completed using (B.3-22f), (B.3-22g), etc., in sequence. The result becomes

$$
\begin{bmatrix}
-4 & 3 & 1 & 1 & 0 & 0 \\
3 & -6 & 2 & 0 & 3 & 0 \\
1 & 2 & -3 & 0 & 1 & {}^{25}\!/\!_3 \\
0 & 0 & 1 & 4 & -10 & -{}^{25}\!/\!_3 \\
0 & 1 & 4 & 0 & {}^{16}\!/\!_3 & -{}^{50}\!/\!_9 \\
1 & {}^{28}\!/\!_3 & {}^{34}\!/\!_3 & 0 & 0 & {}^{11}\!/\!_3
\end{bmatrix}
\qquad \text{(B.3-27)}
$$

$$1 \qquad 13 \quad 40 \quad 11$$

The characteristic equation using (B.3-22i) is then

$$s^3 + 13s^2 + 40s + 11 \qquad \text{(B.3-28)}$$

Q.E.D.

Once the procedure is understood, all intermediate computations are done in the calculator with the results noted directly in (B.3-21). The simplicity of the procedure and the result analogous to (B.3-27) would then be clear.

Besides the ability to check intermediate results, (B.3-20) provides a means of recovering the system mode shape.

An intermediate vector $\{z\}$ is obtained from $[A_{np}]$ augmented as shown in (B.3-29):

$$
\begin{bmatrix}
a_{n+1,n+1} - \omega_i^2 & a_{n+1,n+2} & \cdots & a_{n+1,2n-1} & a_{n+1,2n} \\
-1 & a_{n+2,n+2} - \omega_i^2 & \cdots & a_{n+2,2n-1} & a_{n+2,2n} \\
0 & -1 & \cdots & a_{n+3,2n-1} & a_{n+3,2n} \\
0 & 0 & \cdots & a_{n+4,2n-1} - \omega^2 & a_{n+4,2n} \\
\vdots & \vdots & & \vdots & \vdots \\
0 & 0 & \cdots & -1 & a_{2n,2n} - \omega^2
\end{bmatrix}
\begin{Bmatrix}
z_1 \\ z_2 \\ z_3 \\ z_4 \\ \vdots \\ z_n
\end{Bmatrix} = 0
$$

$$\text{(B.3-29)}$$

Substitute one of the frequencies, ω_i^2, of the system natural motion in (B.3-29) with $z_n = 1$. $z_i(i = 1, n - 1)$ can then be determined by back substitution. The mode shape $\{x\}$ is developed from the product

$$\{x\} = [A_{np}]\{z\} \qquad \text{(B.3-30)}$$

B.4 FACTORING

In this section we try to find the factors or roots of the characteristic equation. There exists a multitude of published techniques for both hand and automatic use. The factors of the characteristic equation are necessary to determine the natural characteristics of the system, the time domain response, section B.5 and frequency response, Section B.6.

Linear Factors. An efficient technique for hand or machine is based on rewriting the characteristic equation

$$a_0 s^n + a s^{n-1} + a_2 s^{n-2} + \cdots + a_n \tag{B.4-1a}$$

as

$$\{\{[(a_0 s + a_1)s + a_2]s + a_3\}s + \cdots \}s + a_n \tag{B.4-1b}$$

where for a particular value of s, the evaluation works outward. This inside-out evaluation may be conveniently arranged in a form known as *synthetic division*. The coefficients, a_i, are written in ascending order, supplying zeros if necessary. The division process may be used for either linear or quadratic factors. The derivative of (B.4-1a) may be evaluated at a particular value of s by a continuation of the process.

If we divide (B.4-1a) by a linear factor $(s + \alpha)$, we generate a quotient and a remainder constant, R. The remainder, R, is the value of the characteristic equation (B.4-1a) at α. Writing (B.4-1a) using only coefficients, inserting zeros where necessary, then dividing by α gives

$$
\begin{array}{r|cccccc}
-\alpha & a_0 & a_1 & a_2 & \cdots & a_{n-1} & a_n \\
        &   & -b_0\alpha & -b_1\alpha & \cdots & -b_{n-2}\alpha & -b_{n-1}\alpha \\
\hline
        & b_0 & b_1 & b_2 & \cdots & b_{n-1} & b_n = R
\end{array}
\tag{B.4-2}
$$

where $b_0 = a_0$, $b_1 = a_1 - a_0\alpha$, etc.

The value b_n would also result if (B.4-1b) was evaluated at $s = -\alpha$. If α were an exact root of (B.4-1a) then the reduced characteristic equation from (B.4-2) would be

$$b_0 s^{n-1} + b_1 s^{n-2} + \cdots + b_{n-1} \tag{B.4-3}$$

and $(s + \alpha)$ would be a factor of (B.4-1a). This may be restated as

$$(s + \alpha)(b_0 s^{n-1} + b_1 s^{n-2} + \cdots + b_{n-1})$$
$$= a_0 s^n + a_1 s^{n-1} + a_2 s^{n-2} + \cdots + a_n \tag{B.4-4}$$

The procedure described may be used to divide the characteristic equation (B.4-1a) by quadratic factors $(s^2 + \alpha s + \beta)$. To perform division, only the coefficients of (B.4-1a) are needed as before, with the coefficients of the quadratic factor written in descending order along the left side with signs reversed. Based on (B.4-1a), the general form for quadratic division is

a_0	a_1	a_2	a_3	$\ldots$	a_{n-2}	a_{n-1}	a_n
$-\alpha$	$-b_0\alpha$	$-b_1\alpha$	$-b_2\alpha$	$\ldots$	$-b_{n-3}\alpha$	$-b_{n-2}\alpha$	
$-\beta$		$-b_0\beta$	$-b_1\beta$	$\ldots$	$-b_{n-4}\beta$	$-b_{n-3}\beta$	$-b_{n-2}\beta$
$b_0 = a_0$	b_1	b_2	b_3	$\ldots$	b_{n-2}	b_{n-1}	b_n

$$\underbrace{\text{Quotient} = \text{reduced characteristic equation}} \qquad \underbrace{\text{Remainder}}$$

$$(B.4\text{-}5)$$

The division process represented by the procedure of (B.4-2) and (B.4-5) may be used to determine the coefficients, s, of the partial fraction expansion of the transfer function and the factors of the characteristic equation. The process of determining these factors is an iterative one of repeated division with improved divisors at each successive trial.

A characteristic equation developed for the temperature distribution within components and assemblies and undamped structural systems have only linear factors. On the other hand, the characteristic equation of a damped structural system will be composed of quadratic factors and depending upon the type of support motion considered, may include some linear factors.

A good starting value for the iterative process of a characteristic equation having linear factors is given by the last two terms of (B.4-1a) as

$$\alpha_0 = a_n/a_{n-1} \qquad (B.4\text{-}6)$$

Following the process of (B.4-2), a better or improved factor is given by

$$\alpha_n = a_n/b_{n-1} \qquad (B.4\text{-}7)$$

Successive values of α_i converge to the lowest factor of the equation. Examination of rate of change of successive α_i may give cause to estimate a better value than can be obtained using (B.4-7).

EXAMPLE B.4-1.   Determine the lowest linear factor of the characteristic equation

$$s^4 + 14s^3 + 67s^2 + 126s + 72 \qquad (B.4\text{-}8)$$

The starting value using (B.4-6) becomes

$$\alpha_0 = {}^{72}\!/_{126} = 0.57 \qquad (B.4\text{-}9)$$

Based on the procedure of (B.4-2), the result of the first iteration becomes

$$
\begin{array}{r|rrrrr}
-0.57 & 1 & 14 & 67 & 126 & 72 \\
& & -0.57 & -7.6551 & -33.82 & \\
\hline
& 1 & 13.43 & 59.34 & 92.17 &
\end{array}
\qquad \text{(B.4-10)}
$$

An improved estimate for the factor is found using (B.4-7) which gives

$$
\alpha_1 = \frac{72}{92.17} = 0.78
\qquad \text{(B.4-11)}
$$

A second iteration of (B.4-2) using (B.4-11) produces the result

$$
\begin{array}{r|rrrrr}
-0.78 & 1 & 14 & 67 & 126 & 72 \\
& & -0.78 & -10.3116 & -44.217 & \\
\hline
& 1 & 13.22 & 56.688 & 81.78 &
\end{array}
\qquad \text{(B.4-12)}
$$

where a further improved factor (B.4-7) becomes

$$
\alpha_2 = \frac{72}{81.78} = 0.88
\qquad \text{(B.4-13)}
$$

Examination of successive differences between (B.4-9), (B.4-11) and (B.4-13), suggests an extrapolated trial $\alpha_3 = 1.0$. A third iteration of (B.4-2) using this value of α_3 indicated no additional improvement occurs. When this condition arises the factor $(s + \alpha_n)$ is a root of the characteristic equation. In this case the lowest linear factor of (B.4-8) was determined as $(s + 1)$. The next lowest linear factor is determined by a continuation of process (B.4-2) on the reduced characteristic equation (B.4-3).

Q.E.D.

The process of (B.4-2) can be accelerated by forming corrections based on the value of the characteristic equation and its derivative at the value of the trial factor α_i. A new trial is determined from

$$
\alpha_{i+1} = \alpha_i - \frac{D(\alpha_i)}{D'(\alpha_i)}
\qquad \text{(B.4-14)}
$$

given $D(\alpha_i)$ = value of the characteristic equation at $s = -\alpha_i$;
$\quad\quad D'(\alpha_i)$ = value of the derivative of the characteristic equation $s = -\alpha_i$.

Equation (B.4-14) is the basis of Newton's method, which was derived from the Taylor expansion by retaining linear terms. Equation (B.4-14) applies for factors which are either real or complex. Complex arithmetic makes the evaluation of improved successive factors more laborious than for linear factors. Use of (B.4-14) causes the correct significant figures of the factor to double each iteration. The convergence is therefore quadratic.

The evaluation of the derivative $D'(\alpha_i)$ is easily accomplished by a continuation of the process (B.4-2) on the reduced characteristic equation. The process applied to (B.4-1a) for linear factors becomes

$$
\begin{array}{c|ccccccc}
-\alpha & a_0 & a_1 & a_2 & \cdots & a_{n-2} & a_{n-1} & a_n \\
& & -b_0\alpha & -b_1\alpha & \cdots & -b_{n-3}\alpha & -b_{n-2}\alpha & -b_{n-1}\alpha \\
\hline
b_0 = a_0 & b_1 & b_2 & \cdots & b_{n-2} & b_{n-1} & b_n = D_{(a)} \\
& -c_0\alpha & -c_1\alpha & \cdots & -c_{n-3}\alpha & -c_{n-2}\alpha & \\
\hline
c_0 = a_0 & c_1 & c_2 & \cdots & c_{n-2} & c_{n-1} = D'(\alpha) &
\end{array}
$$

$$\text{(B.4-15)}$$

A starting value for (B.4-15) is given by (B.4-6) as before.

EXAMPLE B.4-2. Resolve Example B.4-1 using procedure (B.4-15) to determine the lowest linear factor.

Analogous to (B.4-10), the first iteration using (B.4-15) becomes

$$
\begin{array}{c|ccccc}
-0.57 & 1 & 14 & 67 & 126 & 72 \\
& & -0.57 & -7.6551 & -33.82 & -52.54 \\
\hline
& 1 & 13.43 & 59.34 & 92.17 & 19.46 = D \quad \text{(B.4-16)} \\
& & -0.57 & -7.33 & -29.65 & \\
\hline
& 1 & 12.86 & 52.01 & 62.52 = D' &
\end{array}
$$

An improved estimate for the factor using (B.4-14) is given as

$$\alpha_1 = -0.57 - \frac{19.46}{62.52} = -0.88 \qquad \text{(B.4-17)}$$

The second iteration produces the improved factor

$$\alpha_2 = -0.88 - 0.107 = -0.987 \qquad \text{(B.4-18)}$$

A third iteration produces the lowest linear factor: $\alpha_3 = 1.0$. \hspace{1cm} Q.E.D.

The process of (B.4-2), (B.4-5) and (B.4-15) produces a reduced characteristic equation which has the lowest root factored out. The next lowest root is found by repeating the process using the reduced equation. This assures that the roots are found in order of increasing magnitude, giving the best overall accuracy for all roots. Gross errors in the smaller roots can occur if the larger roots are factored first.

Both (B.4-2) and (B.4-15) produce a reduced equation of order $n - 1$ expressed by (B.4-3). Application of (B.4-5) produces a reduced equation of order $n - 2$.

Complex Factors.   The amount of effort to determine complex factors of the characteristic equation can be reduced by first determining linear factors if they exist using any of the methods described. Then an iterative process based on (B.4-5) using a trial quadratic found by truncating the higher-order terms of the characteristic equation is employed. An improved quadratic factor is found using Newton's method (B.4-14) with complex D and D'. The division by the quadratic ($s^2 + \alpha s + \beta$) gives the form of the characteristic equation from (B.4-5) as

$$(s^2 + \alpha s + \beta)q_0(s) + b_{n-1}s + b_n \qquad (B.4\text{-}19)$$

where $q(s)$ is the reduced equation given by

$$q_0(s) = b_0 s^{n-2} + b_1 s^{n-3} + \cdots + b_{n-2} \qquad (B.4\text{-}20)$$

Equation (B.4-19) may be expressed as

$$(s + Z)(s + Z^*)q_0(s) + b_{n-1}s + b_n \qquad (B.4\text{-}21)$$

where Z^* is the complex conjugate of Z. D of (B.4-14) is found by substituting $-Z$ into the linear factor of (B.4-19), i.e.

$$D(Z) = -b_{n-1}Z + b_n \qquad (B.4\text{-}22)$$

Dividing (B.4-20) by the quadratic factor ($s^2 + \alpha s + \beta$) gives a form similar to (B.4-19) as

$$(s^2 + \alpha s + \beta)q_1(s) + c_{n-3}s + c_{n-2} \qquad (B.4\text{-}23)$$

where

$$q_1(s) = c_0 s^{n-4} + c_1 s^{n-5} + \cdots + c_{n-4} \qquad (B.4\text{-}24)$$

Since $Z = a + jb$, the value of D' of (B.4-14) becomes

$$D'(Z) = -2bj(-c_{n-3}Z + c_{n-2}) + b_{n-1}$$
$$= (b_{n-1} - 2b^2c_{n-3}) + j2b(c_{n-3}a - c_{n-2}) \quad \text{(B.4-25)}$$

An application of (B.4-14) then gives

$$Z_{i+1} = Z_i + \frac{D(Z)}{D'(Z)} = (a + jb)_i + (c + jd) \quad \text{(B.4-26)}$$

or
$$Z_{i+1} = (a + c) + j(b + d) \quad \text{(B.4-27)}$$

The improved trial quadratic is determined from $(s + Z_{i+1})(s + Z^*_{i+1})$ as

$$s^2 + 2(a + c)s + [(a + c)^2 + (b + d)^2] \quad \text{(B.4-28)}$$

EXAMPLE B.4-3.   Determine the factors of the characteristic equation

$$s^4 + 400s^3 + 1.617(10^6)s^2 + 3.155(10^8)s + 4.666(10^{11}) \quad \text{(B.4-29)}$$

The quadratic formed by truncating the higher order terms becomes

$$1.617(10^6)s^2 + 3.155(10^8)s + 4.666(10^{11}) \quad \text{(B.4-30a)}$$

or
$$s^2 + 195.1s + 2.886(10^5) \quad \text{(B.4-30b)}$$

Factors of (B.4-30b) are $(s + Z)(s + Z^*)$ where

$$Z = 97.55 + j528.2 \quad \text{(B.4-31)}$$

The coefficients b_i and c_i used in developing an improved quadratic factor are determined by dividing (B.4-29) by (B.4-30b). Using the process defined by (B.4-5) gives

	1	400	$1.617(10^6)$	$3.155(10^8)$	$4.666(10^{11})$
-195.1		-195.1	$-3.997(10^4)$	$-2.514(10^8)$	
$-2.886(10^5)$			$-2.886(10^5)$	$-5.913(10^7)$	$-3.718(10^{11})$
	1	204.9 ·	$1.288(10^6)$	$4.974(10^6) = b_3$	$9.48(10^{10}) = b_4$
-195.1		-195.1		Remainder	
$-2.886(10^5)$			$-2.886(10^5)$		
	1	$9.8 = c_1 (10^6) = c_2$			(B.4-32)

Constants used in Newton's equation (B.4-26) are determined using (B.4-22) and (B.4-25) for order $n = 4$ of the quartic (B.4-29); therefore,

$$D(Z) = -b_3 Z + b_4 = 9.431(10^{10}) - j2.627(10^9)$$

$$D'(Z) = (b_3 - 2b^2 c_1) + j2b(c_1 a - c_2)$$

$$= [4.974(10^6) - 2(528.2^2)9.8] + j2(528.2)[9.8(97.55) - 10^6] \qquad \text{(B.4-33)}$$

$$D'(Z) = -4.943(10^5) - j1.055(10^9)$$

From (B.4-26)

$$\frac{D(Z)}{D'(Z)} = \frac{[9.431(10^{10}) - j2.627(10^9)][-4.943(10^5) + j1.055(10^9)]}{[4.943(10^5)]^2 + [1.055(10^9)]^2}$$

$$= \frac{2.819(10^{18}) + j9.952(10^{19})}{1.114(10^{18})} \qquad \text{(B.4-34)}$$

$$\frac{D(Z)}{D'(Z)} = 2.53 + j89.37$$

so that (B.4-26) using (B.4-31) and (B.4-34) becomes

$$Z_1 = (97.55 + j528.2) + (2.53 + j89.37) = 100.08 + j617.57 \qquad \text{(B.4-35)}$$

This leads directly to the improved quadratic factor (B.4-28) as

$$s^2 + 2(100.08)s + 100.08^2 + 617.57^2 \qquad \text{(B.4-36a)}$$

or $\qquad\qquad\qquad s^2 + 200.16s + 3.914(10^5) \qquad\qquad\qquad$ (B.4-36b)

Dividing (B.4-29) by (B.4-36b) as in (B.4-32) gives

	1	400	$1.617(10^6)$	$3.155(10^8)$	$4.666(10^{11})$
-200.16		-200.16	$-4.0(10^4)$	$-2.373(10^8)$	
$-3.914(10^5)$			$-3.914(10^5)$	$-7.822(10^7)$	$-4.64(10^{11})$
	1	199.84	$1.186(10^6)$	$-2.0(10^4) = b_3$	$2.556(10^9) = b_4$
-200.16		-200.16		Remainder	
$-3.914(10)^5$			$-3.914(10^5)$		
	1	$-0.32 = c_1$	$7.942(10^5) = c_2$		

$$\text{(B.4-37)}$$

The process of determining an improved quadratic factor follows (B.4-33), (B.4-34), (B.4-35). Using (B.4-35), we again obtain

$$D(Z) = 2.558(10^9) + j1.235(10^7)$$
$$D'(Z) = 2.241(10^5) - j9.81(10^8) \tag{B.4-38}$$
$$Z_2 = (100.08 + j617.57) + (-0.012 + j2.608) = 100.07 + j620.2$$

From (B.4-28) the improved quadratic becomes

$$s^2 + 200.14s + 3.947(10^5) \tag{B.4-39}$$

The factor (B.4-39) is of acceptable accuracy as judged by a change in the natural frequency $\sqrt{\beta}$ of 0.42% from the previous factor (B.4-36b).

The reduced characteristic equation which also represents the second factor is a result of dividing (B.4-29) by (B.4-39) using process (B.4-5).

	1	400	$1.617(10^6)$	$3.155(10^8)$	$4.666(10^4)$
-200.14		-200.14	$-4.0(10^4)$	$-2.366(10^8)$	
$-3.947(10^5)$			$-3.947(10^5)$	$-7.888(10^7)$	$-4.665(10^{11})$
	$b_0 = 1$	$b_1 = 199.86$	$b_2 = 1.182(10^6)$	-5522	$6.46(10^2)$

$$\tag{B.4-40}$$

The reduced equation in the form of (B.4-20) is

$$s^2 + 199.86s + 1.182(10^6) \tag{B.4-41}$$

The factored form of (B.4-29) using (B.4-39) and (B.4-41) becomes

$$[s^2 + 200.14s + 3.947(10^5)][s^2 + 199.86s + 1.182(10^6)] \tag{B.4-42}$$

Q.E.D.

Approximate Methods.   An approximate method is available that reduces the effort required to obtain the quadratic factors of characteristic equations of damped physical systems. This method is based on an exact representation of the characteristic equation by quadratic factors with higher order terms of the product discarded. Accuracy of this method increases as the amplification Q_i and separation of the natural modes of the system increases. Only even order systems are considered since real roots are assumed removed. Coefficients α_i of the quadratic factors $(s^2 + \alpha_i s + \beta_i)$ of (B.4-1a) are given by

$$\begin{Bmatrix} \alpha_1 \\ \alpha_2 \\ \cdot \\ \cdot \\ \cdot \\ \alpha_m \end{Bmatrix} = \frac{1}{a_0} \begin{bmatrix} d_1 & & & \\ & d_2 & & \\ & & \cdot & \\ & & & \cdot \\ & & & & d_m \end{bmatrix}^{-1} \begin{bmatrix} B \end{bmatrix} \begin{Bmatrix} a_1 \\ a_3 \\ \cdot \\ \cdot \\ \cdot \\ a_{m-1} \end{Bmatrix} \qquad \text{(B.4-43a)}$$

given

$$d_i = (-1)^m \prod_{j \neq i} (\beta_i - \beta_j) \qquad \text{(B.4-43b)}$$

$$[B] = \begin{bmatrix} (-1)^m \beta_1^{m-1} & (-1)^{m-1} \beta_1^{m-2} & \cdots & \beta_1 & -1 \\ (-1)^m \beta_2^{m-1} & (-1)^{m-1} \beta_2^{m-2} & \cdots & \beta_2 & -1 \\ \cdot & \cdot & & \cdot & \cdot \\ \cdot & \cdot & & \cdot & \cdot \\ \cdot & \cdot & & \cdot & \cdot \\ (-1)^m \beta_m^{m-1} & (-1)^{m-1} \beta_m^{m-2} & \cdots & \beta_m & -1 \end{bmatrix} \qquad \text{(B.4-43c)}$$

Where coefficients are developed from right to left and $\beta_i^r|_{r<1} = 0$, $\beta_i = $ roots of (B.4-1a) with even exponents of s only, i.e.,

$$a_0 s^n + a_2 s^{n-2} + \cdots + a_0 = (s^2 + \beta_1)(s^2 + \beta_2) \cdots (s^2 + \beta_m)$$

$$m = n/2$$

EXAMPLE B.4-4.   Resolve Example B.4-3 using the approximate method of (B.4-43).

Roots of (B.4-29) with even exponents

$$s^4 + 1.617(10^6)s^2 + 4.666(10^{11}) \qquad \text{(B.4-44)}$$

are found using the quadratic equation or (B.4-15) with (B.4-14). The result is expressed as

$$(s^2 + 3.76 \times 10^5)(s^2 + 1.241 \times 10^6) \qquad \text{(B.4-45)}$$

where $\beta_1 = 3.76(10^5)$ and $\beta_2 = 1.241(10^6)$

Equation (B.4-43b) gives

$$d_1 = \beta_1 - \beta_2 = -8.65(10^5)$$
$$d_2 = \beta_2 - \beta_1 = 8.65(10^5)$$

so that (B.4-43a) becomes

$$\begin{Bmatrix} \alpha_1 \\ \alpha_2 \end{Bmatrix} = \begin{bmatrix} 1/d_1 & 0 \\ 0 & 1/d_2 \end{bmatrix} \begin{bmatrix} \beta_1 & -1 \\ \beta_2 & -1 \end{bmatrix} \begin{Bmatrix} a_1 \\ a_3 \end{Bmatrix}$$

$$= \begin{bmatrix} -\dfrac{1}{8.65(10^5)} & 0 \\ 0 & \dfrac{1}{8.65(10^5)} \end{bmatrix} \begin{bmatrix} 3.76(10^5) & -1 \\ 1.241(10^6) & -1 \end{bmatrix} \begin{Bmatrix} 400 \\ 3.155(10^8) \end{Bmatrix}$$

$$\begin{Bmatrix} \alpha_1 \\ \alpha_2 \end{Bmatrix} = \begin{Bmatrix} 190.87 \\ 209.13 \end{Bmatrix}$$

where $a_1 = 400$ and $a_3 = 3.155(10^8)$ from (B.4-29). The factors of (B.4-29) are then given by

$$(s^2 + 190.87s + 3.76 \times 10^5)(s^2 + 209.1s + 1.241 \times 10^6) \quad \text{(B.4-46)}$$

Q.E.D.

Comparison of (B.4-46) with (B.4-42) shows that reasonably accurate factors are obtained using the approximate method. A variance of less than 2.4% in the damped frequencies of Example B.4-3 containing closely spaced natural frequencies and high damping supports the simplicity of this approach.

B.5 TIME RESPONSE

The time response of the transfer function can be found conveniently by looking up the response function in a large table of Laplace transforms and routinely copying the inverse transform term by term. An efficient alternate procedure involves performing a partial fraction expansion of the response function. Then using a short table containing a fundamental set of transforms to recover the time response.

The overall time response of a system is composed of the sum of individual elementary motions which can be evaluated by examining the components of the partial fraction expansion or by plotting the response of each term. Consider the transfer function $x(s)/y(s)$ where both $x(s)$ and $y(s)$ are expressed as polynomials with the degree of $x(s)$ less than that of $y(s)$. The factored transfer function using the procedures of B.4 can be expressed generally as

$$\frac{x(s)}{y(s)} = \frac{K(s + \beta_1)(s + \beta_2) \cdots}{(s + \alpha_1)(s + \alpha_2)(s + \alpha_3) \cdots} \tag{B.5-1}$$

where α_i or β_i may be real, complex or zero. When complex, these factors always appear in conjugate pairs for a physical system. Expressing conjugate pairs as quadratics the form of (B.5-1) and its partial fraction expansion becomes

$$K \frac{(s^2 + c_0 s + d_0)(s^2 + c_1 s + d_1) \cdots (s + \beta_1)(s + \beta_2) \cdots}{(s^2 + a_0 s + b_0)(s^2 + a_1 s + b_1) \cdots (s + \alpha_1)(s + \alpha_2) \cdots}$$

$$= \frac{A_1}{(s + \alpha_1)} + \frac{A_2}{(s + \alpha_2)} + \cdots + \frac{B_0 s + D_0}{(s^2 + a_0 s + b_0)} + \frac{B_1 s + D_1}{(s^2 + a_1 s + b_1)} + \cdots$$

$$\tag{B.5-2}$$

Methods of Procedure B.5-1 may be used to evaluate coefficients A_i, B_i and D_i.

Procedure B.5-1

A. Clear all fractions by multiplying through by the factors in the denominator of the left side of (B.5-2). Two approaches using this form in order of simplicity are:
   1. Substitute the values of s for each linear factor and solve the reduced equation for the coefficient.
   2. Expand the right side arranging the resulting terms in decreasing powers of s. Equate coefficients of corresponding powers of s and solve the resulting system of equations.
B. The Heaviside approach can be used to recover each coefficient for linear factors.
C. Develop even powers of s to recover coefficient of quadratic factors of the form $(s^2 + \alpha^2)$.

Combinations of the methods of Procedure B.5-1 can be used for expressions containing linear, multiple poles or quadratic factors.

EXAMPLE B.5-1.   Determine the partial fraction expansion of

$$\frac{s^2 + 1}{(s + 1)(s + 2)(s + 3)} = \frac{A_1}{(s + 1)} + \frac{A_2}{(s + 2)} + \frac{A_3}{(s + 3)} \qquad \text{(B.5-3)}$$

Using Procedure B.5-1 (A.1) to obtain A_1, multiply through by the factor $(s + 1)$ to obtain

$$\frac{s^2 + 1}{(s + 2)(s + 3)} = A_1 + \frac{A_2(s + 1)}{(s + 2)} + \frac{A_3(s + 1)}{(s + 3)} \qquad \text{(B.5-4)}$$

Setting $s = -1$ which is the value of $s + 1 = 0$, A_1 can be determined directly

$$\frac{(-1)^2 + 1}{(-1 + 2)(-1 + 3)} = A_1 + 0 + 0 \qquad \therefore A_1 = 1 \qquad \text{(B.5-5)}$$

Alternatively, the value of A_1 could have been obtained using the Heaviside approach. Covering the $(s + 1)$ factor in the denominator on the left side of (B.5-3) and evaluating the remaining expression at $s = -1$ give the same form and result as (B.5-5). Applying this procedure for determining A_2 we cover (or mentally remove) the $(s + 2)$ factor and evaluate the remaining expression at $s = -2$ to give

$$A_2 = \frac{s^2 + 1}{(s + 1)(s + 3)} \bigg|_{s=-2} = \frac{(-2)^2 + 1}{(-2 + 1)(-2 + 3)} = -5 \qquad \text{(B.5-6)}$$

The value of A_3 can similarly be determined giving the partial fraction expansion of (B.5-3) as

$$\frac{s^2 + 1}{(s + 1)(s + 2)(s + 3)} = \frac{1}{(s + 1)} - \frac{5}{(s + 2)} + \frac{5}{(s + 3)}$$

$$\text{Q.E.D.}$$

Procedure B.5-1(A.2) can be used to recover the coefficients of both linear and quadratic factors.

EXAMPLE B.5-2.   Determine the partial fraction expansion of

$$\frac{2s + 4}{(s^2 + 1)(s + 1)^2} = \frac{A_1}{(s + 1)^2} + \frac{A_2}{(s + 1)} + \frac{B_0 s + D_s}{(s^2 + 1)} \qquad \text{(B.5-7)}$$

Clearing fractions and collecting coefficients of corresponding powers of s,

$$2s + 4 = (A_2 + B_0)s^3 + (2B_0 + D_0 + A_1 + A_2)s^2$$
$$+ (2D_0 + B_0 + A_2)s + (D_0 + A_1 + A_2) \quad \text{(B.5-8)}$$

Equating coefficients of corresponding powers of s gives the set of equations

$$B_0 \qquad\qquad + A_2 = 0$$
$$2B_0 + D_0 + A_1 + A_2 = 0$$
$$B_0 + 2D_0 \qquad + A_2 = 2 \qquad \text{(B.5-9)}$$
$$D_0 + A_1 + A_2 = 4$$

The coefficients determined from the solution of (B.5-9) are

$$A_1 = 1, \quad A_2 = 2, \quad B_0 = -2, \quad D_0 = 1$$

The method of Example B.5-1 could have been used to determine the coefficient of the multiple pole factor A_1 in (B.5-7). In this case

$$A_1 = \frac{2s + 4}{(s^2 + 1)}\bigg|_{s=-1} = \frac{2(-1) + 4}{(-1)^2 + 1} = 1 \qquad \text{(B.5-10)}$$

Procedure B.5-1(C) could have been used to recover the coefficients B_0 and D_0 of the quadratic factor of (B.5-7). This approach requires factors of the denominator which are even powers of s only. Since the double pole $(s + 1)^2$ has an odd power of s, multiplication of both the numerator and the denominator by $(s - 1)^2$ produces only even powers of s in the denominator. Multiplication of (B.5-7) gives

$$\frac{(2s + 4)(s - 1)^2}{(s^2 + 1)(s + 1)^2(s - 1)^2} = \frac{2s(s^2 - 3) + 4}{(s^2 + 1)(s^2 - 1)^2} \qquad \text{(B.5-11)}$$

The coefficients B_0 and D_0 are determined by evaluating (B.5-11) with the quadratic removed at the value $s^2 = -1$:

$$B_0 s + D_0 = \frac{2s(s^2 - 3) + 4}{(s^2 - 1)^2}\bigg|_{s^2 = -1} = \frac{8s + 4}{4} = -2s + 1 \qquad \text{(B.5-12)}$$

Use of combinations of these various methods can reduce both the number of unknowns and the resulting set of equations for determining the remaining coefficients. The required expansion of (B.5-7) then becomes

$$\frac{2s + 4}{(s^2 + 1)(s + 1)^2} = \frac{1}{(s + 1)^2} + \frac{2}{(s + 1)} + \frac{-2s + 1}{(s^2 + 1)} \qquad \text{(B.5-13)}$$

Q.E.D.

The structure of the transfer function (B.5-1) leads to the development of an additional procedure for reducing the effort associated with the partial fraction expansion. In general, the highest power of s in the numerator is less than that of the denominator. This relationship can be expressed as

$$\frac{x(s)}{y(s)} = \frac{K(s^n + b_0 s^{n-1} + \cdots + b_{n-1})}{(s^m + a_0 s^{m-1} + \cdots + a_{m-1})} \qquad \text{(B.5-14)}$$

The sum of the coefficients of the linear terms of the partial fraction expansion are related to the relationship of n to m in (B.5-14).

Procedure B.5-2

	Exponent Relationship	Sum of the Coefficients of Linear Factors
A.	$n \leq m - 2$	0
B.	$n = m - 1$	1
C.	$n = m$	$b_0 - a_0$
D.	$n = m + 1$	$b_1 - b_0 + a_0 - a_1$

EXAMPLE B.5-3.   Determine the value of expansion coefficient A_3 in Example B.5-1 given $A_1 = 1$ and $A_2 = -5$.

Inspection of (B.5-3) shows that $n = 2$ and $m = 3$ so that procedure B.5-2(B) applies and is stated for (B.5-3) as

$$A_1 + A_2 + A_3 = 1 \qquad \text{(B.5-15a)}$$

or

$$A_3 = 1 - (A_1 + A_2) \qquad \text{(B.5-15b)}$$

Substituting the values given for A_1 and A_2, we obtain

$$A_3 = 1 - (1 - 5) = 5 \qquad \text{(B.5-16)}$$

Q.E.D.

EXAMPLE B.5-4.   Determine the partial fraction expansion of

$$\frac{s + \beta_1}{(s + \alpha_1)(s + \alpha_2)^2} = \frac{A_1}{s + \alpha_1} + \frac{A_2}{s + \alpha_2} + \frac{A_3}{(s + \alpha_2)^2} \qquad (B.5\text{-}17)$$

Using Procedure B.5-1(B) for evaluating A_1 and A_3, we have

$$A_1 = \frac{s + \beta_1}{(s + \alpha_2)^2} \bigg|_{s = -\alpha_1} = \frac{\beta_1 - \alpha_1}{(\alpha_2 - \alpha_1)^2} \qquad (B.5\text{-}18)$$

$$A_3 = \frac{s + \beta_1}{(s + \alpha_1)} \bigg|_{s = -\alpha_2} = \frac{\beta_1 - \alpha_2}{(\alpha_1 - \alpha_2)} \qquad (B.5\text{-}19)$$

Since $n = 1$ and $m = 3$ in (B.5-17), Procedure B.5-2(A) applies therefore:

$$A_1 + A_2 = 0 \qquad (B.5\text{-}20a)$$

or

$$A_2 = -A_1 \qquad (B.5\text{-}20b)$$

Substituting (B.5-20b), (B.5-19) and (B.5-18) into (B.5-17) gives the desired partial fraction expansion:

$$\frac{s + \beta_1}{(s + \alpha_1)(s + \alpha_2)^2} = \frac{(\beta_1 - \alpha_1)}{(\alpha_2 - \alpha_1)^2}\left[\frac{1}{(s + \alpha_1)} - \frac{1}{(s + \alpha_2)}\right] + \frac{(\beta_1 - \alpha_2)}{(\alpha_1 - \alpha_2)(s + \alpha_2)^2}$$

$$(B.5\text{-}21)$$

Q.E.D.

The properties of Procedure B.5-2 are a consequence of using Procedure B.5-1(A.2) on normalized transfer functions (B.5-14). When the transfer functions are not normalized, the values of the sum of coefficients are different than indicated. For example, Procedure B.5-2(B) for the transfer function

$$\frac{x(s)}{y(s)} = \frac{b_0 s^{m-1} + b_1 s^{m-2} + \cdots}{(s + a_0)(s + a_1) \cdots (s + a_n)} \qquad (B.5\text{-}22)$$

becomes

$$A_0 + A_1 + \cdots + A_n = b_0 \qquad (B.5\text{-}23)$$

Procedure B.5-2(A) applies whether the transfer function is normalized or not, as seen in Example B.5-2. The relationship given by this procedure is illustrated in the first equation of the set (B.5-9).

Example B.5-2 also illustrates the use of the expansion coefficient $Bs + D$ for quadratic factors which have imaginary components. The coefficient B represents the linear component of the factor which is added to coefficients of other linear factors when applying Procedure B.5-2. This was also illustrated in Example B.5-2.

The coefficients A_i are readily determined for linear factors which are identical factors of the given transfer function. This was illustrated using the Heaviside approach in (B.5-10) of Example B.5-2 and (B.5-18) and (B.5-19) of Example B.5-4. Other coefficients can be found by using Procedure B.5-2 or by manipulating the partial fraction expansion.

A simple approach when quadratic factors are present follows the same technique of determining coefficients A_i, with these known terms moved to the left side of the equation. The resulting equation can be expressed in a matrix form for determining unknown coefficients A_i, B_i and D_i. This is best illustrated by the next two examples.

EXAMPLE B.5-5.   Resolve Example B.5-2 using the matrix approach.

Since the linear factor of coefficient A_1 of (B.5-7) is identical to a factor of the denominator of the transfer function, it is readily determined as in (B.5-10). Equation (B.5-7) can then be rewritten as

$$\frac{2s + 4}{(s^2 + 1)(s + 1)^2} - \frac{A_1}{(s + 1)^2} = \frac{A_2}{(s + 1)} + \frac{B_0 s + D_0}{(s^2 + 1)} \qquad \text{(B.5-24a)}$$

When each factor of (B.5-24a) is conditioned to give a common denominator, the equation that results is

$$2s + 4 - A_1(s^2 + 1) = A_2(s + 1)(s^2 + 1) + (B_0 s + D_0)(s + 1)^2 \qquad \text{(B.5-24b)}$$

A matrix equation for unknowns A_2, B_0 and D_0 can be written from (B.5-24b), using the technique of Appendix B-1, considering the coefficient of each unknown separately in terms of the constants as powers of s:

$$
\begin{array}{c}
\\
s^3 \\
s^2 \\
s \\
\text{const}
\end{array}
\begin{array}{ccc}
(A_2) & (B_0) & (D_0) \\
\left[\begin{array}{ccc}
1 & 1 & 0 \\
1 & 2 & 1 \\
1 & 1 & 2 \\
1 & 0 & 1
\end{array}\right]
\end{array}
\left\{\begin{array}{c}
A_2 \\
B_0 \\
D_0
\end{array}\right\}
=
\left\{\begin{array}{c}
0 \\
-1 \\
2 \\
3
\end{array}\right\}
\qquad \text{(B.5-25)}
$$

The first column of (B.5-25) gives the coefficient of A_2 as $(s^3 + s^2 + s + 1)$. Since (B.5-25) are over determined (more equations than unknowns), any row can be removed giving a unique set of equation to determine A_2, B_0 and D_0. Removing the second row from (B.5-25) gives

$$\begin{bmatrix} 1 & 1 & 0 \\ 1 & 1 & 2 \\ 1 & 0 & 1 \end{bmatrix} \begin{Bmatrix} A_2 \\ B_0 \\ D_0 \end{Bmatrix} = \begin{Bmatrix} 0 \\ 2 \\ 3 \end{Bmatrix} \qquad \text{(B.5-26)}$$

Solving (B.5-26) using methods of Appendix A.6 or A.7 gives

$$A_2 = 2, \quad B_0 = -2, \quad D_0 = 1$$

Q.E.D.

EXAMPLE B.5-6.   Develop the partial fraction expansion of

$$\frac{s + c}{s(s^2 + as + b)} = \frac{A_1}{s} + \frac{B_0 s + D_0}{s^2 + as + b} \qquad \text{(B.5-27)}$$

Coefficient A_1 is determined directly using Heaviside by covering the s in the denominator while evaluating the remaining equation at s = 0, giving

$$A_1 = \frac{s + c}{s^2 + as + b}\bigg|_{s=0} = c/b \qquad \text{(B.5-28)}$$

Moving the A_1/s term to the left side of (B.5-27) while developing a common denominator for each term gives an equation for determining B_0 and D_0 as

$$s + c - A_1(s^2 + as + b) = (B_0 s + D_0)s \qquad \text{(B.5-29a)}$$

Substituting A_1 from (B.5-28) into (B.5-29a) and arranging coefficients of s similar to the right hand side of the equation gives

$$\left[-\frac{c}{b}s + \left(1 - \frac{ac}{b}\right) \right] s = (B_0 s + D_0)s \qquad \text{(B.5-29b)}$$

The quantities B_0 and D_0 can be found directly from (B.5-29b) by comparing coefficients of corresponding powers of s between the right and left hand side of the equation. The unknowns then become

$$B_0 = -\frac{c}{b}, \quad D_0 = \left(1 - \frac{ac}{b}\right) \qquad \text{(B.5-30a)}$$

The partial fraction expansion then becomes

$$\frac{s + c}{s(s^2 + as + b)} = \frac{c/b}{s} + \frac{(-c/b)s + (1 - ac/b)}{s^2 + as + b} \qquad \text{(B.5-31)}$$

A matrix equation of (B.5-29b) would not give further insight because of the s coefficient correspondence which is available directly. If required, however, the matrix development would have the form

$$
\begin{matrix}
 & (B_0) & (D_0) & & & \\
s^2 & \begin{bmatrix} 1 & 0 \\ 0 & 1 \end{bmatrix} & & \begin{Bmatrix} B_0 \\ D_0 \end{Bmatrix} & = & \begin{Bmatrix} -c/b \\ 1 - ac/b \end{Bmatrix}
\end{matrix}
\qquad \text{(B.5-29c)}
$$

Since the matrix is an identity matrix, (B.5-29c) may be written as

$$\begin{Bmatrix} B_0 \\ D_0 \end{Bmatrix} = \begin{Bmatrix} -c/b \\ 1 - ac/b \end{Bmatrix} \qquad \text{(B.5-30b)}$$

Q.E.D.

The matrix method facilitates expansion of transfer functions which contain several quadratic factors in addition to linear factors of the form represented by (B.5-2). Consider a transfer function with two quadratic factors:

$$\frac{x(s)}{y(s)} = \frac{C_0 s^4 + C_1 s^3 + C_2 s^2 + C_3 s + C_4}{s(s^2 + a_0 s + b_0)(s^2 + a_1 s + b_1)} \qquad \text{(B.5-31)}$$

The coefficient A_0 of the linear factor is readily determined using the Heaviside approach. The coefficients of the quadratic factors are determined using the matrix developed from

$$\frac{x(s)}{y(s)} - \frac{A_0}{s} = \frac{B_0 s + D_0}{s^2 + a_0 s + b_0} + \frac{B_1 s + D_1}{s^2 + a_1 s + b_1} \qquad \text{(B.5-32)}$$

which becomes

$$
\begin{matrix}
 & & & & & & \text{Coef. of powers of } s \\
 & & & & & & \text{of } x(s) - y(s)A_0/s \\
s^4 \\
s^3 \\
s^2 \\
s
\end{matrix}
\begin{bmatrix}
1 & 0 & 1 & 0 \\
a_1 & 1 & a_0 & 1 \\
b_1 & a_1 & b_0 & a_0 \\
0 & b_1 & 0 & b_0
\end{bmatrix}
\begin{Bmatrix}
B_0 \\
D_0 \\
B_1 \\
D_1
\end{Bmatrix}
=
\begin{Bmatrix}
C_0 - A_0 \\
C_1 - (a_1 + a_0)A_0 \\
C_2 - (b_0 + b_1 + a_0 a_1)A_0 \\
C_3 - (a_0 b_1 + a_1 b_0)A_0
\end{Bmatrix}
\qquad \text{(B.5-33)}
$$

The general solution for the coefficients of transfer function (B.5-32) are obtained from (B.5-33) using methods of Appendix A.7, as

$$
\begin{Bmatrix} B_0 \\ D_0 \\ B_1 \\ D_1 \end{Bmatrix} = \frac{1}{|A|} \begin{bmatrix} b_0^2 - a_0(a_1b_0 - a_0b_1) - b_0b_1 & (a_1b_0 - a_0b_1) & (b_1 - b_0) & (a_0 - a_1) \\ -b_0(a_1b_0 - a_0b_1) & b_0(b_0 - b_1) & -b_0(a_0 - a_1) & a_0(a_0 - a_1) - b_0 + b_1 \\ b_1^2 + a_1(a_1b_0 - a_0b_1) - b_0b_1 & -(a_1b_0 - a_0b_1) & -(b_1 - b_0) & -(a_0 - a_1) \\ b_1(a_1b_0 - a_0b_1) & -b_1(b_0 - b_1) & b_1(a_0 - a_1) & -a_1(a_0 - a_1) + b_0 - b_1 \end{bmatrix} \{C\}
$$

$$(B.5\text{-}34)$$

given

$$|A| = (a_1 - a_0)(a_1b_0 - a_0b_1) + (b_0 - b_1)^2$$

$$A_0 = C_4/(b_0b_1)$$

$$
\{C\} = \begin{Bmatrix} C_0 - A_0 \\ C_1 - (a_1 + a_0)A_0 \\ C_2 - (b_0 + b_1 + a_0a_1)A_0 \\ C_3 - (a_0b_1 + a_1b_0)A_0 \end{Bmatrix}
$$

It is left as an exercise for the reader to show that (B.5-34) with $A_0 = 0$ is a solution for the coefficients of the partial fraction expansion of

$$\frac{x(s)}{y(s)} = \frac{C_0s^3 + C_1s^2 + C_2s + C_3}{(s^2 + a_0s + b_0)(s^2 + a_1s + b_1)} \qquad (B.5\text{-}35)$$

EXAMPLE B.5-7. Determine the partial fraction expansion of

$$\frac{s^2 + 2.398s + 579}{(s^2 + 2.507s + 190.771)(s^2 + 2.7s + 781)}$$

$$= \frac{B_0s + D_0}{(s^2 + 2.507s + 190.771)} + \frac{B_1s + D_1}{(s^2 + 2.7s + 781)} \qquad (B.5\text{-}36)$$

Coefficients of the first column of (B.5-34) are not required since $C_0 = 0$ in (B.5-35). The coefficients of (B.5-36) are then determined from (B.5-35) as

$$
\begin{Bmatrix} B_0 \\ D_0 \\ B_1 \\ D_1 \end{Bmatrix} = \begin{bmatrix} -0.004144 & 0.001695 & -0. \\ -0.32340 & 0.000106 & 0.001694 \\ 0.004144 & -0.001695 & +0. \\ 1.3242 & -0.000434 & -0.001694 \end{bmatrix} \begin{Bmatrix} 1 \\ 2.398 \\ 579 \end{Bmatrix} = \begin{Bmatrix} -7.82(10^{-5}) \\ 0.6577 \\ 7.82(10^{-5}) \\ 0.3423 \end{Bmatrix} \qquad (B.5\text{-}37)
$$

where

$$|A| = 0.1936\,(-1442.57) + 34838.7 = 348107.7$$

The required partial fraction expansion in then given by

$$\frac{x(s)}{y(s)} = \frac{0.6577}{s^2 + 2.5078s + 190.771} + \frac{0.3423}{s^2 + 2.7s + 781} \qquad \text{(B.5-38a)}$$

where only the significant coefficient of the expansion has been retained.   Q.E.D

Approximate Method for Quadratic Factors.   Approximate methods can be used for expansion terms containing quadratic factors when the square of the natural frequencies b_i of (B.5-2) are different from one another by a factor of three or more and the system is lightly damped, i.e., $Q_i \geq 2$. In this case, the simpler Heaviside approach, Procedure B.5-1(B), may be used to determine the coefficients. Application of this procedure when these conditions exist is similar to (B.5-11) with factors a_i in the denominator of the transfer function ignored but retained as factors of the expansion.

EXAMPLE B.5-8.   Resolve the transfer function of Example B.5-7 using the approximate method.
  The given transfer meets the conditions since $b_1 > 3 \times b_0$, i.e., $781 > (3 \times 190.7)$ and $Q_0 \simeq \sqrt{190.7/2.5} = 5.5 > 2$. The approximate transfer function and its expansion then become

$$\frac{s^2 + 2.398s + 579}{(s^2 + 190.771)(s^2 + 781)}$$

$$= \frac{B_0 s + D_0}{(s^2 + 2.507s + 190.771)} + \frac{B_1 s + D_1}{(s^2 + 2.7s + 781)} \qquad \text{(B.5-38b)}$$

Coefficients B_i and D_i are determined by evaluating the left side of (B.5-38b). Therefore

$$B_0 s + D_0 = \left.\frac{s^2 + 2.398s + 579}{s^2 + 781}\right|_{s^2 = -190.771} = \frac{2.398s + (579 - 190.77)}{(781 - 190.771)}$$

$$B_0 s + D_0 = 0.00406s + 0.6577$$

$$B_1 s + D_1 = \left.\frac{s^2 + 2.398s + 579}{s^2 + 190.771}\right|_{s^2 = -781} = \frac{2.398s + (579 - 781)}{(190.771 - 781)}$$

$$B_1 s + D_1 = -0.00406s + 0.3422$$

The resulting expansion is given by (B.5-38a), where only the numerically significant coefficients have been retained.   Q.E.D.

The s operator provides a convenient transformation enabling algebraic manipulation and solution of the differential equations of modeled systems. This transformation, which removes the time variable, develops equations easier to solve than the original form. To regain the time variable, an inverse transformation of each term of the partial fraction expansion is required. The use of the s operator follows the well known technique of Laplace transformation.

The transformation of (B.1-3) to (B.1-4) is based on the premise that the initial conditions are zero. This condition is acceptable when determining the natural modes or impulse response characteristics of the system. The impulse response of (B.1-4) is given by its inverse transformation. The natural modes characterize the behavior of the system in the frequency domain whereas the impulse response is particularly useful when the response to random excitation is being studied.

To include initial conditions, the transformation of the first and second derivative in the differential equations of the system contains the additional information, i.e.,

$$\ddot{x}(t) \rightarrow x(s)s^2 - sx_0 - \dot{x}_0 \qquad (\text{B.5-39})$$

Given x_0 and $\dot{x}_0$, the initial displacement and velocity respectively. Similarly, the transformation of the rate of change in temperature $t = dt/d\phi$ as in (B.1-27b) becomes

$$\dot{i}(\phi) = T(s)s - t_0 \qquad (\text{B.5-40})$$

The system time response is obtained from the partial fraction expansion (B.5-2) which includes the contribution of the initial conditions represented by (B.5-39) and (B.5-40). The inverse transformation of any term of the expansion can be developed with the aid of Table B.5-1. This table is by no means complete but does represent the functions obtained from physical systems models of packaged electronic assemblies. Additional tables of transform pairs are found in books by Nixon, Roberts and Doetsh listed in the References.

EXAMPLE B.5-9.   The aluminum chassis shown in Figure B.5-1(a) has an approximate uniformly distributed weight of 40 lb (18.14 kg) with a moment of inertia along the vertical axis of 3 in^4 (1.248 × 10^{-6}M^4). Assuming the chassis and support are initially at rest, determine the resulting vertical motion following suddenly applied support motion described by

$$y(t) = 0.01 \cos 628t \text{ (in)} \qquad (\text{B.5-41a})$$

$$y(t) = 0.254 \cos 628t \text{ (mm)} \qquad (\text{B.5-41b})$$

Table B.5-1. Transform Pairs

Case	$f(s)$	$f(t)$	Parameter Definition
I	u/s	u	
II	$u/(s + \alpha)$	$ue^{-\alpha t}$	
III	$u/(s + \alpha)^2$	$ute^{-\alpha t}$	
IV	$u/(s + \alpha)^n$	$ut^{n-1}e^{-\alpha t}/(n - 1)!$	
V	$u/(s^2 + b)$	$\dfrac{u}{\sqrt{b}} \sin (t\sqrt{b})$	
VI	$us/(s^2 + b)$	$u \cos (t\sqrt{b})$	
VII	$u/(s^2 + as + b)$	$\dfrac{u}{\omega_0} e^{-at/2} \sin \omega_0 t$	$\omega_0 = \sqrt{b - \left(\dfrac{a}{2}\right)^2}$
VIII	$u\left(s + \dfrac{a}{2}\right)\Big/ (s^2 + as + b)$	$ue^{-at/2} \cos \omega_0 t$	$\omega_0 = \sqrt{b - \left(\dfrac{a}{2}\right)^2}$
IX	$us/(s^2 + as + b)$	$ue^{-at/2}\left(\cos \omega_0 t - \dfrac{a}{2\omega_0} \sin \omega_0 t\right)$	$\omega_0 = \sqrt{b - \left(\dfrac{a}{2}\right)^2}$

The equivalent mass and stiffness of the modeled system, Figure B.5-1(b), is determined from case III of Table B.1-1.

$$M_{eq} = 0.504 \frac{w}{g} = 0.504 \left(\frac{40}{386}\right) = 0.052 \frac{\text{lb} \cdot \text{s}^2}{\text{in}}$$

$$= 0.504(18.14) = 9.144 \text{ kg}$$

(B.5-42a)

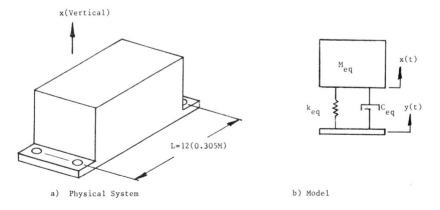

a) Physical System   b) Model

Figure B.5-1. Electronic Assembly and Idealization.

$$k_{eq} = \frac{48EI}{L^3} = \frac{48(10^7)3}{12^3} = 8.333\,(10^5)\,\frac{\text{lb}}{\text{in}}$$

$$= \frac{48(6.894 \times 10^{10})(1.248 \times 10^{-6})}{0.305^3} \qquad \text{(B.5-42b)}$$

$$= 1.456(10^8)\,\text{kg/s}^2$$

The equivalent damping for the modeled system is determined from

$$C_{eq} = \frac{1}{Q}\,\sqrt{K_{eq}M_{eq}} \qquad \text{(B.5-43)}$$

where Q is a function of k_{eq} and M_{eq} given approximately by

$$Q = \left(\frac{1}{2\pi}\,\sqrt{\frac{k_{eq}}{M_{eq}}}\right)^{1/2} \qquad \text{(B.5-44)}$$

Substituting (B.5-44) into (B.5-43), the relationship for determining the damping coefficient becomes

$$C_{eq} = \sqrt{2\pi}\,(k_{eq}M_{eq}^3)^{1/4} \qquad (M/T) \qquad \text{(B.5-45)}$$

Substituting values with consistent units for M_{eq} and k_{eq} from (B.5-42) into (B.5-45)

$$C_{eq} = \sqrt{2\pi}[8.333(10^5)(0.052^3)]^{1/4} = 8.246\,\frac{\text{lb}\cdot\text{s}}{\text{in}} \qquad \text{(B.5-46a)}$$

$$C_{eq} = \sqrt{2\pi}[1.456(10^8)(9.144^3)]^{1/4} = 1447.5\,\frac{\text{kg}\cdot\text{s}}{\text{in}} \qquad \text{(B.5-46b)}$$

Summation of system forces analogous to (B.1-2a) for Figure B.5-1(b) gives

$$0.052\ddot{x} + 8.246\dot{x} + 8.333(10^5)x = 8.333(10^5)(0.01\cos 628t) \qquad \text{(B.5-47a)}$$

or

$$\ddot{x} + 158.6\dot{x} + 1.6025(10^7)x = 1.6025(10^5)\cos 628t \qquad \text{(B.5-47b)}$$

In metric units (B.5-47b) is expressed as

$$\ddot{x} + 158.3\dot{x} + 1.592(10^7)x = 4044.4\cos 628t \qquad \text{(B.5-47c)}$$

The s operator transformation of (B.5-47b) using case VI of Table B.5-1 for the forcing function with zero system initial conditions gives

$$[s^2 + 158.6s + 1.6025(10^7)]x(s) = \frac{1.6025(10^5)s}{s^2 + 3.94(10^5)} \quad \text{(B.5-48a)}$$

or

$$x(s) = \frac{1.6025(10^5)s}{[s^2 + 3.94(10^5)][s^2 + 158.6s + 1.6025(10^7)]} \quad \text{(B.5-48b)}$$

Using the approximate method similar to Example B.5-8 for the partial fraction expansion of (B.5-48b) gives

$$x(s) = \frac{0.01025s}{s^2 + 3.94(10^5)} - \frac{0.01025s}{s^2 + 158.6s + 1.6025(10^7)} \quad \text{(B.5-49a)}$$

In metric units, (B.5-49a) is expressed as

$$x(s) = \frac{2.604(10^{-4})s}{s^2 + 3.94(10^5)} - \frac{2.604(10^{-4})s}{s^2 + 158.3s + 1.592(10^7)} \quad \text{(B.5-49b)}$$

The time domain solution is obtained by performing the inverse transformation on (B.5-49) using both cases VI and IX of Table B.5-1. This operation gives

$$x(t) = 0.01025 \cos(628t) - 0.01025e^{-79.3t}\left(\cos \omega_0 t - \frac{79.3}{\omega_0} \sin \omega_0 t\right)$$

or

$$x(t) = 0.01025 \cos(628t) - 0.01025e^{-79.3t} \cos \omega_0 t \text{ inches} \quad \text{(B.5-50)}$$

where $\omega_0 = \sqrt{1.6025(10^7) - 79.3^2} = 4002.3$ rad/s (637 Hz)

The motion that results from (B.5-50) is the sum of the forced support motion and the natural motion of the chassis as shown in Figure B.5-2(a). The total motion is the sum of the support motion, Figure B.5-2(b), and the chassis motion, Figure B.5-2(c), as indicated by (B.5-50). (Q.E.D.)

The response of a system may also be described by a graphical process known as *convolution*. Convolution may be viewed as a summation of the transient responses of a system due to time varying impulse excitations. This excitation is analogous to a hammer blow representing a force on a physical system. In the study of a system's temperature history, the impulse is characterized by a temperature or heat pulse of very short duration. The dynamic characteristics of the system are defined by its impulse response function. In response to an impulsive input, the initially dormant system springs to life and then gradually

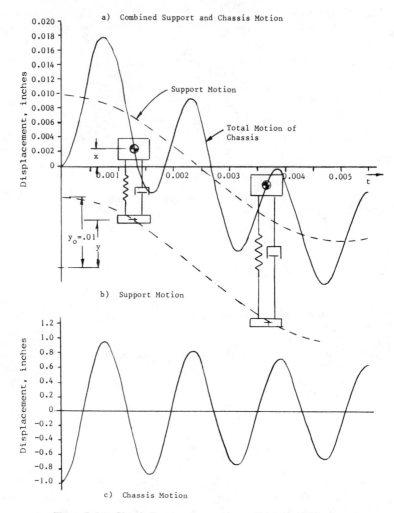

Figure B.5-2. Chassis Response to an Abrupt Sinusoidal Disturbance.

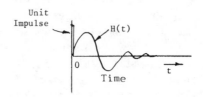

Figure B.5-3. System Impulse Response Function.

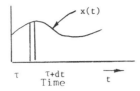

Figure B.5-4. An Excitation Considered as a Series of Pulses.

regains its static equilibrium as time passes. A system response to an impulse is illustrated in Figure B.5-3.

Since the system is initially dormant, $H(t) = 0$ for $t < 0$ before the impulse occurs. The impulse response function $h(t)$ gives the response at time, t, to a unit impulse at $t = 0$, i.e., after a delay of duration t. It follows that $H(t - \tau)$ is the response at time t to a unit impulse at time τ, i.e., after a delay of duration $(t - \tau)$. Consider an excitation $x(t)$ composed of a series of impulses as shown in Figure B.5-4.

An impulse of Figure B.5-4 between τ and $\tau + d\tau$ has the magnitude $x(\tau)d\tau$. The system response at time t to this impulse alone is just the fraction $x(\tau)d\tau/1$ of the unit impulse at $t = \tau$, which is $H(t - \tau)$. The contribution of the shaded area of Figure B.5-4 then contributes $H(t - \tau)x(\tau)d\tau$ to the system responses at time t.

For linear systems, the principle of superposition applies so the total response at t is obtained by summing all the separate responses to all the impulses of $x(t)$ from t to the time of origin, i.e., $t = 0$. This integration process expresses an important input-output relationship of a linear system known as the convolution integral:

$$y(t) = \int_0^t H(t - \tau)x(\tau)d\tau \qquad \text{(B.5-51a)}$$

or

$$y(t) = \int_1^\tau H(\tau)x(t - \tau)d\tau \qquad \text{(B.5-51b)}$$

In the s domain, $H(s)$ is known as the transfer function relating the input $x(s)$ to the output $y(s)$, Figure B.5-5.

Figure B.5-5. Operation of a Transfer Function.

Graphical integration of (B.5-51) is easily performed by four steps in Procedure B.5-3.

Procedure B.5-3

A. Fold either the input or impulse response about the ordinate to obtain $F_1(-\tau)$.
B. Translate the folded image an amount t giving $F_1(t - \tau)$.
C. Multiply the result of step B by the remaining function giving $F_1(t - \tau)\, f_2(\tau)$.
D. Compute the area of the $F_1(t - \tau)F_2(\tau)$ plot versus time from 0 to t.

For some models, expansion by partial fractions may be more direct. The graphical approach develops the response at any time t_i determined by the degree of translation used in step B. Repetition of the process for different t_i's will develop the transient behavior of the model. The graphical approach is of advantage when the input to the system is a complicated function of time.

Consider the impulse response of the component model of Figure B.1-8 which is developed from (B.1-27a) as

$$WC_p \frac{dT}{d\phi} + UT = \delta \tag{B.5-52}$$

With the use of the s operator, the impulse transfer function becomes

$$\frac{T(s)}{\delta(s)} = \frac{1}{WC_p s + U} \tag{B.5-53}$$

Case II of Table B.5-1 then gives the impulse response as

$$F_1(\phi) = \frac{1}{WC_p} e^{-(U/WC_p)\phi} \tag{B.5-54}$$

The inputs to the component of Figure B.1-8 is composed of injected or heat dissipation $q(\phi)$, and the sink or reference temperature $t_s(\phi)$. These inputs may be constant or functions of time. One additional constraint is the initial temperature t_i of the component. These inputs and constraints can be summed and plotted as an equivalent heat load:

$$Q_e(\phi) = [T_s(\phi) - t_i]U + q(\phi) \tag{B.5-55}$$

Equations (B.5-54) and (B.5-55) are shown graphically in Figure B.5-6(a) and Figure B.5-6(b) respectively.

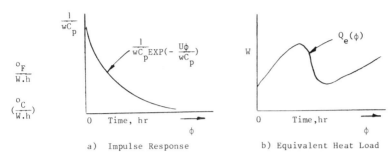

a) Impulse Response   b) Equivalent Heat Load

Figure B.5-6. Transient Characteristics of the Single Node Component Model in Figure B.1-8.

Steps A and B of Procedure B.5-3 is illustrated for the impulse response on Figure B.5-7(a) for time ϕ_1. The product formed in step C is easily determined by superposing Figure B.5-7(a) on Figure B.5-6(b). The area under the product curve for time 0 to ϕ_1 obtained by conventional methods plus the initial temperature t_i gives the response temperature at time ϕ_1.

EXAMPLE B.5-10. The component of Figure B.1-8 is to be tested in an oven with a temperature profile of Figure B.5-8(a). The component conductance U and capacity WC_p are given on the model, Figure B.5-8(b). Determine the component temperature after 1 hour of operation assuming the component was initially soaked at 70°F (21.1°C), dissipating 2W during the test.

Based on the corner temperatures of Figure B.5-8(a), the equivalent heat load at corresponding times given by (B.5-55) become

$$\phi = 0 \quad : Q(0) \quad = (80 - 70)(0.025) + 2 = 2.25W$$

$$\phi = 1h \quad : Q(1) \quad = (160 - 70)(0.025) + 2 = 4.25W$$

$$\phi = 2.5h: Q(2.5) = Q(1)$$

$$\phi = 3h \quad : Q(3) \quad = (40 - 70)(0.025) + 2 = -1.25W$$

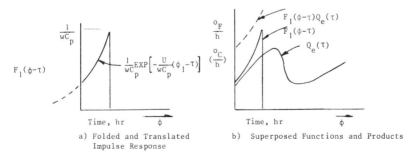

a) Folded and Translated   b) Superposed Functions and Products
Impulse Response

Figure B.5-7. Illustration of Procedure B.5-3.

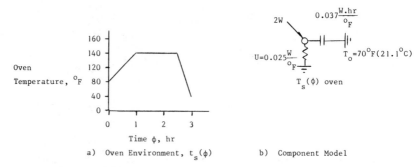

a) Oven Environment, $t_s(\phi)$ b) Component Model

Figure B.5-8. Component Environment and Idealized Model for Example B.5-10.

The equivalent heat load described by these points are plotted in Figure B.5-9(a). The results of steps A and B of Procedure B.5-3 applied to Figure B.5-9(a) are shown superposed on the impulse response on Figure B.5-9(b). The plotted result of step C is illustrated in Figure B.5-9(c).

Integration of the area of Figure B.5-9(c) using the trapezoidal rule for three strips gives $66.9°F$ $(36.7°C)$. The component temperature after one hour of operation then becomes

$$t(1) = t_0 + 66.9 = 136.9°F \ (58.28°C)$$

Q.E.D.

An observation of the graphical approach leads to two conclusions:

1. For arbitrary input, the multiplication and integration steps of Procedure B.5-3 may be structured for automated computation with high speed computers or handheld calculators.

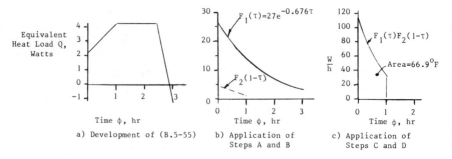

a) Development of (B.5-55)   b) Application of   c) Application of
                                    Steps A and B   Steps C and D

Figure B.5-9. Illustration of Procedure B.5-3.

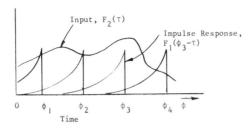

Figure B.5-10. Visualizing the Response at Different Sample Times θ_i.

2. The operations Procedure B.5-3(C) and (D) may be thought of as weighing all past values of the impulse response by the input. This is illustrated in Figure B.5-10 since either the impulse response or input function can be folded as defined by (B.5-51).

The response at any time is dominated by the values of the input at times determined by the impulse. For the impulse response of Figure B.5-10, little effect on the output occurs at times much before the sampling time ϕ_i. This provides a useful visualization of the response characteristics without requiring an evaluation over the entire input history.

The graphical process may also be used to determine the transient response of structural systems.

B.6 FREQUENCY RESPONSE

Considerable insight into the dynamic behavior of a linear system can be obtained from its response to a sine wave input. If the input to the system of Figure B.5-5 is defined as a sine wave of fixed amplitude and frequency given by

$$x(t) = x_0 \sin \omega t \qquad (B.6-1)$$

then the steady state response must also be a sine wave at the same frequency but with a phase difference ϕ, and amplitude determined by the transmission characteristics of the structure. For the system of Figure B.5-5, the steady state response becomes

$$y(t) = y_0 \sin (\omega t - \phi) \qquad (B.6-2)$$

The structure is assumed to be quiescent when excitations (B.6-1) are non-existent. The transfer function for the system is then expressed by the ratio $y_0/$

x_0 at frequency ω. The inputs to the system may be forces, pressures, displacements, velocities, accelerations, etc., taken individually or in some combinations.

A plot of the ratio of magnitudes at a location in the system as a function of frequency can be used to determine the response amplitude for a known excitation of that location. A different location exhibits a different response spectrum unique to its transmission characteristics. The frequency spectrum portrays the structural interrelationship of all the components and elements of the hardware for the range of frequencies examined at a particular location. An understanding of the interrelated motions enables appropriate hardware changes to minimize unwanted coupling and attendant amplifications. A troublesome resonance in the neighborhood of a critical frequency may be eliminated or moved to a frequency of less concern. Design characteristics of components and subsystems within the unit can be defined using information gleaned from frequency spectra of candidate locations.

In some cases, the input may be limited to sinusoidal excitation developed by fans or motors within the enclosure. Small disturbance forces due to rotating mass unbalance may be amplified by an adjoining structure generating unwanted noise or loads. The frequency spectrum is a valuable tool for dealing with these problems.

Many hardware configurations may be represented by the single degree of freedom model of Figure B.1-1. A normalized frequency spectrum for this system is illustrated in Figure B.1-4 for several damped configurations. Resonance depicted by the magnitude of the amplitude ratio occurs when the excitation frequency approaches that of the system. At resonance $s^2 = -k_{eq}/M_{eq}$ in the denominator of (B.1-9) giving

$$\frac{x_m}{x_f} = \frac{cs + k_{eq}}{M_{eq}s^2 + cs + k_{eq}} = s + \frac{k_{eq}}{c} \qquad \text{(B.6-3)}$$

which expressed as an absolute value becomes

$$\left| \frac{x_m}{x_f} \right|_{\Omega=1} = \sqrt{Q^2 + 1} \qquad \text{(B.6-4a)}$$

Equation (B.6-4a) for a lightly damped system is

$$\frac{x_m}{x_f} = \simeq Q \qquad \text{(B.6-4b)}$$

Undamped systems, Q infinite, vibrate indefinitely at their natural frequency when set in motion. Suppose a sinusoidal disturbance excites an undamped system with $\Omega = 1$. Additional energy driving each cycle causes the amplitude to increase eventually to an infinitely large value. Contribution of damping in actual hardware due to friction and pseudo-viscous dissipative mechanisms prevent infinite amplitudes.

The denominator of (B.1-9) represents the magnitude of the transfer functions. The characteristic equation of a multicoordinate system contains many such factors. The nondimensional form of the denominator (B.1-11), illustrated in Figure B.6-1, is commonly used for design and analysis of these systems. Maximum amplification

$$\left| \frac{x_m}{x_f} \right|_{max} = \frac{Q}{\sqrt{1 - \left(\frac{1}{2Q}\right)^2}} \qquad \text{(B.6-5a)}$$

actually occurs at a frequency given by

$$\frac{f}{f_n} = \sqrt{1 - \frac{1}{2Q^2}} \qquad \text{(B.6-5b)}$$

Most electronic equipment, however, exhibits amplifications $Q \simeq \sqrt{f_n}$ so that

$$\left| \frac{x_m}{x_f} \right|_{max} \simeq Q \qquad \text{(B.6-6c)}$$

at frequency

$$\frac{f}{f_n} \simeq 1 \qquad \text{(B.6-6d)}$$

The frequency response of a particular location or component of a multicoordinate system is derived from the transfer function relating response amplitude to the input. Inputs may be defined at any coordinate representing a physical hardware location including the foundation. The transfer function will consist of a ratio of polynomials in the form of (B.5-14).

The frequency response derived from (B.5-14) either numerically or graphically is usually depicted as a log-log plot of the ratio x/y versus frequency. Conventional grid paper with any increment is convenient for manual plotting. A logarithmic scale is easily developed following the guidelines of Figure B.6-2.

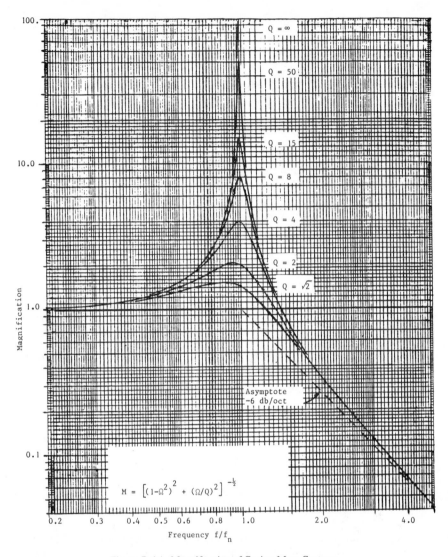

Figure B.6-1. Magnification of Spring Mass Systems.

As an example, assume that a frequency range from 10 to 1000 Hertz (two decades) is to be represented and quarter inch grid paper is available. Each decade may be represented by a 2.5 inch segment which is uniformly divided by the grid into 10 spaces. Tic marks are then drawn at the grid value corresponding to the log increment, following Figure B.6-2. The resulting scale suitably calibrated is shown in Figure B.6-3.

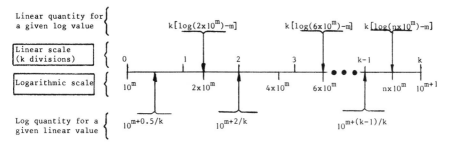

Figure B.6-2. Logarithmic–Linear Scale Correspondence.

Numerical Approach.   Computational effort is significantly reduced by grouping terms with odd and even exponents of s and forming successive products similar to (B.4-1b). This operation expresses (B.5-14) as

$$\frac{x(s)}{y(s)} = K \frac{[(((s^2 + b_1)s^2 + b_3)s^2 + \cdots)s^2 + b_{n-1} + (((b_0 s^2 + b_2)s^2 + b_4)s^2 + \cdots)s]}{(((s^2 + a_1)s^2 + a_3)s^2 + \cdots)\, s^2 + a_{n-1} + (((a_0 s^2 + a_2)s^2 + a_4)s^2 + \cdots)s}$$

$$(B.6-7)$$

or

$$\frac{x(s)}{y(s)} = K \frac{(A_1 + B_1 s)}{A_2 + B_2 s}$$

The magnitude of the response ratio $|x(\omega)/y(\omega)|$ becomes

$$\left| \frac{x(\omega)}{y(\omega)} \right| = K \left[\frac{A_1^2 + (B_1\omega)^2}{A_2^2 + (B_2\omega)^2} \right]^{1/2} \qquad (B.6-8)$$

When damping is neglected, the response ratio (B.6-8) is

$$\left| \frac{x(\omega)}{y(\omega)} \right| = \left| \frac{A_1}{A_2} \right| \qquad (B.6-9)$$

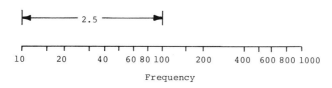

Figure B.6-3.

Equation (B.6-9) artifically increases the amplitudes at resonances but retains good accuracy elsewhere. The correct ratio at resonances could then be determined using (B.6-8).

EXAMPLE B.6-1. Plot the frequency response for the transfer function

$$\frac{x(s)}{y(s)} = \frac{0.0225s^3 + 4.684s^2 + 13.02s + 1800}{0.0121s^4 + 0.039s^3 + 11.679s^2 + 13.02s + 1800} \quad \text{(B.6-10)}$$

Grouping odd and even exponents of s, then arranging the results in the form of (B.6-7) gives

$$\frac{x(s)}{y(s)} = \frac{(4.684s^2 + 1800) + (0.0225s^2 + 13.02)s}{(0.0121s^2 + 11.679)s^2 + 1800 + (0.039s^2 + 13.02)s} \quad \text{(B.6-11)}$$

Which without damping setting $s^2 = -\omega^2$ becomes

$$\frac{x(\omega)}{y(\omega)} = \left|\frac{A_1}{A_2}\right| = \frac{1800 - 4.684\omega^2}{1800 - \omega^2(11.679 - 0.0121\omega^2)} \quad \text{(B.6-12)}$$

Values of (B.6-12) are conveniently organized as follows.

ω	4	8	12	16	20	24	28	32	36	40
A_1	1725	1500	1125	601	−73.6	−898	−1872	−2996	−4270	−5694
A_2	1616	1102	369	−397	−935.6	−912.6	81	2528	6987	14089
$\lvert x(\omega)/y(\omega)\rvert$	1.07	1.36	3.05	1.51	0.078	0.984	23.1	1.18	0.611	0.404

The sign change of A_1 indicates an antiresonance (zero) for $16 < \omega < 20$. The sign changes of A_2 indicates a resonance $12 < \omega < 16$ and another at $24 < \omega < 28$. Additional values of x/y in the frequency ranges stated are required to adequately describe these features.

Values of ω were selected midway between the ranges, i.e., 14, 18 and 26. Values of $\lvert x(\omega)/y(\omega)\rvert$ are plotted as Figure B.6-4 on the logarithmic format developed using Figure B.6-2.

The resonances are estimated at $\omega = 13.5$ and $\omega = 27.5$ by evaluating the sign changes of A_2 and selecting ω midway between the corresponding values. The same method using A_1 puts the antiresonance at approximately $\omega = 19.5$. The values may be used in (B.6-11) to obtain limiting values of $\lvert x/y\rvert$. At $\omega = 13.5$, for example, (B.6-8) gives

$$\left.\left|\frac{x}{y}\right|\right|_{\omega=13.5} = \left[\frac{946.3^2 + (8.92 \times 13.5)^2}{73.4^2 + (5.91 \times 13.5)^2}\right]^{1/2} = 8.8 \quad \text{(B.6-13)}$$

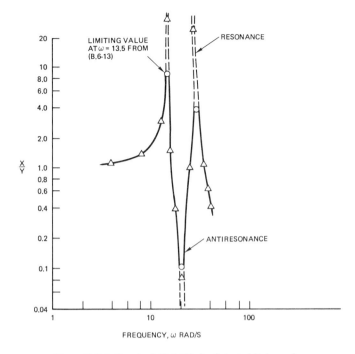

Figure B.6-4. Graph of (B.6-12) for Selected Values of ω.

Limiting values at resonances are shown as circled points on Figure B.6-4, which now defines the frequency response of the given system. The accuracy of ω at resonance requires sufficient halving of the interval defined by the sign changes of A_1 and A_2 for resonance and antiresonance respectively.   Q.E.D.

Graphical Approach. Graphical methods may be used for a factored or unfactored transfer function. This method portrays the frequency response with less data than the numerical method to resolve resonances. One graphical technique uses the factored form of the transfer function defined in Appendix B.4 and the second is based on the unfactored form (B.5-14).

Factored-Graphical Technique. Once the factors of both the numerator and denominator of (B.5-14) are found, the transfer function

$$\frac{x(s)}{y(s)} = \frac{K(s^2 + c_0 s + d_0)(s^2 + c_1 s + d_1) \cdots (s + \beta_1)(s + \beta_2) \cdots}{(s^2 + a_0 s + b_0)(s^2 + a_1 s + b_1) \cdots (s + \alpha_1)(s + \alpha_1) \cdots}$$

$$(B.6\text{-}14)$$

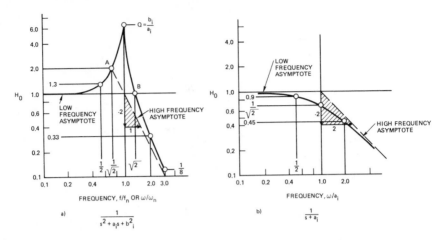

Figure B.6-5. Magnitude Response of Transfer Function Factors.

explicitly defines the resonances, antiresonances and associated peak amplifications. Based on the asymptotic characteristics of Figure B.6-1, the response for each quadratic of (B.6-14) can be quickly sketched. The linear factors are equally simple to sketch. The product of each factor for any ω gives x/y. Normally, only a few points are required to define the frequency spectrum. Asymptotic characteristics for quadratic and linear factors are illustrated in Figure B.6-5. The values for the magnitude of factors in the numerator are equal to the reciprocal to those of Figure B.6-5.

EXAMPLE B.6-2.　Resolve Example B.6-1 using the factored-graphical technique.
Factors of (B.6-10) are determined using the methods of Appendix B.4. The factored result is

$$\frac{x(s)}{y(s)} = \frac{(s + 206.88)(s^2 + 0.926s + 386)}{(s^2 + 0.777s + 193.17)(s^2 + 2.47s + 770.12)} \qquad \text{(B.6-15)}$$

Each factor of (B.6-15) may be normalized independently of the other factors to illustrate clearly each resonance and amplification Q. Performing this operation on (B.6-15) gives

$$\frac{x(s)}{y(s)} = \frac{\left(\dfrac{s}{206.88} + 1\right)\left(\dfrac{s^2}{19.65^2} + \dfrac{1}{21.2}\dfrac{s}{19.65} + 1\right)}{\left(\dfrac{s^2}{13.89^2} + \dfrac{1}{17.89}\dfrac{s}{13.89} + 1\right)\left(\dfrac{s^2}{27.75^2} + \dfrac{1}{11.23}\dfrac{s}{27.75} + 1\right)} \qquad \text{(B.6-16)}$$

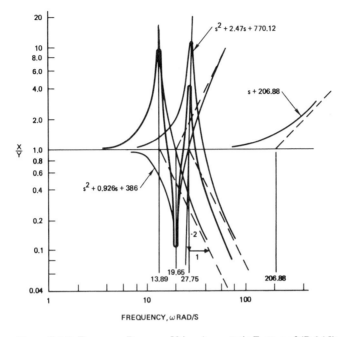

Figure B.6-6. Frequency Response Using Asymptotic Factors of (B.6-15).

One factor in the denominator of equation (B.6-16) indicates that $Q = 17.89$ for the resonant frequency of 13.89 rad/s. Each factor is lightly sketched using asymptotic and point characteristics of Figure B.6-5 or Figure B.6-6.

Products of the ordinate values for fixed frequencies interconnected with the darkened curve defines the frequency response. The process is easily followed and exhibits good accuracy once the factors of the transfer function are found.   Q.E.D.

The points which define the response envelope of Figure B.6-5(a) are accurate for large variations in damping. The effect variations in Q have on the frequency of points A and B are tabulated below

| Q | Values of f/f_n or ω/ω_n | |
	Point A, $H_0 = 2$	Point B, $H_0 = 1$
∞	$1/\sqrt{2} = 0.71$	$\sqrt{2} = 1.41$
5	$1/\sqrt{1.91} = 0.72$	$\sqrt{1.96} = 1.40$
2	$1/\sqrt{1.33} = 0.86$	$\sqrt{1.75} = 1.32$

Notice that the high frequency asymptote intercepts point A on Figure B.6-5(a). Other points on Figure B.6-5 which behave similarly are an octave below and above the frequency of the factor.

Unfactored-Graphical Technique.   The effort required to define the frequency response of a transfer function may be reduced by using (B.5-14) directly.

The quadratic formed by truncating higher order terms in the numerator of (B.5-14) is rewritten as

$$b_{n-2}s^2 + b_{n-1}s + b_n \qquad (B.6\text{-}17)$$

If (B.6-17) is underdamped, $b_{n-2}s^2 = b_n$ at its resonance so that $b_{n-1}s$ becomes dominant at $\omega_1 = \sqrt{b_n/b_{n-2}}$. The frequency ω_1 establishes a hypothetical boundary for examining the lower order terms of the denominator. When (B.6-17) is overdamped, one hypothetical boundary is at $\omega_a = b_n/b_{n-1}$ and a second is at $\omega_b = b_{n-1}/b_{n-2}$. As additional higher order numerator terms of (B.5-14) are considered, other boundaries can be defined, each at a successively higher frequency. Only boundaries in the frequency range dominated by the resonances (zeros) of the denominator are of interest.

On and within each boundary, the dominant numerator term is used with successively higher order denominator terms applicable to the frequency range. As each higher frequency boundary is encountered, the active denominator term applies to each numerator term which is dominant. The process is illustrated in Figure B.6-7 for the transfer function

$$\frac{x(s)}{y(s)} = \frac{b_0 s^3 + b_1 s^2 + b_2 s + b_3}{a_0 s^4 + a_1 s^3 + a_2 s^2 + a_3 s + a_4} \qquad (B.6\text{-}18)$$

The ratio of the dominant numerator and terms of the denominator form asymptotes to which the response curve may be sketched. Figure B.6-5 may be used to add points to guide construction. The intersection where terms such as $b_1/a_2 = b_1/a_0\omega^2$ adjoining a -6 db/oct slope defines a resonance of the system. The amplification at these resonances is developed by extending the response curve to the -3 db/oct asymptote, which represents the damping contribution at the corresponding resonant frequency. At the intersection cited, the resonant frequency would be $\omega = \sqrt{a_2/a_0}$. Resonances found using this technique and the characteristics of the response are approximately correct.

The asymptotes generated by the ratio of individual numerator and denominator terms having values > 1 over a broad range of frequencies influence the construction of response curve. Suppose $b_1/a_1\omega$ is ≤ 1.0 and $b_3/a_3\omega \simeq 2.0$ at $\omega = \sqrt{a_2/a_0}$ in Figure B.6-7. The response at $\omega = \sqrt{a_2/a_0}$ would then be

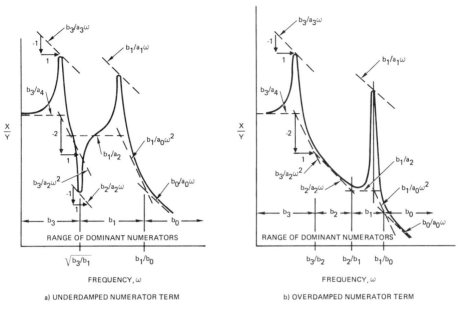

Figure B.6-7. Illustration of Numerator Boundaries and Asymptotes Formed by Ratios of Numerator and Denominator Terms.

sketched to the dominant asymptote $b_3/a_3\omega$. This is illustrated in Example B.6-3. If confusion arises because of several asymptotes of apparent dominance, then use of the numerical approach of (B.6-8) is recommended at the actual resonances in the vicinity of the approximate frequency.

EXAMPLE B.6-3. Determine the frequency response for the transfer function

$$\frac{x(s)}{y(s)} = \frac{0.0209s^2 + 13.02s + 1800}{0.0121s^4 + 0.039s^3 + 11.68s^2 + 13.02s + 1800} \quad \text{(B.6-19)}$$

The normalized numerator of (B.6-19) is

$$\left(\frac{s^2}{293.5^2} + \frac{1}{0.47} \frac{s}{293.5} + 1 \right) \quad \text{(B.6-20)}$$

Since the factor (B.6-20) is overdamped, boundaries are defined by individual numerator terms. Figure B.6-8 depicts the asymptote relationships and the response curve construction.

The amplitude at ω_2 may be checked using (B.6-8) along with the resonance, $\omega_2 = 27.7$ determined by factoring the denominator as in (B.6-16). The 12% difference of

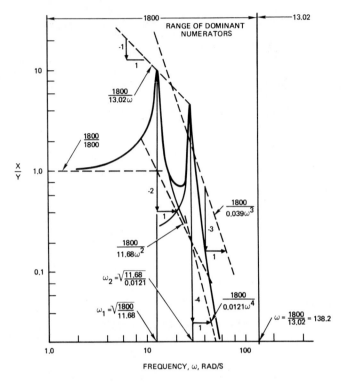

Figure B.6-8. Frequency Response for (B.6-19).

the graphical relative to the factored value of ω_2 can result in dramatic variations of amplitude in (B.6-8).

$$\left(\frac{x}{y}\right)_{\omega=27.7} = \left[\frac{1784^2 + 360.6^2}{38.3^2 + 468.2^2}\right]^{1/2} = 3.87 \qquad (B.6-21)$$

The frequency response is sketched on Figure B.6-8 using the asymptotes in a term-by-term manner with increasing frequency. The rolling off slope at $\omega = 50$ rad/s is asymptotic to the -4 slope (-12 db/oct) as shown.   Q.E.D.

EXAMPLE B.6-4.   Resolve Example B.6-1 using the unfactored-graphical technique. The transfer function

$$\frac{x(s)}{y(s)} = \frac{0.0225s^3 + 4.684s^2 + 13.02s + 1800}{0.0121s^4 + 0.039s^3 + 11.679s^2 + 13.02s + 1800} \qquad (B.6-22)$$

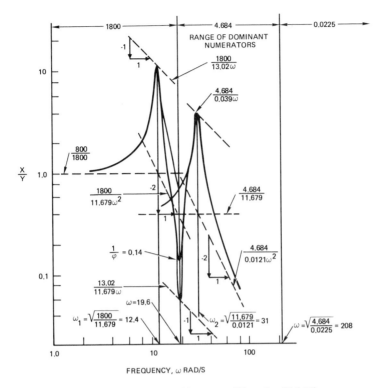

Figure B.6-9. Frequency Response of Equation (B.6-10).

possesses an underdamped quadratic in the numerator in the normalized form

$$\left(\frac{s^2}{19.6^2} + \frac{1}{7.05}\frac{s}{19.6} + 1\right) \tag{B.6-23}$$

Equation (B.6-23) indicates an antiresonance at $\omega = 19.6$ rad/s with $1/Q = 0.14$. The boundaries of the frequency plot domain are defined by numerator terms 1800 and $4.684s^2$ on Figure B.6-9. Asymptote development and the response curve follow the process of Figure B.6-7(a).

The suitability of the response curve using the unfactored-graphical technique can be judged by comparing Figure B.6-9 with Figure B.6-6 or Figure B.6-4.   Q.E.D.

Appendix C

Properties of Materials, Areas, Solids and Shells

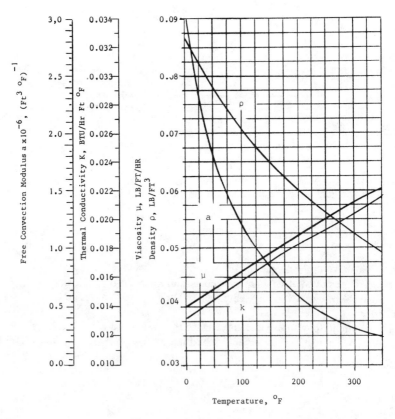

Figure C.1. Properties of Air.

Table C.1. Material Properties

Material	$k, \dfrac{W}{\text{in} \cdot {}^\circ F}$	$\alpha, \dfrac{\text{in}}{\text{in} \cdot {}^\circ F} \times 10^{-5}$	$C_p, \dfrac{W \cdot s}{\text{lb} \cdot {}^\circ F}$	$E, \dfrac{M \cdot \text{lb}}{\text{in}^2}$	Poisson's Ratio, ν	$\rho, \dfrac{\text{lb}}{\text{in}^3}$	$\sigma_{tu}, \dfrac{k \cdot \text{lb}}{\text{in}^2}$	$\sigma_{ty}, \dfrac{k \cdot \text{lb}}{\text{in}^2}$
Metals								
Aluminum 5052	1.95	1.32		9.954	0.334	0.098	34	24
6061	1.97	1.30	253.2	9.954	0.340	0.098	42	36
6061	1.93	1.29	242.5	9.954	0.334	0.10	77	66
Be	2.30	0.68	466.0	42.00	0.10	0.066	86	58
Be-Cu	1.50	2.20	105.5	18.50	0.27	0.297	160	120
Cadmium	1.29	0.73	58.0	9.90	0.30	0.312	11.9	
Copper	5.19	0.93	99.1	17.20	0.326	0.322	40.0	30
Gold	4.31	0.79	32.7	11.10	0.41	0.698	29.8	29.8
Kovar	0.30	0.77	114.2	19.50		0.320	34.4	59.5
Magnesium	1.21	1.47	263.6	6.50	0.35	0.065	39.8	28
Nickel	1.06	0.73	116.9	29.80	0.30	0.320	71.1	50.0
Silver	5.93	1.07	59.0	10.60	0.37	0.380	41.2	41.2
Solder 63/37	0.60	1.28		2.50	0.40		7.0	
Steel Carbon (1010)	0.82	0.64	125.5	30.00	0.292	0.290	70.0	36.0
Stainless (304)	0.22	0.89	126.0	28.40	0.305	0.290	80.0	40
Ceramics								
Alumina (Al_2O_3)	0.473	0.311	234.4	54.0		0.130	25.0	20.0
Beryllia (B_eO)	2.596	0.377	312.0	46.0		0.105	20.0	
Mira	0.005	0.422	216.7	10.0		0.105		5.5
Quartz	0.560	0.0277	199.3	10.4	0.17	0.094	27.9	
Magnesia (MgO)	2.066	0.490		10.0		0.101	12.0	
Plastics								
Epoxy glass (G10) (X/Y)	0.0037	0.55	253.0	2.36	0.12	0.071	25.0	35
(Z)	0.0037	4.00	253.0	2.36	0.12	0.071	25.0	35
Lexan	0.0027	3.75	306.0	0.379		0.047	9.7	9.7
Nylon	0.0033	5.00	400.0	0.217		0.041	11.8	11.5
Teflon	0.0029	5.00	263.0	0.150		0.077		4.0
Mylar	0.0020	0.94	295.0	0.550		0.050	25.0	
Rubber								
Synthetic (med)	0.0042	4.44	448.0		0.033		3.0	

Table C.2. Properties of the Atmosphere

Altitude		Temperature		Pressure in Hg	Pressure Ratio $P/P_{\text{sea level}}$	Density		Density Ratio $\rho/\rho_{\text{sea level}}$
KFT	KM	°F	°C			lb/ft^3	g/cm^3	
0	0	59.000	15.000	29.921	1.000	7.647(−2)	1.226(−3)	1.000
1	3.048(−1)	55.434	13.019	28.856	9.644(−1)	7.426(−2)	1.191(−3)	9.711(−1)
2	6.096(−1)	51.868	11.038	27.821	9.298(−1)	7.210(−2)	1.156(−3)	9.428(−1)
3	9.143(−1)	48.303	9.057	26.817	8.963(−1)	6.998(−2)	1.122(−3)	9.151(−1)
4	1.219	44.738	7.076	25.843	8.637(−1)	6.792(−2)	1.089(−3)	8.881(−1)
5	1.524	41.173	5.096	24.897	8.321(−1)	6.589(−2)	1.056(−3)	8.617(−1)
6	1.829	37.609	3.116	23.980	8.014(−1)	6.393(−2)	1.025(−3)	8.359(−1)
7	2.133	34.035	1.136	23.090	7.717(−1)	6.199(−2)	9.994(−4)	8.107(−1)
8	2.438	30.482	−0.843	22.228	7.429(−1)	6.012(−2)	9.638(−4)	7.861(−1)
9	2.743	26.918	−2.823	21.391	7.149(−1)	5.828(−2)	9.343(−4)	7.621(−1)
10	3.048	23.355	−4.803	20.581	6.878(−1)	5.648(−2)	9.056(−4)	7.386(−1)
15	4.572	5.546	−14.697	16.893	5.646(−1)	4.814(−2)	7.718(−4)	6.294(−1)
20	6.096	−12.255	−24.586	13.761	4.599(−1)	4.077(−2)	6.537(−4)	5.331(−1)
25	7.619	−30.047	−34.470	11.118	3.716(−1)	3.430(−2)	5.500(−4)	4.486(−1)
30	9.144	−47.831	−44.350	8.903	2.975(−1)	2.866(−2)	4.594(−4)	3.747(−1)
35	10.667	−65.606	−54.225	7.060	2.359(−1)	2.375(−2)	3.808(−4)	3.106(−1)
40	12.191	−69.700	−56.500	5.558	1.858(−1)	1.889(−2)	3.029(−4)	2.471(−1)
45	13.715	−69.700	−56.500	4.375	1.462(−1)	1.487(−2)	2.385(−4)	1.945(−1)
50	15.239	−69.700	−56.500	3.444	1.151(−1)	1.171(−2)	1.877(−4)	1.531(−1)
60	18.287	−69.700	−56.500	2.135	7.137(−2)	7.259(−3)	1.164(−4)	9.492(−2)
70	21.335	−67.424	−55.235	1.325	4.429(−2)	4.479(−3)	7.181(−5)	5.857(−2)
80	24.383	−61.977	−52.209	8.273(−1)	2.765(−2)	2.758(−3)	4.421(−5)	3.606(−2)
90	27.431	−56.535	−49.186	5.200(−1)	1.738(−2)	1.710(−3)	2.742(−5)	2.236(−2)
100	30.478	−51.098	−46.166	3.291(−1)	1.099(−2)	1.068(−3)	1.712(−5)	1.396(−2)
150	45.718	19.403	−6.998	4.018(−2)	1.343(−2)	1.112(−4)	1.754(−6)	1.454(−3)
200	60.957	−2.671	−19.262	5.846(−3)	1.954(−4)	1.696(−5)	2.719(−7)	2.217(−4)

Numeric representation $x.xxx\,(-y)$ is condensed version of $x.xxx(10^{-y})$

Table C.3. Radiation Properties of Materials

Material	Temperature °F	°C	Absorptivity α	Emissivity ϵ	α/ϵ
Aluminum Alloy					
Commercial	212	100	0.37	0.09	4.11
Rough Polish	212	100	0.31	0.05	6.20
Brass—Polish			0.22	0.03	7.33
Copper					
Commercial	72	22.2	0.76	0.07	10.86
Rough Polish	72	22.2	0.17	0.03	5.67
Iron—Polish	200	93.3	0.29	0.2	1.45
Nickel					
Electrolytic	100	37.8	0.15	0.06	2.50
Electroless	−60	−51.1	0.45	0.16	2.81
Oxidized			0.66	0.50	1.32
Electroplate on Cu	72	22.2	0.48	0.03	16.0
Stainless Steel					
302 Polish	−60	−51.1	0.3	0.17	
410 (1100F oxidized)	70	21.1	0.76	0.13	5.88
301 Black Oxide	32	0.0	0.89	0.75	1.19
Silver (999 + fine)	300	148.9	0.13	0.03	4.3
Solder					
50-50 on Cu	72	22.2		0.03	
Paint					
White	0	−17.8	0.19	0.91	0.21
White-Epoxy Resin	0	−17.8	0.25	0.88	0.28
Black Lacquer	0	−17.8	0.92	0.71	1.30
Black Epoxy Resin	0	−17.8	0.95	0.89	1.07

Table C.4. Properties of Plane Areas

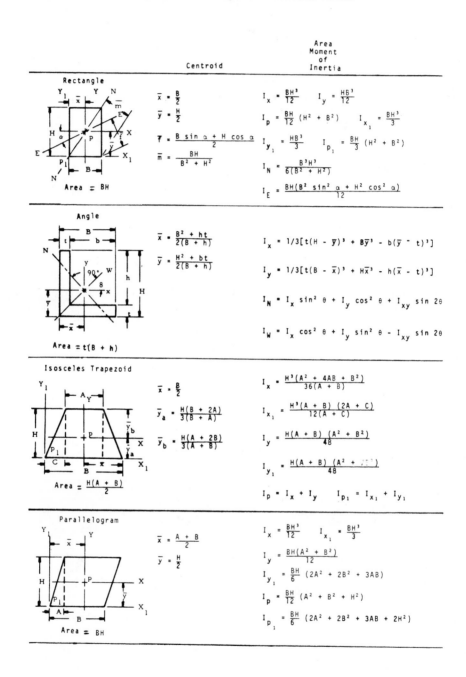

	Centroid	Area Moment of Inertia

Rectangle

$\bar{x} = \dfrac{B}{2}$

$\bar{y} = \dfrac{H}{2}$

$\bar{f} = \dfrac{B \sin \alpha + H \cos \alpha}{2}$

$\bar{m} = \dfrac{BH}{B^2 + H^2}$

Area $= BH$

$I_x = \dfrac{BH^3}{12} \qquad I_y = \dfrac{HB^3}{12}$

$I_p = \dfrac{BH}{12}(H^2 + B^2) \qquad I_{x_1} = \dfrac{BH^3}{3}$

$I_{y_1} = \dfrac{HB^3}{3} \qquad I_{p_1} = \dfrac{BH}{3}(H^2 + B^2)$

$I_N = \dfrac{B^3 H^3}{6(B^2 + H^2)}$

$I_E = \dfrac{BH(B^2 \sin^2 \alpha + H^2 \cos^2 \alpha)}{12}$

Angle

$\bar{x} = \dfrac{B^2 + ht}{2(B + h)}$

$\bar{y} = \dfrac{H^2 + bt}{2(B + h)}$

Area $= t(B + h)$

$I_x = 1/3[t(H - \bar{y})^3 + B\bar{y}^3 - b(\bar{y} - t)^3]$

$I_y = 1/3[t(B - \bar{x})^3 + H\bar{x}^3 - h(\bar{x} - t)^3]$

$I_N = I_x \sin^2 \theta + I_y \cos^2 \theta + I_{xy} \sin 2\theta$

$I_W = I_x \cos^2 \theta + I_y \sin^2 \theta - I_{xy} \sin 2\theta$

Isosceles Trapezoid

$\bar{x} = \dfrac{B}{2}$

$\bar{y}_a = \dfrac{H(B + 2A)}{3(B + A)}$

$\bar{y}_b = \dfrac{H(A + 2B)}{3(A + B)}$

Area $= \dfrac{H(A + B)}{2}$

$I_x = \dfrac{H^3(A^2 + 4AB + B^2)}{36(A + B)}$

$I_{x_1} = \dfrac{H^3(A + B)(2A + C)}{12(A + C)}$

$I_y = \dfrac{H(A + B)(A^2 + B^2)}{48}$

$I_{y_1} = \dfrac{H(A + B)(A^2 + ?\,?)}{48}$

$I_p = I_x + I_y \qquad I_{p_1} = I_{x_1} + I_{y_1}$

Parallelogram

$\bar{x} = \dfrac{A + B}{2}$

$\bar{y} = \dfrac{H}{2}$

Area $= BH$

$I_x = \dfrac{BH^3}{12} \qquad I_{x_1} = \dfrac{BH^3}{3}$

$I_y = \dfrac{BH(A^2 + B^2)}{12}$

$I_{y_1} = \dfrac{BH}{6}(2A^2 + 2B^2 + 3AB)$

$I_p = \dfrac{BH}{12}(A^2 + B^2 + H^2)$

$I_{p_1} = \dfrac{BH}{6}(2A^2 + 2B^2 + 3AB + 2H^2)$

Table C.4. Properties of Plane Areas (*continued*)

	Centroid	Area Moment of Inertia

Obtuse Triangle

Area $= \dfrac{BH}{2}$

$\bar{x} = \dfrac{B + 2C}{3}$

$\bar{y} = \dfrac{H}{3}$

$I_x = \dfrac{BH^3}{36}$ $I_{x_1} = \dfrac{BH^3}{12}$

$I_{x_2} = \dfrac{BH^3}{4}$

$I_y = \dfrac{BH}{36}(B^2 + BC + C^2)$

$I_{y_1} = \dfrac{BH}{12}(B^2 + 3BC + 3C^2)$

$I_p = \dfrac{BH}{36}(H^2 + B^2 + BC + C^2)$

Oblique Triangle

Area $= \dfrac{BH}{2}$

$\bar{x} = \dfrac{B + C}{3}$

$\bar{y} = \dfrac{H}{3}$

$I_x = \dfrac{BH^3}{36}$ $I_{x_1} = \dfrac{BH^3}{12}$

$I_{x_2} = \dfrac{BH^3}{4}$

$I_y = \dfrac{BH}{36}(B^2 + C^2 - BC)$

$I_p = \dfrac{BH}{36}(H^2 + B^2 + C^2 - BC)$

Right Triangle

Area $= \dfrac{BH}{2}$

$\bar{x} = \dfrac{B}{3}$

$\bar{y} = \dfrac{H}{3}$

$I_x = \dfrac{BH^3}{36}$ $I_{x_1} = \dfrac{BH^3}{12}$ $I_{x_2} = \dfrac{BH^3}{4}$

$I_y = \dfrac{B^3 H}{36}$ $I_{y_1} = \dfrac{B^3 H}{12}$

$I_p = \dfrac{BH(B^2 + H^2)}{36}$ $I_{p_1} = \dfrac{BH(B^2 + H^2)}{12}$

Isosceles Triangle

Area $= \dfrac{BH}{2}$

$\bar{x} = \dfrac{B}{2}$

$\bar{y} = \dfrac{H}{3}$

$I_x = \dfrac{BH^3}{36}$ $I_{x_1} = \dfrac{BH^3}{12}$

$I_{x_2} = \dfrac{BH^3}{4}$ $I_y = \dfrac{B^3 H}{48}$

$I_{y_1} = \dfrac{7B^3 H}{48}$ $I_p = \dfrac{4BH^3 + 3B^3 H}{144}$

$I_{p_1} = \dfrac{4BH^3 + 7B^3 H}{48}$

Hollow Circle

Area $= \pi(R^2 - r^2)$

$\bar{x} = \bar{y} = R$

$I_x = I_y = \dfrac{\pi(R^4 - r^4)}{4}$

$I_p = \dfrac{\pi(R^4 - r^4)}{2}$

$I_{x_1} = I_{y_1} = \dfrac{\pi(5R^4 - 4R^2 r^2 - r^4)}{4}$

Table C.5. Properties of Solids and Twin Shells

	Centroid	Weight Moment of Inertia

Rectangular Prism

Volume = ABH

$$\bar{x} = \frac{A}{2}$$

$$\bar{y} = \frac{B}{2}$$

$$\bar{z} = \frac{H}{2}$$

$$I_x = \frac{W}{12}(B^2 + H^2)$$

$$I_{x_1} = \frac{W}{3}(B^2 + H^2)$$

$$I_{x_2} = \frac{W}{12}(B^2 + 4H^2)$$

$$I_y = \frac{W}{12}(A^2 + H^2)$$

$$I_{y_1} = \frac{W}{3}(A^2 + H^2)$$

$$I_z = \frac{W}{12}(A^2 + B^2)$$

$$I_{z_1} = \frac{W}{3}(A^2 + B^2)$$

Hollow Right Circular Cylinder

Volume = $\pi H(R^2 - r^2)$

$$\bar{z} = \frac{H}{2}$$

$$I_x = I_y = \frac{W}{12}\left[3(R^2 + r^2) + H^2\right]$$

$$I_{x_1} = I_{y_1} = W\left[\frac{R^2 + r^2}{4} + \frac{H^2}{3}\right]$$

$$I_z = \frac{W}{2}(R^2 + r^2)$$

Right Circular Cone

Volume = $\frac{\pi R^2 H}{3}$

$$\bar{z} = \frac{H}{4}$$

$$I_x = I_y = \frac{3W}{20}\left(R^2 + \frac{H^2}{4}\right)$$

$$I_{x_1} = I_{y_1} = \frac{W}{20}(3R^2 + 2H^2)$$

Right Rectangular Pyramid

Volume = $\frac{ABH}{3}$

$$\bar{z} = \frac{H}{4}$$

$$I_x = \frac{W}{20}\left(B^2 + \frac{3H^2}{4}\right)$$

$$I_{x_1} = \frac{W}{20}(B^2 + 2H^2)$$

$$I_y = \frac{W}{20}\left(A^2 + \frac{3}{4}H^2\right)$$

$$I_{y_1} = \frac{W}{20}(A^2 + 2H^2)$$

$$I_z = \frac{W}{20}(A^2 + B^2)$$

Table C.5. Properties of Solids and Twin Shells (*continued*)

	Centroid	Weight Moment of Inertia
Isosceles Wedge	$\bar{z} = \dfrac{H}{3}$	$I_x = \dfrac{W}{36}(2H^2 + 3A^2)$
		$I_y = \dfrac{W}{72}(4H^2 + 3B^2)$
Volume $= \dfrac{ABH}{2}$		$I_z = \dfrac{W}{24}(2A^2 + B^2)$
Right Angled Wedge	$\bar{x} = \dfrac{A}{3}$	
	$\bar{y} = \dfrac{B}{2}$	$I_x = \dfrac{W}{36}(2H^2 + 3B^2)$
	$\bar{z} = \dfrac{H}{3}$	$I_y = \dfrac{W}{18}(A^2 + H^2)$
Volume $= \dfrac{ABH}{2}$		$I_z = \dfrac{W}{36}(2A^2 + 3B^2)$
Lateral Cylindrical Shell	$\bar{z} = \dfrac{H}{2}$	$I_x = I_y = \dfrac{W}{2}(R^2 + \dfrac{H^2}{6})$
		$I_z = WR^2$
Surface Area $= 2\pi RH$		$I_{x_1} = I_{y_1} = \dfrac{W}{6}(3R^2 + 2H^2)$
Lateral Surface of a Circular Cone	$\bar{z} = \dfrac{H}{3}$	$I_x = I_y = \dfrac{W}{4}(R^2 + \dfrac{2}{9}H^2)$
		$I_z = \dfrac{WR^2}{2}$
Surface Area $= \pi R\sqrt{R^2 + H^2}$		$I_{x_1} = I_{y_1} = \dfrac{W}{12}(3R^2 + 2H^2)$

Appendix D

Conversions

The metric system used is based on the International System of Units (SI) modified in certain instances to accommodate accepted units in the field of engineering and science. The modifications which refer only to the use of degrees Celsius instead of Kelvin are flagged with an asterisk (*).

Scientific notation is used to express multipliers, accommodating both very large and very small values in the same space. The format consists of a coefficient followed by a capital E, indicating the exponent, along with a positive or negative two digit number. The numbers indicate the power of 10 by which the coefficient must be multiplied to obtain the conversion required.

Example. Convert 3.0 feet per second squared (ft/s^2) to SI units.

From (1), the acceleration
$$= (3.0 \text{ ft/s}^2)(3.048\text{E}-01) = 9.144\text{E}-01 \text{ m/s}^2$$
$$= (3.0 \text{ ft/s}^2)(0.3048) = 0.9144 \text{ m/s}^2$$

(1) Acceleration

Angular (θ/T^2): radian per second squared (rad/s^2) =	
degree per second squared (deg/s^2)	$\times$ 1.745 329 E$-$02
revolution per minute squared (r/min^2)	$\times$ 1.745 329 E$-$03
Linear (L/T^2): meter per second squared (m/s^2) =	
centimeter per second squared (cm/s^2)	$\times$ 1.000 E$-$02
foot per second squared (ft/s^2)	$\times$ 3.048 E$-$01
inch per second squared (in/s^2)	$\times$ 2.540 E$-$02
gravity (G or g)	$\times$ 9.806 650

(2) Area (L^2): square meter (m^2) =

square centimeter (cm^2)	$\times$ 1.000 E$-$04
square foot (ft^2)	$\times$ 9.290 304 E$-$02
square inch (in^2)	$\times$ 6.451 600 E$-$04

(3) Density—mass capacity (M/L^3): kilogram per cubic meter (kg/m^3) =

gram per cubic centimeter (g/cm^3)	$\times$ 1.000 E$+$03
pound per cubic inch (lb/in^3)	$\times$ 2.767 990 E$+$04
pound per cubic foot (lb/ft^3)	$\times$ 1.601 846 E$+$01
slug per cubic foot (slug/ft^3)	$\times$ 5.153 788 E$+$02

(4) Energy (work-thermal-electrical) (ML/T^2) joule: (J) =

British thermal unit (BTU)	$\times$ 1.054 350 E+03
calorie (cal)	$\times$ 4.184
erg	$\times$ 1.000 E+07
foot-pound force (ft·lbf)	$\times$ 1.355 818
foot-poundal (ft·pdl)	$\times$ 4.214 011 E−02
watt second (w·s)	$\times$ 1.000

(5) Energy per area time (M/T^3): watt per square meter (W/m^2) =

British thermal unit per square foot second $(BTU/(ft^2·s))$	$\times$ 1.134 893 E+04
British thermal unit per square foot hour $(BTU/(ft^2·h))$	$\times$ 3.152 481
calorie per square centimeter minute $(cal/(cm^2·min))$	$\times$ 6.973 333 E+02
erg per square centimeter second $(erg/(cm^2·s))$	$\times$ 1.000 E−03
watt per square centimeter (W/cm^2)	$\times$ 1.000 E+04
watt per square inch (W/in^2)	$\times$ 1.550 003 E+03

(6) Flow

Mass flow (M/T): kilogram per second (kg/s) =

pound per second (lb/s)	$\times$ 4.535 924 E−01
pound per minute (lb/min)	$\times$ 7.559 873 E−03
pound per hour (lb/h)	$\times$ 1.259 979 E−04
slug per second (slug/s)	$\times$ 1.459 390 E+01

Volume flow (L^3/T): cubic meter per second (m^3/s) =

cubic foot per second (ft^3/s)	$\times$ 2.831 685 E−02
cubic foot per minute (ft^3/min)	$\times$ 4.719 474 E−04
cubic meter per minute (m^3/min)	$\times$ 1.666 667 E−02

(7) Force (ML/T^2): newton (N) =

dyne (dyn)	$\times$ 1.000 000 E−05
kilogram force (kgf)	$\times$ 9.806 650
kilopond (kp)	$\times$ 9.806 650
kip	$\times$ 4.448 222 E+03
pound force (lbf)	$\times$ 4.448 222
poundal (pdl)	$\times$ 1.382 550 E−01

(8) Force per length (M/T^2): newton per meter (N/m) =

dyne per centimeter (dyn/cm)	$\times$ 1.000 E−03
kilogram force per meter (kgf/m)	$\times$ 9.806 650
pound per foot (lb/ft)	$\times$ 1.459 390 E+01
pound per inch (lb/in)	$\times$ 1.751 268 E+02

(9) Frequence (T^{-1}): hertz (Hz) =

cycle per minute (c/min)	$\times$ 1.666 667 E−02
cycle per second (c/s)	$\times$ 1.000

(10) Heat

Flow rate: watt (W) =

British thermal unit per hour (BTU/h)	$\times$ 2.928 750 E−01
calorie per minute (cal/min)	$\times$ 6.973 333 E−02

calorie per second (cal/s) $\times$ 4.184

joule per second (J/s) $\times$ 1.000

Specific heat capacity*: joule per kilogram degrees Celsius (J/(kg·°C)) =

    British thermal unit per pound (mass) degree $\times$ 4.184   E+03
    Fahrenheit (BTU/lb·°F)

    Calorie per gram degree Celsius (cal/(g·°C)) $\times$ 4.184   E+03

Specific energy: joule per kilogram (J/kg) =

    British thermal unit per pound (mass) (BTU/lb) $\times$ 2.324 444 E+03

    Calorie per gram (cal/g) $\times$ 4.184 000 E+03

Thermal conductance*: watt per square meter degree Celsius (W/(m²·°C)) =

    British thermal unit per second square foot degree $\times$ 2.042 808 E+04
    Fahrenheit (BTU/(s·ft²·°F))

    British thermal unit per hour square foot degree $\times$ 5.674 466
    Fahrenheit (BTU/(h·ft²·°F))

Thermal conductivity*: watt per meter degree Celsius (W/(m·°C)) =

    British thermal unit—foot per hour square foot $\times$ 1.729 577
    degree Fahrenheit (BTU·ft/(h·ft²·°F))

    British thermal unit—inch per hour square foot $\times$ 1.441 314 E−01
    degree Fahrenheit (BTU in/(h·ft²·°F))

    British thermal unit—inch per second square foot $\times$ 5.188 732 E+02
    degree Fahrenheit (BTU·in/(s·ft²·°F))

    Calorie per centimeter second degree Celsius (cal/ $\times$ 4.184   E+02
    (cm·s·°C))

    Calorie per centimeter hour degree Celsius (cal/(cm· $\times$ 1.162 222 E−01
    h·°C))

Thermal resistance*: degree Celsius square
meter per watt (°C·m²/W) =

    degree Fahrenheit hour square foot per British $\times$ 1.762 280 E−01
    thermal unit (°F·h·ft²/BTU)

(11) Inertia

    Area (L⁴): meter⁴ =

      inch⁴ (in⁴) $\times$ 4.162 314 E−07

      foot⁴ (ft⁴) $\times$ 8.630 975 E−03

      millimeter⁴ (mm⁴) $\times$ 1.000   E−12

    Mass (ML²): Kilogram meter squared (kg·m²) =

      pound foot squared (lb·ft²) $\times$ 4.214 011 E−02

      pound inch squared (lb·in²) $\times$ 2.929 397 E−04

      slug foot squared (slug·ft²) $\times$ 1.355 818

(12) Length (L): meter (m) =

    foot (ft $\times$ 3.048 006 E−01

    inch (in) $\times$ 2.540   E−02

    kilometer $\times$ 1.000   E+03

    micrometer (μm) $\times$ 1.000   E−06

(13) Linear mass density (M/L): kilogram per meter (kg/m) =

    pound per foot (lb/ft) $\times$ 1.488 164

    pound per inch (lb/in) $\times$ 1.785 797 E+01

(14) Mass load concentration (M/L^2): kilogram per square meter (kg/m^2) =

  gram per square centimeter (g/cm^2)   × 1.000   E+01

  pound per square inch (lb/in^2)   × 7.030 696 E+02

  pound per square foot (lb/ft^2)   × 4.882 428

(15) Mass (M): kilogram (kg) =

  gram (g)   × 1.000   E−03

  kilogram force-second squared per meter ($kgf \cdot s^2/m$)   × 9.896 650

  pound (avoirdupois) (lb)   × 4.535 924 E−01

(16) Power (ML^2/T^3): watt (W) =

  British thermal unit per hour (BTU/h)   × 2.928 751 E−01

  British thermal unit per second (BTU/s)   × 1.054 350 E+03

  Calorie per second (Cal/s)   × 4.184

  erg per second (erg/s)   × 1.000   E−07

  foot-pound per second ($ft \cdot lb/s$)   × 1.355 818

(17) Pressure-Stress (M/LT^2): pascal (Pa) =

  atmosphere-standard (atm)   × 1.013 250 E+05

  dyne per square centimeter (dyn/cm^2)   × 1.000   E−01

  gram force per square centimeter   × 9.806 650 E+01

  inch of mercury (16°C)   × 3.376 850 E+03

  inch of water (16°C)   × 2.448 400 E+02

  kilogram force per square centimeter (kgf/cm^2)   × 9.806 650 E+04

  kip per square inch (kip/in^2)   × 6.894 757 E+06

  millimeter of mercury (0°C)   × 1.333 220 E+02

  newton per square meter (N/m^2)   × 1.000

  pound per square foot (lb/ft^2)   × 4.786 026 E+01

  pound per square inch (lb/in^2)   × 6.894 757 E+03

  tor (mm Hg)   × 1.333 220 E+02

(18) Torque-bending moment (ML^2/T^2): newton meter ($N \cdot m$) =

  dyne centimeter ($dyn \cdot cm$)   × 1.000   E−07

  kilogram force meter ($kgf \cdot m$)   × 9.806 650

  kip-foot ($kip \cdot ft$)   × 1.355 818 E+02

  pound-foot ($lb \cdot ft$)   × 1.355 818

  pound-inch ($lb \cdot in$)   × 1.129 848 E−01

(19) Velocity

  Angular (θ/T): radian per second (rad/s) =

  revolution per minute (f/min)   × 1.047 198 E−01

  revolution per second (r/s)   × 6.283 185

  Linear (L/T): meter per second (m/s) =

  foot per second (ft/s)   × 3.048   E−01

  foot per hour (ft/h)   × 8.466 667 E−05

  inch per second (in/s)   × 2.540   E−02

  millimeter per second (mm/s)   × 1.000   E−03

(20) Viscosity Kinematic (L^2/T): square meter per second (m^2/s) =

  centistoke (cSt)   × 1.000   E−06

  square foot per second (ft^2/s)   × 9.290 304 E−02

  Dynamic (M/LT): pascal second ($Pa \cdot s$) =

centipoise (cP) $\times$ 1.000   E−03
poundal second per square foot $(pdl \cdot s/ft^2)$ $\times$ 1.488 164
pound force second per square foot $(lbf \cdot s/ft^2)$ $\times$ 4.788 026 E+01
pound per foot-second $(lb/(ft \cdot s))$ $\times$ 1.488 164
slug per foot-second $(slug/(ft \cdot s))$ $\times$ 4.788 026 E+01
(21) Volume-capacity (L^3): cubic meter (m^3) =
    cubic centimeter (cm^3) $\times$ 1.000   E+06
    cubic foot $\times$ 2.831 685 E−02
    cubic inch (in^3) $\times$ 1.638 706 E−05
    cubic millimeter (mm^3) $\times$ 1.000   E−09

Any SI unit which is not listed in the tables can easily be determined by the application of dimensional substitution.

Example: Convert the viscosity $(lb/(min \cdot in))$ to its SI equivalent, $(Pa \cdot s)$.

Applying the relationships listed in the mass and length tables, 1 lb = 0.453 592 kg, 1 inch = 0.0254m and 1 min = 60s. Substitution gives 1 lb/(min·in) = (0.453 592 kg)/(60s $\times$ 0.0254m) = 0.297 633 Pa·s

(22) Temperature

degree Kelvin (K)   = degree Celsius (°C) + 273.15

degree Rankine (°R) = degree Fahrenheit (°F) + 459.67

differential Δ°C   = $\frac{5}{9}$ Δ°F

differential ΔK   = $\frac{5}{9}$ Δ°R

Conversions between °F and °C are performed by an easily remembered relationship based on the common temperature point $t \cdot_C = t \cdot_F = -40$.

$$t \cdot_C = \tfrac{5}{9} (t \cdot_F + 40) - 40 \qquad \text{(D22-1)}$$

$$t \cdot_F = \tfrac{9}{5} (t \cdot_C + 40) - 40 \qquad \text{(D22-2)}$$

Example: Convert 131°F to its degree Celsius equivalent.
Solution: Simply add 40, multiply by the appropriate coefficient ($\frac{5}{9}$), then subtract 40.

$$t \cdot_C = \tfrac{5}{9}(131 + 40) - 40 = 55.°C$$

Example: Convert 71°C to its degree Fahrenheit equivalent.
Solution: Simply add 40, multiply by the appropriate coefficient ($\frac{9}{5}$), then subtract 40.

$$t \cdot_F = \tfrac{9}{5}(71 + 40) - 40 = 159.8°F$$

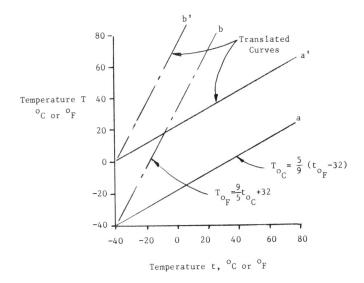

Figure D22-1. Development of Simplified Conversion in (D22-1) and (D22-2).

The conversion technique which is independent of direction ($t \cdot_C \rightarrow t \cdot_F$) or ($t \cdot_F \rightarrow t \cdot_C$) involves the same process of adding 40., multiplying by the coefficient, then subtracting. The coefficient is remembered by the fact that Fahrenheit values are greater than Celsius values (above -40 that is), so ⅘ applies whenever converting from degrees Celsius (°C) to degrees Fahrenheit (°F). Similarly ⅝ applies whenever converting from the "larger" °F to °C.

The relationships (D22-1) and (D22-2) were developed from a 40 degree translation along the ordinate of the linear classic conversion formula which pass through the common point of -40. This is shown in Figure D22-1. With the addition of $+40°$ to the equation for a line a, the equation for line a' in terms of the coefficient (slope) becomes

$$\text{line } a': \quad t \cdot_C = \tfrac{5}{9}(t \cdot_F - 32.) + 40 = \tfrac{5}{9}(t \cdot_F + 40)$$

Similarly for line b':

$$\text{line } b': \quad t \cdot_F = (\tfrac{9}{5}t \cdot_C + 32) + 40 = \tfrac{9}{5}(t \cdot_C + 40)$$

Equations (D22-1) and (D22-2) result from shifting the translated origin back to the original origin.

Index

Index

*The letter T following a page number designates a Table. Similarly, the letter E designates a
design Example, the letter F designates a Figure and the letter P indicates a procedure.